Topics in Safety, Risk, Reliability and Quality

Volume 34

More information about this series at http://www.springer.com/series/6653

Adrian V. Gheorghe · Dan V. Vamanu
Polinpapilinho F. Katina · Roland Pulfer

Critical Infrastructures, Key Resources, Key Assets

Risk, Vulnerability, Resilience, Fragility, and Perception Governance

Springer

Adrian V. Gheorghe
Engineering Management and Systems Engineering
Old Dominion University
Norfolk, VA
USA

Polinpapilinho F. Katina
Engineering Management and Systems Engineering
Old Dominion University
Norfolk, VA
USA

Dan V. Vamanu
Department of Life and Environmental Physics
Horia Hulubei National Institute of Physics and Nuclear Engineering, IFIN-HH
Bucharest—Magurele
Romania

Roland Pulfer
Action4Value GmbH
Harpstedt
Germany

ISSN 1566-0443 ISSN 2215-0285 (electronic)
Topics in Safety, Risk, Reliability and Quality
ISBN 978-3-319-69223-4 ISBN 978-3-319-69224-1 (eBook)
https://doi.org/10.1007/978-3-319-69224-1

Library of Congress Control Number: 2017955249

Printed on acid-free paper

This Springer imprint is published by Springer Nature
The registered company is Springer International Publishing AG
The registered company address is: Gewerbestrasse 11, 6330 Cham, Switzerland

To my children, Anastasia, Alexandra, and Paul

Adrian V. Gheorghe

To Jenny—my beloved wife, and Bogdan, Alin, Aurel—precious children and followers

Dan V. Vamanu

To my brothers, Jeremy, James, Boazin, Obeid, and Bernard, and Ma-Vickie

Polinpapilinho F. Katina

To my entire family for enduring long hours of inattention

Roland Pulfer

Foreword

Recently, I saw two articles posted on the BBC side-by-side. Their titles, 'What If… Conceivable Crises: Unpredictable in 2017, Unmanageable in 2020?' and 'How to Cope with the End of the World' both suggesting that not only is life short, but also it may be getting a lot shorter than we think. So, the appearance of the book you hold in your hands comes as a welcome relief. The threat to the critical infrastructures we rely upon for everyday life has never been greater. Infrastructures supplying food, energy, transportation, health care, information exchange, education, communication, social services, or you name it, these must all function, all the time, in order to our twenty-first century way of life to continue. If anyone of them suffers an extreme event and collapses for an extended period of time, we are all in deep trouble. There is simply no room for error. This book explores exactly what must be done to ensure that these types of systemic failures will not occur.

The range of topics covered in the following pages is breathtaking in its scope. Modeling of critical infrastructures and the types of extreme events that can destroy them occupies a central role in this story. So does the exploration of various ways to govern (i.e., manage and control) these infrastructures so that they can be made resilient to 'unknown unknown' destructive events. Questions of how to measure the vulnerability of the infrastructures are examined with an eye toward how to anticipate when a system is entering into the danger zone. In short, this book offers a one-stop shopping tour of just about every aspect of infrastructure that must be addressed in order for a community, country, company, or organization to feel secure that the systems of daily life will function effectively *all the time.*

I commend the authors for the thoroughness of their investigation into the Critical Infrastructure Problem and urge the reader to take the message presented here seriously.

Vienna, Austria
August 2017

John L. Casti
Author, X-Events: The Collapse of Everything

Preface

We are living in a time when Man—perhaps as a result of a body of knowledge that, more often than not, feeds from itself with only little consideration for the changes around—flatters himself with the belief of being not only more informed and intelligent, but also *wiser* and capable of making ever better decisions—to the extent of moving the entire species into what may seem to emerge as a new epoch —an Anthropocene, of sorts. True indeed: in the years since World War II, remarkable advances in domains such as materials, food, medicine, transportation, and communications have changed lives and habitat in manners effective and significant enough to leave lasting marks on longevity, standards of living and—some would contend—the very human condition at the planetary scale. At the same time, however, as by the curse of an unforgiving balance of evils against goods, a whole series of events have exposed, especially over the past two decades, surprising and transcultural weaknesses in the society foundations, fabric and functioning, and in the much-sought shared world order. From natural disasters of unprecedented violence to industrial accidents of the range of black swans, to terror through self-appointed martyrdom ominous signs are increasingly piling up indicating that something is utterly wrong—with nature, with ourselves and, most probably, with what one does to the other. Organizations as well as individuals that were once considered reliable and trustworthy turn out to be fragile, vulnerable, inconsistent, incompetent when not downright corrupt and, in the end, inconsequential in the face of the turbulence and unpredictability that seem, today, to be the name of the game.

In the face of the recurrent failure to identify and implement effective solutions via traditional, disciplinary approaches, one thought commonly attributed to Albert Einstein, looms large upon us: *We cannot solve our problems with the same thinking we used when we created them.* In all honesty, one should confess that there is a long way from noticing the need above to understanding what exactly this means; and the more—should be done; and how. The only pervasive feeling is that any attempt to make sense of the issues, must involve *thinking out of the box* and originating unconventional approaches that may hopefully integrate their organizational, managerial, social, political, cultural, and simply human aspects and their interplay. Moreover, when targeting a problem identified in a given system, we should perhaps also 'look sideways and open our minds to other places' that display similarities in structural patterns and dynamics; and thus, conjure the power of the analogies in the attempt to reveal an underlying 'business logic' of different facts and phenomena, which would make those intelligible. For a belief that authors of the present text hold is that most of the other ways of confronting the threats and anxieties of our times can only serve to further boost what Joseph Tainter refers to as a *Runaway Train*. In a Runaway Train model, a society is seen as a complex system '*impelled along a path of increasing complexity, unable to switch directions, regress, or remain static. When obstacles impinge, it can continue in only the direction it is headed, so that catastrophe ultimately results*' (Tainter 1988, p. 59).

Taking a Runaway Train from its fatal course to another timeline is nothing less than changing history. And even though 'ending' the Cold War and alleviating an all-out nuclear confrontation had the semblance of being just that, it only took a couple of decades to realize that we are all far away from being off the hook. And the explanation is almost inescapable: to switch the tracks takes more than one lever to be acted upon—one has to pull many levers; and all of these should be critically important, effective, and working in harmony to the same end. The landscape for this train might as well involve seemingly inescapable effects of globalization, the shift of individual markets, and a sheer number of reactive government regulations where nothing can be anticipated, let alone planned for. It is in this context that this book offers a number of thoughts on topics that have emerged in the last three to two decades and fast in migrating from the academic debate to the forefront of the political agendas: ***the critical infrastructures***.

To these authors, *critical infrastructures* are systems 'so vital and ubiquitous that their incapacity or destruction would not only affect the security and social welfare of any nation, but would also cascade [send disruptive waves] across borders' (Gheorghe et al. 2007, p. 6). While history, cultural differences, and current realities may trim in specific ways, the lists of critical infrastructures and their designations, the inventory would generally cover—without being limited to—the Chemical Sector; Commercial Facilities Sector; Communications Sector; Critical Manufacturing Sector; Dams Sector; Defense Industrial Base Sector; Emergency Services Sector; Energy Sector; Financial Services Sector; Food and Agriculture Sector; Government Facilities Sector; Healthcare and Public Health Sector; Information Technology Sector; Nuclear Reactors, Materials, and Waste Sector; Sector-Specific Agencies; Transportation Systems Sector; and Water and Wastewater Systems Sector (USDHS 2016).

The motivation for this research is two-fold. First, our problems as discussed are not likely to go away anytime soon. In fact, both the literature (see, e.g., Ansoff 1984; Cohen and Ahearn 1980; Martin 2006; Rasmussen and Batstone 1989; Richardson 1994; Tainter 1988; and Weick 1988) and the news reports continue to remind us that violence, crime and war, lawlessness, mismanagement, natural disasters, and gross depletion of resources—to name just a few matters of concern, are increasingly frequent and severe. At the center of it all is the public well-being. The book attempts to delve into this complex and sometimes controversial issue by highlighting *the need for good governance* involving concepts of, among others, vulnerability, resilience, and fragility in addressing *what is good for the many*. Second, there is a need to 'do something' about all these, for doing nothing can only let the Runaway Train slide down toward the predictable catastrophic dead end ahead.

Our 'doing something' comes in the form of offering *research models* as tools to understand, that is—to diagnose and predict, the behavior of the complex techno-socio-economic systems at hand in the debate. Faithful to our own beliefs, most of the models embody analogies with emblematic models in physics, with which the critical infrastructures—as well as the society itself and its paraphernalia—share the profile of what is known as 'many-body systems' featuring 'cooperative phenomena' and 'phase transitions'—the latter usually felt as disruptive occurrences. Inevitably, our models are rather educational in nature and scope. Although originated over a number of years now in an academic environment, they are the products of authors' propensity *to place the analytics in general, and the visual analytics in particular at the fingertips of the real-life-business actors*—policy makers, financiers and insurers, industry managers, emergency responders, and the like.

These models are instrumental in understanding how the world operates and in terms could inform the means for managing complex situations, instituting change, and empowering people society's well-being. The challenge at hand is enormous: systems, their element, and dependencies as well as interdependencies are often unknown, not measured properly, data privacy, and data protection are frequently misunderstood and confused with transparency, systems are not well documented, just to name a few. As a result, living in a 'system of systems' world, let alone 'managing and controlling' is a daunting endeavor, leaving bruised, those who are brave.

Interestingly, and as a solid proof that every subject matter has, eventually, it's right time, it is about now that the following *Top 10 Business Intelligence Trends for 2016* are formally recognized: (i) governance and self-service analytics become best friends, (ii) visual analytics becomes common language, (iii) data product chain becomes democratized, (iv) data integration gets exciting, (v) advanced analytics is no longer just for analysts, (vi) cloud data and cloud analytics take off, (vii) analytics center of excellence becomes excellent, (viii) mobile analytics stands on its own, (ix) people begin to dig into IoT [Internet of Things] data, and (x) new technologies rise to fill the gaps (Tableau 2016).

As such, this book may prove to be a useful read for a variety of readers interested in navigating the foggy waters of ambiguities and uncertainties of the twenty-first century. At one end, business leaders and policy makers may find it insightful in matters of say, investments in critical infrastructures, key resources, and key assets (CIKRKA). At the other end are our dear graduate students, who often work on the testing grounds of theories and models that, in time, may turn into real-life applications—it is only fair that we plant seeds now in their minds.

With this audience in mind, seventeen chapters and seven appendixes have been developed.

In Chap. 1, the underlying notions of critical infrastructures, key resources, and key assets are introduced, with emphasis on relevance to the topics of safety and well-being in the twenty-first century. The elements of space, undersea, and belowground are discussed as exemplary cases of new, complex, and critical theaters of action, each combining in a natural fashion the three-fold condition of critical infrastructure, key resource, and key asset. In addition, this chapter establishes a need for extending the traditional risk approach by going beyond mere probabilities and consequences, to the insightful concepts of vulnerability, resilience, and fragility.

Chapter 2 deals with a conceptualization of governance, first introduced in Chap. 1. Specifically, a need for a flexible many-faceted governance strategy for a diverse set of stakeholders is articulated. To that end, this chapter covers models for structural, operational, managerial, and national vulnerabilities in a 'community' setting that enable the dynamic capability to master organizational inefficiently.

Chapter 3 addresses the concept of hysteresis. The chapter looks at the importance of hysteresis as well as its implementation in a quantitative model—QVA—for an assessment of cooperative behavior, given the tendency to resist stress and maintain system state configuration.

Chapter 4 provides a 'system of systems' model of the world. This is done using readily and publicly available data from the *CIA's World Factbook.*

Chapter 5 explores the applicability of cellular automata as a viable approach for assessing risk and vulnerability with emphasis on three aspects: forest model, fire model, and smoke.

Chapter 6 is concerned with the application of QVA: *Quantitative Vulnerability Assessment* model to a nuclear reactor vulnerability assessment.

Chapter 7 offers a game approach to dealing with an emerging threat to space systems (i.e., satellites) is presented including scenario model that could be used to shot down such systems.

Chapter 8 offers insights into methods and tools that can be used by managers, political pundits, policy makers, scientists, and even hackers, to make decisions, even without having full knowledge of complex situations.

Chapter 9 offers insights into a model for assessing the vulnerability of territorial kind due to emissions. The procedural agenda for the model is discussed along with break points for chemical and radioactive release.

Chapter 10 provides a *System Resilience Governance Profile* developed for the management of complex situations along with its governance architecture as well as its calculations. A model-driven approach to resilience, complementing, System Resilience Governance Profile is provided in Chap. 11. This model, dynamic in nature, offers utility in understanding current, future, and intermediate situations involving critical infrastructures.

Chapters 12–15 are application-oriented chapters supporting theories presented in the preceding chapters. These applications range from applications at the national level—Switzerland (Chaps. 12 and 13), to specific systems—the cases for Sihl Dam (Chap. 14), to airflow dispersion in complex terrains in urban areas (Chap. 15), to regional vulnerability—Germany and EU (Chap. 16). The last chapter (Chap. 17) offers insights into proposed research along methodological, epistemological, ontological, and nature of man.

The book also contains a set of appendices are meant to complement the book chapters. These include a hierarchical holographic vulnerability assessment model, notes on an emerging domain of *Complex System Governance*, an expert-oriented tutorial for systems many bistable entities, a mix-game elaborating QVA model, a listing of systems theory-based pathologies, lexicography for threat index, and introductory notes on VulPet—a software platform for assessing vulnerability in petrochemical plants. A glossary of terms, based on the 2013 National Infrastructure Protection Plan (USDHS 2013), concludes the matter.

Acknowledgements

Beyond authors' toiling, this book is a measurable expression of their intense intellectual interaction and cross-fertilization of ideas with several distinguished colleagues and partners-in-mind from the Academia and the World of Business of many denominations. Most especially our gratitude goes to the following:

Professor Wolfgang Kröger—ETH Zürich, Switzerland; Dr. Ioannis Papazoglou—National Center for Scientific Research 'DEMOKRITOS', Greece; Adolf Dörig—Dörig + Partner AG, Switzerland; Prof. Radu Cornel—University Politehnica of Bucharest, Romania; Jürg Birchmeier—ETH Zürich, Switzerland; Prof. Charles Keating—National Centers for System of Systems Engineering, Old Dominion University, USA; Marcelo Masera—Directorate for Energy, Transport, and Climate, Institute for Energy and Transport, Joint Research Centre of the European Commission, the Netherlands, and Prof. Liviu Muresan—EURISC Foundation, Romania.

Authors acknowledge the support of the Integrated Risk Governance Project—IHPD/Future Earth under grant number: 2010DFB20880; 2012DFG20710.

Authors are also grateful to graduate students and young researchers at Horia Hulubei National Institute of Physics and Nuclear Engineering, IFIN-HH Bucharest, Romania and ETH Zürich, Laboratorium fur Siecherheistanalytik, Zürich, Switzerland for many mutually seminal discussions and participative enthusiasm.

Norfolk, USA — Adrian V. Gheorghe
Bucharest-Magurele, Romania — Dan V. Vamanu
Norfolk, USA — Polinpapilinho F. Katina
Harpstedt, Germany — Roland Pulfer

References

Ansoff, H. I. (1984). *Implanting strategic management*. Englewood Cliffs: Prentice-Hall.

Cohen, R. E., & Ahearn, F. L. (1980). *Handbook for mental health care of disaster victims*. Hutchinson: Johns Hopkins University Press.

Gheorghe, A. V., Masera, M., De Vries, L., Weijnen, M., & Kröger, W. (2007). Critical infrastructures: The need for international risk governance. *International Journal of Critical Infrastructures, 3*(1/2), 3–19. http://doi.org/10.1504/IJCIS.2007.011543

Martin, J. (2006). *The meaning of the 21st century: A vital blueprint for ensuring our future*. New York, NY: Riverhead Books.

Rasmussen, J., & Batstone, R. (1989). Why Do Complex Organisational Systems Fail? World Bank Environmental Working Paper, 20.

Richardson, B. (1994). Socio-technical disasters: Profile and prevalence. *Disaster Prevention and Management: An International Journal, 3*(4), 41–69.

Tableau. (2016). Top 10 Business Intelligence Trends for 2016. Retrieved September 27, 2016, from http://www.tableau.com/learn/whitepapers/top-10-business-intelligence-trends-2016

Tainter, J. A. (1988). *The collapse of complex societies*. New York, NY: Cambridge University Press.

USDHS. (2013). NIPP 2013: Partnering for critical infrastructure security and resilience. U.S. Department of Homeland Security, Washington, D.C. Retrieved from http://www.dhs.gov/xlibrary/assets/nipp-ssp-national-monuments-icons.pdf

USDHS. (2016). Critical Infrastructure Sectors. Department of Homeland Security. Retrieved September 27, 2016, from https://www.dhs.gov/critical-infrastructure-sectors

Weick, K. E. (1988). Enacted sensemaking in crisis situations. *Journal of Management Studies, 25* (4), 305–317. http://doi.org/10.1111/j.1467-6486.1988.tb00039.x

About the Book

In the face of increasing failures, comments attributed to Albert Einstein loom large upon us: '*we cannot solve our problems with the same thinking we used when we created them.*' A pervasive feeling is that any attempt to make sense of current terrain of complex systems, *must involve thinking out-and-above the box* and originating unconventional approaches that integrate organizational, managerial, social, political, cultural, and human aspects and their interplay.

The present textbook offers research-based models and tools for diagnosing and predicting behavior of complex techno-socio-economic systems in the domain of critical infrastructures, key resources, key assets and the open bazaar of space, undersea, and belowground systems. These models embody emblematic models in Physics, within which the critical infrastructures, as well as the society itself and its paraphernalia, share the profile of *many-body systems* featuring *cooperative phenomena and phase transitions*—the latter usually felt as disruptive occurrences.

The book and its models place emphasis on the analytics of the real-life-business actors, among others, policy makers, financiers and insurers, industry managers, and emergency responders.

Contents

About the Authors

Adrian V. Gheorghe currently serves as Professor of Engineering Management and Systems Engineering and is the Batten Endowed Chair on System of Systems Engineering with the Department of Engineering Management and Systems Engineering at Old Dominion University (Norfolk, Virginia, USA).

Professor Gheorghe holds a M.Sc. in Electrical Engineering from the Faculty of Power Engineering, Bucharest Polytechnic Institute (Bucharest, Romania), a Ph.D. in Systems Science/Systems Engineering from City University (London, U.K.), an MBA from Academy of Economic Studies (Bucharest, Romania), and a M.Sc. Engineering-Economics, Bucharest Polytechnic Institute, (Bucharest, Romania).

Professor Gheorghe serves as Senior Scientist with the European Institute for Risk and Communication Management (Bucharest, Romania) and Vice President World Security Forum (Langenthal, Switzerland). He has worked with different organizations including International Atomic Energy Agency (IAEA), International Institute for Applied Systems Analysis, and Joint Research Centre of the European Commission.

His profile includes editorship for several *international scientific journals, including International Journal of Critical Infrastructures and International Journal of System of Systems Engineering)* and scientific board memberships, including International Journal of Global Energy Issues. He has published several books, including '*Emergency Planning Knowledge*' (VdF Verlag 1996), '*Integrated Risk and Vulnerability Management Assisted by Decision Support Systems: Relevance and Impact on Governance*' (Springer, 2005), and '*Critical infrastructures at risk: Securing the European electric power system*' (Springer, 2006). Consistently addressing the topical area, his recent book is titled '*Critical infrastructures: Risk and vulnerability assessment in transportation of dangerous goods—Transportation by road and rail*' (Springer, 2016).

Dan V. Vamanu is a Senior Researcher with Horia Hulubei National Institute of Physics and Nuclear Engineering (Bucharest, Romania). Holding a Ph.D. in Theoretical Physics (1978). He originally worked in Quantum Theory of Solids, publishing in the field of elementary excitations—phonons and magnons in thin films and impure crystals. He was called to public service in the years of the first major *World Oil Crisis* and served as chief planner, executive manager and international communicator for Romania's programme on new and renewable sources of energy, being active in the United Nations' NRSE Conference process (1980–1986), the World Energy Conference (1974–1983), the World Energy Fair Symposia (1981–1982) in the USA and as UNESCO liaison officer.

Since 1986, in the trail of the Chernobyl events, Dr. Vamanu developed an interest in computer-assisted radiological assessment, nuclear emergency management, risk and vulnerability analysis in complex systems including critical infrastructures and decision support solutions. He rejoined Horia Hulubei National Institute of Physics and Nuclear Engineering (IFIN-HH) in 1990 and substantively contributed to assertions of topics in institute's profile. His work (http://www.nipne.ro/research/publications/148-publications.html) was noted by the Romanian Academy (Horia Hulubei Prize for Physics, 2010) and foreign institutions including the International Atomic

Energy Agency, Nuclear Energy Agency NEA-OECD Data Bank, U.S. Nuclear Regulatory Commission – where he worked as a foreign assignee (1994–1996), French Institute of Radioprotection and Nuclear Safety CEA-IRNS, Swiss Federal Institute of Technology (ETH Zürich, Switzerland), Paul Scherrer Institute and the SwissRe Group (Switzerland), World Institute for Disaster Risk Management, World Bank, Old Dominion University (Norfolk, Virginia, USA). By appointment of these, he acted as a consultant, project member, research contributor and IT provider.

He co-authored a review on thin magnetic films in the American Encyclopaedia *Solid State Physics* (vol. 27, Academic Press), a book—*Emergency Planning Knowledge* (VdF Verlag, 1996) and two monographic volumes—*Integrated Risk and Vulnerability Management Assisted by Decision Support Systems: Relevance and Impact on Governance* (Springer, 2005); and *Critical Infrastructures at Risk: Securing the European Electric Power System* (Springer, 2006). Dr. Vamanu is a member of the Editorial Board of the *International Journal of Critical Infrastructures and the International Journal of System of Systems Engineering.*

Polinpapilinho F. Katina currently serves as a Postdoctoral Researcher at the National Centers for System of Systems Engineering and is an Adjunct Assistant Professor in the Department of Engineering Management and Systems Engineering at Old Dominion University (Norfolk, Virginia, USA).

Dr. Katina holds a B.Sc. in Engineering Technology, a M.E. in Systems Engineering and a Ph.D. in Engineering Management and Systems Engineering, all from Old Dominion University (Norfolk, Virginia, USA). He received additional training from, among others, Environmental Systems Research Institute (Redlands, California), the University of Edinburgh (Edinburgh, UK), and Politecnico di Milano (Milan, Italy). He serves on the Board of Directors for the *International Society for Systems Pathology* (Claremont, California).

His areas of research include critical infrastructure protection, energy systems (Smart Grids), engineering management, decision-making under uncertainty, complex system governance, infranomics, systems engineering, systems of systems engineering, system pathology, and systems theory.

Dr. Katina has co-authored more than 60 peer-reviewed papers to international journals, including *International Journal of Critical Infrastructure Protection, International Journal of Critical Infrastructures, Requirements Engineering, and INCOSE's Insight* and conferences, including ASEM, IEEE, IISE, and WEFTEC. He served as a Guest Editor for the *International Journal of Critical Infrastructures* (2014) and *International Journal of System of Systems Engineering* (2015).

He edited his first book, *Infranomics: Sustainability, Engineering Design and Governance* (Springer, 2014). His second book is titled *Critical Infrastructures: Risk and Vulnerability Assessment in Transportation of Dangerous Goods—Transportation by Road and Rail* (Springer, 2016).

Roland Pulfer has more the 25 years of entrepreneur experience. He founded Action4Value GmbH (Harpstedt, Germany) and currently serves as Strategy Officer at Action4Value GmbH where he conducts research in areas of vulnerability analysis, risk 4.0, and conformance management.

Roland Pulfer holds a Ph.D. in Industrial Engineering and is active in various roles for innovative software companies in domains of enterprise digitization (risks, controls, processes, organization, and infrastructure), holistic search technology, cyber defence, and conformance monitoring.

Dr. Roland Pulfer is Member of the Technical Supervisory Board for Old Dominion University (Norfolk, Virginia, USA). He sits on several editorial boards for scientific journals included the *International Journal of Critical Infrastructures*. He has published several books in the field of enterprise development including *Business (re-) engineering: Best Practices:*

The TopEaseTM Approach (2001) and *Control Your Business: The Balance between Principles and Pragmatism* (2006). Dr. Pulfer holds several patents, 2 in Switzerland and 5 in United States of America, in areas of modelling complex system, model driven enterprise transformation and system resilience governance.

Part I
The Foundations of the Wellbeing

Chapter 1
Critical Infrastructures, Key Resources, and Key Assets

Abstract This introductory chapter articulates the topics of critical infrastructure, key resources, and key assets (CIKRKA). It offers the reader a clear definition of the notions, explain their relevance and also the need to develop intelligible and robust models to diagnose and predict risks, vulnerability, resilience, fragility, and perception. This chapter reveals space, undersea, and belowground as three privileged sectors of human endeavor were critical infrastructures, resources, and assets coexist in the guise of complex systems that tend to assume a leading position in the overall, global CIKRKA system of systems. This chapter sets the stage for the remainder of the book

1.1 Critical Infrastructures

A review of the literature suggests that the performance of and quality of life in the modern society depends, to a large degree, on the quality of its infrastructure. This feeling is shared across governments, industry, as well as academia (Katina and Keating 2015). Allegedly, at the core of this opinion is the concern about the *well-being* of the People. The well-being can be graded from 'high' to 'low' in terms of the social, economic, psychological, spiritual, or medical state of an individual or group of people. A 'high' level of well-being suggests in some sense that the individual or group's condition deserves a positive appreciation, while a 'low' well-being is associated with a negative appraisal.[1]

It should be obvious that a high level of well-being is desirable. Arguably, the efforts to push well-being toward ever higher levels are manifest in the changes occurring in, or imposed upon, our modern society. People increasingly demand, among others, fairness, safety, security, long-term sustainability, and in-time quality services (Casti 2012; Thissen and Herder 2003b; Northrop and Connor 2013; Tainter 1988). For some authors, such demands might be indicative about man's coming to the realization of a painful deficit in being in tune with the surroundings, or rather the 'universe' (Li 2013). Apart from other, metaphysical entanglements,

[1] http://www.thefreedictionary.com/wellbeing.

A.V. Gheorghe et al., *Critical Infrastructures, Key Resources, Key Assets*, Topics in Safety, Risk, Reliability and Quality 34,
https://doi.org/10.1007/978-3-319-69224-1_1

this realization could be linked to increasingly frequent and severe manifestations of risks and vulnerabilities as well as threats that confront humanity in the twenty-first century. In the wake of natural disasters such as Hurricane Katrina, or man-engineered criminal acts such as the 9/11 attacks, people have become acutely aware of their dependency on a number of *amenities* without which life becomes anything from difficult to unbearable. On a more profound level, however, it would immediately become evident that all amenities are products of a long and miscellaneous series of 'physical' and 'virtual' *systems* including the raw materials and processing industries (chemicals, etc.); energy suppliers, transporters, and distributors; nuclear reactors and their fuel cycle industry; dams and other works of land engineering; food and agriculture enterprises; banking and insurance; commercial outlets; communication facilities; defense industrial bases; emergency services; government facilities, health care, and public health; information technology; waste management and treatment industries; and many others (Gheorghe 2004; Kröger and Zio 2011; Masys 2015; Obama 2013). This immensely varied, complicated, and loosely structured yet highly interactive on the inside machinery to generate and maintain the well-being was somehow expeditiously baptized 'infrastructure' encompasses a broad range of subsystems involving physical sectors (e.g., roads and electrical systems) as well as virtual, or 'soft' systems involving information and telecommunications as well as supervisory control and data acquisition (SCADA) systems (GAO 2004). This view of infrastructure systems comes in tune with the European Council stand, according to which critical infrastructures 'consist of those physical and information technology facilities, networks, services and assets which, if disrupted or destroyed, would have a serious impact on the health, safety, security or economic wellbeing of citizens or the effective functioning of governments in the Member States' (European Council 2004, p. 3). Table 1.1 provides a collection of more or less formal definitions for the critical infrastructures. Failure of such systems—regardless of the source—anthropic or natural—can have an alarming impact and consequences on the public well-being, which extends to individuals, business, government, as well as the environment. It comes as no surprise that some authors suggest that 'failure of these infrastructures…is one of the most important vulnerabilities of modern society' (Thissen and Herder 2003b, p. 1). As an example, it was pointed out, Hurricane Katrina, which made landfall in New Orleans (Louisiana) on August 29, 2005, was the costliest and one of the five deadliest hurricanes to ever strike the USA'S with an estimated damage of over $100 billion dollars (Knabb et al. 2011; Townsend 2006).

Throughout human history, there were always systems that failed (Sandage 2006). However, there is a sense that failure of systems is increasing with a higher level of calamity (Ansoff 1984; Cohen and Ahearn 1980; Richardson 1994; Weick 1988). Thus, the Rasmussen and Batstone (1989) assertion that: 'the frequency and magnitude of organizational failures and the subsequent impacts are increasing at an alarming rate' (Rasmussen and Batstone 1989, p. ii) should stand true. This conjecture is supported by many researchers, including an annual analysis by Swiss

Table 1.1 Defining features for critical infrastructures

Critical infrastructure perspective themes	Author(s)
…systems and assets, whether physical or virtual, so vital to the United States that the incapacity or destruction of such systems and assets would have a debilitating impact on security, national economic security, national public health or safety, or any combination of those matters	US Congress (2001) p. 115, Stat. 401
…consist of those physical and information technology facilities, networks, services and assets which, if disrupted or destroyed, would have a serious impact on the health, safety, security or economic well-being of citizens or the effective functioning of governments in the Member States	European Council (2004, p. 3)
…organizational and physical structures and facilities of such vital importance to a nation's society and economy that their failure or degradation would result in sustained supply shortages, significant disruption of public safety and security, or other dramatic consequences	Germany Federal Ministry of the Interior, FRG (2009, p. 4)
…are so vital that their incapacity or destruction would have a debilitating impact on the defense or economic security of the United States	Clinton (1996, p. 37347)
…large scale, man-made systems that function interdependently to produce and distribute essential goods (such as energy, water and data) and services (such as transportation, banking and health care). An infrastructure is termed critical if its incapacity or destruction has a significant impact on health, safety, security, economics and social well-being	Zio (2016, p. 3)
Infrastructures are critical because they provide services that are vital to one or more broad governmental or societal functions or attributes. This can be related to survivability of citizens as far as the safety of their life is concerned, or to their quality of life	Gheorghe et al. (2006, p. 5)
The domain of critical infrastructures deals with engineering systems which are characterized by a high degree of technical complexity, social intricacy, and elaborate processes, aimed at fulfilling important functions in the society	Gheorghe and Katina (2014, p. 195)
…current approach to CI protection and mitigation primarily focuses on large malicious and cataclysmic events of terrorism, cyber-attacks, and natural events… [there is] … need to understanding the slow, evolving, and inane events that could accumulate into significant events over-time	Calida and Katina (2012, p. 87)

Reinsurance.[2] This phenomenon has created, at least in the domain of critical infrastructures, a need for grading the operating state of an infrastructure system as 'under threat,' 'vulnerable,' 'operable,' 'inoperable,' as well as a need to develop means to strengthen the security of such systems. Understanding the status of an infrastructure system is one of the key aspects of infrastructure systems research (USDHS 2013). In fact, the *International Risk Governance Council* notes that infrastructures 'have been, remain and will probably always be subject to changes of different degrees and speed, with technological development and market liberalization as the current drivers' (IRGC 2006, p. 49). Subsequently, it makes sense to try and understand and monitor the status of infrastructures and identify the changes that might influence such systems. The changes in question can take different forms; rapid technological changes driven by information and telecommunications (ITC); institutional changes involving a shift from public to private partnerships; increasing complexity as a result of interplay between technology, behavior and policies; as well as increasing concerns for a sustainable planet (Thissen and Herder 2003b; IRGC 2006).

Complexity in this case conforms to Sussman's (2005) definition of a system composed of a group of related units (subsystems) for which the degree and nature of the relationship are imperfectly known. When a system is in this state, its behavior and structural patterns are always in constant flux making it difficult to understand and ascertain any useful knowledge (Sousa-Poza et al. 2008). Nonetheless, efforts were made in regard to understanding 'relationships.' The term 'relationship' is used to refer to 'mutually reliant relations between systems' and is often assimilated with interdependency in infrastructure research. Debatably, efforts to understand complexity in infrastructures can begin with understanding interdependencies. The Rinaldi et al. (2001) research may provide a starting point into exploration of interdependencies. Six categories of interdependency are suggested by various researchers (Dudenhoeffer et al. 2006; Katina et al. 2014; Rinaldi et al. 2001). Table 1.2 provides a summary of types of infrastructure interdependencies that could be used in ascertaining system relationships.

It is widely accepted that stronger interdependencies across critical infrastructure systems, especially the heavy reliance on ITC, 'have increased the potential vulnerabilities to physical and cyber threats and potential consequences resulting from the compromise of underlying systems or networks' (USDHS 2013, p. 8). It would appeal to reason, as suggested by Calida and Katina (2012), that in a network of interdependent systems, a system cannot be expected to operate as an isolated entity, this implying that even a seemingly isolated system or its operations can affect operations of another system far from its point of origin. This is especially the case when critical infrastructure systems cross physical national boundaries.

In the present research, and certainly within the context of interdependencies, two corollaries are submitted:

---

[2]More information can be found at http://www.swissre.com/.

Table 1.2 Type of infrastructure interdependencies

Interdependency type	Relevant themes	Implications for infrastructure development
Physical interdependency	This is a relationship that 'arises from the physical linkage between the inputs and outputs of two agents [where the] commodity produced or modified by one infrastructure (an output) is required by another infrastructure for it to operate (an input)' (Rinaldi et al. 2001, p. 15) such as drinking water and electricity	A consideration of the direct and physical influence of external systems including outputs, product, goods, and services to a system of interest. For example, an operator of a water system should be concerned with risks in electrical grid, since availability of clean drinking water is physically dependents on electricity used in water treatment
Cyber interdependency	A relationship based on ubiquitous and pervasive use of information and communications technologies (ICT). Many critical systems provide essential goods and services with the help of control systems such as supervisory control and data acquisition (SCADA) systems that remotely monitor and control operations used coded signals over ICT communication channels (Katina et al. 2016a; Rinaldi et al. 2001)	A consideration of cyber interdependency could enable one to examine the nature of reliance of ICT within a given scenario. This analysis might include an overview of a cyber aspects a system including an articulation of relation between internal and external systems, processes monitored and controlled, types of SCADA architectures deployed (i.e., 1st generation, 2nd generation, 3rd generation, 4th generation), and cyber-related risks as well as countermeasures
Geographical interdependency	This is a relationship that exists when different infrastructure systems share the same environment such as electrical power lines that share the same corridor with a bridge (DiSera and Brooks 2009; Katina et al. 2014)	This involves a consideration of geographical interdependencies associated with a need for common environment that typically enables coupling of infrastructure systems. Coupling creates a situation in which an attack on one is an attack to all. For example, a destruction of a bridge affects electricity transmission, if there is a shared corridor
Logical interdependency	A logical interdependency exists between infrastructures if the state of each infrastructure depends on the state of the other via some mechanism that is neither physical, cyber, nor geographical (Rinaldi 2004) such as power deregulation policy	An exploration into 'other mechanisms' beyond physical, cyber, and geography. Other mechanism could involve the role of time, space, perception, and geo-politics
Policy and/or procedural interdependency	This is a 'hidden' and not-so-obvious relationship that only becomes apparent after a	Attempts to feedforward and development of scenarios that might offer insights into how quality of

(continued)

Table 1.2 (continued)

Interdependency type	Relevant themes	Implications for infrastructure development
	change, in form of a policy and/procedure that takes effect in one part of the system. For example, several regulations that were issued in the wake of 9/11 attacks affected all air transport systems, changing the flying experience (Mendonca and Wallace 2006)	goods and services could be influenced by changes in policy at national, state, regional, and local levels. The intent of such efforts is the discovery of possible direct effects of changes as well as 'unintended' consequences on critical infrastructures
Societal interdependency	Societal interdependency is a situation in which infrastructure operations are affected by public opinion. For example, after 9/11 attacks, air traffic was reduced due to the public's evaluation of travel safety, resulting in job cuts and bankruptcies (Dudenhoeffer et al. 2006; Katina et al. 2014)	This analysis involves examination of public opinion on critical topics as they relate to infrastructure goods, services, and operations. The intent of such efforts is an attempt to understand impact of infrastructure operations. This might include, for example, understanding public perception of emerging concepts, for example, Smart Grids

- First, 'know' your system. This 'knowing' might involve articulation of the system as an infrastructure with a well-defined purpose along with stakeholders such as owners, operators, and clients (Blanchard and Fabrycky 2006).
- Second, include the 'environment' of the infrastructure in the analysis. In systems research, the environment is often projected as anything beyond the system itself. According to Skyttner (2005), the environment is actually anything 'outside of the direct control of the system and [includes] any phenomenon influencing the processes and behaviors of the system' (Skyttner 2005, p. 63).

These corollaries are essential in creating a matrix instrumental in the discovery and measuring interdependencies (Setola 2010). Figure 1.1 provides an example of an assessment matrix as suggested by IRGC (2006, p. 50). The figure illustrates infrastructure dependencies. The red-colored boxes represent a high degree of dependence, while the green-colored boxes represent a low infrastructure dependence. The yellow-colored boxes represent an in-between relationship. The split in colors projects a transition phase in a form of a trend. It should be noted that in the absence of a more detailed assessment and analysis, the categorization of interdependencies along with an assessment matrix could be used as a starting point for risk governance strategies.

These strategies might inspire the development of a strategic approach for infrastructure protection. In fact, different parameters such as likelihood of threat, vulnerabilities, and uncertainty consequences associated with terrorist activities, natural disasters, and accidents could be incorporated in similar approaches (Bush et al. 2005). This may open a discussion on methods and tools for modeling

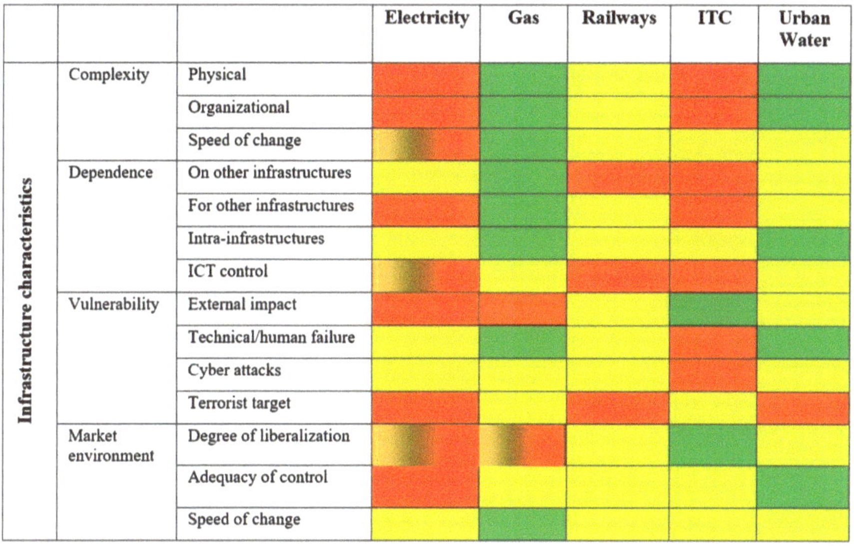

Fig. 1.1 Assessment matrix for infrastructures, modified from IRGC (2006)

structures and simulating the behavior of infrastructure systems (Fletcher 2002; Niemeyer 2004; Nozick et al. 2004). There is a trend to use modeling and simulation (M&S) techniques to reveal infrastructure dependencies, within specific sectors (Calida and Katina 2015; Kröger 2008). These techniques include, but are not limited to, network topology (Eusgeld et al. 2009), graph theory (Garrett et al. 2011), fully coupled blast effects modeling (McMichael et al. 2009), and Multi-area Thevenin Equivalent [MATE] (Rahman et al. 2011). For example, *Critical Infrastructure Modeling System, CIMS,* is an agent-based approach that can be used to model an infrastructure, its elements, and relationships, in order to understand individual component behavior (Dudenhoeffer et al. 2006; Permann 2007). CIMS provides means to model cascading effects and consequences in infrastructure systems through observation of emergent infrastructure behavior. CIMS can build a model from a simple underlying bitmap, satellite photographs, maps, or charts.

At this point, it must be stressed that there are still challenges associated with modeling infrastructure systems as 'wholes.' It should be noted that the term 'whole' is used to suggest a 'sector' in a domain of critical infrastructure such as transportation system. As suggested by Eusgeld et al. (2008), modeling and simulating such a system with its interdependent systems are a difficult endeavor. It is easy to attribute such difficulty to, among other factors, the availability, accessibility, and validity of data and the exchangeability between modeling tools (Eusgeld et al. 2008). Beyond that, however, Pederson et al. (2006) provide a more radical and compelling response: There is a lack of a well-developed framework that could be used across all critical infrastructure sectors. The two commonly used frameworks, high-level architecture (HLA) and the distributed interactive simulation (DIS), Pederson et al. (2006) suggest, have multiple disadvantages and are very limited in applications. Moreover, while current M&S techniques are useful in physical interdependencies, 'doing so on a large scale is a resource challenge' (Pederson et al. 2006, p. 18). Our intent is not to undermine the utility of M&S tools and techniques, rather, we point at a need for robust methods and tools capable of holistic analysis of infrastructure systems. Such efforts are especially of critical importance in areas where interests converge on interdependent infrastructure systems at regional, national, and international scales.

There are further complications within the present discourse. The first one is ontological in nature: There is no clear line of demarcation indicating what is 'critical' and what is not. It appears that as changes occur in society, over time, different infrastructure systems, previously non-critical, are identified as critical. This observation, made explicit by Katina and Keating's (2015) research in comparing the Patriot Act of 2002 and the PCCIP of 1996, suggests that we might be on a course of not granting the due attention to infrastructure systems essential to public well-being. Avoiding such misconduct would call upon us to embrace a broadening of what constitutes critical infrastructures. Attempts to broaden the current perspectives are, in fact, the basis for the remainder of this book.

The second difficulty adds a deontological or ethical dimension to our conceptual problems. The embarrassment begins the moment one realizes that discussing critical infrastructures from the angle of 'securing the well-being of people' may mean skidding into an attitude fraught with arrogance. For indeed, what is 'well-being,' for whom?

For some, it is securing water, food, and shelter—period.

For others, it is securing water, food, shelter, health care, and education.

Again, for others, it is securing water, food, shelter, health care, education, a safe and sound natural environment, equal opportunities, human rights, and, in a word, freedom.

And for each category, 'infrastructure' would have a different meaning, its contents ever incrementing while carrying with it also the criticality threshold.

Probing into the complexities of the issue above would require another book—and one definitely worth writing. Not for us, here and now. At the risk of losing two-thirds of the potential readers worldwide, who would prefer reading 'the other book,' we will stick to only the technical side of the matters and talk critical infrastructures in their most heavy, sophisticated, arrogant, and vulnerable version—the one that, if poorly managed, can take us all down.

First, let us address two key aspects of critically important systems: key resources and key assets.

1.1.1 Key Resources

The general concept of critical infrastructures, as previously mentioned, has penetrated and became customary in many sectors. Along with it, two other terms make headway in the literature concerning critical infrastructures: 'key assets' and 'key resources.' At a fundamental level, one needs to know that both key assets and key resources are worth protecting, similarly to 'critical infrastructures.' The US Patriot Act of 2001 (US Congress 2001) provides one of the most widely used definitions for critical infrastructures. This document influenced subsequent acts including the Homeland Security Act of 2002 which introduces the concept of 'key resources.' A key resource was defined as 'a publicly or privately controlled resource essential to the minimal operations of the economy and government' (USDHS 2002, p. 116 STAT. 2141).

In highlighting the importance of key resources, the Homeland Security Act of 2002 called for a comprehensive assessment of their vulnerabilities including risk assessment aimed at types of risk, probability of attack occurrence, feasibility, as well as efficacy of possible countermeasures. Interestingly, this act did not offer any specific examples of key resources (Moteff and Parfomak 2004). However, Bennett (2007, p. 54) suggests that the 'destruction of a key resource would not endanger vital systems but could cause large-scale injury, death, or destruction of property and/or profound damage to our national prestige and confidence.' Clearly, these definitions underscore the importance of understanding the nature of key resources including risks, threats, and vulnerabilities.

1.1.2 Key Assets

Nearly a year after the Homeland Security Act of 2002, the President of the USA introduced the *National Strategy for the Physical protection of Critical Infrastructures and Key Assets*. The strategy introduced the concept of 'key assets.' Key assets were defined as 'individual targets whose destruction would not endanger vital systems, but could create local disaster or profoundly damage our Nation's morale or confidence' (Bush 2003, p. 7). There are three categories associated with key assets, described in Table 1.3. Similar to 'key resources,' 'key assets' can be distinguished from 'critical infrastructures' based on a 'scale of damage.' In the *National Strategy for the Physical protection of Critical Infrastructures and Key Assets*, it is noted that key assets while 'alone may not be vital to the continuity of critical services on a national scale, but an attack on any one of them could produce, in the worst case, significant loss of life and/or public health and safety consequences' (Bush 2003, p. 7). The importance of key assets (e.g., historical attractions and monuments) relates to the fact that such assets are seen as being part of a heritage of a given society and as such, an attack on these, is also an attack on the society's psyche and lifestyle.

In summary, the following are the most salient features of the conceptual triad of 'critical infrastructures,' 'key resources,' and 'key assets' (Katina and Keating 2015):

Table 1.3 Three main categories of key assets

Key asset category	Definition	Examples
Category I	This category comprises the diverse array of national monuments, symbols, and icons that represent our nation's heritage, traditions and values, and political power. Such asset attracts large number of tourists and frequent media attention	Prominent historical attractions, monuments (e.g., African Burial Ground), cultural icons, and centers of government and commerce
Category II	This category includes facilities and structures that represent national economic power and technological advancement. These assets tend to house significant amounts of hazardous materials, fuels, and chemical catalysts that enable important production and processing functions	Specialized medical and pharmaceutical laboratories, nuclear power and chemical plants, and hydro electrical dams. Failures in such systems can have significant impact on public health and confidence, and the economy
Category III	This category includes such structures as prominent commercial centers, office buildings, and sports stadiums, where large numbers of people regularly congregate to conduct business or personal transactions, shop, or enjoy a recreational time	Entertainment and media (e.g., motion picture studios, broadcast media), public assembly (e.g., arenas, and stadiums), and sports leagues (e.g., professional sports leagues)

- First and foremost, incapacitating/crippling/destroying 'critical infrastructures,' 'key assets,' or 'key resources' can have a significant impact on public health and safety, public confidence, and the economy. In fact, continued operations of such systems support daily societal activities, to the point that such systems are normally taken for granted.
- The concepts 'critical infrastructures,' 'key resources,' and 'key assets' encompass elements of physical, 'hard' systems, such as roads and highways, and 'soft' systems that control hard systems, such as SCADA, and, therefore, the distinction between a 'critical' and non-critical system can be difficult when addressing varied stakeholders.
- These systems are all prone to be confronted with naturally occurring disruptive events such as earthquakes and hurricanes as well as with threats posed by man including human error, accidents, and malicious attacks.
- Critical infrastructures, key resources, and key assets (CIKRKA) largely operate into the open. The fact makes them easy targets for attacks from several agents seeking to exploit system vulnerabilities.
- The ever more intense use of information technologies and easy access to powerful computing continue to close the 'gap' among infrastructure systems, so that the number of systems operating as 'isolated systems' is diminishing, and networked configurations tend to prevail. In such networks, cascading failures are always possible.

- As we move deeper into the twenty-first century, the society will continue to be shaped by social changes involving a higher demand for quality and safer products, goods, and services, the inevitable globalization, and private–public governance policies, among others. In this environment, the way we decide to view and use concepts such as protection, management, and control will need to be thought over again, especially in relation to infrastructure systems and the evolving canvas of threats enshrouding them.
- In our search for measures to provide protection and security, efforts should not be limited to those infrastructures deemed as 'critical.' Appropriate protection and security should also cover 'key resources' and 'key assets.' In fact, a successful malicious attack could instill fear and therefore affect the morale of infrastructure, resource, or asset owners/managers (de Silva 2016).

These themes suggest a need for robust approaches for dealing with infrastructure systems. In the case of risk management for infrastructures, the *National Infrastructure Protection Plan* (NIPP) developed a framework, originally introduced in 2006 and modified in 2013. The framework, illustrated in Fig. 1.2, supports a decision-making process involving collaborative efforts of infrastructure owners and operators in dealing with various risks. It is not meant as a prescriptive approach for managing risk; rather, it is meant to be an 'organizing construct' for thinking and doing risk management in CIKRKA (USDHS 2013).

It should be noted that the solution proposed considers three elements of critical infrastructures: The physical, cyber, and human aspects that are dealt with in a sequence of five key phases as described in Table 1.4.

In summary, the NIPP risk management framework 'enables the critical infrastructure community to focus on those threats and hazards that are likely to cause harm, and employ approaches that are designed to prevent or mitigate the effects of those incidents. It also increases security and strengthens resilience by identifying and prioritizing actions to ensure continuity of essential functions and services and support enhanced response and restoration' (USDHS 2013, p. 16).

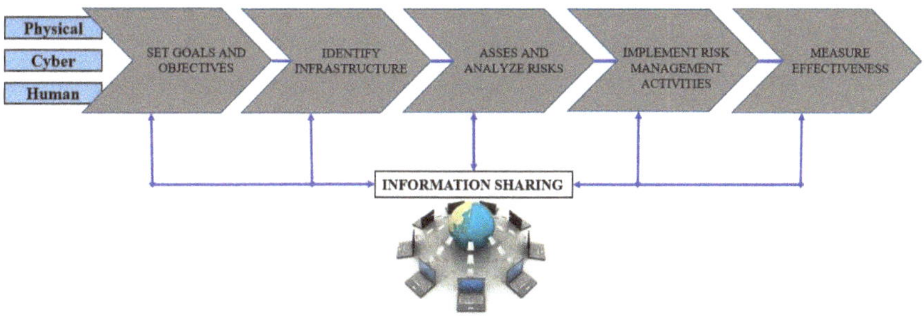

Fig. 1.2 NIPP critical infrastructure risk management framework, adapted from USDHS (2013)

1.2 Terminology

The introductory notes of this book suggested the importance of extending traditional concepts to the novel situations confronting the analysts. It is only reasonable that we provide 'working' definitions of such concepts in the context of the current research. The reader is directed to the *Glossary of Terms* for a more comprehensive and extensive listing and definition of related ideas.

1.2.1 Risk

The term *risk* enjoys many definitions accepted to various degrees, and no definition unanimously accepted. It is present, used and debated in the literature for years now (Holton 2004; Knight 1921; Komljenovic et al. 2016). Risk is typically defined in terms of *probability* of occurrence of an event and the magnitude of the expected *consequences* (ASCE 2009). Additionally, risk is often associated with *uncertainty*. In terms of system life cycle, risk is associated with uncertainty and opportunities related to cost, schedule, and performance (INCOSE 2011). The INCOSE handbook notes that 'every new system or modification of an existing system is based on pursuit of opportunity' and that 'Risk is always present in the life cycle of systems…' due to technical factors (INCOSE 2011, p. 214). In decision making, risk is associated with probabilities of *unknown outcomes* and uncertainty (Gibson et al. 2007). Risk has also been defined as 'the *potential* that something will go wrong as a result of one or a series of events' (Blanchard 2008, p. 344) and is equated to 'a *probability* event' (Garvey 2009, p. 33).

Several definitions imply that *risk* is that which happens without one's planning, anticipation, or intent. Moreover, Hill (2012) posits that the notion of risk can also be subjective and for the most part, a *mental construct*. Yet another perspective suggests that there can be different *levels of risk* (Gheorghe et al. 2000). Arguably, critical infrastructures can be *exposed* to different types of risks since they do not operate in isolation in relation to internal and external interfaces. Furthermore, it is reasonable to assume that critical infrastructures are always *under threat* from naturally occurring events such as flooding, drought, pandemics, as well as malicious attacks. In assessing risk for infrastructure systems, the objective can be associated with a risk classification as 'accepted/acceptable' or 'unaccepted/unacceptable.' This is typically done to enable decision making, allocation of scarce resources, and policy formulation and/or revision. Moreover, Vamanu et al. (2016) suggest that there are several possible angles in assessing risk, varying based on targeted industry, and analyst concerns including, among others, environmental, ecological, and public health. It stands to reason that some domains, such as the nuclear, aerospace, oil, rail, and the defense, might have well-defined risk assessment approaches and methods because of the long-standing availability of statistical data rooted in historical accounts and a consolidated safety culture. However, there

Table 1.4 Phases and objectives of NIPP risk management framework, adapted from USDHS (2013)

Phase	Objectives
Set infrastructure goals and objectives	At the national level, this phase calls for the establishment of a set of broad national goals for critical infrastructure security and resilience. Owners and operators of critical infrastructure at regional entities identify objectives and priorities for critical infrastructure consistent with national priorities, national goals, and sector objectives, tailored and scaled to their operational and risk environments and available resources
Identify infrastructure	To manage critical infrastructure risks effectively, assets, systems, and networks that are essential for continued operation must be identified, considering associated dependencies and interdependencies. It was observed that not all actors involved in this process (governments, industry, academia) view infrastructure systems from the same perspective (Katina and Keating 2015). This includes differences in appraising infrastructure 'criticality' at the national, regional, and sector levels (USDHS 2013)
Assess and analyze risks	Critical infrastructure risks are assessed in terms of threat, vulnerability, and consequence. In the context: • 'Threats' are agents and/or circumstances having the potential to harm life, information, operations, the environment, and/or property • 'Vulnerability' represents a physical feature or operational attribute that renders an entity open to exploitation or susceptible to a given hazard • 'Consequence' is an effect of an event, incident, or occurrence There is a broad range of methodologies that can be used in risk assessment enabling informed decision making
Implement risk management activities	This phase is concerned with prioritizing activities to manage critical infrastructure risk based on the criticality of the affected infrastructure, costs, and potential for risk reduction. Implementation activities include: • Identify, deter, detect, disrupt, and prepare for threats and hazards • Reduction of vulnerabilities • Mitigating the consequences
Measure effectiveness	Once risk management activities are implemented, the next phase involves an evaluation of the effectiveness of risk management efforts within sectors and at national, state, local, and regional levels by developing metrics for both direct and indirect indicator measurement. This phase serves, among others, as the basis for rearticulation of vision and national and regional goals and assessing progress

is a need for new methods, tools, and techniques for assessing new and emerging threats such as those associated with cybersecurity in cyber-physical systems. In spite of this rich—if sometimes confusing—conceptual landscape, central to risk remains a construct defined in terms of probability and consequence, with variations, and visualized with a *risk matrix* conducive to a mitigative policy based on a corrective intervention *As Low as Reasonably Achievable* (ALARA).

1.2.2 Vulnerability

Like risk, the concept of *Vulnerability* has many definitions accepted to various degrees, and no definition unanimously accepted (Katina et al. 2014; Song 2005; Vamanu et al. 2016). In fact, 'vulnerability' was long considered as being closely similar to risk, if only with a broader interpretation. However, Song (2005) notes that some authors make a clear distinction between vulnerability and risk. For example, Turner et al. (2003) depict *vulnerability* as a degree to which a system, subsystem, or system component is likely to experience harm due to exposure to a hazard, either a perturbation or a stress/stressor.

In Einarsson and Rausand (1998) as well as Holmgren et al. (2001), vulnerability is defined as the properties of a system that may weaken or limit its ability to survive and perform its mission in the presence of threats that originate both within and outside the system boundaries. Song's (2005) research establishes a critical difference between vulnerability and the degree of vulnerability: *vulnerability is the susceptibility and resilience/survivability of the community/ system and its environment to hazards. Susceptible comprises two aspects: exposure and sensitivity; survivability mainly comprises robustness, reliability, redundancy, and adaptation four aspects* (Song 2005, p. 15). The degree of vulnerability is the numerical index of the vulnerability based on different criteria, usually in the range 0–100% (Song 2005).

Aven's definition appears to be consistent with other research when invoking 'manifestation of the inherent states of the system that can be subjected to a natural hazard or be exploited to adversely affect that system' (Aven 2011, p. 515). Regardless of diverging perspectives on vulnerability definition, there is consensus on the need to consider vulnerability in system assessment—a clear indication that, so far, 'something was missing' for the analyst's peace of mind. Recognizing the need, the *International Risk Governance Council* stipulates that vulnerability is a viable area of research, especially for coupled critical infrastructure systems. This might be attributed to 'basic weaknesses, such as over-complexity and traded-off security factors, and [the fact that such systems] face multiple threats, including exposure to natural hazards and malicious attacks' (IRGC 2007, p. 4). The more essential is the vulnerability issue in critical infrastructures when one remembers that most infrastructures operate out into the open and are therefore exposed to different elements.

In distinguishing between vulnerability and risk, Song (2005) directs attention to the differences in the manner of analysis associated with the two concepts. In risk assessment, one might select a particular stress (or threat, hazard) of concern and seek to identify consequences for a variety of system properties. In contrast, in vulnerability assessment, one selects a particular system (or component) and

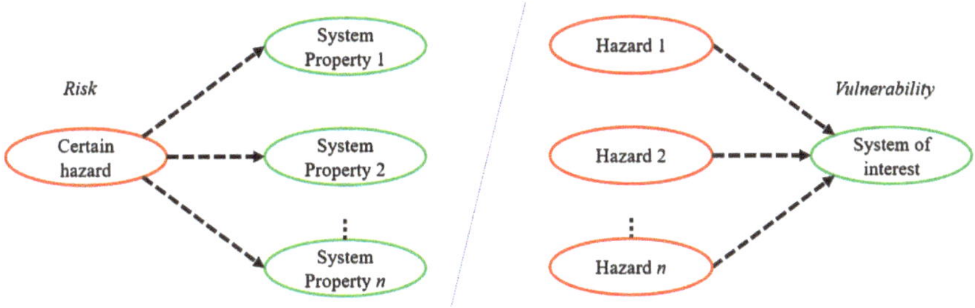

Fig. 1.3 Difference between objectives of risk assessment and vulnerability assessment, modified from Song (2005)

examines how it can be affected by a variety of stressors. Obviously, such an analysis will involve identification of means to reduce vulnerability (Tokgoz and Gheorghe 2013). In Song's clear-cut words, vulnerability describes 'inherent characteristics of a system that create the potential for harm but are independent of the risk of occurrence of any particular hazard' (Song 2005, p. 19). Figure 1.3 depicts the difference between risk and vulnerability. Apart from the manner of assessment, differences can also be pointed at in the scope of the analysis as well as in emphasis (Einarsson and Rausand 1998; Song 2005). For those interested in a step-by-step procedure for identifying, prioritizing, and managing risks, a framework, Hierarchical Holographic Vulnerability Assessment, is presented in Appendix A.

Again, like with risk, there exist different models for dealing with vulnerability. Further discussions on these can be found elsewhere (see, e.g., Vamanu et al. 2016). In the present text, the *quantitative vulnerability assessment* (QVA) takes the central role. The QVA is a result of a warranted analogy with *quantitative risk assessment* (QRA)—a term coined in the closing decade of the past century and having made quite a career in the community of risk and safety managers, worldwide. Like its risk-related counterpart, QVA is about expressing its object—vulnerability—in numbers, in a scientifically defendable and practically meaningful way.

In one approach to QVA, commented upon in this book, vulnerability is described as 'a predictive quantity reflecting system's selective stress reaction toward respective threat' (Vamanu et al. 2016, p. 95). Details aside, it suffices to say that model's 'assumption zero' was that critical or otherwise complex real-life structures and systems can be accommodated within the concept of a multi-component, multi-indicator system; the parts of which would show some kind of collective behavior by virtue of their interactions, as well as some susceptibility to external factors acting upon the structure as a whole. To quantify the vulnerability of such a generic system, the model rests on two control variables and an equation of state. The model input includes an arbitrarily large number of indicators accounting for the system internal processes (fast-varying) as well as external forces (slow-varying) assumed to uniformly act upon the components of the entire structure. The model output is a membership fraction qualifying the integrity of the system in terms of operability defined in terms of the proportion of 'operable' versus 'inoperable' states of the system. The solution is directly conducive to the definition of a *two-parameter phase space* of the system, where three *vulnerability basins* may be identified: (i) system stable—low vulnerability; (ii) system and vulnerability—critical; and (iii) system unstable—high vulnerability. Moreover, a 0–100 *Vulnerability Scale* and the means to measure the respective *Vulnerability Index* were offered as an operational expression of a QVA.

Talking 'basins of vulnerability' is to highlight an important issue: *How tolerable may one deem the vulnerability*. This implies that an operator or user in a system, or the society, plays a major role in determining the vulnerability level at which they are comfortable. Thus, a model aimed at capturing the concept of vulnerability needs to be developed based on user belief system. In so doing, it is reasonable to expect that a low (or acceptable) vulnerability may turn to 'critical' or even high (or unacceptable)—which duly introduces in the assessment a requisite element of subjectivity—the stakeholder perception. Therefore, QVA can be seen as a method to diagnose vulnerability, as well as an approach to dynamically monitor the time evolvement of the vulnerability as the indicators and the perception change.

Through the years, the generic model described was implemented in a number of applications targeting different systems and critical infrastructures. Like with many other vulnerability assessment models meanwhile developed, central to a sound QVA approach was *to see and treat vulnerability as an evolving system state*, entailing such notions as *susceptibility, resilience, and fragility*. Quantified, vulnerability could be graded, mapped and managed, to the extent of even conceiving policies for systems, including critical infrastructures, *As Resilient as Reasonably Achievable* (ARARA).

1.2.3 Resilience

No matter how sophisticated in intentions and scope, the vulnerability analysis of a system will always rest on two queries: (i) How *susceptible* is the system to suffer from threats, if these materialize; (ii) to what degree can a system *recover* from an effective threat hit, and how fast—which in the context translates as 'how *resilient* the system is.' True enough, both susceptibility and resilience can be convincingly quantified. Thus, *susceptibility* may be taken as the amount of damage per unit of hit intensity, expressed in, for example, monetary terms; whereas resilience may be seen as a Cartesian product of an *acceptable* percent recovery of system operability by the time lapse to reach that percentage. The fact feeds the allegation that vulnerability may just be 'a function of susceptibility and resilience'—something like $V = f(S, R)$ as suggested by Song (2005), which, while basically true, may, however, deprive the concept of vulnerability from a lot of the complexities that have made it worth recognition and appreciation.

Leaving aside a scholastic debate of the issue, one may rather focus on the accepted perception that resilience is, fundamentally, about system *ability to withstanding a threat/attack* (Gheorghe and Katina 2014; Martin-Breen and Anderies 2011), drawing attention to the fact that, in effect, 'to withstand an attack' can almost univocally be equivalent to *recover from the hit up to an acceptable, fractional level from its original, nominal operability*. The concept of resilience is important in the realm of critical infrastructures for several reasons, among which one may stand out: *It is possible that a system be susceptible, and yet has a high resilience*. And, from a vulnerability standpoint, the later feature is overwhelmingly consequential.

The nature of critical infrastructures, especially from the point of view of goods and services, calls for the need to recover from effective disruptions to the largest possible (affordable) extent and as soon as possible. This is, in part, an implicit recognition that hazards and threats will eventually materialize. Failure to return to some degree of normality can easily lead to crisis and have a debilitating impact on people. Especially, severe are such effects in interdependent systems, as in the case of the California Electricity Crisis of 2000 and 2001, when the state of California suffered large-scale blackouts, collapse of several companies, and eventual political turmoil having roots in electricity supply shortages caused by market manipulations, illegal shutdown of pipelines, and capped retail electricity prices (Sweeney 2002).

Second, in dealing with infrastructure systems, it may be of interest to realize that being susceptible to threats does not automatically reduce the expected performance of a given infrastructure. In fact, it is possible to have a susceptible system that has a high level of resilience and, thereby, an acceptable level of vulnerability. This would certainly imply specific mechanisms to enhance system capabilities to withstand hazard impacts and solutions to promptly provide the required recovery resources—provisions that should best be considered from the system design phase, on a direct *and indirect* cost/benefit base.

Indeed, in a broader sense, interest in infrastructure resilience can be concerned with evaluating existing resilience mechanisms—technical, logistic, and operational, relating to selected potential disasters. Such considerations should provide the basis for developing prioritization and benefit trade-offs for investments to increase resilience, preparedness, and response capabilities for a given system of interest, and even regions. Moreover, prioritization and ranking can also serve as the basis for the following (Song 2005):

- Identifying most critical scenarios related to the concerned unwanted situations.
- Ranking the vulnerable points to provide a foundation for allocation of limited resources and establishment of mitigation plans.
- Determining the dominating vulnerabilities of the system for further infrastructure assessment.

In such analysis, *sensitivity* and *attractiveness* are surrogate measures for the likelihood of disruptive events happening (see e.g., Song 2005 and Vamanu et al. 2016). For sensitivity, one needs to examine whether or not the vulnerable points are on critical stress under various internal or external unfavorable conditions. For attractiveness, one assumes that the more vulnerable a point, the more it makes an 'attractive' target for adversities—be these of natural (random, unintentional) or anthropic (intentional, malevolent) origin. Various conditions in the natural and anthropic/societal environment of the system should be examined to determine their potential to represent adversities for an infrastructure.

One set of features seems to increasingly gain prominence in the strive to ever improve infrastructure resilience, namely the defensive properties of the system of interest. An investigation on this topic offers insights into a number of aspects including system protective characteristics, maintenance capability, deterrence, detection, delay, adaptability, robustness, reliability, redundancy, and availability of warning systems. The following recommendations stand out as particularly relevant in the context (Katina et al. 2016b):

- Adopt a variety of mechanisms intended to preemptively boost protection measures of a system of interest.
- Take a reactive approach to the threats to the system of interest. This could be done through implementation of several measures that could be used to resist attacks on a system.
- Take steps to ensure a capability to preserve or improve the operability of a system of interest despite attempts to distort it.
- Consider means to discourage attacks on a system in order to secure operability.
- Develop the ability to identify concealed threats that could affect system operability.
- Develop abilities and capabilities to impede an attacker from penetrating into a system, physically or otherwise.
- Develop the ability to respond to and recover from a disruptive event as soon as possible.
- Design systems having the ability to detect and delay threats as well as having a capability to alert about intrusions (Klump and Kwiatkowski 2010).

1.2.4 Fragility

Johnson and Gheorghe (2013) note that a threat posed by a hazard to a system, typically associated with uncertainty of occurrence and degree of impact, is typically presented from an exogenous viewpoint, with the main focus being the hazard itself. In such cases, risk analysis appears to be one of the best approaches. In contrast, when the concern is the system itself, the consideration must shift to include exogenous *and* endogenous aspects, which would definitely require a recourse to deal with concepts of, among others, vulnerability, resilience, and *fragility* (Johnson and Gheorghe 2013). In describing the fragility of a system, from the perspective of adaptive complex systems, Johnson and Gheorghe posit the following about a dynamic environment (Johnson and Gheorghe 2013, pp. 160–161):

> …a host of things are always changing: conditions, constraints, treats, opportunities, and so on. The ability to make internal adjustments in response to, or in anticipation of, external environmental changes, is the essence of being adaptive. In less complex systems, these changes take place based on pre-established rules in the system…[however] Complex adaptive systems…are not only responsive to environmental dynamics; they have the ability to learn from experiences.

Subsequently, it has been suggested that the manner in which complex systems respond to hazards can be characterized on a continuum ranging from ‘fragile’ to ‘robust’ to ‘anti-fragile’ (Johnson and Gheorghe 2013). In this case, fragility indicates the possibility that the system be degraded by stress/threat. Robust means that the system remains unchanged when under stress/threat, while anti-fragility means the ability to improve with stress (Taleb 2014). On the surface, one could submit that vulnerability and fragility appear to describe a similar consideration of a system. However, there is a critical distinction for those interested in examining system *failures,* relating to why a system can become vulnerable or fragile. Johnson and Gheorghe (2013) plainly offer a response: ‘Vulnerable systems fail because of their degree of exposure to a stress [hazard] of a specific nature, while fragile systems fail because they are easily broken regardless of the nature of stress they are exposed to’ (Johnson and Gheorghe 2013, p. 161).

More interesting is the ‘anti’-fragility. Taleb (2014) argues that to a certain degree, the ability of a system to withstand stress is a function of some deliberate, intentional exposure to small stressing events. In effect, the thesis of anti-fragility is that small stresses can actually strengthen a system and therefore offer the ability to protect a system from extreme stress. While this idea might sound counterintuitive at the first glance, a further examination of the concept can yield surprising results, for example, when considering morphogenesis in the biological process that causes organisms to develop shape (Becvar and Becvar 1999), or the function of role models in society (Herzfeld 2001). To a certain extent, exposure stress, especially at an early stage, is beneficial in preparing systems for future stressful events.

At this point, one might ponder more carefully on infrastructure fragility or robustness. In consideration of the potential benefits of anti-fragility, it becomes evident that such systems might need to be exposed, intentionally to low-level stressing events in order to increase their ability to withstand higher-level or even extreme stresses. Obviously, such endeavors would involve sophisticated methods, tools, and fine-tuning techniques to properly manage fragility and increase the anti-fragility levels. However, with a few exceptions (see, e.g., Comfort et al. 2005; Johnson and Gheorghe 2013), there is a lack of models dedicated to the task. Certainly, such approaches might take advantage of advanced simulation techniques (e.g., quantum cellular automata and parallel computing) to simulate stress and its effects on critical infrastructures in order to bolster infrastructure anti-fragility.

1.2.5 Perception

The role of perception cannot be overemphasized when dealing with any given ‘situation,’ whether good or bad. Concerning a risk, Hill (2012) notes that analyst’s perception on the probability of occurrence of an event and the bearing of its potential consequences are fundamental for one's understanding and valuing, among others, threat, vulnerability, and consequence. Notice that perception is intrinsic to an individual and is therefore related to deep-seated fundamental assumptions such as one's beliefs and predispositions. In an examination of a well-known disaster, Hurricane Katrina, Katina (2016) offers a sobering finding: *It is possible to make decisions and take actions contrary to what is expected*; the 'expected' being, in that case, the evacuation of the area prone to disaster. Why one *might not* evacuate, in spite of the high probability of occurrence of an event and heavy in consequences, is bound to remain hidden in the perception of the individual(s) at the decision helm during the events. At this point and in the case of risk, it becomes evident that risk perception is a 'subjective judgment about the severity of a risk scenario to an asset; [it] may be driven by sense, emotion, or personal experience' (Hill 2012, p. 20). This idea is also supported by Reason's (1990) research, suggesting that making decisions and taking actions are related to mental cognitive processes. Weber and Hsee (1998) would go further on, contending that perception can vary based on one's background. Their survey seemed to indicate that contrary to their American counterparts, Chinese students had a tendency to engage in 'risky' behavior since their culture emphasizes collectivism and interdependence in family and the community as a whole. An extension of this thinking might involve philosophies and ideologies that drive national agendas. Drawing from Beck (2006), Gheorghe (2005a, b), and Katina (2016), Table 1.5 provides, perhaps as an exaggeration, key differences related to perception of different aspects in the USA and the European Union.

In the case of Hurricane Katrina, the literature also reveals a number of reasons motivating the non-evac conduct on behalf of the residents (Eisenman et al. 2007; Katina 2016; Seed et al. 2008; Townsend 2006). Specifically, Elder et al. note that a 'collective memory of past hurricanes combined with distrust of authorities led to minimization of their perceived risk associated with Hurricane Katrina' (Elder et al. 2007, p. S113). Moreover, evidence exists that there was a perception that a Katrina-like hurricane could *not* hit New Orleans (Townsend 2006). These examples are simply selected to emphasize the need to pay attention due to perceptions in the assessment of risks and vulnerabilities related to CIKRKA.

Several models are available for collecting and analyzing perception, including Social Amplification of Risk Framework, Cultural Theory Model, and Psychometric Model. The Social Amplification of Risk Framework is an interdisciplinary approach that combines psychology, sociology, anthropology, and communications. The Cultural Theory Model suggests that people choose to worry about certain risk scenarios based on their social engagements (Sjöberg 1999). The Psychometric Model is largely based on measurement of knowledge, perception, abilities, or personality characteristics (Slovic et al. 1979). In the field of critical infrastructures, key resources, and key assets, such models may prove useful in ranking infrastructure systems.

Table 1.5 Diverging perceptions—USA and Europe

Area of interest	Differences	
Country/region	USA	EU
Dominant philosophy	• *Laissez-faire*: Generally, genetically modified foods (GMF) are considered safe, as long there is no evidence to the contrary	• *Precautionary principle:* policymakers are induced to reject GMFs in favor of more research to establish the fact that they are safe
Homeland security	• Win war • It is a war issue • Set global alliance for global fight • We must reestablish borders • We must reorganize the government • Concerned with consequences • We must quantify risks • We must define business standards • R&D spending $4 billion • Concerned with competitiveness of US corporations	• Win peace • It is a law enforcement issue • The neighborhood vision • We must strive for a borderless society • Not sure who has authority • Concerned with probability • We must protect civil liberties • EU is fragmented • Striving for bilateral cooperation under national responsibility • R&D spending: €1 billion

1.2.6 Governance

There is one definition of 'governance.' Many of the various perspectives are driven by the nature of systems and interests and system operations, or rather in-operability in such systems. Following Calida (2013) and the subsequent works (Calida and Keating 2014; Keating et al. 2014), Table 1.6 is drawn to depict the multitude of perspectives for governance.

Clearly, there is a large spectrum of perspectives on governance. These perspectives are sufficient to offer insights into three related and basic governance attributes: *direction*, *oversight*, and *accountability,* as depicted in Fig. 1.4.

There is a good relationship between themes of governance and critical infrastructure topics of, among others, risk, vulnerability, resilience, fragility, and perception. For instance, sustaining a coherent identity and vision that supports consistent decision, action, interpretation, and strategic priorities involves understating possible risks, vulnerabilities, and the related concepts. Clearly, governance is a topic for any 'typical' infrastructure system (i.e., water and energy) but even more for open bazaar of space, undersea, and belowground.

At this point, it is perhaps proper to take a step back and ponder about what we have thus far:

- We now know what *critical infrastructures, key resources, and key assets* (CIKRKA) are and have hopefully fathomed their bearing in keeping the world on track.

Table 1.6 Different perspectives on governance

Governance type	Description	Proponents
Process-centric	A governing arrangement where one or more public agencies directly engage non-state stakeholders in a collective decision-making process that is formal, consensus-oriented, and deliberative and that aims to make or implement public policy or manage public programs or assets	Ansell and Gash (2008, p. 544)
	…social turbulence kept within bounds, and change steered in desired directions… preserves order and continuity, but not necessarily the maintenance of the status quo	Dunsire (1990, p. 18)
Structure-centric	…the totality of conceptual ideas about these Interactions…(these in relation to the act of governing)	Kooiman (2003, p. 79)
	…the activity of coordinating communications in order to achieve collective goals through collaboration	Willke (2007, p. 10)
State-centric	…the process through which state and nonstate actors interact to design and implement policies within a given set of formal and informal rules that shape and are shaped by power…	The World Bank (2017, p. 3)
Hybrid	…the reflexive self-organization of independent actors involved in complex relations of reciprocal interdependence, with such self-organization being based on continuing dialogue and resource-sharing to develop mutually beneficial joint projects and to manage the contradictions and dilemmas inevitably involved in such situations	Jessop (2003, p. 142)
	…interdependence between organizations… continuing interactions between network members, caused by the need to exchange resources and negotiate shared purposes…game-like interactions, rooted in trust and regulated by rules of the game negotiated and agreed by network participants…a significant degree of autonomy; they are self-organizing	Rhodes (2007, p. 1246)
Corporate governance	…the system of checks and balances, both internal and external to companies, which ensures that companies discharge their accountability to all their stakeholders and act in a socially responsible way in all areas of their business activity	Brennan and Solomon (2008, p. 890)
New public management	…the means for achieving direction, control, and coordination of wholly or partially autonomous individuals or organizations on behalf of interests to which they jointly contribute	Lynn et al. (2000, p. 235)

(continued)

Table 1.6 (continued)

Governance type	Description	Proponents
Public policy	…the ways in which stakeholders interact with each other in order to influence the outcomes of public policies	Bovaird (2005, p. 220)
	…the processes and institutions, both formal and informal, that guide and restrain the collective activities of a group	Keohane and Nye (2000, p. 12)
International security	…the emergence and recognition of principles, norms, rules and behavior that both provide standards of acceptable public behavior and that are followed sufficiently to produce behavioral regularities	Keohane and Nye (1989)
Social and political	Governance denotes the structures and processes which enable a set of public and private actors to coordinate their interdependent needs and interests through the making and implementation of binding policy decisions in the absence of a central political authority	Krahmann (2003, p. 11)
	…arrangements in which public as well as private actors aim at solving societal problems or create societal opportunities, and aim at the care for the societal institutions within which these governing activities take place	Kooiman (2000, p. 139)
Earth	…the interrelated and increasingly integrated system of formal and informal rules, rule-making systems, and actor-networks at all levels of human society (from local to global) that are set up to steer societies towards preventing, mitigating, and adapting to global and local environmental change and, in particular, earth system transformation, within the normative context of sustainable development	Biermann et al. (2009)

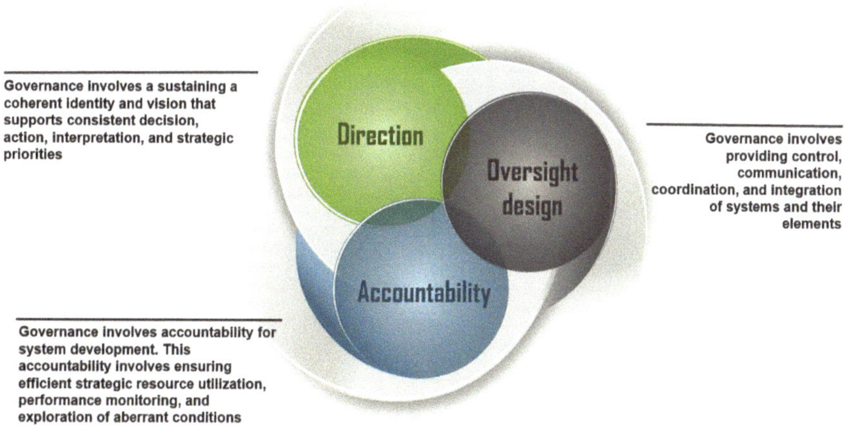

Fig. 1.4 Three relevant themes for governance

- We circumscribed the issue with the CIKRKA triad to *securing its continual, sound, and safe operations as a 'sine qua non' condition of societal and personal well-being*.
- We have discovered that CIKRKA is 'one of those things' that cannot be univocally defined via an *inventory*; this not only because of the tremendous variety of the triad's contents, rather, also because *the contents are essentially and permanently trimmed by the perception of the observers, analysts, and managers involved*.

The corollary of the above may be that CIKRKA may never be treated as a library of nicely indexed items, but rather as an open bazaar, where what makes sense in order to make it manageable is to try and identify and, hopefully, control the *commonalities* in the properties of the immensely varied stocks and parts sharing the market space and of the operators. And we also believe that implementing such a vision-directing paradigm requires an additional, adequate level of abstraction to capture CIKRKA into an intelligible, as generic as feasible and practical modeling, simulation and visualization framework that should decisively integrate the unavoidable concepts of vulnerability, resilience, fragility, and perception.

In the sequel, we will table in the American sense (postpone the debate) the paradigm of an open bazaar, regardless of how tempting it may be; and rather table in the British sense (start considering) what we feel, in light of developments on record over the past decade, to constitute a natural and increasingly consequential, if less visible, complement of the bazaar's departmental structure: the critical infrastructures of space; undersea; and belowground as illustrated in Fig. 1.5.

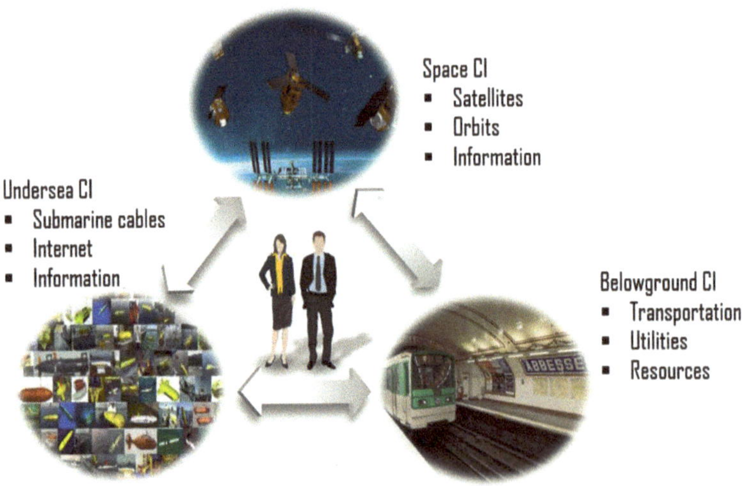

Fig. 1.5 Augmenting CIKRKA: the space, the undersea, and the belowground

1.3 Open Bazaar Newcomers: The Space, Undersea, and Belowground

As indicated already, while all infrastructures are, or eventually become, visible in the sense of capturing the public interest, an infrastructure qualifies critical only when people will effectively feel that the respective system's failure would represent a clear and present danger of crippling his or her 'well-being.' Since years now satellite TV, the offshore oil and the deep-down richness of the planet were taken for granted. Moreover, as long as the services of such origins were only incipient or marginal in both kind and volume, all things were definitely normal—and never 'critical.'

Well, not anymore.

1.3.1 The Space Critical Infrastructure

Space systems including unmanned air systems, satellites, rockets, space probes, planetary stations, ground stations, and among others are rapidly becoming key enablers for a number of commercial, scientific, and military applications, being currently and increasingly embedded in the functioning of societies, economies, as well as in lifestyles and governance processes (Francis 2010; Gheorghe and Yuchnovicz 2015; Hesse and Hornung 2015; Schmieer 2015). In a globalized twenty-first century society, space systems offer vital services ranging from cheap, constant, and instantaneous communications with worldwide coverage, to navigation systems, remote sensing, threat assessment, and early warning systems involving, among others, Global Positioning System (GPS), GLObal NAvigation Satellite System (GLONASS), Galileo, and BeiDou Navigation Satellite System.

The current deals of space systems also include financial transactions (e.g., algorithm-banking or robobanking) and industrial processes (e.g., supervisory control and data acquisition—SCADA), Earth observation for scientific study and daily conveniences like weather forecasts, but also command and control capabilities during emergency and crisis situations. Hesse and Hornung (2015) note that 'usability of innovative space applications seems to be almost unlimited, for example in domains such as meteorology, research on climate change, telemedicine, disaster management, and precision farming' (Hesse and Hornung 2015, p. 188).

Our concern for space systems stays primarily with the aspects of *risk, vulnerability, resilience, fragility*, and even *perception*. Not only these issues are applicable to space systems, but also they can have a real impact on societies. Take, for example, the case of the meteorite impact in Chelyabinsk, Russia. The event took place concurrently with the coincidental flyby of an asteroid of the same caliber as the one responsible for the 1908 Tunguska event which flattened 2000 km^2 of forest, thus highlighting the urgency of documenting and dealing with potentially dangerous space objects. The growth of the awareness on the potential dangers that traverse our planetary neighborhood has only added to the urgency of the need to address the vulnerabilities of the space systems, some of which are in charge not only with our well-being, but also with our physical existence.

Succinctly, we submit that the space critical infrastructure (SCI) represents *a system of systems* which encompasses hardware, workforce, environment, facilities, business and organizational entities, and multi-directional interactions essential for the maintenance of vital societal functions, health, safety, security, economic or social well-being of people, the destruction or disruption of which would have a significant impact on, virtually, any given nation. Arguably, in this vision, a satellite can represent a critical infrastructure, the information it handles—a key resource, while its orbit can be seen as a key asset. While the continued operation of SCI could be taken for granted, the failure can have real consequences. Moreover, such consequences can affect the operability of other systems in interdependence with the SCI, including the undersea and belowground systems.

1.3.2 The Undersea Critical Infrastructure

Similar to SCI, undersea systems, with an emphasis on submarine cables, provide vital societal functions related to health, safety, security, and economic or social well-being of people, whose destruction or disruption would have a significant impact on nations. The importance of undersea cables can be illustrated from several angles, ranging from communications to politics and intelligence. Some estimates suggest that 99% of all transoceanic Internet communications is carried through undersea cables (Starosielski 2015). In a recent article in *Computerworld UK*, Charlotte Jee notes that 'undersea cables, rather than satellites, carry almost all transoceanic Internet data nowadays' (Jee 2016). Not only do these cables are transmit data, but also the rate of transmission is faster and cheaper compared to satellites (Brown 2015). This is why you are able to Skype or WebEx with colleagues, friends, and family across the world, shop online from the comfort of your home, and stream that favorite movie.

Beyond the obvious utility of undersea cables, a historical account of this critical system provides a revealing fact relating to *espionage and politics*. It suffices to say that undersea cables and state sovereignty are intricately linked (Khatri 1996). There has always been international intrigue regarding undersea cables. As reported in the *Times* in 1959, the US Navy boarded and searched a Russian fishing trawler off the coasts of Newfoundland on suspicions of tampering with the undersea cables (Miller 2015). During the Cold War, the US Navy, the Central Intelligence Agency, and National Security Agency ran a joint mission with the objective of wiretapping on Soviet underwater communication lines (Brown 2015; Hoffman 2010; Miller 2015). Most recently, there has been a call for 'neutrality of Internet' after concerns emerged over the spying on Dilma Rousseff and wiretapping on Angela Merkel (Baker 2014). To guarantee the neutrality of the Internet, Dilma Rousseff, Herman Van Rompuy

(European Council President), and Jose Manuel Barroso (European Commission President) issued a joint statement calling for a future installation of a fiber-optic submarine cable linking Brazil to Europe directly (Baker 2014).

The aspects above highlight sensitive issues and elements of criticality of the undersea cables infrastructure (UCI). Moreover, on the factual side, undersea cables appear definitely confronted with risks and vulnerabilities (Miller 2015; Starosielski 2015). For example, a 2007 Internet service disruption in Vietnam lasting several months was a result of pulling two fiber-optic cables by scrap metal salvagers (Miller 2015). In 2013, the cutting of the undersea cables North of Alexandria, Egypt, resulted in a 60% drop in Internet services (Al-Youm 2013). Strategies such as 'security through obscurity' appear obsolete. Perhaps the time is ripe for resilience-oriented measures and strategies addressing anti-fragility, which would certainly require an improved, more comprehensive perception on the different facets of UCI.

1.3.3 The Belowground Critical Infrastructure

Consider this, around the world and as of October 2014, there are 55 countries and over 140 cities hosting 160 metro systems (UITP 2014). This belowground transportation system is used by over 150 million people a day (UITP 2014). This mode of transportation is only an element of systems that are linked to the belowground critical infrastructure (BCI) spanning an array of industries including dot com, oil/gas, and utilities. A clear majority of critical mineral resources are found belowground including coltan, gold, tungsten, and tin. Tantalum from coltan is used to manufacture tantalum capacitors, used in electronic products including cell phones, computers, and other electronic devices.

The oil and gas industry is part of the BCI. In Europe alone, notes a report by the European Parliament, that gas demand of the member states has substantially increased during the past 50 years, a fact echoed in the system of pipelines and storage facilities. Due to uncertainty in production, 'underground gas storage facilities are used to balance demand and supply' (EP 2009, p. 8). Interestingly, about 25% of the energy used in the USA in 2013 came from natural gas. Obviously, the natural gas sector of the energy industry includes the production, processing, transportation, distribution, and storage (Katina and Unal 2015). In Fig. 1.6, the left side illustrates the complexity of the oil pipeline network in Europe, according to a map published by the European Parliament (EP 2009, p. 35). The right side of the figure maps the US underground natural gas storage facilities as of 2015 as offered by the Energy Information Administration (EIA 2008).

Of course, belowground utilities involve more than pipelines and storage tanks for gas and oil. Water, gas, and sewage, among others, travel via a network of belowground utilities to, for example, treatment plants. In the New York area alone, 'Six natural gas—and fuel oil-fired steam generating facilities in Manhattan, Brooklyn, and Queens can collectively produce over 10 million pounds of steam

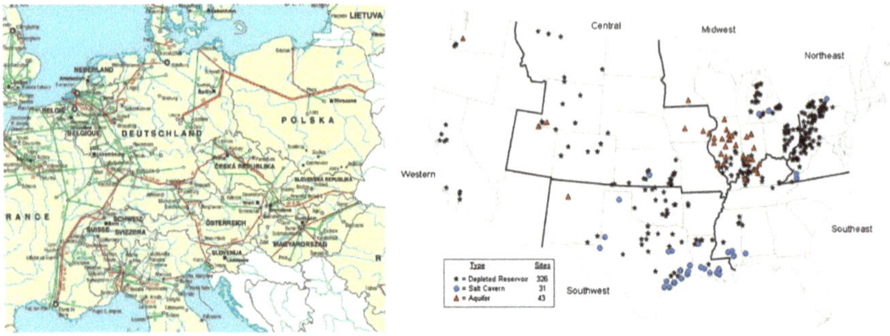

Fig. 1.6 Network of gas pipelines in Europe (left) and belowground gas storage facilities in the USA (right), modified from EP (2009) and EIA (2008), respectively

per hour, either cogenerating this steam along with electricity, or producing steam alone in massive boilers. A network of 105 miles of belowground pipes transports this steam to customers' (NYC 2007, p. 110). It is important to also note that natural gas is responsible for 65% of the heating needs throughout the city of New York (NYC 2007).

A fact worth remembering is that belowground systems are complex, and in most cases, we do not think about them until they fail (NYC 2007). Case in point, a gas explosion, on March 26, 2015, in a Manhattan neighborhood, New York, as a result of an illegal tap into a gas main, killed two and injured nineteen, with the ensuing fire destroying three adjacent buildings. And more, the Flint water crisis in Flint, Michigan: corrosive Flint River water caused a leak from aging pipes to leach into the water supply, entailing extremely elevated levels of a heavy metal neurotoxin. Thousands of children were exposed to drinking contaminated water that may later cause serious health problems.

Probing questions might involve asking, for example, was the Flint water crisis an isolated problem? What is the probability that a similar event will occur in City X, and what about the associated consequences? These types of questions conjure the concepts of vulnerability, resilience, fragility, and perception. Undoubtedly, there are major challenges in ensuring continued and steady supply of belowground resources as well as maintaining and protecting the BCIs. Think of the relationship between regional conflicts and the so-called resource trails (Humphreys et al. 2007), or you perhaps you would rather recall the role of uranium mined in the Congo, in shaping the World War II outcome and the world thereafter (Williams 2016). In cases like these, the European Commission recognizes the need to reduce the vulnerabilities associated with the different systems, including the belowground (EP 2009). In other cases, such as the hydraulic fracturing or the emerging initiative of belowground CO_2 sequestration, comprehensive assessments and strategic plans to address economic effects, public policy, health, and environmental impact may lag behind the expectations of significant segments of the public opinion (see, e.g., Litovitz et al. 2013).

1.4 Remarks

In concluding this chapter, let it be said that augmenting the CIKRKA 'Open Bazaar' by bringing to attention the space, undersea, and belowground realms was motivated not only by a quest for completeness—futile, anyway, in the face of an anthropic universe in continual expansion—but also by the drive of seeing the world through a different set of lenses—one that filters out cheap enthusiasms and unilateral interpretations of 'progress' while encouraging a more analytical and many-sided assessment, where all parts are listened to and no concern is dismissed out of hand, no matter how esoteric its substance may seem. That is why, along with tangible *threats* and rationally accepted *risks* one talks, here and in the sequel, the rather elusive at a first glance *vulnerabilities*, *resilience,* and *fragility*—all under the inevitable and highly relevant censorship of the *perceptions*.

There is, of course, a vast difference between seeing and doing something. To these authors, *doing* means developing methodologies, models, methods, and tools for intervention and governance. A starting point might be a good governance 'steering' a system to enable the realization of the desired outcomes. This realization might as well have genesis in elements of 'direction,' 'oversight design,' and 'accountability' as they relate to governance. This is an approach that reverberates through the remainder of this book as we address the conceptualization of the governance notion, models for assessment, and offer working examples.

References

Al-Youm, A.-M. (2013, March 27). *Internet saboteur caught, says Telecom Egypt CEO.* Retrieved September 15, 2016, from http://www.egyptindependent.com/news/internet-saboteur-caught-says-telecom-egypt-ceo.

Ansell, C., & Gash, A. (2008). Collaborative governance in theory and practice. *Journal of Public Administration Research and Theory, 18*(4), 543–571.

Ansoff, H. I. (1984). *Implanting strategic management.* Englewood Cliffs: Prentice-Hall.

ASCE. (2009). *Guiding principles for the nation's critical infrastructure.* Reston: American Society of Civil Engineers.

Aven, T. (2011). On some recent definitions and analysis frameworks for risk, vulnerability, and resilience. *Risk Analysis, 31*(4), 515–522.

Baker, J. (2014, February 25). *EU mulling investment in South America-Europe undersea cable.* Retrieved September 15, 2016, from http://www.pcworld.com/article/2101820/eu-mulling-investment-in-south-americaeurope-undersea-cable.html.

Beck, U. (2006). Living in the world risk society. *Economy and Society, 35*(3), 329–345.

Becvar, D. S., & Becvar, R. J. (1999). *Systems theory and family therapy: A primer* (2nd ed.). Lanham, MD: University Press of America.

Bennett, B. T. (2007). *Understanding, assessing, and responding to terrorism: Protecting critical infrastructure and personnel.* Hoboken, NJ: Wiley.

Biermann, F., Betsill, M. M., Gupta, J., Kanie, N., Lebel, L., Liverman, D., et al. (2009). *Earth system governance: People, places and the planet. Science and implementation plan of the earth system governance project* (No. Earth System Governance Report 1, IHDP Report 20). Bonn: IHDP: The

Earth System Governance Project. Retrieved from http://www.earthsystemgovernance.org/sites/default/files/publications/files/Earth-System-Governance_Science-Plan.pdf.
Blanchard, B. S. (2008). *System engineering management*. Hoboken: Wiley.
Blanchard, B. S., & Fabrycky, W. J. (2006). *Systems engineering and analysis* (4th ed.). Upper Saddle River, NJ: Pearson—Prentice Hall.
Bovaird, T. (2005). Public governance: Balancing stakeholder power in a network society. *International Review of Administrative Sciences, 71*(2), 217–228.
Brennan, N. M., & Solomon, J. (2008). Corporate governance, accountability and mechanisms of accountability: An overview. *Accounting, Auditing & Accountability Journal, 21*(7), 885–906.
Brown, D. W. (2015). *10 facts about the Internet's undersea cables*. Retrieved September 15, 2016, from http://mentalfloss.com/article/60150/10-facts-about-internets-undersea-cables.
Bush, B., Dauelsberg, L., LeClaire, R., Powell, D., DeLand, S., & Samsa, M. (2005). *Critical infrastructure protection decision support system (CIP/DSS) Project overview. Presented at the International System Dynamics Conference, Boston, MA*. Retrieved from http://www.systemdynamics.org/conferences/2005/proceed/papers/LECLA332.pdf.
Bush, G. W. (2003). *The National strategy for the physical protection of critical infrastructures and key assests*. Washington, DC: The White House. Retrieved from https://www.dhs.gov/xlibrary/assets/Physical_Strategy.pdf.
Calida, B. Y. (2013). *System governance analysis of complex systems* (Ph.D.). Old Dominion University, United States—Virginia.
Calida, B. Y., & Katina, P. F. (2012). Regional industries as critical infrastructures: A tale of two modern cities. *International Journal of Critical Infrastructures, 8*(1), 74–90.
Calida, B. Y., & Katina, P. F. (2015). Modelling the 2008 financial economic crisis: Triggers, perspectives and implications from systems dynamics. *International Journal of System of Systems Engineering, 6*(4), 273–301.
Calida, B. Y., & Keating, C. B. (2014). System governance: Emergence of practical perspectives across the disciplines. In A. V. Gheorghe, M. Masera, & P. F. Katina (Eds.), *Infranomics* (pp. 269–296). Geneva, Switzerland: Springer International Publishing.
Casti, J. (2012). *X-events: Complexity overload and the collapse of everything*. New York, NY: William Morrow.
Clinton, W. J. (1996). Executive order 13010: Critical infrastructure protection. *Federal Register, 61*(138), 37345–37350.
Cohen, R. E., & Ahearn, F. L. (1980). *Handbook for mental health care of disaster victims*. Hutchinson: Johns Hopkins University Press.
Comfort, L. K., Ko, K., & Zagorecki, A. (2005). Coordination in rapidly evolving disaster response systems: The role of information. In T. Terano, H. Kita, T. Kaneda, K. Arai, & H. Deguchi (Eds.), *Agent-based simulation: From modeling methodologies to real-world applications* (pp. 208–219). Tokyo: Springer Tokyo.
de Silva, E. (Ed.). (2016). *National security and counterintelligence in the era of cyber espionage*. Hershey, PA: IGI Global.
DiSera, D., & Brooks, T. (2009). *The geospatial dimensions of critical infrastructure and emergency response*. Pipeline and Gas Journal, *236*(9), 1–4.
Dudenhoeffer, D., Permann, M. R., & Manic, M. (2006). CIMS: A framework for infrastructure interdependency modeling and analysis. In *Proceedings of the 38th conference on Winter simulation* (pp. 478–485). Monterey, California: Winter Simulation Conference. https://doi.org/10.1109/WSC.2006.323119.
Dunsire, A. (1990). Holistic governance. *Public Policy and Administration, 5*(1), 4–19.
EIA. (2008). *U.S. underground natural gas storage facilities, Close of 2007*. Retrieved September 16, 2016, from https://www.eia.gov/pub/oil_gas/natural_gas/analysis_publications/ngpipeline/undrgrndstor_map.html.
Einarsson, S., & Rausand, M. (1998). An approach to vulnerability analysis of complex industrial systems. *Risk Analysis, 18*(5), 535–546.
Eisenman, D. P., Cordasco, K. M., Asch, S., Golden, J. F., & Glik, D. (2007). Disaster planning and risk communication with vulnerable communities: Lessons from Hurricane Katrina. *American Journal of Public Health, 97*(Supplement_1), 109–115.

Elder, K., Xirasagar, S., Miller, N., Bowen, S. A., Glover, S., & Piper, C. (2007). African Americans' decisions not to evacuate New Orleans before Hurricane Katrina: A qualitative study. *American Journal of Public Health, 97*(Supplement 1), S124–S129.

EP. (2009). *Gas and oil pipelines in Europe* (No. PE 416.239 (IP/A/ITRE/NT/2009-13)). Brussels: European Parliament. Retrieved from http://www.europarl.europa.eu/document/activities/cont/201106/20110628ATT22856/20110628ATT22856EN.pdf.

European Council. (2004). *Communication from the commission to the council and the European Parliament: Critical infrastructure protection in the fight against terrorism* (pp. 1–11). Brussels, Belgium: Commission of the European Communities. Retrieved from http://eur-lex.europa.eu/LexUriServ/LexUriServ.do?uri=CELEX:52004DC0702:EN:NOT.

Eusgeld, I., Henzi, D., & Kröger, W. (2008). *Comparative evaluation of modeling and simulation techniques for interdependent critical infrastructures* (p. 50). Zurich, Switzerland: ETH Zurich.

Eusgeld, I., Kröger, W., Sansavini, G., Schläpfer, M., & Zio, E. (2009). The role of network theory and object-oriented modeling within a framework for the vulnerability analysis of critical infrastructures. *Reliability Engineering & System Safety, 94*(5), 954–963.

Fletcher, D. R. (2002). *Spatial information technologies in critical infrastructure protection: A research agenda in CIP* (pp. 1–8). Santa Barbara, CA: University of California. Retrieved from http://www.ncgia.ucsb.edu/ncrst/research/cip/CIPAgenda.pdf.

Francis, M. S. (2010). UAS uses, capabilities, grand challenges. In *Encyclopedia of aerospace engineering* (pp. 1–11). New York: NY: Wiley.

FRG. (2009). *National strategy for critical infrastructure protection* (pp. 1–18). Berlin, Germany: Federal Ministry of the Interior. Retrieved from http://www.bmi.bund.de/cae/servlet/contentblob/598732/publicationFile/34423/kritis_englisch.pdf.

GAO. (2004). *Critical infrastructure protection: Challenges and efforts to secure control systems* (pp. 1–47). Washington, DC: US Government Accountability Office.

Garrett, R. K., Anderson, S., Baron, N. T., & Moreland, J. D. (2011). Managing the interstitials, a system of systems framework suited for the Ballistic Missile Defense System. *Systems Engineering, 14*(1), 87–109.

Garvey, P. R. (2009). *Analytical methods for risk management: A systems engineering perspective*. Boca Raton: Chapman & Hall/CRC.

Gheorghe, A. V. (2004). Risks, vulnerability, sustainability and governance: A new landscape for critical infrastructures. *International Journal of Critical Infrastructures, 1*(1), 118–124.

Gheorghe, A. V. (2005a). *Critical infrastructures protection from [(systems) engineering] to {(system of systems) engineering}*. Norfolk: ETH Zurich.

Gheorghe, A. V. (Ed.). (2005b). *Integrated risk and vulnerability management assisted by decision support systems: Relevance and impact on governance* (Vol. 8). Dordrecht, The Netherlands: Springer.

Gheorghe, A. V., & Katina, P. F. (2014). Editorial: Resiliency and engineering systems—Research trends and challenges. *International Journal of Critical Infrastructures, 10*(3/4), 193–199.

Gheorghe, A. V., Masera, M., Weijnen, M. P. C., & De Vries, J. L. (Eds.). (2006). *Critical infrastructures at risk: Securing the European electric power system* (Vol. 9). Dordrecht: Springer.

Gheorghe, A. V., Mock, R., & Kröger, W. (2000). Risk assessment of regional systems. *Reliability Engineering and System Safety, 70*(2), 141–156.

Gheorghe, A. V., & Yuchnovicz, D. (2015). The space infrastructure vulnerability cadastre: Orbital debris critical loads. *International Journal of Disaster Risk Science, 6*(4), 359–371.

Gibson, J. E., Scherer, W. T., & Gibson, W. F. (2007). *How to do systems analysis*. Hoboken, NJ: Wiley-Interscience.

Herzfeld, M. (2001). *Anthropology: Theoretical practice in culture and society*. Malden: Blackwell Publishers.

Hesse, M., & Hornung, M. (2015). Space as a critical infrastructure. In K.-U. Schrogl, P. L. Hays, J. Robinson, D. Moura, & C. Giannopapa (Eds.), *Handbook of space security* (pp. 187–201). New York: Springer, New York.

Hill, K. N. (2012). *Risk quadruplet: Integrating assessments of threat, vulnerability, consequence, and perception for homeland security and homeland defense* (Ph.D.). Old Dominion University, United States—Virginia.

Hoffman, D. E. (2010). *The dead hand: The untold story of the cold war arms race and its dangerous legacy*. New York: Anchor Books.

Holmgren, A., Molin, S., & Thedéen, T. (2001). *Vulnerability of complex Infrastructure; power system and supporting digital communication system*. Presented at the 5th International Conference on Technology, Policy, and Innovation, Utrecht, the Netherlands: LEMMA Publishers.

Holton, G. A. (2004). Defining risk. *Financial Analysts Journal, 60*(6), 19–25.

Humphreys, M., Sachs, F. D., & Stiglitz, J. E. (Eds.). (2007). *Escaping the resource curse*. New York: Columbia University Press.

INCOSE. (2011). *Systems engineering handbook: A guide for system life cycle processes and activities* (H. Cecilia, Ed.) (3.2 ed.). San Diego, CA: INCOSE.

IRGC. (2006). *White paper on managing and reducing social vulnerabilities from coupled critical infrastructures* (pp. 1–68). Geneva: Switzerland.

IRGC. (2007). *Managing and reducing social vulnerabilities from coupled critical infrastructures* (pp. 1–16). Geneva: International Risk Governance Council. Retrieved from http://www.irgc.org/IMG/pdf/IRGCinfra_site06.11.07-2.pdf.

Jee, C. (2016, June 30). *10 things you didn't know about Google's undersea internet cable*. Retrieved September 15, 2016, from http://www.computerworlduk.com/galleries/infrastructure/10-top-fun-facts-about-googles-undersea-internet-cable-3619557/.

Jessop, B. (2003). Governance and metagovernance: On reflexivity, requisite variety, and requisite irony. In H. P. Bang (Ed.), *Governance, as social and political communication* (pp. 142–172). Manchester, England: Manchester University Press.

Johnson, J., & Gheorghe, A. V. (2013). Antifragility analysis and measurement framework for systems of systems. *International Journal of Disaster Risk Science, 4*(4), 159–168.

Katina, P. F. (2016). Individual and societal risk (RiskIS): Beyond probability and consequence during Hurricane Katrina. In A. J. Masys (Ed.), *Disaster forensics: Understanding root cause and complex causality* (pp. 1–23). Geneva: Springer International Publishing.

Katina, P. F., & Keating, C. B. (2015). Critical infrastructures: A perspective from systems of systems. *International Journal of Critical Infrastructures, 11*(4), 316–344.

Katina, P. F., Keating, C. B., & Gheorghe, A. V. (2016a). Cyber-physical systems: Complex system governance as an integrating construct. In H. Yang, Z. Kong, & M. D. Sarder (Eds.), *Proceedings of the 2016 Industrial and Systems Engineering Research Conference*. Anaheim, CA: IISE.

Katina, P. F., Keating, C. B., Zio, E., & Gheorghe, A. V. (2016b). A criticality-based approach for the analysis of smart grids. *Technology and Economics of Smart Grids and Sustainable Energy, 1*(1), 14. doi:10.1007/s40866-016-0013-2.

Katina, P. F., Pinto, C. A., Bradley, J. M., & Hester, P. T. (2014). Interdependency-induced risk with applications to healthcare. *International Journal of Critical Infrastructure Protection, 7* (1), 12–26.

Katina, P. F., & Unal, R. (2015). Application of fuzzy sets in decision analysis for prioritising critical energy infrastructures. *International Journal of Decision Sciences, Risk and Management, 6*(1), 1–15.

Keating, C. B., Katina, P. F., & Bradley, J. M. (2014). Complex system governance: Concept, challenges, and emerging research. *International Journal of System of Systems Engineering, 5* (3), 263–288.

Keohane, R., & Nye, J. (1989). *Power and interdependence*. New York: NY: Harper Collins.

Keohane, R., & Nye, J. (Eds.). (2000). *Governance in a globalizing world*. Washington: DC: Brookings Institution.

Khatri, F. I. (1996). *Optical solution propagation and control* (Ph.D.). Massachusetts Institute of Technology, United States—Massachusetts.

Klump, R., & Kwiatkowski, M. (2010). Distributed IP watchlist generation for intrusion detection in the electrical smart grid. In T. Moore & S. Shenoi (Eds.), *Critical infrastructure protection IV* (pp. 113–126). New York: Springer, Berlin Heidelberg.

Knabb, R. D., Rhome, J. R., & Brown, D. P. (2011). *Tropical cyclone report Hurricane Katrina 23–30 August 2005* (pp. 1–43). Washington, DC: National Oceanic and Atmospheric Administration. Retrieved from http://www.nhc.noaa.gov/data/tcr/AL122005_Katrina.pdf.

Knight, F. H. (1921). *Risk, uncertainty, and profit*. Boston, MA: Hart, Schaffner & Marx; Houghton Mifflin Co.

Komljenovic, D., Gaha, M., Abdul-Nour, G., Langheit, C., & Bourgeois, M. (2016). Risks of extreme and rare events in asset management. *Safety Science, 88*, 129–145. doi:10.1016/j.ssci.2016.05.004.

Kooiman, J. (2000). Societal governance: Levels, models and orders of social-political interaction. In J. Pierre (Ed.), *Debating governance: Authority, steering and democracy* (pp. 138–166). Oxford, UK: Oxford University Press.

Kooiman, J. (2003). *Governing as governance*. London: UK: SAGE Publications Ltd.

Krahmann, E. (2003). Conceptualizing security governance. *Cooperation and Conflict, 38*(1), 5–26.

Kröger, W. (2008). Critical infrastructures at risk: A need for a new conceptual approach and extended analytical tools. *Reliability Engineering & System Safety, 93*(12), 1781–1787.

Kröger, W., & Zio, E. (2011). *Vulnerable systems*. London, UK: Springer-Verlag.

Li, J. (2013). The visible hand: From struggling survival to viable sustainability. In *Proceedings of the 56th Annual Meeting of the ISSS* (pp. 1–19). San Jose, CA: International Society for the Systems Sciences. Retrieved from http://journals.isss.org/index.php/proceedings56th/article/view/1959.

Litovitz, A., Curtright, A., Abramzon, S., Burger, N., & Samaras, C. (2013). Estimation of regional air-quality damages from Marcellus Shale natural gas extraction in Pennsylvania. *Environmental Research Letters, 8*(1), 014017. doi:10.1088/1748-9326/8/1/014017.

Lynn, L., Heinrich, C., & Hill, C. (2000). Studying governance and public management: Challenges and prospects. *Journal of Public Administration Research and Theory, 10*(2), 233–261.

Martin-Breen, P., & Anderies, J. M. (2011). *Resilience: A literature review* (p. 64). New York: The Rockefeller Foundation. Retrieved from http://www.rockefellerfoundation.org/blog/resilience-literature-review.

Masys, A. (Ed.). (2015). *Disaster management: Enabling resilience*. New York, NY: Springer International Publishing.

McMichael, L. D., Noble, C. R., Margraf, J. D., & Glascoe, L. G. (2009). *Assessing the vulnerability of large critical infrastructure using fully-coupled blast effects modeling (p. 12)*. Presented at the 5th Congress on Forensics Engineering, Washington, D.C. Retrieved from https://e-reports-ext.llnl.gov/pdf/371487.pdf.

Mendonca, D., & Wallace, W. A. (2006). Impacts of the 2001 world trade center attack on New York City critical infrastructures. *Journal of Infrastructure Systems, 12*(4), 260–270.

Miller, G. (2015, October 29). *Undersea internet cables are surprisingly vulnerable*. Retrieved September 15, 2016, from https://www.wired.com/2015/10/undersea-cable-maps/.

Moteff, J., & Parfomak, P. (2004). *Critical infrastructure and key assets: Definition and identification*. Washington: DC: Congressional Research Service. Retrieved from http://www.fas.org/sgp/crs/RL32631.pdf.

Niemeyer, K. (2004). Simulation of critical infrastructures. *Information and Security. An International Journal, 15*(2), 120–143.

Northrop, R. B., & Connor, A. N. (2013). *Ecological sustainability: Understanding complex issues*. Boca Raton, FL: CRC Press.

Nozick, L. K., Turnquist, M. A., Jones, D. A., Davis, J. R., & Lawton, C. R. (2004). Assessing the performance of interdependent infrastructures and optimizing investments. In *System sciences, 2004. Proceedings of the 37th Annual Hawaii International Conference on* (pp. 1–7). Big Island, Hawaii.

NYC. (2007). *A stronger, more resilient New York* (pp. 106–130). New York: City of New York. Retrieved from http://www.nyc.gov/html/sirr/downloads/pdf/final_report/Ch_6_Utilities_FINAL_singles.pdf.
Obama, B. H. (2013). *Critical infrastructure security and resilience*. Washington, D.C.: The White House. Retrieved from http://www.fas.org/irp/offdocs/ppd/ppd-21.pdf.
Pederson, P., Dudenhoeffer, D., Hartley, S., & Permann, M. (2006). *Critical infrastructure interdependency modeling: A survey of US and international research* (No. INL/EXT-06-11464) (p. 125). Idaho Falls, ID: Idaho National Laboratory. Retrieved from http://www.inl.gov/technicalpublications/Documents/3489532.pdf.
Permann, M. R. (2007). Toward developing genetic algorithms to aid in critical infrastructure modeling. In *2007 IEEE Conference on Technologies for Homeland Security* (pp. 192–197). Woburn, MA. https://doi.org/10.1109/THS.2007.370044.
Rahman, H. A., Marti, J. R., & Srivastava, K. (2011). Quantitative estimates of critical infrastructures' interdependencies on the communication and information technology infrastructure. *International Journal of Critical Infrastructures, 7*(3), 220–242.
Rasmussen, J., & Batstone, R. (1989). Why do complex organisational systems fail? World Bank Environmental Working Paper, No. 20.
Reason, J. (1990). *Human error*. Cambridge, UK: Cambridge University Press.
Rhodes, R. A. W. (2007). Understanding governance: Ten years on. *Organization Studies, 28*(8), 1243–1264.
Richardson, B. (1994). Socio-technical disasters: Profile and prevalence. *Disaster Prevention and Management: An International Journal, 3*(4), 41–69.
Rinaldi, S. M. (2004). Modeling and simulating critical infrastructures and their interdependencies. In *Proceedings of the 37th Hawaii International Conference on System Sciences* (pp. 1–8). Big Island, Hawaii. https://doi.org/10.1109/HICSS.2004.1265180.
Rinaldi, S. M., Peerenboom, J., & Kelly, T. K. (2001). Identifying, understanding, and analyzing critical infrastructure interdependencies. *IEEE Control Systems, 21*(6), 11–25.
Sandage, S. A. (2006). *Born losers: A history of failure in America*. Cambridge: Harvard University Press.
Schmieer, M. (2015). *Inventory of space based assets (infrastructures and services) to be classified as part of a critical infrastructure*. Ispra: Institute for the Protection and Security of the Citizen, European Commission Joint Research Center.
Seed, R. B., Bea, R. G., Abdelmalak, R. I., Athanasopoulos-Zekkos, A., Boutwell, G. P., Briaud, J. L., et al. (2008). New Orleans and Hurricane Katrina: Introduction, overview, and the east flank. *Journal of Geotechnical and Geoenvironmental Engineering, 134,* 701.
Setola, R. (2010). How to measure the degree of interdependencies among critical infrastructures. *International Journal of System of Systems Engineering, 2*(1), 38–59.
Sjöberg, L. (1999). Risk perception in Western Europe. *Ambio, 28*(6), 543–549.
Skyttner, L. (2005). *General systems theory: Problems, perspectives, practice* (2nd ed.). Singapore: World Scientific Publishing Co., Pte. Ltd.
Slovic, P., Fischhoff, B., & Lichtenstein, S. (1979). Rating the risks. *Environment, 21*(3), 14.
Song, C. (2005). *A methodological framework for vulnerability assessment for critical infrastructure systems, hierarchical holographic vulnerability assessment (HHVA) (Thesis)*. Zürich: ETH Zürich.
Sousa-Poza, A. A., Kovacic, S., & Keating, C. B. (2008). System of systems engineering: An emerging multidiscipline. *International Journal of System of Systems Engineering, 1*(1/2), 1–17.
Starosielski, N. (2015). *The undersea network*. Durham: Duke University Press.
Sussman, J. M. (2005). *Perspectives on intelligent transportation systems*. New York, NY: Springer.
Sweeney, J. L. (2002). *The California electricity crisis*. Stanford: Hoover Institution Press.
Tainter, J. A. (1988). *The collapse of complex societies*. New York, NY: Cambridge University Press.
Taleb, N. N. (2014). *Antifragile: Things that gain from disorder* (Reprint edition). New York: Random House Trade Paperbacks.

The World Bank. (2017). *World development report 2017: Governance and the law (No. 112303)*. Washington: DC: The World Bank. Retrieved from http://elibrary.worldbank.org/doi/book/10.1596/978-1-4648-0950-7.

Thissen, W. A., & Herder, P. M. (2003b). *Critical infrastructures: State of the art in research and application*. Boston, MA: Kluwer Academic Publishers.

Tokgoz, B. E., & Gheorghe, A. V. (2013). Resilience quantification and its application to a residential building subject to hurricane winds. *International Journal of Disaster Risk Science, 4*(3), 105–114.

Townsend, F. F. (2006). *The federal response to Hurricane Katrina: Lessons learned*. Washington, DC: US Government Printing Office.

Turner, B. L., Kasperson, R. E., Matson, P. A., McCarthy, J. J., Corell, R. W., Christensen, L., et al. (2003). A framework for vulnerability analysis in sustainability science. *Proceedings of the National Academy of Sciences, 100*(14), 8074–8079.

UITP. (2014). *Statistics brief: World metro figures*. Brussels: L'Union Internationale des Transports Publics. Retrieved from http://www.uitp.org/sites/default/files/cck-focus-papers-files/Local_PT_in_the_EU_web%20(2).pdf.

US Congress. (2001). *Uniting and strengthening America by providing appropriate tools required to intercept and obstruct terrorism (USA PATRIOT ACT) Act of 2001* (No. 147) (p. 115 Stat. 271–402). Washington, DC: 107th Congress. Retrieved from http://www.gpo.gov/fdsys/pkg/PLAW-107publ56/content-detail.html.

USDHS. (2002). *Homeland Security Act of 2002* (No. Public Law 107–296) (p. 116 Stat. 2135–2321). Washington, DC: 107th Congress. Retrieved from https://www.dhs.gov/homeland-security-act-2002.

USDHS. (2013). *NIPP 2013: Partnering for critical infrastructure security and resilience*. Washington, D.C.: U.S. Dept. of Homeland Security. Retrieved from www.dhs.gov/xlibrary/assets/nipp-ssp-national-monuments-icons.pdf.

Vamanu, B. I., Gheorghe, A. V., & Katina, P. F. (2016). *Critical infrastructures: Risk and vulnerability assessment in transportation of dangerous goods—Transportation by road and rail* (Vol. 31). Cham, Switzerland: Springer International Publishing.

Weber, E. U., & Hsee, C. (1998). Cross-cultural differences in risk perception, but cross-cultural similarities in attitudes towards perceived risk. *Management Science, 44*(9), 1205–1217.

Weick, K. E. (1988). Enacted sensemaking in crisis situations. *Journal of Management Studies, 25* (4), 305–317. doi:10.1111/j.1467-6486.1988.tb00039.x.

Williams, S. (2016). *Spies in the Congo: America's atomic mission in World War II*. New York: PublicAffairs.

Willke, H. (2007). *Smart governance: Governing the global knowledge society*. Frankfurt, Germany: Campus Verlag GmbH.

Zio, E. (2016). Critical infrastructures vulnerability and risk analysis. *European Journal for Security Research*, 1–18.

Chapter 2
Governance Vulnerability Facets

Abstract In this chapter, several models supporting the notion governance for vulnerability assessment are presented. These include structural vulnerability, operational vulnerability, managerial vulnerability, and relational vulnerability. These notions are presented in view of *Quantitative Vulnerability Assessment* (QVA), which is a method to diagnose vulnerability in complex systems with a focus on strategies that could be undertaken for sustained system development. Theory supporting QVA is presented as well as the general means of transportability of the application are presented.

2.1 Strategic Approach for Dealing with Diverse Stakeholders

It is obvious that organizations in the twenty-first century operate under conditions of ambiguity, complexity, emergence, interdependence, and uncertainty (Flood and Carson 1993; Katina et al. 2014; Skyttner 2005). Regardless of your system of interest: health care, energy, transportation, security, etc.: you organizations operate under increasing lack of clarity and situational understanding, it has many richly and dynamically interacting stakeholders, systems, and subsystems with behavior difficult to predict, analysts and stakeholders might lack the ability to deduce behavior, structure, and performance of the constituent elements, and there is a likelihood that your system is influenced and it influenced the state of interconnected systems. Using this backdrop, one can argue that many of our systems, critical to public well-being, operate in the open and have a good chance of failing. The former, which is the subject of this book, for the most part, is referred to as vulnerability. There are several models that address the concept of vulnerability. These models are the subject of the remainder of this chapter along with strategic measures that enable system development and sustainability. Appendix B has been prepared to address 'governance' at a more general level.

A.V. Gheorghe et al., *Critical Infrastructures, Key Resources, Key Assets*, Topics in Safety, Risk, Reliability and Quality 34,
https://doi.org/10.1007/978-3-319-69224-1_2

2.2 Angles and Targets of Vulnerability

2.2.1 *Structural Vulnerability, System Stability, and Hysteresis*

The ever-increasing complexity associated with technology permeates system structures and patterns. The behavior of high-tech-addictive modern society in conjunction with the collective, contagious anxiety and the unrest brought about by hesitant and confusing reshufflings in the world order and the globalization, places vulnerability of critical infrastructures on top of the agenda of all consequential establishments. Governments, defense industries, private organizations, banking systems, natural catastrophes, technical failures, and accidents as well as terrorists tend to merge into a collage defining the landscape of challenges for the twenty-first century. Our attempt to put some order to this 'mess' comes in the form of a model for vulnerability assessment. The goal is to contribute to a management toolset for critical infrastructures management that places emphasis on strategies for sustainable development under present conditions. In particular, this chapter addressed the following topics: (i) quantification of the concept of vulnerability, (ii) making vulnerability an operational concept for sustainable development strategies, and (iii) enabling systems engineering as an approach to vulnerability management. This is done with the aim of arriving at a methodological approach for vulnerability estimating in critical infrastructures different levels (i.e., local and regional) and means to measure potential impact on system sustainable development.

There is a scarcity of practical approaches to quantify vulnerability in critical infrastructures. In the present text, the proposed model, practical and sound, offers: (i) a two-parameter description of vulnerability and the respective equation of state of the system: 'operable' and 'inoperable,' (ii) a division of the two-parameter phase space of the system into 'vulnerability basins,' and (iii) a scale of 0–100 'vulnerability' and the means to measure the respective 'vulnerability index.' In essence, the proposed method can offer the ability to diagnose current system vulnerability. The method uses an extensive set of indicators involving internal and external elements with the capability to dynamically monitor the time evolvement of the vulnerability as change occurs. Appendix C is reserved for an in-depth discussion on how to arrive at the compact analytic solution for the equation of systems with many component systems. Certainly, this involves hysteresis in which the current state of a system might depend on its history as found in ferromagnetic and ferroelectric materials as evidenced in thermostats and Schmitt triggers to prevent unwanted frequent switching. The aim of the model is to operationalize the concept of vulnerability in the context of multi-dimensional indicators of sustainability.

2.2.1.1 QVA: The Basic Assumptions

Quantitative Vulnerability Assessment (QVA) is a result of a warranted equivalence with *Quantitative Risk Assessment* (QRA)—coined within the closing decade of the past century and having made quite a career in the community of risk and safety managers worldwide (Gheorghe and Vamanu 2004a, b; Vamanu et al. 2016). Like its risk-related counterpart, QVA is about expressing its object—vulnerability—in numbers, in a scientifically defendable and practically meaningful way. Unlike QRA, QVA has to face an even more difficult task, for at this time there is no agreed 'closed formula' for vulnerability, whereas for risk, one does have a formula: *risk* of a disruptive event equals the *probability* of occurrence of an event times the *measure of event consequences* powered to a subjective *consequence perception exponent*.

At the root of this dis-symmetry is common semantics. Without excessively elaborating, let it be noted that such a popular reference as the Webster's new explorer encyclopedic dictionary (Merriam-Webster 2006) retains, in the entry for 'risk,' the instrumental ingredients of the formula. Table 2.1 attempts to draw out these differences.

In QRA, the task is to take a well-substantiated *noun* to a number. In QVA, the task is to take an *adjective*, reflective of a virtuality (i.e., *open to*...) to a number. To achieve this, four assumptions are made:

Assumption 1 First, one needs to adapt an operational definition for vulnerability as openness of a system—openness to losing its design functions, and/or structural integrity, and/or identity under the combined interplay of two sets of factors (*U* and *V*), where *U* is risk-featuring factor while *V* is management response-featuring factor. All factors are supposed to be eventually quantifiable by appropriate indicators. *U* factors involve risks that the system is prone to (i.e., the disruptive developments). These include (i) elements *internal* to the system, and/or (ii) reflective to the *processes* that the system hosts, (iii) to the performance of a system of interest. We refer to these as *fast-variable indicators* since they are on the move, constantly. *V* factors involve slow-variable indicators external to the system. These influence system and capability of the system's management to react/respond to internal developments.

Assumption 2 The method carries the assumption that the measurable and monitored indicators (i.e., parameters) can be aggregated such that control variables of *U* and *V* can be obtained. This then suggests that *U* and *V* are membership functions of the fuzzy set theory (Christen et al. 1995; Katina and Unal 2015). Accordingly, if

Table 2.1 A basic dis-symmetry of risk and vulnerability

Risk (noun)	Vulnerable (adjective)
The *chance* of injury, damage, or loss; dangerous change; hazard; the *degree* of probability of loss	Open to being physically or emotionally wounded; open to attack or damage

X_i, $i = 1, 2, \ldots, n$ are the normalized indicators contributing in the definition of U, then one has:

$$U(X_1, X_2, \ldots, X_n) = \min\left(1, \left(X_1^p + X_2^p + \cdots + X_n^p\right)^{\frac{1}{p}}\right) \tag{2.1}$$

where X_i are obtained from the physical indicators Y_i as:

$$X_i = A\log_{10}(Y_i) + B, \quad i = 1, 2, \ldots, n \tag{2.2}$$

The constants A and B are, in turn, derived from the assumed knowledge of two pairs of values for the normalized and physical indicators: $X_i^{(1)} = 0.2$ and $X_i^{(2)} = 0.6$.

$$A\ \log_{10}\left(Y_i^{(1)}\right) + B = X_i^{(1)} \tag{2.3}$$

$$A\ \log_{10}\left(Y_i^{(2)}\right) + B = X_i^{(2)}$$

Wherefrom

$$A = \left(X_i^{(2)} - X_i^{(1)}\right) \Big/ \left(\log_{10}\left(Y_i^{(2)}\right) - \log_{10}\left(Y_i^{(1)}\right)\right)$$
$$B = \left(X_i^{(2)}\log_{10}\left(Y_i^{(1)}\right) - X_i^{(1)}\log_{10}\left(Y_i^{(2)}\right)\right) \Big/ \left(\log_{10}\left(Y_i^{(1)}\right) - \log_{10}\left(Y_i^{(2)}\right)\right) \tag{2.4}$$

A similar set of equations would be given for $V(X_1, X_2, \ldots, X_n)$.

Assumption 3 Once U and V are determined, one then assumes that these make the aggregated control variables of a two-state, multi-component system (see Chap. 7 in Vamanu et al. 2016). By design, the solution adopted for modeling vulnerability comes close to the Bragg–Williams approximation. According to this approach, the membership fractions in a two-state system can be obtained based on probabilities of individual transitions between the two states. The interplay of the actual 'physical' and potentially numerous system indicators will result in variations of the aggregated parameters (U and V), which in turn drives the system 'state' in and out of a region of instability (Vamanu et al. 2016). In a conventional sense, an *operable* system may thereby appear as: (i) stable and therefore featuring a low vulnerability, (ii) critically unstable (i.e., vulnerable), or (iii) unstable and thereby featuring a high vulnerability. Beyond these, the system may only be found *inoperable*. A schematic of structural vulnerability is presented in Fig. 2.1.

Assumption 4 As given above, it is not possible to create a *Vulnerability Scale* based on the assessment of the system state in the U space and V space. The following is adapted: (i) measuring the *Vulnerability Index* is done using Euclidian distance of the state of U and V to the cusp line in the $U \geq 0$, $V \geq 0$ region of the (U, V) plane, and (ii) normalizing the index such that, everywhere on the cusp line,

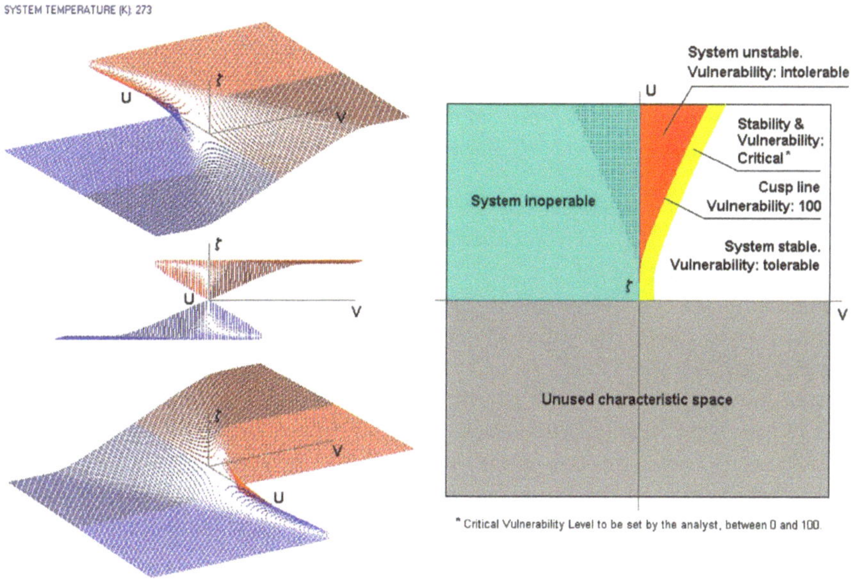

Fig. 2.1 Schematics of the QVA machine; left: characteristic of system (i.e., collection of real solutions of the 'equation of state' Eqs. 2.6 and 2.16. Adapted from Gheorghe and Vamanu (2004b)

including its $V \rightarrow 0$ portion, the *Vulnerability Index* must be equal to 100, the assumed maximum.

Subsequently, if D is the said distance to the cusp line, then the *Vulnerability Index*, V_{scale}, on the 0–100 *Vulnerability Scale* is:

$$V_{\text{Scale}} = 100\left(1 - \frac{D}{15}\right) \tag{2.5}$$

where the (U, V) field has been conventionally limited to $0 \leq U \leq 15$, $0 \leq V \leq 15$.

Since no analytic solution for the equation of the cusp line is readily available, distance, D, is evaluated up to the Bézier interpolation of a sufficient number of (U, V) knots on the cusp. The knots are determined as median points on the positive V-axis, for every positive U, between the last V that provides three solutions to the system's equation of state (i.e., equation of 'characteristic' in the sense of Thom (1975, 1983), namely:

$$\text{th}\left(\frac{U \cdot \zeta + V}{\theta}\right) = 2\zeta \tag{2.6}$$

and the first V, larger than the preceding, that provides only one solution. Symbol th stands for the hyperbolic tangent of the ensuing argument. As argued in the next section, the region of the characteristic's topological foil featuring a single solution to the equation of state is the *region of system stability*, whereas the region featuring three solutions, of which only two can normally be accessed, is the *region of system instability*.

The rough equivalence of (i) *system Instability…Highest/Intolerable Vulnerability* and (ii) *system Stability … Lower/Tolerable Vulnerability* is assumed. In the sense of the definition, within the region of instability, the *Vulnerability Index* is supposed to be uniformly 100, while it would gradually decrease away from the edge of the instability region. In a basic 'simple' computer-assisted QVA exercise, and for the sake of an example, Fig. 2.2 depicts a system described by 30 U-type generic indicators and 20 V-type indicators.

The geometric (Euclidian) distance of the state point in the (U, V) plane to the cusp line is taken as a measure of the vulnerability. The measure is normalized such that vulnerability is 100 everywhere on the cusp line and its analytic continuation as V is equal to 0.

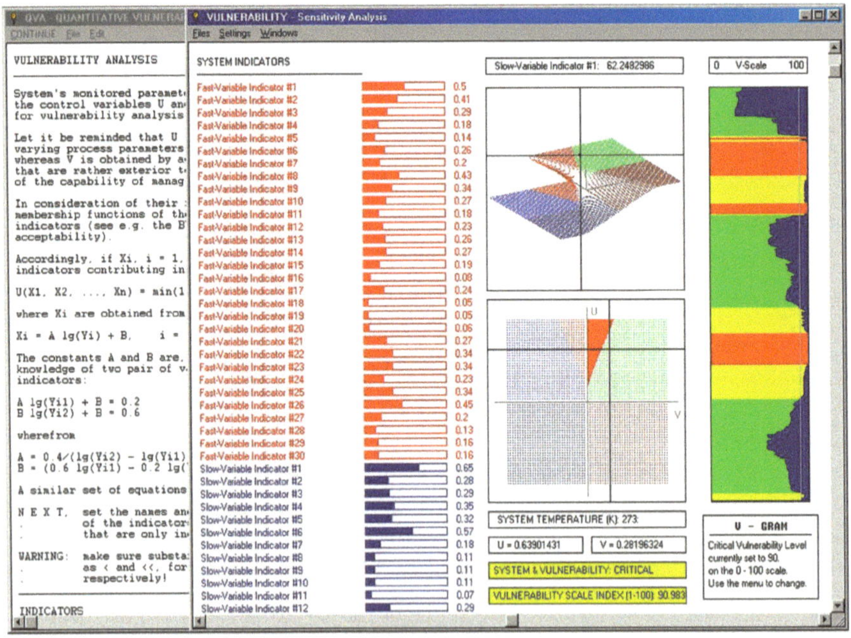

Fig. 2.2 An example of a computerized QVA exercise involving 30 U-type and 20 V-type indicators. V-Gram represents a histogram of vulnerability over time on the scale of 0–100. It can be placed on record and then called back for analysis, adapted from Gheorghe and Vamanu (2014)

2.2.1.2 QVA Modeling: Vulnerability and Stability in Multi-component Systems

Let us assume a system consisting of a large number, M, of elemental constituents or *members*. Elemental is taken in context as in 'atomic' sense such that a member should be seen as complex and fully connected to its environment, and yet indivisible as in a 'black box.' System members interact with each other with, in principle, a varying intensity. To describe the interaction, a *coupling constant*, or *intrinsic parameter*, U, is assumed to be either known or inferable. However, it is also recognized that members state can also be influenced by factors exterior to the system, as issue being accountable via an *influence field* or *extrinsic parameter*, V.

A member of the system may assume only two distinct *states*, state 1 and state 2. The generic 'states' may be seen as opposite in respect of a given criterion of judgment as in normal–abnormal, up–down, and pro–con, although this is not *always* the case. The only condition of essence is that states 1 and 2 are distinguishable from each other. At any given time, t, let M_1 members be in state 1 and M_2 members be in state 2. Since only two states are possible, one has:

$$M_1 + M_2 = M \tag{2.7}$$

The overall state of the system may then be described via the pair of numbers (M_1, M_2), while the system dynamics, or 'motion' in its state space, will follow from variations in M_1 and M_2 that should be consistent with Eq. (2.7). The smallest transitions in the state of the system would obviously involve alterations by one unit in the numbers of members:

$$(M_1-1, M_2+1)\underset{w_{21}\ \rightarrow}{\overset{\leftarrow\ w_{12}}{\rule{2em}{0.4pt}}}(M_1, M_2)\underset{\leftarrow\ w_{12}}{\overset{w_{21}\ \rightarrow}{\rule{2em}{0.4pt}}}(M_1+1, M_2-1) \tag{2.8}$$

Assume that the respective transitions are governed by the probabilities W_{12} and $W_{21,}$ respectively, as indicated in the relationship (Eq. 2.8). Admission of the process (Eq. 2.8) leads also to the recognition of a *function of distribution* of the system's states, $f(M_1, M_2)$, that would obey the master Eq. (2.9):

$$\begin{aligned}\partial f(M_1,\ M_2,\ t)/\partial t = {} & w_{21}(M_1-\ 1,\ M_2+\ 1)\cdot f(M_1-\ 1,\ M_2+\ 1)\\ & + w_{12}(M_1+\ 1,\ M_2-\ 1)\cdot f(M_1+\ 1,\ M_2-\ 1)\\ & - (w_{21}(M_1,\ M_2) + w_{12}(M_1,\ M_2))\cdot f(M_1,\ M_2)\end{aligned} \tag{2.9}$$

The state (M_1, M_2) of the system can alternatively be described by the *membership fraction*

$$\zeta = (M_1 - M_2)/(2M), \tag{2.10}$$

defined such that if all system members are in state 1, then ζ = 1/2, whereas if all members are in state 2, then ζ = −1/2.

Upon that, one notes that the master Eq. (2.9) involves the following states:

$$\begin{array}{ll} (M_1, M_2) & \zeta \\ (M_1 - 1, M_2 + 1) & \zeta - 1/M \\ (M_1 + 1, M_2 - 1) & \zeta + 1/M \end{array}$$

so that Eq. (3.3) may be rewritten as:

$$\begin{aligned} \partial f(\zeta)/\partial t = {} & w_{21}(\zeta - 1/M)f(\zeta - 1/M) + w_{12}(\zeta + 1/M)f(\zeta + 1/M) \\ & - (w_{21}(\zeta) + w_{12}(\zeta))f(\zeta) \end{aligned} \tag{2.11}$$

The initial assumption that the number, M, of system members is large allows one a series expansion of all quantities in the second member of Eq. (2.11). Restricting the expansion to the second order in (1/M), one obtains:

$$\partial f/\partial t + \partial J/\partial \zeta = 0. \tag{2.12}$$

Equation (2.12) is a continuity (i.e., conservation) equation for the state distribution function f, involving the 'current'

$$J = (1/M)(w_{21} - w_{12}) \cdot f - (1/(2M_2))\partial((w_{21} + w_{12}) \cdot f)/\partial z \tag{2.13}$$

Looking for the stationary states of the system, one assumes now:

$$\partial f/\partial t = 0, \tag{2.14}$$

which leaves one with the equation

$$\partial J/\partial \zeta = 0. \tag{2.15}$$

having as solution

$$J = \text{constant and, in particular, } J = 0 \tag{2.16}$$

Using the expression (2.13) of the current J, Eq. (2.16) can immediately be integrated to give:

$$f(\zeta) = \text{const} \cdot \frac{\exp\left[2M_1 \int\limits_{-1/2}^{\zeta} \frac{w_{21}(\xi) - w_{12}(\xi)}{w_{21}(\xi) + w_{12}(\xi)} \mathrm{d}\xi\right]}{w_{21}(\zeta) + w_{12}(\zeta)} \tag{2.17}$$

The constant in Eq. (2.17) can be determined setting the $f(\zeta)$ to be normalized to 1:

$$\int_{-1/2}^{1/2} f(\zeta)d\zeta = 1 \tag{2.18}$$

To normalize, that is, to fully determine the distribution function $f(\zeta)$, one needs to make an assumption on the analytical form of the transition probabilities, w_{12} and w_{21}. The following expressions would correspond to the notion that the transitions are a cooperative phenomenon:

$$\begin{aligned} w_{12}(\zeta) &= wM_1 \cdot \exp(-U \cdot \zeta + V/\theta) \\ w_{21}(\zeta) &= wM_2 \cdot \exp(U \cdot \zeta + V/\theta) \end{aligned} \tag{2.19}$$

where U is the coupling constant (intrinsic parameter) and V is the influence field (extrinsic parameter) that were previously introduced, while θ is a generalized 'temperature' of the system.

One makes now the natural assumption that the values of the membership fraction ζ that make the distribution function $f(\zeta)$ reach its extremes would make the space of possible states (the 'characteristic') of the system. Taking the expressions (2.19) of the transition probabilities into Eq. (2.17), and requesting that the condition

$$\partial f(\zeta)/\partial\zeta = 0 \tag{2.20}$$

be fulfilled, one has:

$$\mathrm{cth}((U \cdot \zeta + V)/\theta) = (1/2 - 1/(U/\theta - 2M))/\zeta, \tag{2.21}$$

where cth denotes the hyperbolic cotangent function, $\mathrm{cth}(x) = (\exp(x) + \exp(-x))/(\exp(x) - \exp(-x))$. Using again the fact that the number of members, M, in the system is large, the second term in the parenthesis in the right-hand side of Eq. (2.21) is ignored, so that, finally, the space of system states (U, V, ζ) is given by the equation:

$$\mathrm{th}((U \cdot \zeta + V)/\theta) = 2\zeta \tag{2.22}$$

where th denotes the hyperbolic tangent function, $\mathrm{th}(x) = (\exp(x) - \exp(-x))/(\exp(x) + \exp(-x))$. Depending on the degree of interaction between system constituents (members), reflected in the coupling constant U, and on the external influence on all system members—reflected in the field V, and also taking into consideration the temperature, θ, of the system, Eq. (2.22) may display the following number of real solutions ζ that may be related to the overall system condition shown in Table 2.2.

Figure 2.3 renders the situation. The boxes in the left-hand side present the cuspidal foil $\zeta_{sys} = \zeta(U, V)$, also known as system's 'characteristic,' seen in perspective. The (U, V) plane on the right-hand side is color-coded to emphasize the different basins of the system's 'phase space.'

Table 2.2 Overall system conditions associated with real solutions

Number of real solutions	System conditions
1	Stable. Smooth transitions in population membership, between state 1 and state 2
	Low and/or acceptable vulnerability
3; of which 2 are identical	Critical. Sharp transitions in membership between states 1 and 2 are possible. Either state 1 or state 2 may suddenly become improbable
	System is critically vulnerable
3; all different from each other	Unstable. Sharp transitions in membership between states 1 and 2 are possible. Frequency of occurrence of states 1 and 2 are comparable. Though Eq. (22) has three real roots, the intermediate root is generally taken as having no physical meaning and is therefore discarded
	System is dangerously/unacceptably vulnerable

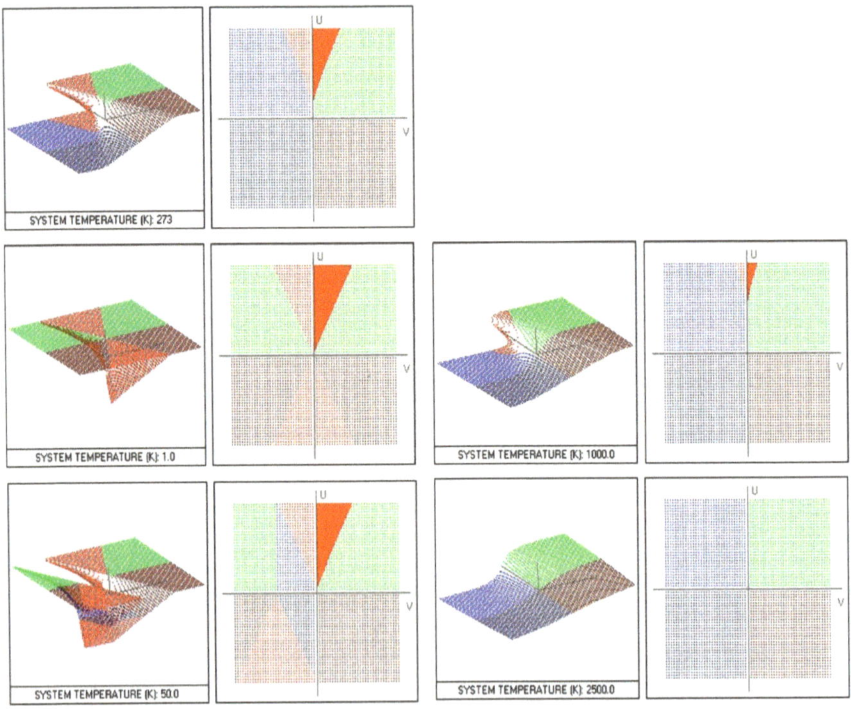

Fig. 2.3 Generic system 'characteristics' at different temperatures, figures adapted from Gheorghe and Vamanu (2004b)

As it turns out, the aspect of the foil expressing the topology of the system's space of states would vary with the generalized 'temperature' θ. The concrete details would, of course, depend on the scaling adopted for the 'energy'-wise parameters involved. Indeed, drawing further upon the physical analogy behind the model one would have $\theta = k_B T$, with k_B a 'Boltzmann' constant relating to the energy per degree of freedom of a system member and T being the 'absolute temperature.' Likewise, with the Ising model of ferromagnetics in mind, the coupling constant U would be reminiscent of the pair exchange energy, while V would bring to mind an external magnetic field casting its influence on all the 'spins' that make up the system. For practical purposes, the exercise attempted has adopted a 'Boltzmann' constant equal to 1/273, while preserving the absolute 'temperature' scale, where the 0 centigrade would correspond to T = 273.15. That would take parameters U and V in the convenient range of 1–15.

On this parameter scaling, at a 'normal' temperature of 273 K, and higher up, the system 'characteristic' would show one instability region in the positive U range, whereas at lower temperatures, a second instability region would progressively manifest itself in the negative U and V range, until, very close to 0 K. The two instability regions would almost connect with each other at $U = 0$ and $V = 0$. At this stage into the exercise, it is perhaps too early to speculate on the significance of such occurrences. More careful thinking should go, for example, into establishing whether negative *Us* (i.e., anti-ferromagnetic states) should at all be accepted, which would indicate a spontaneous antagonism of individual system members. At any rate, the current QVA model, relying on the definitions (Eqs. 2.1 through 2.4) for the indicators, would make use of only the $U \geq 0$, $V \geq 0$ quadrant of the (U, V) plane.

In this section, thus far, QVA is presented as an approach for cooperative behavior in multi-component systems. It is addressed along the line '*to whom it may concern,*' primarily targeting readers with a background in Physics that feel like getting more enlightened about and confident in QVA. If one is interested in the physical analogy of the model as well as logical and calculational flow, along with model assumptions, equations, and the respective notations and remarks on potential system collapse, interdependence, and temperature effects, then the authors suggest acquaintances with 'Appendix D' in Vamanu et al. (2016).

Limitations of QVA

The apparent association of the method proposed by catastrophe theory (CT) (Thom 1975, 1983; Zeeman 1977) may give rise to some discomfort in some segments of the critical readers, given the never-exhausted controversy around the meaning and the value associated with CT. A comprehensive coverage of the issues associated with CT is found in Thom (1983). Aware of those issues, present authors find it appropriate to note that given the way the QVA model was proposed, key objections fall off target. Table 2.3 is drawn to address these objections.

Notice that the last criticism, or rather an objection, steams from susceptibility of the QVA to experimental control—a previous objection, which paves the way to counter the fourth possible objection. Lingering one moment longer in the realm of

Table 2.3 A summary of CT criticisms and comparative QVA strengths

CT Objections	QVA-related responses
'CT may be reproached for being an abstract schema independent of physical reality' (a paraphrase from Thom 1983)	Using as conceptual background for the QVA model, the archetype of order–disorder phenomena and phase transitions in multi-component (many-body) systems, and tracking the theoretical apparatus back to some concrete solutions such as the Bragg–Williams approximation to the Ising model—which, in turn, covers great many cases of 'physical reality'—would largely free the proposed QVA approach from the objection
'CT is in itself purely qualitative and it simultaneously ignores considerations of scale and the quantitative laws of classical Physics' (a paraphrase from Thom 1983)	The QVA model, method, algorithm, and computer code as described are patent proof to the contrary: The statistical mechanics-inspired tools have provided for an effective quantitative approach to vulnerability, including a *Vulnerability Index* and *Scale*. If the scaling conventions may indeed be said as being user-defined, and thereby arbitrary, the topology of the phase space behind these is, on the other hand, univocal and indisputable—within the given model terms
'CT is not susceptible to experimental control' (a paraphrase from Thom 1983)	QVA *is*, undoubtedly, susceptible to experimental control and was created precisely with purpose in mind. The computer codes that were designed to implement the method are only the soft expression of a machine that may eventually take the hard form of a 'black box.' The code is designed to take in input by the dozens, of the physical indicators of a given system and delivering output multimedia including video and sound, which serves as an 'alarm' warning on system evolving vulnerability status
'(There is) … the thorny problem of uniqueness of models in CT: if one has two models, *M*, *M′*, in competition (on the same system), can one always find a model *M″* that covers both?' (a paraphrase from Thom 1983)	With QVA, the uniqueness issue is solved by (i) fairly admitting that models are *not* unique, a model *M* for a system *S* being fully determined by its collection of *U*- and *V*-type indicators, in both numbers and nature, and (ii) emphasizing that the *appropriateness* of a model—the only criterion of interest in accepting it for practical purposes—has to be settled by *experimentation*

Thom's (1983) comments, let us note that, the way it is proposed, the QVA may well make proof of some convenient 'ontological range', which Thom (1983) describes as 'the manner in which the phenomena [can] take place and in which it describes their underlying mechanisms' (Thom 1983, p. 111). The 'phenomenon' in the case of QVA is the *coherent convergence of dozens of internal, fast-varying, and external, slow-varying, system features, expressed in as many physically different indicators, into an identifiable and quantifiable vulnerability state.*

At this point, we suggest an '*Assumption Zero*' of this research: *any critical infrastructure can be accommodated within the concept of a multi-component, multi-indicator system the parts of which would show some kind of collective behavior by virtue of their interacting, as well as some susceptibility to external factors acting upon the system components.*

While the current proposal should be seen as only a test of feasibility, further developments may consolidate a fully operational QVA methodology. In an attempt to make present concepts more 'fun,' researchers developed a 'mix game' approach to concepts outlined in this section—see Appendix D.

2.2.2 Operational Vulnerability and System Dynamics in Phase Portraits

In a study commissioned by the Swiss Federal Department of Defense, Civil Protection and Sports, and Directorate for Security Policy, by a research team of the SwissFederal Institute of Technology (Gheorghe 2004), it was shown that the dynamics of a three-body component business system could be modeled using ordinary differential equations for a generalized measure of component 'productions' of X, Y, and Z such that:

$$
\begin{aligned}
\mathrm{d}X/\mathrm{d}t &= a1 + a2X + a3X^2 + a4XY + a5XZ + a6Y + a7Y^2 + a8YZ + a9Z + a10Z^2\\
\mathrm{d}X/\mathrm{d}t &= a11 + a12X + a13X^2 + a14XY + a15XZ + a16Y + a17Y^2 + a18YZ + a19Z + a20X^2\\
\mathrm{d}X/\mathrm{d}t &= a21 + a22X + a23X^2 + a24XY + a25XZ + a26Y + a27Y^2 + a28YZ + a29Z + a30Z^2
\end{aligned} \tag{2.23}
$$

In this equation, coupling coefficients ai may be nil. It is equally evident that the Euler solving of the system patterns above can also have the topologies of (1) unbounded states and (2) bounded states. The classification for bounded states can include fixed point, limit cycle, and strange attractor.

In turn, qualitative differences between attractor configurations can be captured by two indicators: the Lyapunov exponent, L, and the fractal (in effect, correlation) dimension, F. L is the largest of two quantities that are defined based on comparing successive Euler iterations of the solutions of system (Eq. 2.23):

$$
\begin{aligned}
Xn + 1 &= Xn + hF(Xn, Yn, Zn) \\
Yn + 1 &= Yn + hG(Xn, Yn, Zn) \\
Zn + 1 &= Zn + hH(Xn, Yn, Zn)
\end{aligned}
\tag{2.24}
$$

and which account for the propensity of the solutions to either coalesce over a bounded topological variety or diverge to infinity. L is, therefore, relating to system's *stability*. F, on the other hand, relates to the degree the phase space is occupied by point states of the system: The larger the degree of occupancy, the larger the F. It is conjectured that F relates to two apparently conflicting qualities of the system: the *predictability* and the *maneuverability*. There is a built-in assumption that a system whose phase space pattern occupies more of the space foil-like or bulk configurations of higher F is likely to offer more space of maneuver for the coupling coefficients that describe the exchanges between system components, and yet, on the other hand, it is more difficult to point at a space region where the system state my find itself, at any time.

On the contrary, a string-like (lower F) configuration in the phase space makes the inference of the system whereabouts easier—a higher predictability—whereas the maneuverability is, comparatively, lower.

2.2.2.1 The Rules

Speculating over the features above may result in the following classification of situations that can be monitored as it evolves in time because of fluctuations or otherwise intentional (programmed) evolutions in the coupling (exchange) coefficients ai that can be used in a dashboard for monitoring business systems:

```
---------------
'Stability

if diagnose = "Fixed Point" then
    Current Stability State = "CALM"
          if Previous Stability State = "CALM" then Stability Trend = "CONSTANT"
          if Previous Stability State = "NORMAL" then Stability Trend = "IMPROVING"
          if Previous Stability State = "ACCEPTABLE" then Stability Trend = "IMPROVING"
          if Previous Stability State = "UNACCEPTABLE" then Stability Trend = "IMPROVING"
end if
if diagnose = "Limit Cycle" then
        Current Stability State = "NORMAL"
          if Previous Stability State = "CALM" then Stability Trend = "DETERIORATING"
          if Previous Stability State = "NORMAL" then Stability Trend = "CONSTANT"
          if Previous Stability State = "ACCEPTABLE" then Stability Trend = "IMPROVING"
          if Previous Stability State = "UNACCEPTABLE" then Stability Trend = "IMPROVING"
end if
if diagnose = "Strange Attractor" then
    Current Stability State = "ACCEPTABLE"
          if Previous Stability State = "CALM" then Stability Trend = "DETERIORATING"
          if Previous Stability State = "NORMAL" then Stability Trend = "DETERIORATING"
          if Previous Stability State = "ACCEPTABLE" then Stability Trend = "CONSTANT"
          if Previous Stability State = "UNACCEPTABLE" then Stability Trend = "IMPROVING"
end if
if diagnose = "Unbounded" then
    Current Stability State = "UNACCEPTABLE"
          if Previous Stability State = "CALM" then Stability Trend = "DETERIORATING"
          if Previous Stability State = "NORMAL" then Stability Trend = "DETERIORATING"
          if Previous Stability State = "ACCEPTABLE" then Stability Trend = "DETERIORATING"
          if Previous Stability State = "UNACCEPTABLE" then Stability Trend = "CONSTANT"
end if

'Predictability

if diagnose = "Unbounded" then Current Predictability State = "UNACCEPTABLE"
if diagnose = "Fixed Point" then Current Predictability State = "CALM"
if diagnose = "Limit Cycle" then Current Predictability State = "NORMAL"
if diagnose = "Strange Attractor" then
    if F<=D/4 then
          Current Predictability State = "CALM"
          if Previous Predictability State = "CALM" then Predictability Trend = "CONSTANT"
          if Previous Predictability State = "NORMAL" then Predictability Trend = "IMPROVING"
          if Previous Predictability State = "ACCEPTABLE" then Predictability Trend = "IMPROVING"
          if Previous Predictability State = "UNACCEPTABLE" then Predictability Trend = "IMPROVING"
    end if
    if F>D/4 and F<=D/2 then
          Current Predictability State = "NORMAL"
          if Previous Predictability State = "CALM" then Predictability Trend = "DETERIORATING"
          if Previous Predictability State = "NORMAL" then Predictability Trend = "CONSTANT"
          if Previous Predictability State = "ACCEPTABLE" then Predictability Trend = "IMPROVING"
          if Previous Predictability State = "UNACCEPTABLE" then Predictability Trend = "IMPROVING"
    end if
    if F>D/2 and F<=3*D/4 then
          Current Predictability State = "ACCEPTABLE"
          if Previous Predictability State = "CALM" then Predictability Trend = "DETERIORATING"
          if Previous Predictability State = "NORMAL" then Predictability Trend = "DETERIORATING"
          if Previous Predictability State = "ACCEPTABLE" then Predictability Trend = "CONSTANT"
          if Previous Predictability State = "UNACCEPTABLE" then Predictability Trend = "IMPROVING"
    end if
    if F>3*D/4 then
          Current Predictability State = "UNACCEPTABLE"
          if Previous Predictability State = "CALM" then Predictability Trend = "DETERIORATING"
          if Previous Predictability State = "NORMAL" then Predictability Trend = "DETERIORATING"
          if Previous Predictability State = "ACCEPTABLE" then Predictability Trend = "DETERIORATING"
          if Previous Predictability State = "UNACCEPTABLE" then Predictability Trend = "CONSTANT"
    end if
end if
'Maneuverability

if diagnose = "Unbounded" then
    Current Maneuverability State = "UNACCEPTABLE"
          if Previous Maneuverability State = "CALM" then Maneuverability Trend = "DETERIORATING"
          if Previous Maneuverability State = "NORMAL" then Maneuverability Trend = "DETERIORATING"
          if Previous Maneuverability State = "ACCEPTABLE" then Maneuverability Trend = "DETERIORATING"
          if Previous Maneuverability State = "UNACCEPTABLE" then Maneuverability Trend = "CONSTANT"
    end if
if diagnose = "Fixed Point" then
    Current Maneuverability State = "UNACCEPTABLE"
          if Previous Maneuverability State = "CALM" then Maneuverability Trend = "DETERIORATING"
          if Previous Maneuverability State = "NORMAL" then Maneuverability Trend = "DETERIORATING"
          if Previous Maneuverability State = "ACCEPTABLE" then Maneuverability Trend = "DETERIORATING"
          if Previous Maneuverability State = "UNACCEPTABLE" then Maneuverability Trend = "CONSTANT"
    end if
```

```
if diagnose = "Limit Cycle" then
    Current Maneuverability State = "ACCEPTABLE"
          if Previous Maneuverability State = "CALM" then Maneuverability Trend = "DETERIORATING"
          if Previous Maneuverability State = "NORMAL" then Maneuverability Trend = "DETERIORATING"
          if Previous Maneuverability State = "ACCEPTABLE" then Maneuverability Trend = "CONSTANT"
          if Previous Maneuverability State = "UNACCEPTABLE" then Maneuverability Trend = "IMPROVING"
    end if
if diagnose = "Strange Attractor" then
    if F<=D/8 then
    Current Maneuverability State = "ACCEPTABLE"
          if Previous Maneuverability State = "CALM" then Maneuverability Trend = "DETERIORATING"
          if Previous Maneuverability State = "NORMAL" then Maneuverability Trend = "DETERIORATING"
          if Previous Maneuverability State = "ACCEPTABLE" then Maneuverability Trend = "CONSTANT"
          if Previous Maneuverability State = "UNACCEPTABLE" then Maneuverability Trend = "IMPROVING"
    end if
if F>D/8 and F<=D/4 then
    Current Maneuverability State = "NORMAL"
          if Previous Maneuverability State = "CALM" then Maneuverability Trend = "DETERIORATING"
          if Previous Maneuverability State = "NORMAL" then Maneuverability Trend = "CONSTANT"
          if Previous Maneuverability State = "ACCEPTABLE" then Maneuverability Trend = "IMPROVING"
          if Previous Maneuverability State = "UNACCEPTABLE" then Maneuverability Trend = "IMPROVING"
    end if
if F>D/4 then
    Current Maneuverability State = "CALM"
          if Previous Maneuverability State = "CALM" then Maneuverability Trend = "CONSTANT"
          if Previous Maneuverability State = "NORMAL" then Maneuverability Trend = "IMPROVING"
          if Previous Maneuverability State = "ACCEPTABLE" then Maneuverability Trend = "IMPROVING"
          if Previous Maneuverability State = "UNACCEPTABLE" then Maneuverability Trend = "IMPROVING"
    end if
end if

'System Condition

if Current Stability State = "UNACCEPTABLE" then Current System Condition = "UNACCEPTABLE"

if Current Stability State = "ACCEPTABLE" then
    if ((Current Predictability State = "UNACCEPTABLE") or (Current Maneuverability State =
"UNACCEPTABLE")) then
          Current System Condition = "UNACCEPTABLE"
    end if

    if ((Current Predictability State = "ACCEPTABLE") or (Current Maneuverability State =
"ACCEPTABLE"))
and ((Current Predictability State<>"UNACCEPTABLE") and (Current Maneuverability
State<>"UNACCEPTABLE")) then
          Current System Condition = "ACCEPTABLE"
    end if

    if ((Current Predictability State = "NORMAL") and (Current Maneuverability State = "NORMAL")) then
          Current System Condition = "NORMAL"
    end if

    if ((Current Predictability State = "NORMAL") and (Current Maneuverability State = "CALM")) then
          Current System Condition = "NORMAL"
    end if

    if ((Current Predictability State = "CALM") and (Current Maneuverability State = "NORMAL")) then
          Current System Condition = "CALM"
    end if

    if ((Current Predictability State = "CALM") and (Current Maneuverability State = "CALM")) then
          Current System Condition = "CALM"
    end if
end if

    if Current Stability State = "NORMAL" then
        if ((Current Predictability State = "UNACCEPTABLE") or (Current Maneuverability State =
"UNACCEPTABLE")) then
          Current System Condition = "ACCEPTABLE"
    end if

    if ((Current Predictability State = "ACCEPTABLE") or (Current Maneuverability State =
"ACCEPTABLE"))
and ((Current Predictability State<>"UNACCEPTABLE") and (Current Maneuverability
State<>"UNACCEPTABLE")) then
          Current System Condition = "NORMAL"
    end if

    if ((Current Predictability State = "NORMAL") and (Current Maneuverability State = "NORMAL")) then
          Current System Condition = "CALM"
    end if

    if ((Current Predictability State = "NORMAL") and (Current Maneuverability State = "CALM")) then
          Current System Condition = "CALM"
```

```
    end if

    if ((Current Predictability State = "CALM") and (Current Maneuverability State = "NORMAL")) then
            Current System Condition = "CALM"
    end if

    if ((Current Predictability State = "CALM") and (Current Maneuverability State = "CALM")) then
            Current System Condition = "CALM"
    end if
end if

    if Current Stability State = "CALM" then
        if ((Current Predictability State = "UNACCEPTABLE") or (Current Maneuverability State =
"UNACCEPTABLE")) then
            Current System Condition = "NORMAL"
        end if

    if ((Current Predictability State = "ACCEPTABLE") or (Current Maneuverability State =
"ACCEPTABLE"))
and ((Current Predictability State<>"UNACCEPTABLE") and (Current Maneuverability
State<>"UNACCEPTABLE")) then
            Current System Condition = "CALM"
    end if

    if ((Current Predictability State = "NORMAL") and (Current Maneuverability State = "NORMAL")) then
            Current System Condition = "CALM"
    end if

    if ((Current Predictability State = "NORMAL") and (Current Maneuverability State = "CALM")) then
            Current System Condition = "CALM"
    end if

    if ((Current Predictability State = "CALM") and (Current Maneuverability State = "NORMAL")) then
            Current System Condition = "CALM"
    end if

    if ((Current Predictability State = "CALM") and (Current Maneuverability State = "CALM")) then
            Current System Condition = "CALM"
    end if
end if

    if Previous System Condition = "CALM" then
            if Current System Condition = Previous System Condition then System Condition Trend =
"CONSTANT"
            if Current System Condition<>Previous System Condition then System Condition Trend =
        "DETERIORATING"
    end if

    if Previous System Condition = "NORMAL" then
            if Current System Condition = Previous System Condition then System Condition Trend =
        "CONSTANT"
            if Current System Condition = "CALM" then System Condition Trend = "IMPROVING"
            if Current System Condition = "ACCEPTABLE" then System Condition Trend = "DETERIORATING"
            if Current System Condition = "UNACCEPTABLE" then System Condition Trend = "DETERIORATING"
    end if

    if Previous System Condition = "ACCEPTABLE" then
            if Current System Condition = Previous System Condition then System Condition Trend =
    "CONSTANT"
            if Current System Condition = "CALM" then System Condition Trend = "IMPROVING"
            if Current System Condition = "NORMAL" then System Condition Trend = "IMPROVING"
            if Current System Condition = "UNACCEPTABLE" then System Condition Trend = "DETERIORATING"
    end if

    if Previous System Condition = "UNACCEPTABLE" then
            if Current System Condition = Previous System Condition then System Condition Trend =
    "CONSTANT"
            if Current System Condition<>Previous System Condition then System Condition Trend =
    "IMPROVING"
end if
-----------
```

The overall system condition would result to ‘calm,’ ‘normal,’ ‘acceptable,’ or ‘unacceptable,’ depending on the interplay of the factors described. A trend can also be identified, in consideration of the precedence in system’s conditions: ‘constant,’ ‘improving,’ or ‘deteriorating.’ The code offers one possible interface; we shall refer to the ‘dashboard’ to demonstrate the concept. Figure 2.4 depicts a standard view of the ‘dashboard’ with the left-hand side offering daily monitor of business system components. The right-hand side of the dashboard is for those who might be interested in the patterns behind the performance of the business system. Additional views are presented in Figs. 2.5, 2.6, and 2.7.

Several ‘illustrational’ scenarios were developed using this model. In these scenarios, different system conditions were developed and data used for random (uncorrelated) variation of the system model control parameters in the coupling (exchange) coefficients. Figures 2.8, 2.9, and 2.10 are the products of mockup runs. Figure 2.8 depicts a case for *increased predictability* by the narrowing of the occupied phase space—a consequence of uncorrelated variations in the system model’s control parameters. Notice that the resulting overall *system condition* is ‘normal.’

A case for increased *maneuverability* by the widening of the occupied phase space—a consequence of uncorrelated variations in the system model’s control parameters—is depicted in Fig. 2.9. Notice that the resulting overall system condition is ‘acceptable.’ Figure 2.10 depicts a case for an unacceptable *diminishing of*

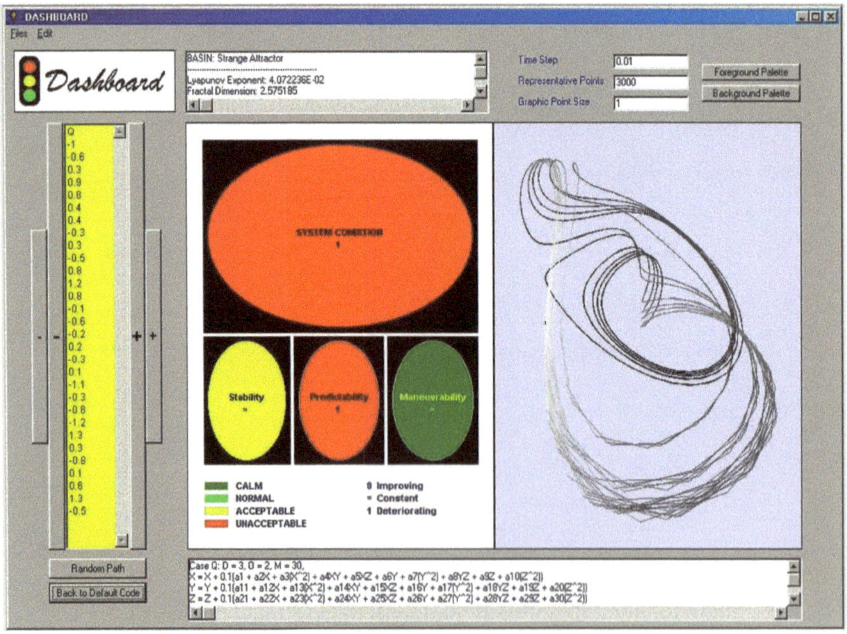

Fig. 2.4 A standard ‘Dashboard’ view indicating system conditions, adapted from Gheorghe and Vamanu (2006)

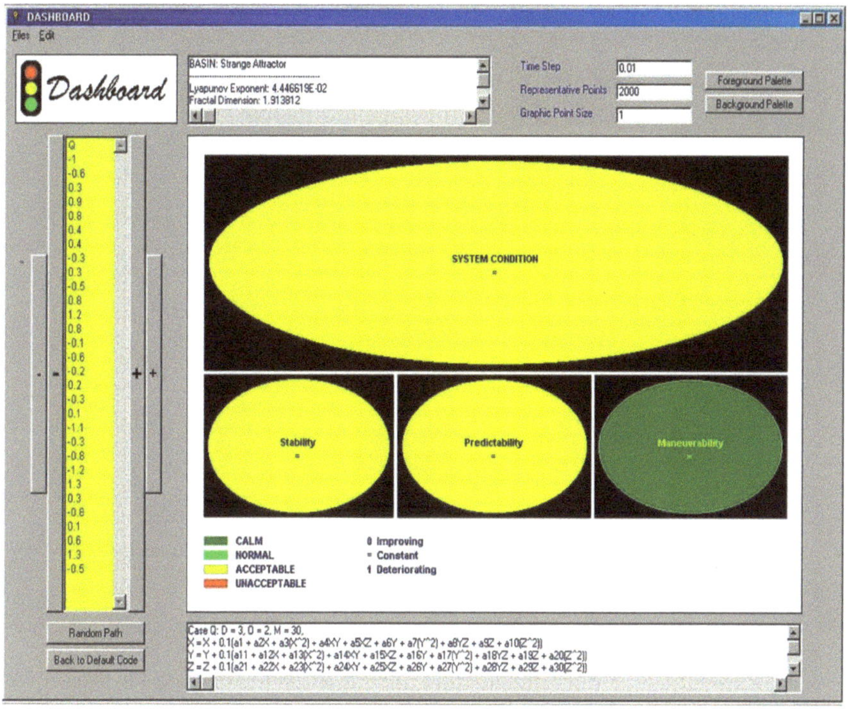

Fig. 2.5 'Dashboard' indicating an 'acceptable' system condition, adapted from Gheorghe and Vamanu (2006)

the predictability by the excessive spread-out of the system state trajectory flow over the phase space—a consequence of uncorrelated variations in the system model's control parameters. Notice that this results in an overall system condition of 'unacceptable.'

It is only fair at this stage to recognize that the scaling of 'system condition' is subject to a considerable arbitraries. The scaling of practical values must depend, extensively, on stakeholder perspective on the matter. The analyst can influence the auditing and numerical experiments; however, the stakeholders of the system should be the primary influencers. Furthermore, there might be a need to implement more refined notions and indicators of the chaos theory.

The concept of a 'dashboard' and the language used in this section are borrowed from the Basel Committee's debate on business operational risks and vulnerability (Doerig 2000; Romeike and Maitz 2001). In particular, Doerig (2000, p. 74) suggests that the dashboard approach 'is intended to provide senior management with a simple overview of operational risk levels and directional trends at the highest reporting aggregation level per business unit.'

The debate over the 'dashboard' approach to risk management has established several competing methods with varying degrees of complexity, sophistication, and feasibility operational risk evaluation in the banking sector. These methods include

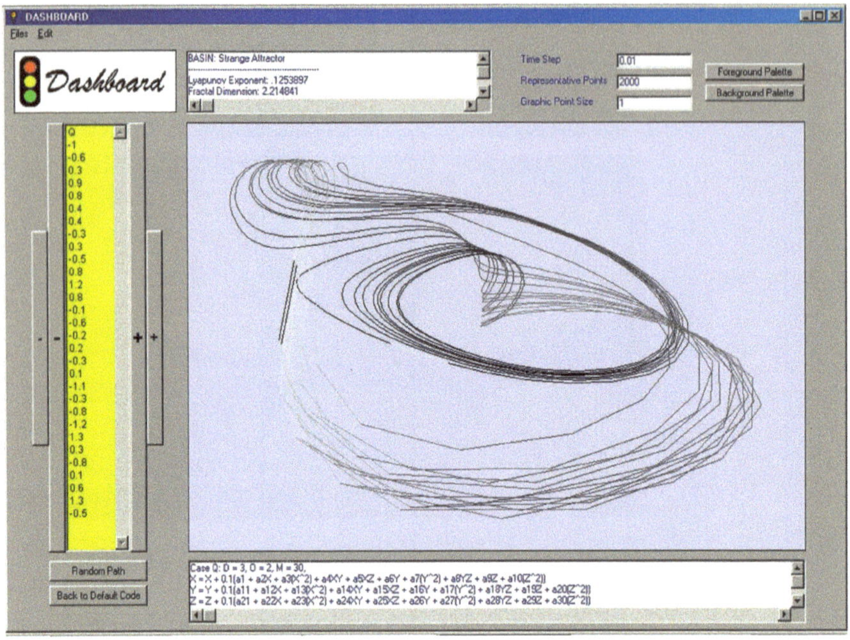

Fig. 2.6 Element of the 'Dashboard' indicating patterns of the business system, adapted from Gheorghe and Vamanu (2006)

'*Basic Indicator Approach*,' '*Standardised Approach*,' '*Internal Measurement Approach*,' and '*Loss Distribution Approach*' (Doerig 2000; Romeike and Maitz 2001). Certainly, there is still room for complementary approaches offering insights into assessment, new and emerging risks and vulnerabilities.

Subsequently, we suggest that approach suggested in this section's series of demonstrations could be used to investigate the 'motions' or exchanges in business structures and components to unveil intelligible structure(s) in the motion itself.

2.2.3 Managerial Vulnerability and Consensual Analytical Hierarchies

There are several methods associated with describing vulnerability (Nilsson et al. 2001) including *Index Method* as suggested in Vamanu et al. (2016). However, such methods do not appear to offer a complete picture of on local authority's actual risk level or ability to manage the risks—the municipal vulnerability/robustness. In this section, an attempt is made, therefore, to design a method that could be used to comprehensively do vulnerability analysis at a municipal level and yet easily updated to account for changes that might occur. To this end, we suggest the following advantages for this method:

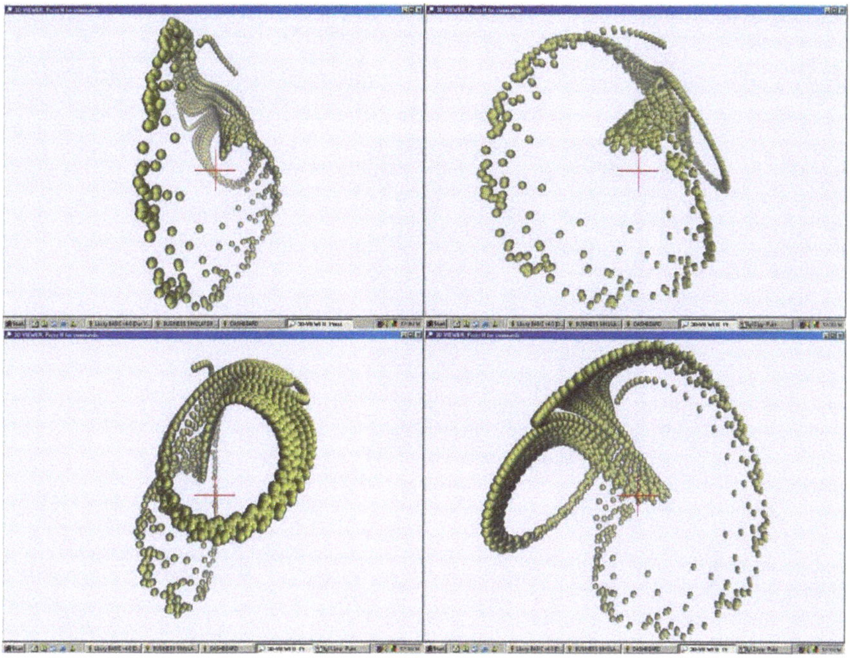

Fig. 2.7 Depicting system's phase space pattern as seen in 3D fashion, from four viewing angles, adapted from Gheorghe and Vamanu (2006)

- It gives a chance for assessing the individual local authority's ability to manage the current risks.
- The vulnerability/robustness can be assessed as a result of generally defined acceptance criteria. These are determined as an upper and lower limit. The result of this is vulnerability is divided up into three areas: one area where vulnerability is unacceptable, another where vulnerability can be tolerated, if all the financially possible efforts have been fulfilled, and a third where vulnerability is generally acceptable. Figure 2.11 depicts these areas.
- The provided vulnerability model could also be used as a basis for distributing financial means to the local municipal authorities.

We now describe the main parts of the model. There are five main parts to this model:

(I) The definition of current hazards and damage types. This model uses five different damage types: loss of life, damage to person, absence owing to illness, damage to the ecological system, and damage to property. Notice that these damage types are offered here randomly. Interestingly, several other damage types can be used including those associated with psychological trauma—a type of damage to the mind that occurs because of a severely distressing event. It should be evident that further work is required in this area to choose as damage types as well as possible consequences.

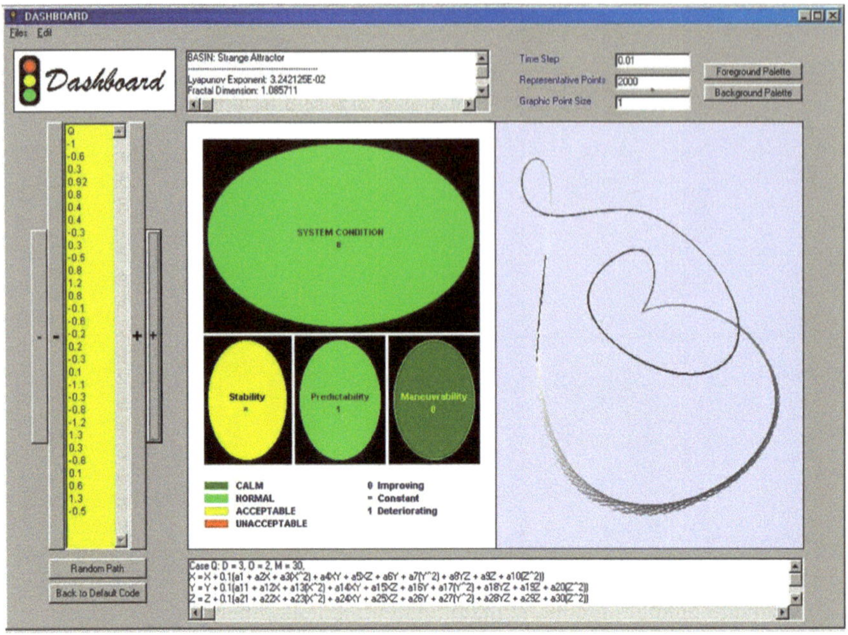

Fig. 2.8 A 'normal' system condition despite 'deteriorating' phase space, adapted from Gheorghe and Vamanu (2006)

(II) Damage and probability are divided up into four classes with the Swiss system as role model with index values of 1, 2, 4, and 4 with 1 corresponding to least harmful and 4 being the most dangerous.

(III) A systematic risk inventory is carried out for all examined municipal hazards, natural and malicious, which are given an index value, on a scale of 1–4, for each type of damage types as well as probability.

(IV) For all hazards, the existing damage indexes are multiplied by the probability index. The product of the damage index and probability index is summed up over relevant damage types (least 1, maximum 5). The sum is named Z_i. The maximum value of Z_i for specific hazard can be, for example, 80. This value can be obtained from a damage class 4, a probability class 4, and 5 damage types. The individual values for Z_i give the municipal danger profile; the sum of Z_i, gives a measurement of the collective risks and corresponds to the local authority's risk value.

(V) An inventory is made of the resources for risk management. For each defined hazard (danger), the capability to manage the risk is described using two coefficients: α_i and β_i where index i describes the actual hazard. The value of α_i and β_i can for each of them amount to 1, but the total value can never be more than 1 (perfectly managed hazard).

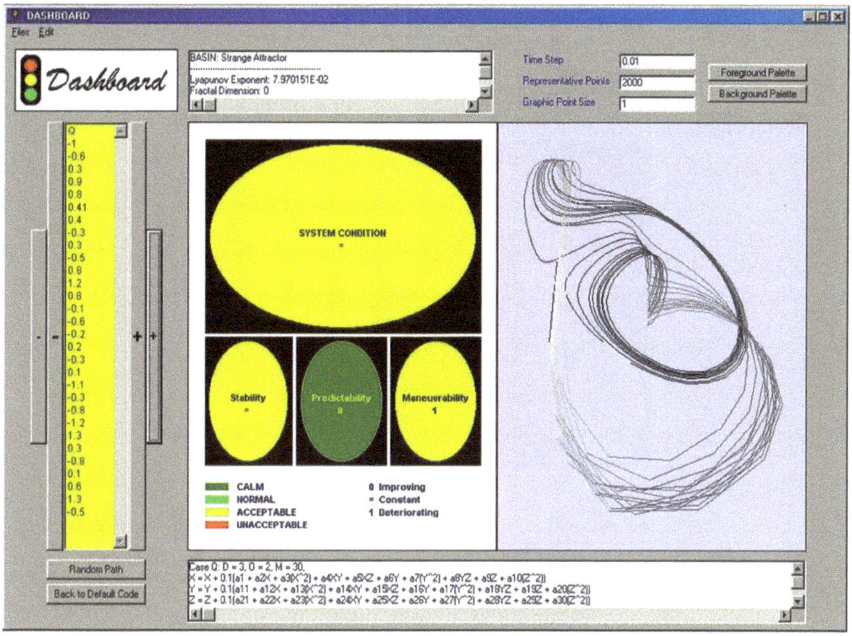

Fig. 2.9 An ‘acceptable’ system condition despite ‘deteriorating’ maneuverability phase space, adapted from Gheorghe and Vamanu (2006)

2.2.3.1 Deriving α_i

α_i describes the general hazard-independent ability to eliminate risks, counteract losses, intrusion, and damage as well as limit damage consequences. The checklists suggested in Lagbo-Bergqvist and Lexén (2000) could be used as a guide for setting value of α_i. In practice, the task is to set a value for the municipality for each of the five parameters: loss of life, damage to person, absence owing to illness, damage to ecological system, and damage to property (see Tables 2.4, 2.5, and 2.6), summing it up and finally normalizing. A structured method is used throughout this process based on multiple-criteria decision analysis (MCDA) method to derive α_i values.

2.2.3.2 Deriving β_i

β_i offers the states for specified hazard i, such as a factory plant, natural disaster (e.g., flooding), or level of the resources (actions) directly linked to this hazard. In this case, a hazard is taken as a ‘danger.’ Once again, this assessment can be made with the help of existing checklists for technical, administrative safety control, competing gaming, and techniques in modeling and simulation, among others. Each one of these actions must be specific of the hazard and, thus, not included in the calculation of the corresponding α_i value.

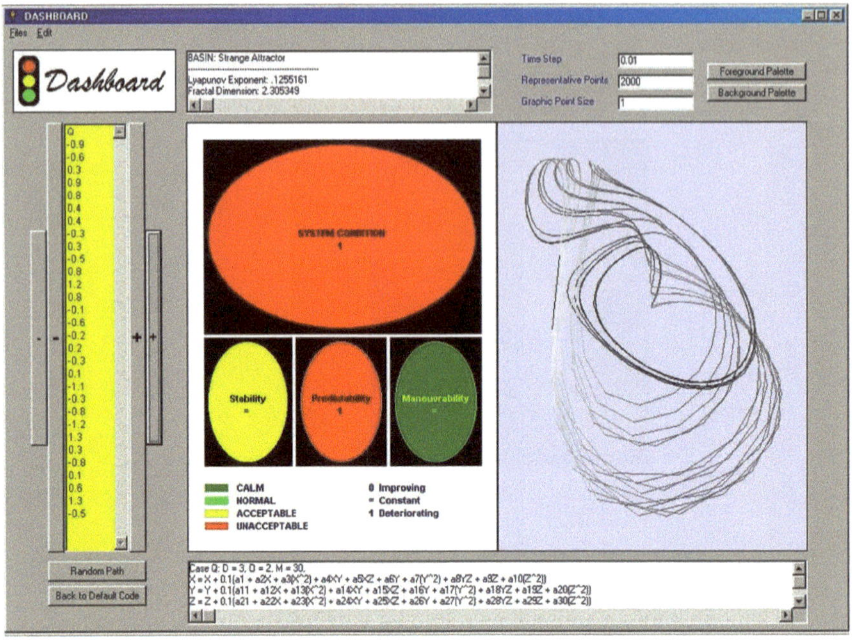

Fig. 2.10 An 'unacceptable' system condition with 'deteriorating' predictability phase space, adapted from Gheorghe and Vamanu (2006)

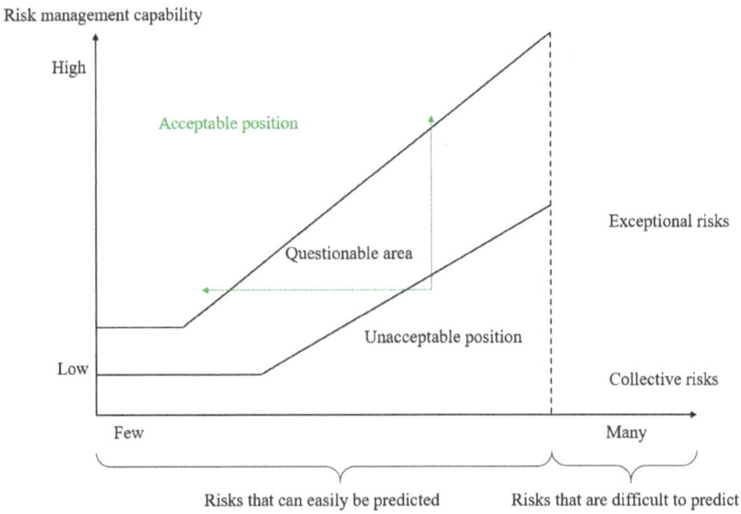

Fig. 2.11 Illustration of vulnerability of municipality as well overall risks and risk management ability

Table 2.4 A classification of accidents affecting the population

Class	Lives	Personal injuries	Personal injury days
1	1–3	1–5	1–999
2	4–10	6–20	1000–49,000
3	11–50	21–100	50,000–499,000
4	>50	>100	>500,000

Table 2.5 A classification of damages to the ecological system

Class	Ecological system (km^2)
1	0–0.1
2	0.1–1
3	1–10
4	>10

Table 2.6 A classification of damages to property

Class	Consequence (Mkr)
1	<1
2	<50
3	<500
4	>500

When determining values of α_i and β_i, such things as existing safety cultures must be evaluated and quantified (e.g., see Warren 2015). A safety control must be carried out partly in the administrative situation and partly from a systematic point of view. Within both areas, methods have been developed for safety control. For example, safety, health, and the environment (SHE) model offers a checklist consisting of 145 points assessed and placed on a scale from 0 to 10 approach (Kemikontoret 1996). Another is Katina's 'pathological' issues, a total of 83, that could hinder organizational performance (Katina 2015a, b). Such models can function as a basis for assessing α_i and β_i, but presumably continued development work is required to select suitable points that are most practical for use on a regular basis.

VI. Presentation of the result. In this phase, the results are presented based on the including criteria for acceptable vulnerability.

An example of the application of the presented model is discussed in the following section, discussing a fictitious municipality's vulnerability inventory. The municipality has 14 hazards of importance. The result is 14 Z_i values ($0.0 < Z_i < 80$) and 14 values of α_i and β_i (for the sake of simplicity, α_i and β_i would not amount to more than 0, 5; that is $0.0 \leq \alpha_i, \beta_i \leq 0.5$).

First, we can conclude the following:

- If we draw profiles with the 14 values for Z, $\alpha_{i,}$ and $\beta_{i,}$ respectively, we obtain the municipality's risk profile stating if and where efforts should be taken (see Fig. 2.13).

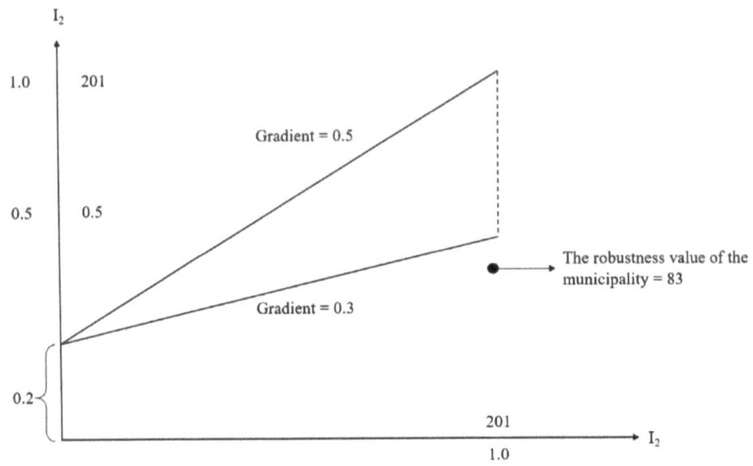

Fig. 2.12 Value of the individual municipality vulnerability management in relation to acceptance criteria

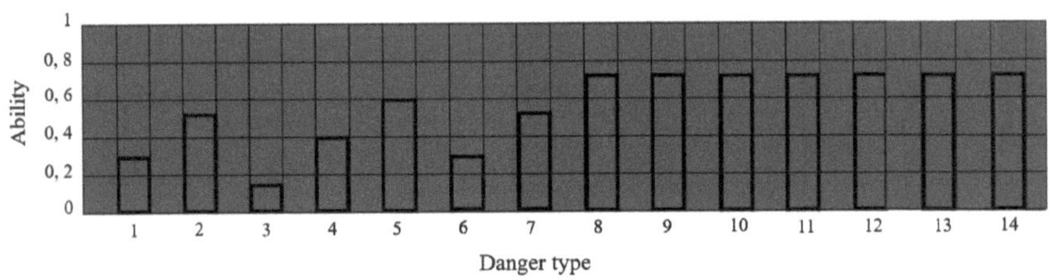

Fig. 2.13 An indication of how far the general and object-specific risk management resources stretch to deal with different risks

- If we sum up the 14 Z values, we get the collective risk value of the municipality (see Table 2.9).

Three indexes are defined with the following values for the investigated local authority:

- I_1 = Maximum possible danger level (Z_{max}) = $14 \times 80 = 1120$
- I_2 = Current risk level $\sum_1^{14} Z_i = 201$
- I_3 = Current vulnerability management capability level $\sum_1^{14} (\alpha_i + \beta_i)Z_i = 83$

The quotient I_2/I_1 gives the relative threat level with the given hazards that have been discovered in the inventory. Of greater interest is the quotient I_3/I_2 that is a direct measurement of how well the local authority manages its vulnerability situation. The nearer to the value of 1, the more robust and resilient a municipality is considered. The value of 1 equals to completely robustness. Acceptance criteria can be directly used against this quotient as a municipality independent means of measurement, and the quotient I_3/I_2 thus becomes an important measurement of the local authority's vulnerability in relation to other local authorities and relative to a given nationwide value.

An alternative way of using the calculated indexes is illustrated in Fig. 2.12 which also illustrates a method of defining a more general applicable acceptance criterion. Horizontal and vertical axes are graded from 0 to 1.0 where 1.0 is equal to the value of the index I_2 (=201). A *lower acceptance curve*, could, for example, be stated with the starting point 0.2 on the *y*-axis and with a slope = 0.3. An *upper acceptance curve* could be given with the same starting point but with the slope = 0.5. If the local authority's I_3 value (the level of the current capability to manage its vulnerability) comes under the lower limit curve, the result would not be acceptable. If I_3 comes over the upper limit curve, the state of the system would certainly be acceptable. An I_3 value between the curves indicates that an improvement, built on an analysis of cost *efficiency*, must be carried out.

The following notes apply:

- **Note 1**: The starting point on the *y*-axis of 0.2 is motivated by the fact that there is always a certain level of risk management, irrespective of the size and number of hazards.
- **Note 2**: By considering an individual local authority and using the relative values of I_2 and I_3, we will always be on the line $I_2 = 1$.

In the present approach, it is quite fine to do away with the relative values by only using the absolute values. In this case, the numbers provided by local authorities can be used. Also, one should differentiate the diagrams based on the classification stipulated by the Swedish Association of Local Authorities (Kemikontoret 1996); otherwise, the solution of the figure could return as 'not acceptable.' The method could be sorted in under all application classes 1, 2, 3, and 4, of the Swiss system, concerning their areas of use.

2.2.3.3 Model Application: Calculating Vulnerability Management Capability Index

The following is an example indicating how an index for vulnerability management capability is calculated for a fictive local authority.

Step 1: Definition of hazard and damage types

In this case, the hazard types of concern involve:

- Threats against municipal services
- Natural catastrophes
- §43 factory buildings
- Hazards associated with information technology safety are excluded.

The following damage types are considered:

- Population—divided into three separate damage types (see Table 2.4)
- Ecological systems (see Table 2.5)
- Property (see Table 2.6).

Step 2: Classification of damage types and probability

Undoubtedly, population can be affected in terms of death and personal injury. Time can also be an important factor when dealing with injury, especially the size of an injury. The effects can be, for example, measured in terms of number of days a person is affected (i.e., personal injury days). Table 2.4 is an attempt to classify accidents into different classes. It should be noted that exactly how such a classification should be made is not clear at this stage. However, authors can speculate that such classification could vary from system to system and from nation to nation. In effect, the context of operation could affect this classification.

An important step, when defining a risk, is to determine the likelihood (i.e., probability) of occurrence of the risk event. In this case, Table 2.7 demonstrates a classification of the probability associated with the damage types.

Step 3: Inventory of hazards

From this classification, we gain the following values associated with the municipality in question:

(A) *Provision systems* (*water, electricity, sewage*)

A vulnerability analysis of the municipal service systems means that indexes for the different damage types and probability are given values (Table 2.8) as stated below:

The three Z values for risk are: $Z_1 = 20$, $Z_2 = 28$, and $Z_3 = 30$. The total of 78 is placed in relation to theoretically possible maximum value of 240.

(B) *Natural disasters* (*severe*)

We consider a municipality with several severe natural risk events, including blizzards, flooding, landslides, and forest fires.

For blizzards

Personal days (off work) index = 3, property damage index = 2, probability class index = 3, Z_i value = $(3 + 2) \times 3 = 15$

For flooding

Personal days (off work) index = 3, ecological system index = 3, property damage index = 4, probability class; index = 4, Z_i value = $(3 + 3 + 4) \times 4 = 40$

Table 2.7 Probability classes for the above damage type

Class	Probability P/year
1	$P < 10^{-3}$/year
2	10^{-3}/year $< P < 10^{-2}$/year
3	10^{-2}/year $< P < 10^{-1}$/year
4	$P > 10^{-1}$/year

Table 2.8 Index and probability for different municipal service systems

Damage type	Municipal services systems		
	Water	Electricity	Sewage
Lives	0	0	0
Injured	3	4	4
Personal injury days	1	3	3
Property	1	0	3
Environment	5	7	10
Probability class	4	0.4	0.3
Total	20	28	30

Table 2.9 Calculation of municipal vulnerability management capability (robustness)

Hazard types	Risk value, Z	α	β	$\sum(\alpha+\beta)$	Vulnerability management capability (robustness) = $\sum(\alpha+\beta) \times$ risk value
Water	20	0.1	0.2	0.3	6
Electricity	28	0.2	0.3	0.5	14
Sewage	38	0.1	0.1	0.2	6
Snow	15	0.1	0.3	0.4	6
Flooding	40	0.1	0.5	0.6	24
Landslide	18	0.1	0.2	0.3	5
Forestry	14	0.2	0.3	0.5	7
§43-object					
N_1	4	0.1	0.3	0.4	=15 for all the objects in total
N_2	3	0.1	0.3	0.4	
N_3	8	0.1	0.3	0.4	
N_4	2	0.1	0.3	0.4	
N_5	5	0.1	0.3	0.4	
N_6	7	0.1	0.3	0.4	
N_7	7	0.1	0.3	0.4	
	$\sum = 201$				$\sum = 83$

For landslides

Index for personal days (off work) index = 2, ecological system index = 2, property damage index = 2, probability index = 3, Z_i value $(2 + 2+2) \times 3 = 18$

For forest fires

Index for personal days (off work) index = 1, ecological system index = 3, property damage index = 3, probability index = 2, Z_i value = $(1 + 3 + 3) \times 2 = 14$

(C) *Industry (including public buildings) and other 43 industrial buildings*

We assume that there are seven industrial buildings and that each building has a probability index of 1 meaning that severe accident happens no more than once in every 1000 years in each of the industrial buildings.

Further, we assume an aggregate scale index 0, 4, 3, 8, 2, 5, 7, 7 for the seven industrial buildings. The total Z value will be 36. Thus, the value of 36 is to be compared to a theoretical maximum value of 560.

Step 4: Inventory of resources for risk management

It is possible to divide up risk management resources, in a local municipal authority, into two parts: First are those that can be considered general in a local authority, α_i, and second are those that are linked to the individual object or phenomenon, β_i, including the 43 industrial buildings. How though do we determine α_i, and β_i? First, we assume that the maximum number of resources needed to deal with the risk level amounts to value of 1. After that, for the sake of simplicity, we divide the resources at random into two similar parts, giving an interval within which it is possible, for a local authority, to find general and object-specific resources.

Therefore, we can assume that:

- For general risk management capability, $0.0 < \alpha_i < 0.5$, and
- For object-linked or the phenomenon linked to the risk management capability, $0.0 < \beta_i < 0.5$

In Table 2.9, the assumed values of α_i and β_i are used. By adding these to each other, we gain the total number of resources (maximum 1) that exist in a local authority. If this percentage value is multiplied by the risk value for the different threat types, a new value is gained which states the level of robustness regarding the risk level.

For each hazard (danger type), the value of the vulnerability management capability (robustness) can be placed against a corresponding risk value. This is done to show the capability of the municipality stack-up against a given risk in a specific area. An example of this 'stackness' is provided in Fig. 2.13. For example, one can conclude that municipal ability to deal with danger type 5 is only at 0.6.

In Table 2.8, we obtain vulnerability management capability value of 83 by adding items on the most right-hand column. It is now possible to summarize the result of the calculations already done as three individual indexes:

- The maximum possible danger or threat level; total maximum sum of the earlier part steps in *Step* 2 = 1080 = I_1
- The current risk level = 201 = I_2
- The current vulnerability management capability level = 83 = I_3

The different indices can be presented in two ways:

1. Relative diagram to determine acceptance. It must be drawn for each municipality individually.
2. Absolute values I_2, I_3, and I_3/I_2 which can apply for a whole nation.

The acceptance criteria, in accordance with point 1 above, are illustrated in the diagram above. The lower acceptance line starts in $(0.2 \times 201) = 40.2$ and ends

$40 + (0.3 \times 201) = 100.5$ and the upper acceptance line ends in $40 + (0.5 \times 201) = 140.5$. The variables consist of starting point and gradient on the lower and upper line. The starting point is determined by defining a baseline for basic services (e.g., all local authorities must have a security coordinator). The exact gradient lines for municipal authorities are determined through several large-scale calibrating studies.

This section offers an approach to assessing local authority's ability to manage risk events using a measure of vulnerability capability that involves lower and upper acceptance curves. The presented model defines hazards and damage types, probabilities associated with such hazards, and the available resources that can be used to deal with such threats. Authors submit that the model offers utility at a local level as well as the national level and can be used in connection with different risks and well as known checklists.

2.2.4 *Relational Vulnerability and System Penetrability*

The topic of assessing the vulnerability in critical infrastructures is becoming extremely important, under the stringent needs for protecting them against malicious, technical, and natural disasters. A few attempts have been made to give an adequate working framework to the concept of vulnerability. However, these efforts do not fully reflect the stringent needs to quantify the vulnerability and then offer systematic steps for (1) an agreed upon criteria for vulnerability acceptance—the framing of this issue should get you to realize that is no such a thing as systems free of vulnerability—and (2) vulnerability economics, implying the fact that vulnerability of systems could decrease by allocating resources at different stages of system evolution. What is suggested in this section is considering vulnerability assessment of critical infrastructures by addressing the aspect of their *complexity*. In defining and measuring complexity of such systems, the concept of graphs is needed and used. In this case, the vulnerability, referring to nature of the system itself, is seen as the capacity of a system made of people, hardware, software, organizational, and management procedures being *penetrated*. The degree of vulnerability is then supported by the capability of the system performing its designed functions.

This section addresses a special line of thought, setting the task of *taking a straightforward approach to complexity as a source of vulnerability*. The practical goal is to attach a relevant metric to the internal connectivity of multi-component systems so that this is turned to account from a quantitative vulnerability assessment (QVA) oriented standpoint.

2.2.4.1 Models of Relational Vulnerability

Since the promotion of the concept—complexity-induced vulnerability—requires a versatile *modus operandi*, able to accommodate a variety of user-defined, convincing applications, a generic model was sought. The reference in hand were the graphs, as a comprehensive expression of multi-component systems and their internal connectivity—*ergo*, 'complexity,' in the parochial sense adopted. There are several assumptions associated with models in question starting with assumption zero:

Assumption 0: The operational representation of a multi-component system is a graph.

Here is the spelt-out equivalence. The members (i.e., constituents, parts) of the system are the graph's ***knots***. The interactions of the members are represented by directed knot ***links***, and the graph is customized to a system by attaching to knots a set of ***features***, appropriately quantified and normalized on a vulnerability-relevant scale. ***Knots*** are, generically, the irreducible components or 'atoms' of a system and are the subjects of the analysis. Depending on the nature of the targeted system, 'knots' may be employees, departments, subsidiaries, contractors, parts in an engineered machinery, circuitry, plant, member-states of an alliance, etc., *or collections of these*, showing a sufficient degree of coherence to play a coordinated part in the overall system's internal interaction game.

Links connect knots such that exchange/trade information, energy, and/or substance are possible. In effect, links define a system by way of its exchange boundaries. Links enter the model by *Connection Lists* attached to each knot in the graph of the system in question. Normally, exchanges between knots proceed under an authority rule, or otherwise said—in hierarchic fashions. That is why links are *directed*, so that, in the sense of the model, knot A may have knot B on its connection list, while knot B may not necessarily have knot A on its connection list. Links are of critical importance in evaluating, among others, the security efficiency, efficacy, and sustainability of the system.

Features are meant to characterize the knots. Depending on what the 'knot' is (i.e., employees, departments, subsidiaries, contractors, parts in an engineered machinery, etc.), 'features' should be selected providing maximum relevance concerning the objective of the analysis (e.g., security, efficiency, efficacy, etc.). In the current model, 'features' enter the quantitative vulnerability assessment through *values* and *weights*. The feature 'values' are attached to the system 'knots' and provided as a decimal number in the range 1 through 9. It is assumed to be in direct proportion to the degree of *vulnerability relevance* that the feature may attain for different knots. In a way of a random example, in an embassy, within feature 'position,' a desk clerk might be given a value of 3 compared to a value of 9 that could be given to a cipher officer. Contrastingly, feature 'weights' compares features in terms of their *relative* vulnerability relevance such that feature '*Clearance*' is more vulnerability relevant than feature '*Qualification*.' Weights are entered by the user as arbitrary numbers and are eventually normalized by the code over a span of 0.0 to 1.0. In this case, the 'weights' are meant to discriminate among 'features' placing these in perspective as far as *importance* and play their part in quantitative evaluations involving the 'knots' and their 'features.' The model and its algorithms have been implemented in a software tool, DOMINO, as part of a decision support system. Four screenshots of DOMINO are presented in Fig. 2.14 for a system with 100 knots and features.

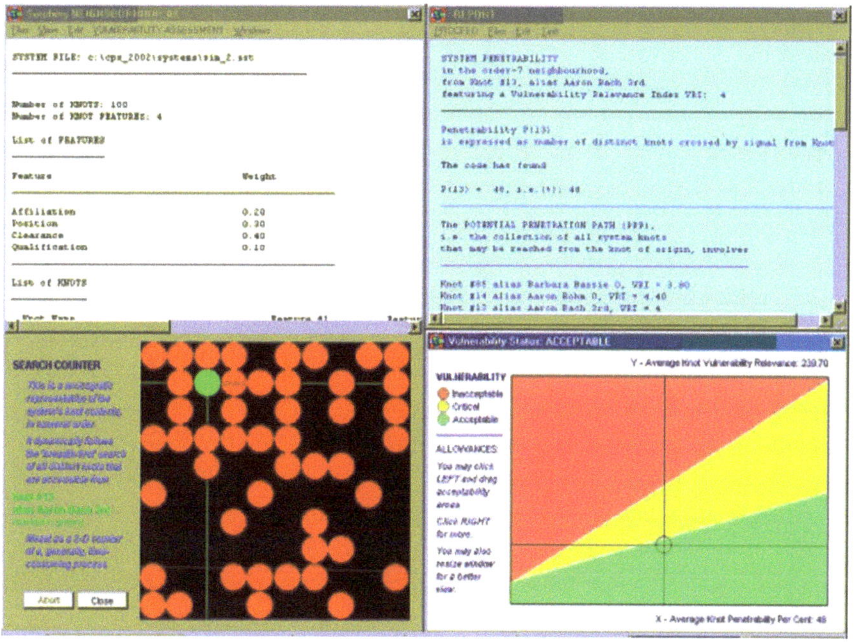

Fig. 2.14 DOMINO—DSS features for complexity vulnerability assessment, adapted from Gheorghe and Vamanu (2004a)

2.2.4.2 Connectivity as Penetrability

There are, basically, two possible interpretations of the meaning of internal connectivity. These meanings are not contradictory, rather complementary and in fact intertwined. One is a *benign,* while the other is *cautious* interpretation. As per the benign interpretation, the more extensive and multi-lateral the exchanges among system parts or constituents, the better. In this instance, the system is considered 'functional,' 'lively,' 'active,' and 'dynamic'—terms often associated with the promise of high productivity, efficacy, alert response to inputs, and high profits. Complementarity to this view is the perspective of looking for lack of connectivity which would reveal chances of inherent defects in the systems (e.g., a short circuit in the control room of a nuclear power plant), an accidental instruction (i.e., a static discharge in a highly relevant computer circuitry), or a foul play (i.e., a fatal virus dropped in a company database), and which could be initiated at one specific knot in the system and having a higher chances to propagating throughout the system, thus having the potential to impair larger system segments. In this interpretation, a higher connectivity rhymes with a higher vulnerability. All epistemological (e.g., see Fernandez 2009; Flood and Carson 1993) and ethical debate left aside, this research exercise will try and reconcile the two stands, which would generate the operational assumptions as indicated in Table 2.10.

Table 2.10 Operational assumptions for connectivity as penetrability

Assumptions	Description	Implications
Assumption I	A higher internal connectivity in a system is a desirable quality only to the extent that the cumulated vulnerability relevance of the connected knots is tolerable	This allegation expresses the fact that not all 'knots' (i.e., system constituents have the same vulnerability relevance. The involvement in exchanges—of information, energy, substance of *some*—could be more meaningful and attention-catching, than others. When assumption I is considered, it produces the following consolidating findings
Assumption 2	The higher the vulnerability relevance of the knots involved in the exchange path of any knot of origin, including the relevance of the knot of origin itself, the higher the vulnerability induced in the overall system by the respective knot of origin	
Assumption 3	The higher the cumulated vulnerability relevance of the system's knots, the higher the system vulnerability itself	

Upon a consideration of these assumptions, one might be attempted to characterize a system vulnerability in terms of its 'complexity.' To this end, we suggest two distinct, if not completely independent, parameters: (a) system's penetrability and (b) connectivity's vulnerability relevance. System's penetrability is a quality that may have metrics such as the number (e.g., average number) of knots that can be accessed starting from a (any) given knot in the system. Connectivity's vulnerability relevance depends on penetrability as defined above along with vulnerability relevance grades as assigned to knot features. In an *X–Y* plane underlined by these parameters, one may conduct a meaningful appraisal of *'vulnerability tolerance,'* as a means of understanding and recognizing that vulnerability of a part of life, be it functional or structural, and that it can be inherent, has unavoidable drawbacks, or otherwise limitation, of all *negentropic* systems.

2.2.4.3 Quantifying Vulnerability Relevance of Penetrability

With this said, one must recall that the objective function of the investigation may easily be written. Let's consider:

N_k be the number of knots, $K_{i,}$ $i = 1, 2,\ldots, N_{k,}$ in the graph **G** representing a multi-component system,

N_f be the number of vulnerability-relevant knot features F_j, $j = 1, 2,\ldots, N_f$,
$W(F_j)$ be the weights of the features F_j, $j = 1, 2,\ldots, N_f$, where

$$0 \leq W(F_j) \leq 1 \tag{2.25}$$

$G(F_j, K_i)$ be the value (grade) of the feature F_j of knot K_i, where

$$1 \leq G(F_j, K_i) \leq 9 \tag{2.26}$$

Then, one has:
The individual vulnerability relevance, $V_k(K_i)$, of knot K_i:

$$V_k(K_i) = \sum_{j-1}^{N_j} W(F_j).G(F_j, K_i) \tag{2.27}$$

(a) The search-path (breadth-first) vulnerability relevance, $V_p(K_i)$, of knot K_i and all the knots that can be accessed either directly or via other knots, into the system (index 'p' for 'path'):

$$V_p(K_i) = V_k(K_i) + \sum_{m-1}^{N_j}{}' V_k(K_m) \tag{2.28}$$

with $V_k(\cdot)$ given by Eq. (2.27). The sign $'$ in Eq. (2.28) emphasizes the limitation of the sum to only those knots that can be, directly or indirectly, accessed starting from knot K_i.

(b) The maximum possible vulnerability relevance of a system's knot:

$$V_{\max} = \mathbf{max}(V_k(K_i)) \cdot N_k = 9 \times SW(F_j) \cdot N_k = 9 \times 1 \times N_k \tag{2.29}$$

obtained in consideration of expressions (2.25), (2.26), and Eqs. (2.27), (2.28).

(c) The average vulnerability relevance per knot of system:

$$V_{\text{avg}} = \left(\sum_{i=1}^{N_i} Vp(K_i) \right) \Big/ N_K \tag{2.30}$$

with $V_p(\cdot)$ given by Eq. (2.28).

One may also define:

(d) The penetrability of the system from knot K_i:

$$P(K_i) = \text{number of distinct knots that can be accessed from } K_i, \tag{2.31}$$

both directly and via other knots, plus 1—the knot of origin

(e) The *Maximum System Penetrability*, obviously given by

$$P_{\max} = N_k \tag{2.32}$$

f. The *Average System Penetrability*, per knot, given by:

$$P_{\text{avg}} = \left(\sum_{i=1}^{N_i} p(K_i) \right) \Big/ N_K \tag{2.33}$$

At this point, it is possible to visualize the issue in hand, complexity-induced vulnerability as depicted in Fig. 2.15.

2.2.4.4 Tolerability of Vulnerability

Since, as indicated, the only meaningful issue in QVA, quantitative vulnerability assessment, is '*How tolerable the vulnerability of this system is*,' a discussion may be conducted:

(a) In the X–Y plane featuring

$$\begin{aligned} X &= P(K_i)/P_{max}, \\ Y &= V_p(K_i)/V_{max}, \end{aligned} \tag{2.34}$$

with the quantities involved given by Eqs. (2.28), (2.29) and (2.31), (2.32), respectively, and

(b) In the X-Y plane featuring

$$\begin{aligned} X &= P_{\text{avg}}/P_{\max}, \\ Y &= V_{\text{avg}}/V_{\max}, \end{aligned} \tag{2.35}$$

with the quantities involved given by Eqs. (2.30), (2.33) and (2.31), (2.32), respectively. While the approach (2.35) would indeed qualify a system's connectivity (i.e., 'complexity'), overall vulnerability relevance, the approach (2.34) has

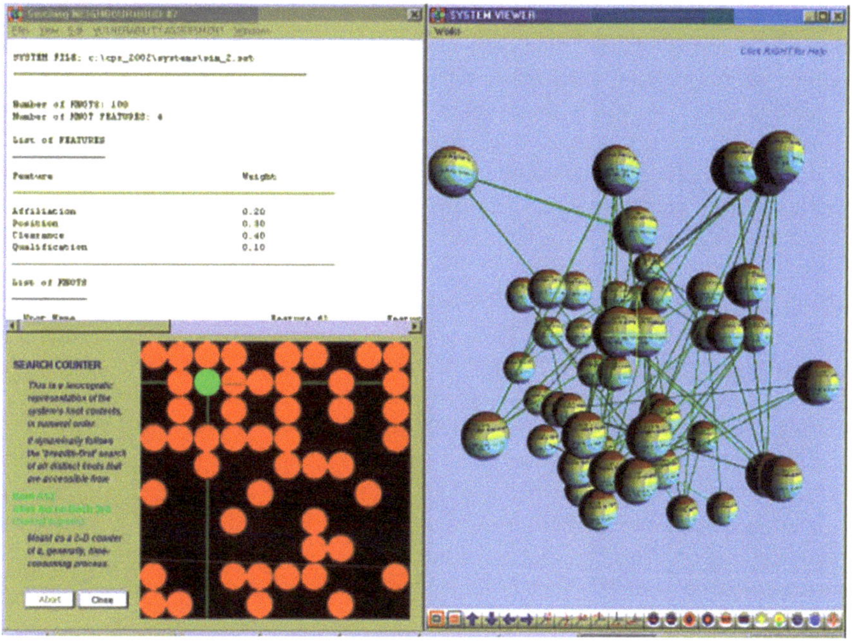

Fig. 2.15 Visualizing complexity-induced vulnerability by DOMINO software, adapted from Gheorghe and Vamanu (2004a)

also merits in signaling extremes, or 'vulnerability spikes,' originating in knots that would deserve special attentions. The *X–Y* space, as defined above, is divided, generally, into three basins: (a) **acceptable vulnerability** basin, indicated by the green area; (b) **critical vulnerability** basin, indicated by the yellow area; and (c) **unacceptable vulnerability** basin, indicated by the red area.

The *X* and *Y* parameters are not to be taken as completely independent of each other, the configuration of the basins remains debatable, and, on this account, the code makes provisions enabling the user to interactively redefine the basins. The default configuration proposed by the code associated with this model ***assumes an acceptable vulnerability at 0-penetrability***. Such a scheme may be termed as *'over-confident.'* It reflects a 'non-guilty-until-otherwise-proved' presumption, or attitude, in the sense that each and every constituent of a system carries, by design, a 'vulnerability relevance.' However, there is an irrefutable reality that one cannot build a healthy system, company, circuit, alliance, etc., resting on the assumption that it is *bound* to be unsafe or malicious. The opposite attitude, assuming an unacceptable vulnerability, even at 0-penetrability, could be termed 'paranoiac' with, however, no derisive connotation. In-between, a 'cautious' or 'conservative' attitude may also be identified, assuming complete uncertainty on vulnerability at 0-penetrability, that is, $Y = 0$ for $X = 0$.

User may position him/herself in respect of the above, by the mouse-driven action of shifting the basin divides. In DOMINO, both the initial left-hand-side gap and the aperture of the 'critical vulnerability' area can be fine-tuned based on response to user's beliefs. This type of analysis, however, introduces a requisite element of subjectivity (i.e., stakeholder perception of vulnerability) since it is possible to have a system in the basin of 'acceptable' vulnerability that might be in a 'critical' or even 'unacceptable' basin. Figure 2.16 depicts full information structure from DOMINO involving the three basins of vulnerability along with relevance to vulnerability assessment for each knot in a structure of interest.

2.2.4.5 Supplementary Model Resources

The evaluation of system vulnerability based on system's internal connectivity, introduced in this chapter, relies heavily on statistical connotations. However, recall that the newcomers of SCI, UCI, and BCI, as well as the likes, might not have standing and readily available statistical data rooted in historical accounts. Certainly, this might be true when attempting to address emerging areas of research such as cybersecurity in cyber-physical systems, blockchain technology, algo- and robot techniques. This suggests a need for application of approaches that are non-statistical in nature: *scenario-based inves*tigations. A scenario-based investigation could provide insights; for example, at any moment in time, a system monitor may be interested in whether knot B could be reached by signals emitted at

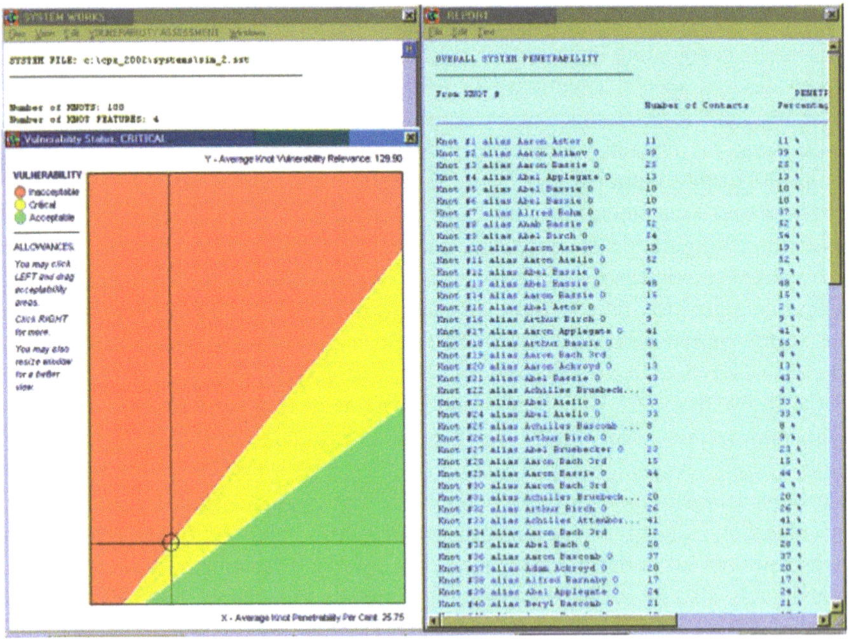

Fig. 2.16 An example of the three vulnerability basins along with knots along with system architecture, adapted from Gheorghe and Vamanu (2004a)

knot A. The *route* through which such a contact proceeds may also be relevant, as well as the *cumulated 'vulnerability relevance'* of all knots involved in the process.

Normative, or ceiling, values for the vulnerability relevance burden of every individual knot, of knot pairs, or groups of knots, may equally be contemplated within a vulnerability-conscious system management. Although the current model places emphasis on the *distinct* number of knots that can be reached from a given knot of origin, the *frequency* of the intermediate knots being visited by the signal started at origin, until it reaches the destination knot, may present importance, in a vulnerability, or foul play scenario, inquiry. DOMINO enables one to study such investigations. Future research expects to include other features that might be of interest to system monitors. A proposed and interesting area of research is system theory-based pathologies (Katina 2015a, b, 2016a, b; Keating and Katina 2012; Troncale 2013). A pathology is defined as a circumstance, condition, factor, or pattern that acts to limit system performance, or lessen system viability, such that the likelihood of a system achieving performance expectation is reduced (Keating and Katina 2012). Simply stated, these are *organizational diseases*. The current state of research has yielded over 80 systems theory-based pathologies classified in terms of dynamics of a system, system goals/missions, information flow, processes, regulation, resources, systemic structures, and understanding (Katina 2015b). Obviously, there remain questions of relating pathologies to vulnerability, levels of pathologies, how pathologies can affect a system (e.g., penetrability), and the economics of pathologies. Appendix E provides an in-depth classification of systems theory-based pathologies.

2.3 Remarks

In this chapter, a new metric for system vulnerability by considering their complexity, as a new and comprehensive measure, is introduced. The applicability of this model is rather generic, and it needs to be adapted to various applications in case. The model has been complemented by the design and implementation of a decision support system, named DOMINO, which exhibits various features related to the process of measuring and assessing the vulnerability of sociotechnical systems.

References

Christen, P., Bohnenblust, H., Seitz, S. (1995). How to compare harm to the population with damage of the environment? A quantitative multi-attribute approach for risk analysis based on fuzzy set theory. In J. J. Mewis, H. J. Pasman, E. E. De Rademaeker (Eds.), *Proceedings of the 8th international symposium* (pp. 691–704). Antwerp: Elsevier Science BV.

Doerig, H.-U. (2000). *Operational risks in financial services: An old challenge in a new environment*. London: Institut international d'etudes bancaires.

Fernandez, G. W. (2009). *Epistemological beliefs and teacher efficacy* (Ph.D.). University of Virginia, United States, Virginia.

Flood, R. L., Carson, E. R. (1993). *Dealing with complexity: An introduction to the theory and application of systems science*. New York: Plenum Press.
Gheorghe, A. V. (2004). *The hidden faults: Towards a standing method to assess Switzerland's vulnerabilities*. Zurich, Switzerland: Laboratory of Safety Analysis, ETH Zurich.
Gheorghe, A. V., Vamanu, D. V. (2004a). Complexity induced vulnerability. *International Journal of Critical Infrastructures, 1*(1), 76–84.
Gheorghe, A. V., Vamanu, D. V. (2004b). Towards QVA—Quantitative Vulnerability Assessment: A generic practical model. *Journal of Risk Research, 7*(6), 613–628.
Gheorghe, A. V., Vamanu, D. V. (2006). Risks in business design for critical infrastructures: the "DASHBOARD" concept. *International Journal of Critical Infrastructures, 2*(1), 70–82.
Katina, P. F. (2015a). Emerging systems theory–based pathologies for governance of complex systems. *International Journal of System of Systems Engineering, 6*(1/2), 144–159.
Katina, P. F. (2015b). *Systems theory-based construct for identifying metasystem pathologies for complex system governance* (Ph.D.). Old Dominion University, United States, Virginia.
Katina, P. F. (2016a). Metasystem pathologies (M-Path) method: Phases and procedures. *Journal of Management Development, 35*(10), 1287–1301. doi:10.1108/JMD-02-2016-0024.
Katina, P. F. (2016b). Systems theory as a foundation for discovery of pathologies for complex system problem formulation. In A. J. Masys (Ed.), *Applications of Systems Thinking and Soft Operations Research in Managing Complexity* (pp. 227–267). Geneva, Switzerland: Springer International Publishing.
Katina, P. F., Pinto, C. A., Bradley, J. M., Hester, P. T. (2014). Interdependency-induced risk with applications to healthcare. *International Journal of Critical Infrastructure Protection, 7*(1), 12–26.
Katina, P. F., Unal, R. (2015). Application of fuzzy sets in decision analysis for prioritising critical energy infrastructures. *International Journal of Decision Sciences, Risk and Management, 6* (1), 1–15.
Keating, C. B., Katina, P. F. (2012). Prevalence of pathologies in systems of systems. *International Journal of System of Systems Engineering, 3*(3/4), 243–267. doi:10.1504/IJSSE.2012.052688.
Kemikontoret, (1996). *Administrativ SHM—revision*. Stockholm: Association of Swedish Chemical Industries.
Lagbo-Bergqvist, E., Lexén, R. (2000). *Vägen till bättre styrning av säkerhetsarbetet i kommuner och landsting*. Stockholm: Svenska kommunförbundet Landstingsförbundet.
Merriam-Webster. (2006). *Webster's new explorer encyclopedic dictionary*. Springfield, MA: Federal Street Press.
Nilsson, J., Magnusson, S., Hallin, P., Lenntorp, B. (2001). *Models for vulnerability auditing and distribution of governmental economical means at the local authority level*. Lund, Sweden: LUCRAM: Lund University Centre for Risk Analysis and Management.
Romeike, F., Maitz, J. (2001). *Operational risk*. London: CSC Financial Services EMEA.
Skyttner, L. (2005). *General systems theory: Problems, perspectives, practice* (2nd ed.). Singapore: World Scientific Publishing Co., Pte. Ltd.
Thom, R. (1975). *Structural stability and morphogenesis*. Reading, MA: Westview Press.
Thom, R. (1983). *Mathematical models of morphogenesis*. (W. M. Brooks, D. Rand, Trans.). New York: Halsted Press.
Troncale, L. (2013). Systems processes and pathologies: Creating an integrated framework for systems science. *INCOSE International Symposium, 23*(1), 1330–1353.

Vamanu, B. I., Gheorghe, A. V., Katina, P. F. (2016). *Critical infrastructures: Risk and vulnerability assessment in transportation of dangerous goods—transportation by road and rail* (Vol. 31). Cham, Switzerland: Springer International Publishing.
Warren, J. H. (2015). *Safety culture monitoring: A management approach for assessing nuclear safety culture health performance utilizing multiple-criteria decision analysis* (Ph.D.). Old Dominion University, United States, Virginia.
Zeeman, E. C. (1977). *Catastrophe theory: Selected papers*. London: Addison-Wesley.

Part II
Governance Modeling, Simulation, and Visualization

Chapter 3
A Physical Analogy for Resilience and Vulnerability

Abstract This chapter sheds lights on concepts presented in Section 2.2.1—hysteresis. Specifically, this chapter shows how the concept of hysteresis is implemented in QVA model for assessment of cooperative behavior, the tendency to resist stress and maintain system state (configuration and performance level) against driving stress.

3.1 An Analogy in Hysteresis

The general consensus is that any system (e.g., a energy system) will consist of *parts*, P_i, $i = 1, 2, \ldots, M$, preferably interacting parts. Once defined, the 'parts' may be seen as individual, atomic (indivisible) components, that:

- usually come in large numbers (M);
- are coupled with each other with a strength that may conveniently be expressed as a generic, coupling 'energy,' ε_{ij}, $i = 1, 2, \ldots, M$, $j = 1, 2, \ldots, M$;
- respond to external stresses, or influences ('fields'), H, each system part featuring an 'energy'

$\mu_i H$, of coupling with the 'field' via a coupling strength μ_i.

In the context, the notion of 'part' embraces a virtually unlimited variety of representations. For an energy system, these may include anything from mines, mills, wells, pipes, power stations, switchyards, transmission lines, distribution facilities, control rooms, dispatching centers, IT assistance facilities—their subassemblies included over entire fuel cycles, to key workers, working units, enterprises, companies, regulators, and political pressure entities. In a first, rough approximation parts either *do function* as per intent and design, or *do not function*, their state being thereby describable via a variable, S, that may assume two values only: $S = 1$ indicating a functional part; or $S = -1$ indicating a dysfunctional part—which accommodates systems within the *Ising mode* (Gheorghe and Vamanu 2008), ubiquitous in Physics and well beyond. Parts may switch from a functional to a dysfunctional state, and conversely, the process being assumed to be, in the

A.V. Gheorghe et al., *Critical Infrastructures, Key Resources, Key Assets*, Topics in Safety, Risk, Reliability and Quality 34,
https://doi.org/10.1007/978-3-319-69224-1_3

final analysis, reversible, and *probabilistic* in nature (Hopkinson and Williams 1912).

Observant to the natural systems that are coherent enough—within their boundaries of definition—to feature a certain autonomy, or quasi-isolation of their own in respect with the remaining environment, the overall behavior of our model–system may be thought to be governed by *a variational principle*, applicable to system's total energy. According to such a principle, in a steady state of the system, the individual states of the parts are such that the system 'energy,' which is given by Eq. (3.1) is a minimum for any given *temperature*:

$$E = -(1/2)\sum_{ij} \varepsilon_{ij} S(i)S(j) - H\sum_{i} \mu_i S(i) \tag{3.1}$$

The first term in Eq. (2.1) denotes the total *internal* 'energy' of the system of interacting parts, whereas the second term features the total 'energy' imparted to the parts by their coupling to the external, compelling 'field' H.

Physicists will immediately note that, in a textbook rendering of an Ising or a Heisenberg model—that are at the origin of our analogy—the normal assumption is that both the coupling ('exchange') energy, $\varepsilon_{ij,}$ and the field-coupling constant, μ_i do *not* depend on the parts *i, j*—a fact that has to do with the assumption that all parts are identical (and in effect indiscriminate) to each other. In this respect, Eq. (3.1) is a generalization to a many-body system of *nonidentical* parts.

In applying the notion above, note that any part-*i* state-flip (from functional, 1, to dysfunctional, −1, or vice versa) entails a change in system's energy, of

$$\Delta E = -S(i)\left(\sum\nolimits_j' \varepsilon_{ij} S(j) + \mu_i H\right) \tag{3.2}$$

where $\sum_j'$ indicates a sum that, in practice, extends over a certain, neighborhood of part *i*—while in principle it may extend over *all* the agents other than *i*.

Following the Ising model philosophy (see e.g., the discussion in references (Gheorghe and Vamanu 2004, 2008; Sprott 1993), a part's behavior is governed by the following set of rules, consistent with the assumptions above:

$$\begin{aligned} &\textit{Rule } 1: \text{If } \Delta E <= 0, \text{then the part would always undergo a state-flip.}\\ &\textit{Rule } 2: \text{If } \Delta E > 0, \text{then part flips state only with a probability,} \end{aligned} \tag{3.3}$$

$$P = \exp(-\Delta E/(k_B T)) \tag{3.4}$$

with T a 'system temperature,' and k_B a 'Boltzmann constant,' conveniently taken as 1. However, *Rule 2* is recommended in practice [e.g., see Metropolis et al. (1953) and Sprott (1993)]. Under this recommendation:

Let r be a (computer-generated) random number, $r >= 0$ and $r < 1$.

Then,

if $r <= P$ [P given by (3.4)] then *do* flip;
else, *do not* flip.

Under these terms, for any 'temperature' T there will, in principle, be M_1 system parts that would be functional and $M_2 = M - M_1$ parts that would be dysfunctional, so that one may define a *system performance fraction*, ζ as:

$$\zeta = (M_1 - M_2)/(2M) \tag{3.5}$$

Definition (3.5) places performance fraction ζ between (−0.5) and (+0.5), and favors the following assessment rule:

A system featuring $\zeta >= 0$ is *mostly functional, whereas*
A system featuring $\zeta < 0$ is *mostly dysfunctional*

And the value judgment placed on a policy/strategy relates to *an assessment of the extent the managed system is kept mostly functional.* It is deemed that the *macroscopic* behavior of a system, normally expressed via variations in a number of indicators of definition perceived as relevant, is a result of system's *microscopic, cooperative* behavior, primarily characterized by the performance fraction, ζ.

The Eqs. (3.1)–(3.5) are, in actual fact, implemented in the current application, meant to provide a graphic expression to one possible manner of characterizing *resilience and performance in systems.*

In essence, the game:

- simulates a system made of user-specified number of parts that interact, both mutually and with external stress fields, at given strengths;
- induces in the system the afore-described microscopic process at part level, cyclically stressing the system by external fields;
- thereby obtains system performance fraction ζ as a function of the applied stress H (See Figs. 3.1, 3.2, 3.3, and 3.4).

3.2 Hysteresis Modeling

The results will systematically indicate an overly important feature of large systems showing cooperative behavior: *their tendency to resist stress and maintain their state (configuration and performance level) against the driving stress applied—an effect known as hysteresis*, a common knowledge in, for instance, the theory and practice of magnetic phenomena and materials and beyond: for a starter, do a *Wikipedia* search for 'Hysteresis.' Notice that the system can have different alert states as described in the DEFense readiness CONdition (DEFCON) as used by the US Armed Forces.

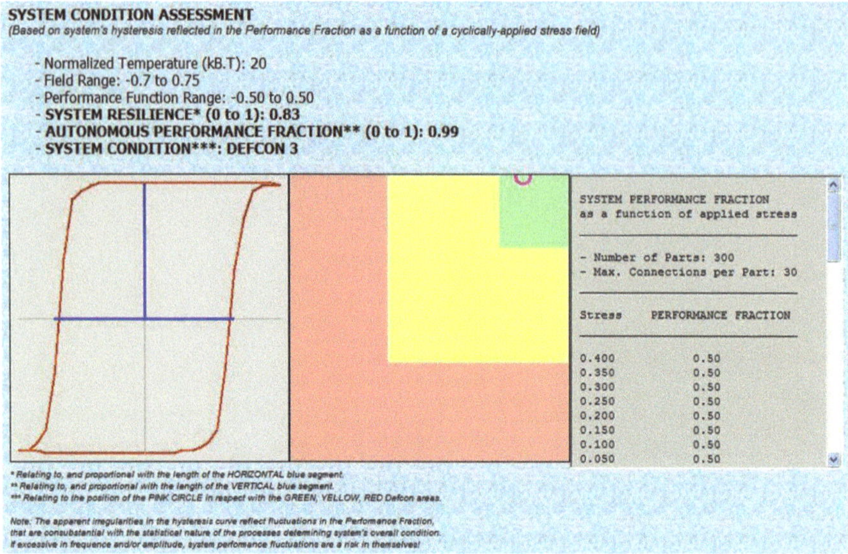

Fig. 3.1 Hysteresis in a 300-part, 30-part links/part system; normalized temperature 20 units—DEFCON 3, adapted from Gheorghe and Vamanu (2009)

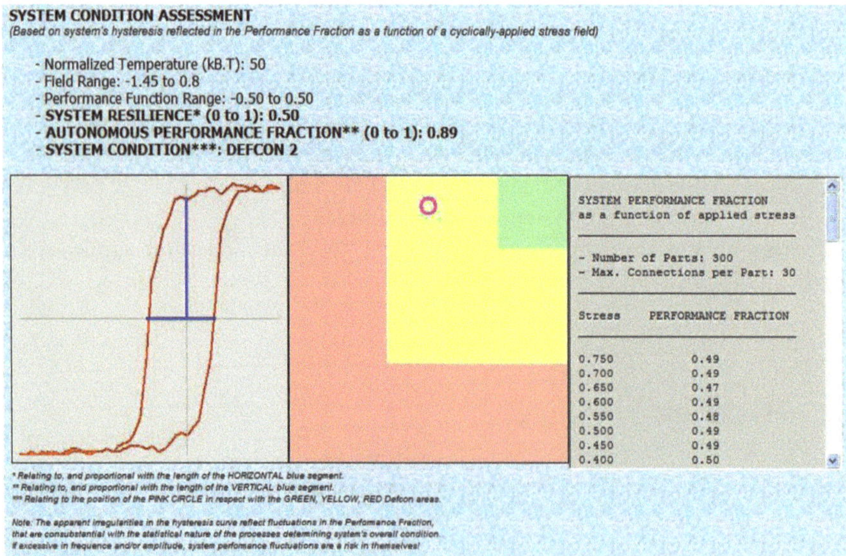

Fig. 3.2 Hysteresis in a 300-part, 30-part links/part system; normalized temperature 50 units—DEFCON 2, adapted from Gheorghe and Vamanu (2009)

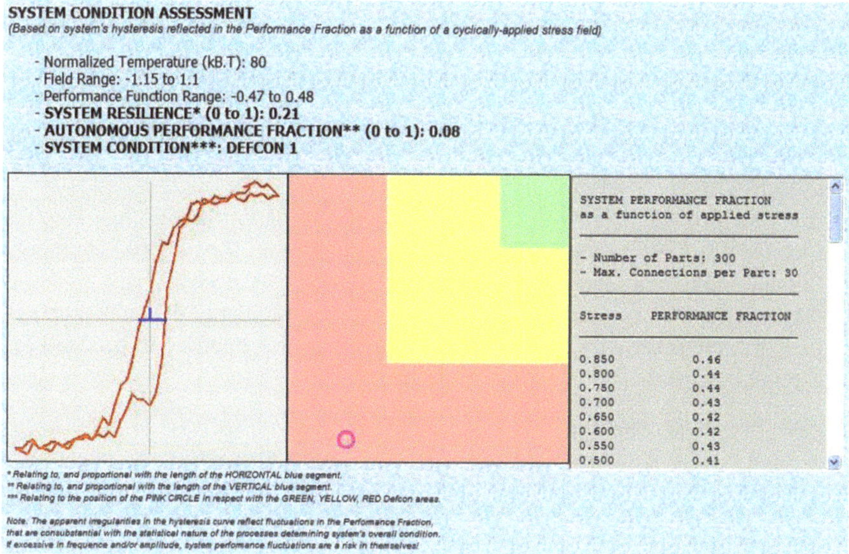

Fig. 3.3 Hysteresis in a 300-part, 30-part links/part system; normalized temperature 50 units—DEFCON 1, adapted from Gheorghe and Vamanu (2009)

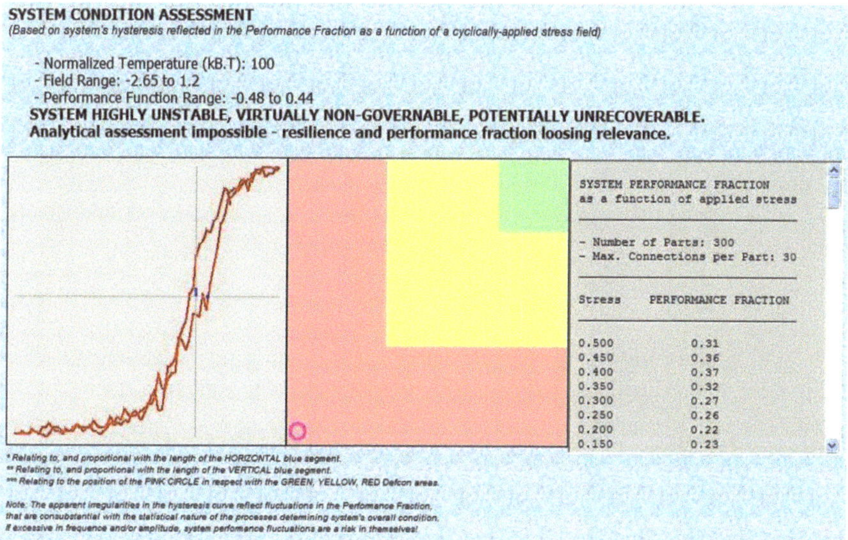

Fig. 3.4 Hysteresis in a 300-part, 30-part links/part system; normalized temperature 100 units, adapted from Gheorghe and Vamanu (2009)

Notice that in Fig. 3.4, the system is 'unstable' and virtually 'non-governable' and potentially 'unrecoverable.'

Additionally, and in plain terms:

- if a system is dominantly functional, then it tends to maintain its level of functionality (performance) despite applied stresses threatening to make parts dysfunctional,
- if a system is dominantly dysfunctional, then it tends to maintain low levels of functionality (performance) despite the applied stresses (i.e., efforts) attempting to make parts functional again; and, perhaps more strikingly,
- the transition from a dominantly functional to a dominantly dysfunctional system, and the other way around, tends to be abrupt (as opposed to gradual) and essentially depends on system's 'temperature.'

It has been suggested that 'the reluctance to changes in the level of performance under applied stress, of large systems featuring cooperative, statistical phenomena that animate their interconnected parts is *resilience*, (Gheorghe et al. 2011). A natural measure of the resilience, in the present discussion, could be described as *the distance of the intersections of the hysteresis cycle with the abscissa* (see Fig. 3.1). Expressed in units of the applied stress (field), this quantity may be termed—by analogy with the Theory of Magnetism—a *'Coercive Force,' or 'Coercivity.'* Further along the analogy, the maximum value of the performance function ζ, measured on the ordinate axis for a nil-stress may be termed *Remnant Performance Level* as opposed to *remnant magnetization* or *remanence*. An alternative, and perhaps a more appropriate term in the context may be *Autonomous Performance Fraction* (APF), indicating a desirable feature of complex systems: their capability to sustain operations even when most of the 'positive stress' (financial, logistic, etc.) required to set the system in motion has been tempered, or withdrawn (Gheorghe et al. 2011).

In such terms, a system deemed '*in good order*,' or '*condition*' should display both,

- a high *resilience*—indicating a good resistance to the effects of negative stresses; and
- a high *autonomous performance fraction*—indicating an acceptable level of performance even in the absence of a positive stress to maintain it.

This finding leaves one with the need to employ in the representation of the system condition *the Cartesian product* of the said quantities in an X–Y plane, one choice being to place the resilience on the X-axis and the APF on the Y-axis. This manner of visualizing/monitoring a systems' condition would immediately call to mind the defense drill that deals with readiness for appropriate response in threatening conditions in terms of 'DEFCONs.' In the context, one may, for instance, leave to the gamer the definition of boundaries between, say, three 'DEFCONs' of incremental degree of severity, the most severe featuring *the lowest system resilience*, OR *the lowest autonomous performance fraction* (APF).

The current module of the ROSTREC Arcade platform (See Muresan 2010) plays with some basic parameters defining a system, namely:

- the *number* of parts and their *susceptibility* (reactivity) to applied stress, assumed to, generally, differ from part to part;
- the *number* of links of every part in the system with other parts, in either physical and/or logical a sense, and the strength of the respective links—that also may differ from one link to the other while remaining, however, reciprocal for any given pair of parts; and
- the 'temperature' of the system—the net effect of which, in a purely algebraic sense is to diminish in bulk, by the same factor, all part susceptibilities and link strengths, which turns out to result in quite dramatic effects on resilience and performance fraction.

As the gamer stretches these parameters within the allowed limits (essentially resulting from the computing power of the average-price desk-/lap/tops) he/she will get:

- a variety of *hysteresis loops*, each providing an indeed graphic expression of the system condition via system's resilience and APF; and
- an X–Y map of APF vs. resilience, for a comparative analysis in terms of DEFCONs, of the consequences of different choices, or evolutions with system's parameters.

3.3 Remarks

After some enduring exercises, one might end up with a 'feeling' of how large systems behave. To these 'feelings,' present authors suggest that on the one hand:

- large and internally coherent systems tend to show a higher level of resilience and Autonomous Performance Fraction. Contrastingly, the level of resilience for small and poorly coherent systems tends to be low and thin of the characteristics of stable and fluctuation-free operations and regimes.
- systems that are subject to poor, negligent, lax management, and governance in terms of, among others, maintenance, monitoring, updating, corporate spirit, truthful self-assessment, and ethics, which translates as 'disorder,' or 'higher temperatures' tend to show degraded resilience and/or performance fractions, down to complete collapse.

On the other hand, the following remarks are also suggested:

- *Highly resilient systems—systems that have a high-grade tend to be… highly vulnerable*! Their vulnerability relates to the near-ideal shape of their hysteresis cycle: quasi-rectangular and covering a large expanse in the performance versus stress in the form of an X–Y plan. This remark is based on the fact that such a shape may encourage a feeling that 'things are all right.' Regardless of the cause of shape, be it negligence, or external circumstances, a prolonged recession, for example, could be seen as 'normal.'

Interestingly, residual positive stress, normally known as 'production and maintenance costs' (e.g., financial, logistic, intelligence), can move the system into the negative stress realm. The system could find itself into dangerously close to the edge, that if reached by a mere further, apparently insignificant decrement or fluctuation, *will take the entire system down into a full-fledged collapse*. Oddly enough, what we have referred to as a 'feeling' that all things are all right as it related to 'systems theory' concept of punctuated equilibrium (Gould and Eldredge 1986) where the long periods of stasis as suggested in Katina (2015) could create a false sense of 'safeness.' Unfortunately, the feeling of safeness tends to lead to system designs that:

- lack virtually any complete and credible early warning systems. Thus, a system might stay stable at a high(est) level of performance although its environment is clearly deteriorating,
- the brutality of the collapse (the steep slope of the hysteresis) that would dramatize the entire scenario; and—perhaps more importantly,
- the remarkably long and costly way to a full system recovery (see the length of the lower hysteresis cycle plateau.)

However, all is not bad. The examined analogy suggests a need to create recovery points in design of complex systems. However, that remains a point of further research as to how to establish recovery points based on hysteresis. Moreover, literature suggests that there can be types of hysteresis (Mayagoitia 1991). There remains an issue of implications of such types of man-made systems.

References

Hopkinson, B., & Williams, G. T. (1912). The elastic hysteresis of steel. *Proceedings of the Royal Society A: Mathematical, Physical and Engineering Sciences, 87*(598), 502–511. doi:10.1098/rspa.1912.0104.

Gheorghe, A., Muresan, L., Celac, S., Caceu, S., Degeratu, C., Lenes, L., et al. (2011). An energy security strategy for Romania: Promoting energy efficiency and renewable energy sources. In A. V. Gheorghe & L. Muresan (Eds.), *Energy security: International and local issues, perspectives, and critical energy infrastructures* (pp. 337–412). New York: Springer Science & Business Media.

Gheorghe, A. V., & Vamanu, D. V. (2004). Towards QVA—Quantitative Vulnerability Assessment: A generic practical model. *Journal of Risk Research, 7*(6), 613–628.

Gheorghe, A. V., & Vamanu, D. V. (2008). Mining intelligence data in the benefit of critical infrastructures security: Vulnerability modelling, simulation and assessment, system of systems engineering. *International Journal of System of Systems Engineering, 1*(1), 189–221.

Gheorghe, A. V., & Vamanu, D. V. (2009). Resilience and vulnerability in critical infrastructure systems—A physical analogy. *International Journal of Critical Infrastructures, 5*(4), 389–397.

Gould, S. J., & Eldredge, N. (1986). Punctuated equilibrium at the third stage. *Systematic Zoology, 35*(1), 143–148.

Katina, P. F. (2015). *Systems theory-based construct for identifying metasystem pathologies for complex system governance* (Ph.D.). Old Dominion University, USA, Virginia.

Metropolis, N., Rosenbluth, A. W., Rosenbluth, M. N., Teller, A. H., & Teller, E. (1953). Equation of state calculations by fast computing machines. *The Journal of Chemical Physics, 21*(6), 1087–1092.

Mayagoitia, V. (1991). The five types of porous structures and their hysteresis loops. *Studies in Surface Science and Catalysis 62,* 51–60, Elsevier. Retrieved from http://linkinghub.elsevier.com/retrieve/pii/S0167299108613087.

Muresan, L. (2010). *Energy security and critical infrastructure protection strategy for Romania and the regional perspective*. Odessa: EURISC Foundation, Romania. Retrieved from http://www.energycharter.org/fileadmin/DocumentsMedia/Events/20100727-PESAIITBSATROMC_S2_LMuresan.pdf.

Sprott, J. C. (1993). *Strange attractors: Creating patterns in chaos*. New York: M&T Books.

Chapter 4
System of Systems Governance

Abstract This chapter elaborates in the concepts of governance, risk, and vulnerability assessment in system of systems. The lens of system of systems (SOS) is used to look into the nature of our natural world. This is done using readily and publicly available data from USA's Central Intelligence Agency (CIA) *World Factbook*. The developed model suggests the possibility of measuring (and even predict) the vulnerability of nations based on a number of indicators.

4.1 Framework for System Governance

Governance is fundamentally related to regulation that enables realization of desired long- and short-term goals (Katina 2015). Schneider and Bauer (2007) espouse that 'if a "problem" is defined as the difference between a preferred state and an undesired status quo, the function of governance is "problem-solving" in the sense of moving to desired states' (2007, p. 11). One would be mistaken to assume that there is one type of governance. It has been suggested that the very concept of 'governance' varies with contexts despite appearing to emerge from Greek and Latin languages caring connotations of 'art of steering' and 'governing.' A recent survey on the concept suggests that there exist types of governance in different disciplines and practices (Calida 2013, 2016). Certainly, this suggests that there can be different articulations of governance for different systems, systems of systems, and even for us, different infrasystems including Space Critical Infrastructure, Undersea Critical Infrastructure, and Belowground Critical Infrastructure.

In this chapter, an attempt is made to offer insights into governing a type of system known as system of systems. Fundamentally, a system of system is a collection of task-oriented or dedicated systems that pool their resources and capabilities together to create new and more integrated systems with functionality and performance beyond any single complex system (Keating et al. 2003; USAF SAB 2005). Methodological approach for defining, abstracting, modeling, and analyzing system of systems problems the engineering of system of systems. Both, system of systems (SoS) and system of systems engineering (SoSE), suggest

A.V. Gheorghe et al., *Critical Infrastructures, Key Resources, Key Assets*, Topics in Safety, Risk, Reliability and Quality 34,
https://doi.org/10.1007/978-3-319-69224-1_4

a need to look beyond individual systems. Many have suggested looking at the concepts of metasystem (Carter 2015; Djavanshir et al. 2009; Gheorghe and Masera 2014) as it relates to 'beyond' capabilities of single systems and aspects of 'co-ordination' and 'integration' (Katina et al. 2014). Table 4.1 provides a summary of typical characterization of system of systems.

Arguably, these characteristics described may of the systems mentioned in the introductory notes of the present research. Additionally, and of interests is what Sousa-Poza et al. (2008) suggest are the problem domains for system of systems and the implication for understanding such systems. A summary is offered in Table 4.2. Subsequently, it should be evident that knowledge and understanding at the level of system of systems and system of systems engineering are different than at the individual system level.

In this chapter, we attempt to apply a 'system of system' lens to our natural world. At this level of lens, certainly, some details might disappear (e.g., a car accident in Moscow, North Dakota). This is not to suggest that such an incident is not relevant. Rather, the focus of the system of system (SoS) at the world level is at different level of inquiry and concern. In the present case, the SoS shall be comprised of nations with different attributes and thus fit the many-body system as described in Sect. 2.2.1.

Table 4.1 Characteristics of system of systems

Characteristic	Description
Operational independence of systems	Disintegrating systems of systems into constituent systems would not render the constituent system inoperable. Rather, each constituent system can operate independently (Maier 1996)
Managerial independence of systems	Constituent systems (making up a system of systems) can be separately acquired and can be managed independently (Maier 1996)
Evolutionary development	Systems of systems evolve over time, with component systems capabilities added, removed, or modified as needs change and experience is gained (Maier 1996)
Emergent behavior	Systems of systems have emergent capabilities and properties that do not reside in the component systems (Maier 1996)
Geographical distribution of systems	Systems of systems are comprised of constituent complex systems geographically distributed with the ability to readily exchange information (Maier 1996)
Networks of systems	Networks define connectivity between independent systems in the SoS through rules of interaction (DeLaurentis 2005)
Heterogeneity of systems	Constituent systems are of significantly different nature, with different elementary dynamics that operate on different time scales (DeLaurentis 2005)
Trans-domain study	This is a proposition suggesting that effective study of SoS requires unifying knowledge across fields of study: engineering $\cup$ economy $\cup$ policy $\cup$ operations (DeLaurentis 2005)

Table 4.2 An operational landscape for system of systems

Characteristic	Landscape description
Holistic problem space	SoS problem space requires consideration of technical, human/social, managerial, organizational, policy, and political dimensions (Sousa-Poza et al. 2008)
Ambiguity	Problem domain for SoS involves difficulty in clearly demarking SoS problem boundaries, as well as their interpretation (Sousa-Poza et al. 2008)
Uncertainty	Systems of systems problems are not tightly bound, and flexing as additional knowledge of the situation is developed (Sousa-Poza et al. 2008)
Highly contextual	Consideration of circumstances, conditions, factors, and patterns that give meaning and purposes to systems of systems (Sousa-Poza et al. 2008)
Emergence	SoS behavioral and structural patterns, their interpretations, knowledge, understanding, and conditions are always in constant flux (Sousa-Poza et al. 2008)
Non-ergodicity	A phenomenological condition of having no defined states or discernible transitions between states (Sousa-Poza et al. 2008)
Non-monotonicity	A condition in which increases in knowledge are not reciprocated by increases in understanding. Under this condition, decisions are defeasible or tentative (Sousa-Poza et al. 2008)

4.1.1 SOSE Model of the World[1]

In the present context, the System of System (SoS) is a composition of 278 ***systems*** —the World States or countries. The aim is to see the world as a SoS in relation to the protection of critical infrastructures at national level using aspects of quantitative vulnerability assessment. In this case, the system of interest is a collection of critical infrastructures, resources, and political-related indicators. These systems are comprised of subsystems, subsystem sections, and subsection indicators. A working *macroscopic hypothesis* is that a system is (i.e., equivalent to) its set of measurable indicators. The indicator choice is governed by both availability and appropriateness. Authors set a key working assumption: *the most appropriate indicators for vulnerability assessment should be sought in relation to intelligence sources.*

In this regard, *CIA's World Factbook* (2000–2006) is used as primary source indicator, an epoch in world's vulnerability history relaying information about a variety of critical infrastructure systems, associated resources, demographic, and political context. The agency is a civilian foreign intelligence service of the US federal government, tasked with gathering, processing, and analyzing national security information from around the world, primarily through the use of human intelligence. This information is the basis for input into creation of the different models in the present chapter.

[1]This chapter is based on an earlier version of research by Adrian V. Gheorghe and Dan Vamanu published in the *International Journal of System of Systems Engineering*, 'Mining intelligence data in the benefit of critical infrastructures security: vulnerability modeling, simulation and assessment, system of systems engineering,' Vol. 1, Nos. ½, 2008.

4.1.1.1 National Scale Vulnerability

After 9/11, the term 'Homeland Security' emerged as symbolic designation for the USA's nationwide effort to respond to the risks and vulnerabilities that the USA faced. Comparable to the ministry of the interior or home ministry, the US Department of Homeland Security (DHS) emerged with missions revolving around counterterrorism, border security, immigration and customs, cybersecurity, and disaster prevention and management. Arguably, DHS is involved in setting governance rules at the business and human relations at different rules. These levels include: domestic, interface, and international. The *domestic front* involves governance rules within the USA while the *interface* level suggests the interface of the USA and the rest of the world. At the international level, the concern is the manner of understanding and conducting business outside the USA.

As a matter of awareness, and especially after 9/11, numerous presidential directives have been issued. These are directed, among others, toward protecting infrastructure and deterring deliberate attacks as well as increasing national resilience (and thus reducing vulnerability) (Pederson et al. 2006; Katina and Keating 2015). But in the wake of such tragedy and the responses, have we—the nation, the world—changed as a whole? Consider the following illustrations:

> On May 25, 1961, then President, John F. Kennedy proposed in an address to the US Congress a project involving 'landing a man on the Moon and returning him safely to the Earth' by the end of the 1960s

One might suggest, criticality thinking of the issue at hand, that the *project* putting man on the moon and returning safely involved:

- A *doctrine*, mitigating in an unprecedented manner the classic American antagonisms between isolationism and global interventionism, honestly delineated from both and yet unavoidable and thus borrowing from both.
- The *justification for an evolved democracy,* where individual's supremacy and freedoms are to be served by enhanced levels of societal discipline and a strengthened awareness on the value of civic spirit, and community solidarity and coherence, particularly in times of crisis.
- A *procedural revolution,* keyed on monitoring and accountability, with lesser than usual concern for intrusiveness, in a highly standardized and stereo typified society, affecting—again to unprecedented levels—practically all motions and exchanges, whether of people or of values, including—and sometimes paramount so—the information and communications.
- *A legal ethics revolution*, bending the so far sacred stance 'not guilty until otherwise proved' into 'not guilty yet potentially guilty, until otherwise proved'—which, some would contend, may entail monumental consequences on the American way of life.
- A *governance ethics revolution*, resting on the paradoxical and repelling—if otherwise truthful to real life—postulate according to which people can *and should* be served by being governed *even despite themselves,* with the corollary

that the new times and challenges would warrant an enhanced inventory of ways and means to implement 'a government with the people, for the people.'

The said 'project' clearly had far deeper implications. By way of comparison, a closer look at 9/11, especially the pre 9/11 and post 9/11, with respect to the aftermath responses (i.e., immediate responses and domestic reactions) and effects (i.e., health, economic, cultural, and government policies) suggests far greater implications for society as a whole. Touching almost all facets that define 'what is America to me,' the concept of homeland security may as well turn out to be *the very model of the twenty-first Century American Society* and on a purely humane plan, *the price for being the best.* Like all projects, doctrines, justifications, and revolutions, homeland security revolves around a central theme of concern, and in the case of present research efforts, a perceived contemporary view of *being vulnerability.*

Since long intimated by many sensitive spirits, more recently argued by several educated minds, occasionally signaled by selected governance agents and political engineers, America's vulnerability had passed largely overlooked. Sometimes dismissed outright, and perhaps rightly so in the face of much evidence: unmatched military might, economic power, political dominance, and some might argue civilization—some prefer the term 'cultural' influence, and gloriously winning against the anti-communist crusade and the 'end' of Cold War in favor of the Free World. One might think that there should have been a realization that the Land of Brave might feature 'hidden faults' with high seismic potential—figuratively speaking—initiated by internal common sense stating: '*He, who has the most to lose, is the most vulnerable.*'

This repugnant reality for the USA's vulnerability manifested, rather violently, in the 'unexpected' strategy of *9/11.* And—due condolences observed—the *tragedy behind the tragedy* is that, undoubtedly connected as it is, to America's vulnerability (and the world in general) in the post-bipolar world, *9/11* is, however, no more relevant to the latter than the dangle of the warning bells, as it were for 'the flood' in Noah's day...Warning bells do you no good if you can't obey them.

In this chapter, it is argued *vulnerability (American or otherwise) is consubstantial to its endowments, merits, and performance as far as nature, people, infrastructure, economy, political system, and culture.* There can be a price for being the best. And, if there is a consolation to the finding above, *all* countries—or call them nations, societies, and communities—endure the same fate. The chapter is meant to elaborate on the sense of 'we are what we are' in this world in terms of nature, people, infrastructure, economy, political system, and culture, *and this caries a vulnerability tag to it.*

4.1.1.2 The SoS Concept for the World

If we are to see the world as a SoS, there is a need to define the individual systems that comprise the SoS. In our current case, the SoS is comprised of 278 systems (i.e., the World States or countries). The system as described above can have several subsystems, including, among others, suprastructures, core structure, infrastructure, and resources. At this point, we can then conceive all subsystem sections and subsections as defined using indicators. Figure 4.1 depicts the world as a system of systems with 278 countries, along with subsystems and indicators.

Recall present macroscopic hypothesis: *a system is (equivalent to) its set of measurable indicators*. The indicator choice is governed by both availability and appropriateness. Key working assumption: The most appropriate indicators for vulnerability assessment should be sought in relation to intelligence sources. Hence: use as primary indicator source the CIA World Factbook, is reasonable. Researchers selected yearly issues 2000–2006 covering a telling epoch in world's vulnerability history. Table 4.3 is a listing of countries obtained from CIA World Factbook representing systems within the system of systems conceptualization.

In a similar fashion, authors obtained subsystems of the systems above. Table 4.4 lays out the subsystems of interests with much emphasis being placed on infrastructure, *Economy, Demographics, Nature, Politics*, and *Culture*.

At this point, it is possible to describe the indicators associated with the subsystems. These indicators are described in Table 4.5 below in terms of codes:

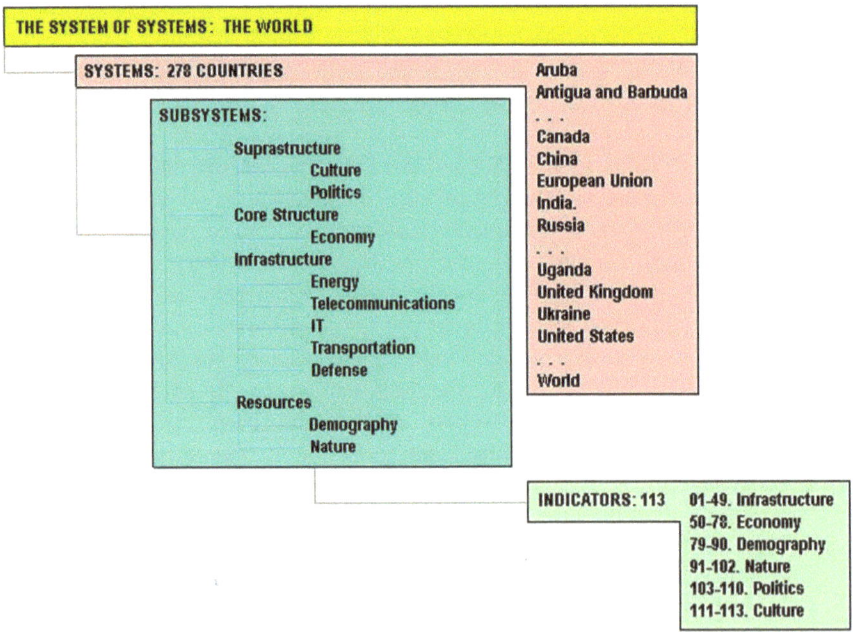

Fig. 4.1 Conceptualizing the world as a system of systems, adapted from Gheorghe and Vamanu (2008)

Table 4.3 A partial listing of countries (systems) of the system of systems world

File: Total Area, km2
Primary data version: CIA World Factbook 2006

Aruba: 193 Antigua & Barbuda: 443 UAEmirates: NA Afghanistan: 647,500 Austria: 83,858 Anguilla: 102 Akrotiri: NA Antarctica: 14,000,000 Bahrain: 665 Congo, Democratic Republic of the: 2,345,410 China: 9,596,960 Chile: 756,950 Cayman Islands: 262 Cocos Islands: 14 Cameroon: 475,440 Comoros: 2170 Colombia: 1,138,910 Northern Mariana Islands: 477 Coral Sea Islands: 18 Costa Rica: 51,100	Central African Republic: 622,984 Cuba: 110,860 Cape Verde: 4033 Cook Islands: 240 Cyprus: 9250 Denmark: 43,094 Djibouti: 23,000 Dominica: 754 Jarvis Island: 4.5 Lithuania: 65,200 Liberia: 111,370 Slovakia: 48,845 Palmyra Atoll: 11.9 Liechtenstein: 160 Lesotho: 30,355 Paraguay: 406,750 Pitcairn Islands: 47 Peru: 1,285,220 Paracel Islands: 9 Spratly Islands: 18 Pakistan: 803,940 Poland: 312,685 Panama: 78,200 Portugal: 92,391	Papua New Guinea: 462,840 Palau: 458 Spain: 504,782 Serbia & Montenegro: 102,350 Saint Lucia: 616 Sudan: 2,505,810 Svalbard: 62,049 Sweden: 449,964 South Georgia and the South Sandwich Islands: 3903 Syria: 185,180 Switzerland: 41,290 United Arab Emirates: 82,880 Trinidad and Tobago: 5128 Tromelin Island: 1 Thailand: 514,000 Tajikistan: 143,100 Turks and Caicos Islands: 430 Tokelau: 10 Tonga: 748 Togo: 56,785 Sao/Tome & Principe: 1001	Tunisia: 163,610 East Timor: 15,007 Turkey: 780,580 Tuvalu: 26 Taiwan: 35,980 Turkmenistan: 488,100 Tanzania: 945,087 Uganda: 236,040 UK: 244,820 US Pacific Island Wildlife Refuges: NA Ukraine: 603,700 USA: 9,629,091 Burkina Faso: 274,200 Uruguay: 176,220 Uzbekistan: 447,400 Yemen: 527,970 Zambia: 752,614 Atlantic Ocean: 76,762,000 Zimbabwe: 390,580

Table 4.4 Listing of subsystems for selected systems in a system of systems world

Subsystems	Description
SUBSYSTEM: INFRASTRUCTURES	1. Electricity - production_ _U_I_INFRASTRUCTURES_ 2. Electricity - production by source fossil fuel_ _U_D_INFRASTRUCTURES_ 3. Hydro_ _U_I_INFRASTRUCTURES_ 4. Nuclear_ _U_I_INFRASTRUCTURES_ 5. Other_ _U_I_INFRASTRUCTURES_ 6. Electricity - consumption_ _U_D_INFRASTRUCTURES_ 7. Electricity - exports_ _U_I_INFRASTRUCTURES_ 8. Electricity - imports_ _U_D_INFRASTRUCTURES_ 9. Oil - production_ _U_I_INFRASTRUCTURES_ 10. Oil - consumption_ _U_D_INFRASTRUCTURES_ 11. Oil - exports_ _U_I_INFRASTRUCTURES_ 12. Oil - imports_ _U_D_INFRASTRUCTURES_ 13. Oil - proved reserves_ _U_I_INFRASTRUCTURES_ 14. Natural gas - production_ _U_I_INFRASTRUCTURES_ 15. Natural gas - consumption_ _U_D_INFRASTRUCTURES_ 16. Natural gas - exports_ _U_I_INFRASTRUCTURES_ 17. Natural gas - imports_ _U_D_INFRASTRUCTURES_ 18. Natural gas - proved reserves_ _U_I_INFRASTRUCTURES_ 19. Irrigated land_ _U_I_INFRASTRUCTURES_ 20. Telephones - main lines in use_ _U_I_INFRASTRUCTURES_ 21. Telephones - mobile cellular_ _U_I_INFRASTRUCTURES_ 22. Domestic_Telephones _U_I_INFRASTRUCTURES_ 23. International_Telephones _U_I_INFRASTRUCTURES_ 24. Radio broadcast stations_ _U_I_INFRASTRUCTURES_ 25. Radios_ _U_I_INFRASTRUCTURES_ 26. Television broadcast stations_ _U_I_INFRASTRUCTURES_ 27. Televisions_ _U_I_INFRASTRUCTURES_ 28. Internet hosts_ _U_I_INFRASTRUCTURES_ 29. Internet Service Providers (ISPs)_ _U_I_INFRASTRUCTURES_ 30. Internet users_ _U_I_INFRASTRUCTURES_ 31. Airports_ _U_D_INFRASTRUCTURES_ 32. Airports - with paved runways total_ _U_D_INFRASTRUCTURES_ 33. Airports - with unpaved runways total_ _U_I_INFRASTRUCTURES_ 34. Heliports_ _U_I_INFRASTRUCTURES_ 35. Pipelines_ _U_D_INFRASTRUCTURES_ 36. Railways total_ _U_I_INFRASTRUCTURES_ 37. Roadways total_ _U_I_INFRASTRUCTURES_ 38. Paved_ _U_D_INFRASTRUCTURES_ 39. Waterways_ _U_I_INFRASTRUCTURES_ 40. Merchant marine total_ _U_D_INFRASTRUCTURES_ 41. by type_ _U_I_INFRASTRUCTURES_ 42. Foreign-owned_ _U_D_INFRASTRUCTURES_ 43. Registered in other countries_ _U_I_INFRASTRUCTURES_ 44. Ports and terminals_ _U_D_INFRASTRUCTURES_ 45. Military branches_ _U_I_INFRASTRUCTURES_ 46. Manpower available for military service males age 16-49_ _U_I_INFRASTRUCTURES_ 47. Manpower fit for military service males age 16-49_ _U_I_INFRASTRUCTURES_ 48. Military expenditures - dollar figure_ _U_I_INFRASTRUCTURES_ 49. Military expenditures - percent of GDP_ _U_D_INFRASTRUCTURES_

(continued)

Table 4.4 (continued)

Subsystems	Description
SUBSYSTEM: ECONOMY	50. Imports - partners_ _U_I_ECONOMY_ 51. GGDP (purchasing power parity)_ _V_I_ECONOMY_ 52. GGDP (official exchange rate)_ _V_I_ECONOMY_ 53. GDP - real growth rate_ _V_I_ECONOMY_ 54. GGDP - per capita (PPP)_ _V_D_ECONOMY_ 55. GDP - composition by sector agriculture_ _V_D_ECONOMY_ 56. Labor force_ _V_I_ECONOMY_ 57. Labor force - by occupation agriculture_ _V_I_ECONOMY_ 58. Industry_Labor force _V_D_ECONOMY_ 59. Services_Labor force _V_I_ECONOMY_ 60. Unemployment rate_ _V_D_ECONOMY_ 61. Population below poverty line_ _V_D_ECONOMY_ 62. Inflation rate (consumer prices)_ _V_D_ECONOMY_ 63. Investment (gross fixed)_ _V_I_ECONOMY_ 64. Budget revenues_ _V_I_ECONOMY_ 65. Expenditures_ _V_D_ECONOMY_ 66. Public debt_ _V_D_ECONOMY_ 67. Agriculture - products_ _V_I_ECONOMY_ 68. Industries_ _V_I_ECONOMY_ 69. Industrial production growth rate_ _V_D_ECONOMY_ 70. Current account balance_ _V_I_ECONOMY_ 71. Exports_ _V_I_ECONOMY_ 72. Exports - commodities_ _V_I_ECONOMY_ 73. Exports - partners_ _V_I_ECONOMY_ 74. Imports_ _V_D_ECONOMY_ 75. Imports - commodities_ _V_D_ECONOMY_ 76. Reserves of foreign exchange and gold_ _V_I_ECONOMY_ 77. Debt - external_ _V_D_ECONOMY_ 78. Economic aid - donor_ _V_I_ECONOMY_
SUBSYSTEM: DEMOGRAPHY	79. Population_ _V_I_DEMOGRAPHY_ 80. Age structure 0-14 years_ _V_I_DEMOGRAPHY_ 81. Median age total_ _V_D_DEMOGRAPHY_ 82. Population growth rate_ _V_I_DEMOGRAPHY_ 83. Birth rate_ _V_I_DEMOGRAPHY_ 84. Death rate_ _V_D_DEMOGRAPHY_ 85. Net migration rate_ _V_D_DEMOGRAPHY_ 86. Infant mortality rate total_ _V_D_DEMOGRAPHY_ 87. Life expectancy at birth total population_ _V_I_DEMOGRAPHY_ 88. Total fertility rate_ _V_I_DEMOGRAPHY_ 89. HIV AIDS - adult prevalence rate_ _V_D_DEMOGRAPHY_ 90. Ethnic groups_ _V_D_DEMOGRAPHY_
SUBSYSTEM: NATURE	91. Area total_ _V_I_NATURE_ 92. land_ _V_I_NATURE_ 93. water_ _V_I_NATURE_ 94. Land boundaries total_ _V_D_NATURE_ 95. border countries_ _V_D_NATURE_ 96. Coastline_ _V_D_NATURE_ 97. continental shelf_ _V_I_NATURE_ 98. Natural resources_ _V_I_NATURE_ 99. Land use arable land_ _V_I_NATURE_ 100. permanent crops_ _V_I_NATURE_ 101. Natural hazards_ _V_D_NATURE_ 102. Environment - current issues_ _V_D_NATURE_

(continued)

Table 4.4 (continued)

Subsystems	Description
SUBSYSTEM: POLITICS	103. Disputes - international_ _V_D_POLITICS_ 104. Maritime claims territorial sea_ _V_D_POLITICS_ 105. Exclusive fishing zone_ _V_D_POLITICS_ 106. Environment - international agreements party to_ _V_I_POLITICS_ 107. Dependent areas_ _V_D_POLITICS_ 108. Political parties and leaders_ _V_I_POLITICS_ 109. Political pressure groups and leaders_ _V_D_POLITICS_ 110. International organization participation_ _V_I_POLITICS_
SUBSYSTEM: CULTURE	111. Religions_ _V_D_CULTURE_ 112. Languages_ _V_D_CULTURE_ 113. Total population_Literacy _V_I_CULTURE_

Table 4.5 Code description associated with subsystems of the world as a system of systems

Indicator codes	Description of codes
_U:	Type-U indicator in the sense of the vulnerability model, referring to the targeted subsystem, in this case, the Infrastructure
_V:	Type-V indicator in the sense of the vulnerability model, referring to subsystems other than the targeted subsystem, in this case, the Economy, Demography, Nature, Politics, and Culture
_D:	Direct intuitive relationship with vulnerability, in the sense that an increment in the indicator entails an increment in vulnerability, and vice versa;
_I:	Inverse intuitive relationship with vulnerability, in the sense that an increment in the indicator entails a decrement in vulnerability, and vice versa;
_INFRASTRUCTURE/ECONOMY/ DEMOGRAPHY/POLITICS/ CULTURE:	Subsystem affiliation of indicator

4.1.1.3 The Vulnerability Model: Microscopic Hypothesis of a Physical Analogy

Assume that a system consists of *agents* (individual, atomic components), that:

- come in large numbers, M,
- agents interact with each other, exchanging an energy, ε,

- respond uniformly to external influences (fields), H,
- each exchanging an energy μH.

For the current representation of 'systems' as countries, the notion of an 'agent' embraces an almost unimaginable variety of representations—from the ordinary citizens (up to billions) down to the parts in a computer (a few tens), a nuclear reactor (thousands), a cardiac pacemaker (a dozen?), a passenger plane (thousands again), etc., and up to business companies (hundreds?), governmental institutions (tens), and media factors (tens to hundreds). Clearly, the actual number of agents and their nature is immaterial. The key point we are making is that agents are *numerous*. Not surprisingly, these agents can either *function* as designed or *not function* as designed. Therefore, the state of these agents could be described via a variable, S, that could assume two values such that:

- $S = 1$ indicating a functional agent; or
- $S = -1$ indicating a dysfunctional agent—which accommodates systems within the *Ising model* (Huang 1963; Vamanu et al. 2003) which is ubiquitous Physics and other fields.

These agents may switch from functional to dysfunctional state and conversely. There is an assumption that this process is essentially *probabilistic* in nature. The overall behavior of a system is governed by *a variational principle*, according to which, in a steady system state, agent's individual states are such that the system energy:

$$E = -\left(\frac{1}{2}\right)\varepsilon \sum_{ij} S(i)S(j) - \mu H \sum_{i} S(i) \tag{4.1}$$

is a minimum for any given *temperature*.

In applying the above, note that any agent i state-flip (from functional, 1, to dysfunctional, -1, or vice versa) entails a change in energy of the system such that:

$$\Delta E = -S(i)\left\{\varepsilon \sum_{j}{}' S(j) + \mu H\right\} \tag{4.2}$$

where $\sum_j'$ indicates a sum that, in practice, extends over a certain, usually close, neighborhood of agent *I*—while in principle, it may extend over all the agents other than i.

Following Ising as suggested by Gheorghe and Vamanu (2004), agent behavior is governed by the following set of rules, consistent with the assumptions above:

$$\text{Rule 1 : if } \Delta E \leq 0, \tag{4.3}$$

then agent always undergoes a state-flip

$$\text{Rule 2}: \text{if } \Delta E > 0, \tag{4.4}$$

then agent flips state only with a probability, $P = \exp\left(\frac{-\Delta E}{k_B T}\right)$

with T a *sui generis* 'system temperature,' and k_B a *sui generis*, a Boltzmann constant, conveniently taken as 1. In practice, see Sprott (1993) and Metropolis et al. (1953). Authors recommend Zeeman (1977) and Gilmore (1981) for implementation of Rule 2.

Accordingly, let r be a (computer-generated) random number,

$$r > = 0 \text{ and } r < 1 \tag{4.5}$$

then, if $r <= P$ (P is given by Eq. (4.4) then *do* flip; otherwise *do not* flip. From a microscopic perspective, the overall state of a system can be described by *the fraction of agents that do function*, ζ, defined as follows:

Let the system consist of M agents, of which M_1 agents are in a functional state ($S = 1$), and M_2 agents are dysfunctional ($S = -1$). Then one has, obviously,

$$M = M_1 + M_2 \tag{4.6}$$

and one may define the fraction of functional agents, or 'the *membership fraction*' as

$$\xi = \frac{M_1 - M_2}{2M}, \tag{4.7}$$

a definition that places ζ between (−0.5) and (+0.5), and sets the assessment rule: 'a system featuring $\zeta > = 0$ is *mostly functional, whereas a* system featuring $\zeta < 0$ is *mostly dysfunctional.*' And, the value judgment placed on a policy/strategy relates *to an assessment of the extent the managed system is kept mostly functional.* The model assumes that the variable M is essentially large, and for all practical purposes, constant. The macroscopic behavior of a system, expressed via variations in its indicators of definition, $I(k)$, $k = 1, 2, \ldots, n_I$ is a result of its microscopic behavior characterized by the membership fraction, ζ.

To infer a relationship between $I(k)$ and ζ, let it be noted that for $S(i) = 1$, performing the sum Σ over M_1 terms $S(j)$ equal to 1 and M_2 terms equal to −1, and in consideration of the definition in Eq. (4.7) of ζ, one has:

$$\Delta E = -\varepsilon(M_1 - M_2) - \mu H = -\varepsilon 0.2M\zeta - \mu H = U\zeta + V \tag{4.8}$$

Similarly, for $S(i) = -1$,

$$\Delta E = \varepsilon(M_1 - M_2) + \mu H = \varepsilon 0.2M\zeta + \mu H = -(U\zeta + V) \tag{4.9}$$

Here, since ε is the exchange (pairing) energy for any pair of interacting agents, $U = 2M\varepsilon$ relates to *the total interaction energy of the system's agents*—a quantity featuring the *internal* dynamics of the system. In turn, energy $V = \mu H$ features *the coupling of agents to the external field H* (uniform influence on agents). Equations (4.8) and (4.9) provide a consistent interpretation of the microscopic, state-flip probability (4.4) in terms of macroscopic, *overall system transition.* Thus, the probability of a system transition from an overall state is characterized as (M_1, M_2)—that is, M_1 functional and M_2 dysfunctional, agents, to an overall state characterized as ($M_1 - 1$, $M_2 + 1$) is;

$$P_{12} = \exp\left(-\frac{(U\zeta + V)}{k_B T}\right) \tag{4.10}$$

whereas, the probability of a system transition from an overall state characterized as (M_1, M_2) to an overall state characterized as ($M_1 + 1$, $M_2 - 1$) is;

$$P_{21} = \exp\left(\frac{(U\zeta + V)}{k_B T}\right) \tag{4.11}$$

To completely seal the gap between the microscopic and the macroscopic visions on the system, one calls to mind the microscopic meaning of U and V (see Eqs. 4.8 and 4.9). As argued, U features the *internal dynamics of the system* (pairing interaction intensity/energy); it would be natural to associate it to a construct made of the system indicators that refer to what is consider *system's chief driving force in matters of vulnerability*. In this respect, authors' primary choice is —*the infrastructure indicators*, coded _U_ in the list given in the preceding table. Likewise, since V features *system's coupling to external influences* (uniform effect of external fields on system agents), it is natural to build it from all indicators *other than* the infrastructure-related ones that include the *economy, demography, nature* (including resources and hazards), *political system features*, and *culture*—all to the extent made available by the primary data source—the *CIA World Factbook* series, 2000–2006. As for *how*, namely, *effectively* expressing the described loose, vague, and *fuzzy* relationships, the *Fuzzy Set theory* (Dhar 1979; Katina and Unal 2015; Zadeh 1965) may offer the natural answer, in the form of two *generalized distances*:

$$U = \left(\sum_k I(k; U)^{p_U}\right)^{\frac{1}{p_U}} \tag{4.12}$$

$$V = \left(\sum_k I(k; V)^{p_V}\right)^{\frac{1}{p_V}} \tag{4.13}$$

In Eqs. (4.12 and 4.13), notation *I*(*k*; *U*) for actually the indicator *I*(*k*) is meant to emphasize the _U (internal, infrastructure-related) nature of the indicator, while *I*(*k*; *V*) signals a _V (external, field-wise) type indicator. p_U and p_V, known as 'fuzzy exponents,' may assume real values from near-zero to, typically, a few tens. A 'phase space' for the system can thus accommodate the system states, and dynamics: essentially 3-dimensional—a state being defined as the triplet (U, V, ζ), it would, however, appear at macroscopic levels as only 2-dimensional, via its projection on the (U, V)-plane—the one that is 'visible' via the indicators. While for $p_U = 2$ and $p_V = 2$, the generalized distances in the U, V plane become true Euclidian distances; there may be no intuitive equivalent for other values of the fuzzy exponents. Nonetheless, p_U and p_V are instrumental in *shaping the distribution of system states* (i.e., country 'positions') *in the phase space*, thus providing *a unique tool for model calibration* in respect of the SoS (the world system of systems) ensemble. All requisites are now in place for *an analysis of the topology* of the (U, V, ζ) phase space. This draws upon a natural *rate (balance, master) equation* that can be written for the function of distribution, $f(M_1, M_2, t)$, *or* $f(\zeta, t)$, of probability of occupancy of the system states in the (U, V, ζ) space:

$$\begin{aligned}\partial f(M_1,M_2,t)/\partial t = {} & P_{21}(M_1-1,M_2+1)\cdot f(M_1-1,M_2+1)\\ & + P_{12}(M_1+1,M_2-1)\cdot f(M_1+1,M_2-1)\\ & - (P_{21}(M_1,M_2)+P_{12}(M_1,M_2))\cdot f(M_1,M_2)\end{aligned} \tag{4.14}$$

that describes the obvious transient process toward a system's steady state:

$$(M_1-1,M_2+1)\underset{w_{21}\;\rightarrow}{\overset{\leftarrow\;w_{12}}{\rule{2em}{0.4pt}}}(M_1,M_2)\underset{\leftarrow\;w_{12}}{\overset{w_{21}\;\rightarrow}{\rule{2em}{0.4pt}}}(M_1+1,M_2-1) \tag{4.15}$$

P_{21} and P_{12} are the transition probabilities given by Eqs. (4.10) and (4.11). In terms of ζ, Eq. (4.14) reads:

$$\begin{aligned}\partial f(\zeta)/\partial t = {} & P_{21}(\zeta-1/M)\,f(\zeta-1/M)+P_{12}(\zeta+1/M)\,f(\zeta+1/M)\\ & -(P_{21}(\zeta)+P_{12}(\zeta))\,f(\zeta)\end{aligned} \tag{4.16}$$

The assumption that the number M of system agents is large allows a series expansion of all quantities in the 2nd member of Eq. (4.16). Restricting the expansion to the 2nd order in ($1/M$) one obtains:

$$\partial f/\partial t+\partial J/\partial\zeta = 0. \tag{4.17}$$

Equation (4.17) is a continuity (conservation) equation for the state distribution function f, involving the 'current'

$$J = (1/M)(P_{21} - P_{12}) \cdot f - (1/(2M2))\partial((P_{21} + P_{12}) \cdot f)/\partial z \tag{4.18}$$

Looking for the stationary states of the system, one takes now

$$\partial f/\partial t = 0, \tag{4.19}$$

which leaves us with the equation

$$\partial J/\partial \zeta = 0. \tag{4.20}$$

having as solution J = constant and, in particular,

$$J = 0. \tag{4.21}$$

In the form:

$$(1/M)(P_{21} - P_{12}) \cdot f - (1/(2M))\partial((P_{21} + P_{12}) \cdot f)/\partial z = 0,$$

that employs Eqs. (4.18) and (4.21) can immediately be integrated to give

$$f(\zeta) = \text{const} \cdot \frac{\exp\left[2M_1 \int\limits_{-1/2}^{\zeta} \frac{P_{21}(\xi) - P_{12}(\xi)}{P_{21}(\xi) + P_{12}(\xi)} \mathrm{d}\xi\right]}{P_{21}(\zeta) + P_{12}(\zeta)} \tag{4.22}$$

To determine the normalizing constant in Eqs. (4.21) and (4.22), the following equation is used.

$$\int\limits_{-1/2}^{1/2} f(\zeta)\mathrm{d}\zeta = 1 \tag{4.23}$$

Taking the expressions (4.10) and (4.11) of the transition probabilities in Eqs. (4.22) and (4.23) and requesting that

$$\partial f(\zeta)/\partial \zeta = 0, \tag{4.24}$$

one has:

$$\operatorname{cth}((U \cdot \zeta + V)/\theta) = (1/2 - 1/(U/\theta - 2M))/\zeta, \tag{4.25}$$

where cth denotes the hyperbolic cotangent function, $\operatorname{cth}(x) = (\exp(x) + \exp(-x)) / (\exp(x) - \exp(-x))$.

Using again the fact that the number of agents M in the system is large, the second term in the parenthesis in the right-hand side of Eq. (4.25) is ignored, so that, finally, the space of system states (U, V, ζ) is given by the equation:

$$\text{th}((U \cdot \zeta + V)/\theta) = 2\zeta \tag{4.26}$$

where th denotes the hyperbolic tangent function, $\text{th}(x) = (\exp(x) - \exp(-x))/(\exp(x) + \exp(-x))$.

Depending on the degree of interaction between system agents, reflected in the internal variable *U*, and on the external influence on all system members—reflected in the external variable *V*, and also taking into consideration the normalized temperature $k_B T$ of the system, the Eq. (2.19) may display a number of real solutions ζ that may relate to the overall system condition. Table 4.6 depicts the number of real solutions and system conditions.

This is a good time to say that we are interested in rendering a topology of the space that could enable assessing evolvement of the world vulnerability. Figures 4.2, 4.3, and 4.4 render the topology of the space phase, in this case: 'assessing the evolvement of the World vulnerability over the years 2000–2006, assuming (subsystem) infrastructures as vulnerability's driving force.' Figure 4.2 is a rendering of the SoS world in a 2D space of *U* and *V* parameters.

Similarly, the world's SoS 3D rendering is possible. This rendering includes a third dimension of ζ in this chapter (Sect. 3.1.1.3). Figure 4.3 is a depiction of the world as a SoS in a 3D space.

In Fig. 4.4, the front view is against the *U*-axis. The same graphic is rendered below with 60° slant view. In both Figs. 4.3 and 4.4, the red-black emphasized area (i.e., the physical basin). The green color represents the system (country) state-points.

At this point, we know the basics of the SoSE model as well as the conceptualization of the QVA model. The reminder of this chapter provides the results of the model.

Table 4.6 System conditions

Number of real solutions	System condition
1	*Stable*. Smooth transitions in population membership, between functional ($S = 1$) and dysfunctional ($S = -1$) states. *Low and/or acceptable vulnerability*
3, of which 2 are identical	*Critical*. Sharp transitions in membership between states 1 and −1 are possible. Either state 1 or state 2 may suddenly become improbable. *System is critically vulnerable*
3, all different from each	*Unstable*. Sharp transitions in membership between states 1 and −1 are possible. Frequency of occurrence of states 1 and 2 are comparable. Though Eq. (2.19) has three real roots, the intermediate root is taken as having no physical meaning and is therefore discarded. *System is dangerously and has in unacceptably vulnerable*

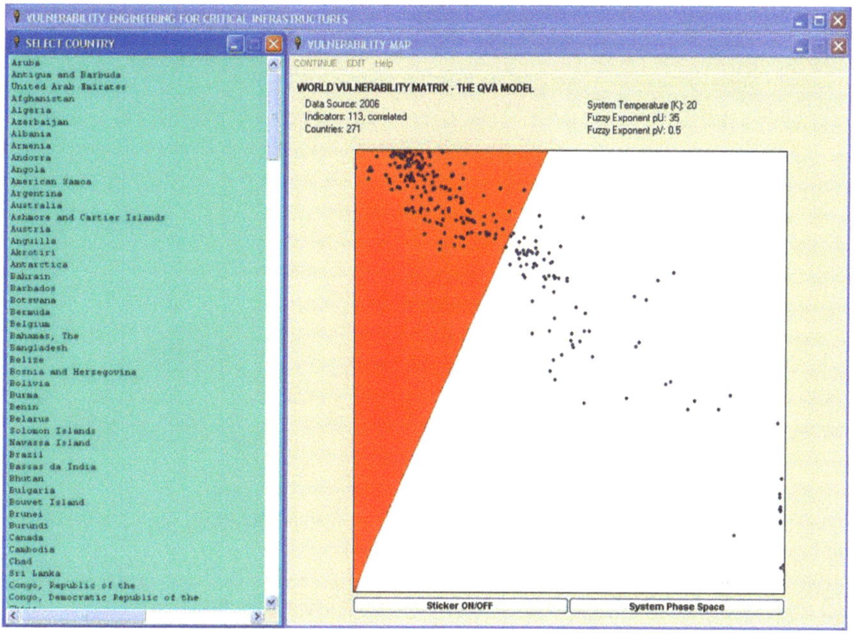

Fig. 4.2 Rendering the world as a SoS in a 2D phase space using *U* and *V*, adapted from Gheorghe and Vamanu (2008)

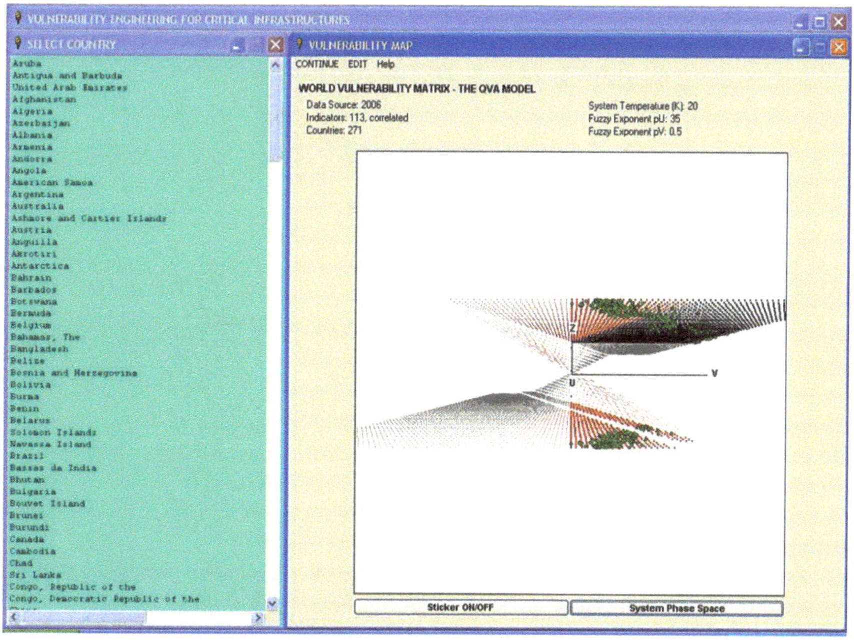

Fig. 4.3 World's SoS in a 3D phase space using *U*, *V*, and ζ with front view against *U*-axis, adapted from Gheorghe and Vamanu (2008)

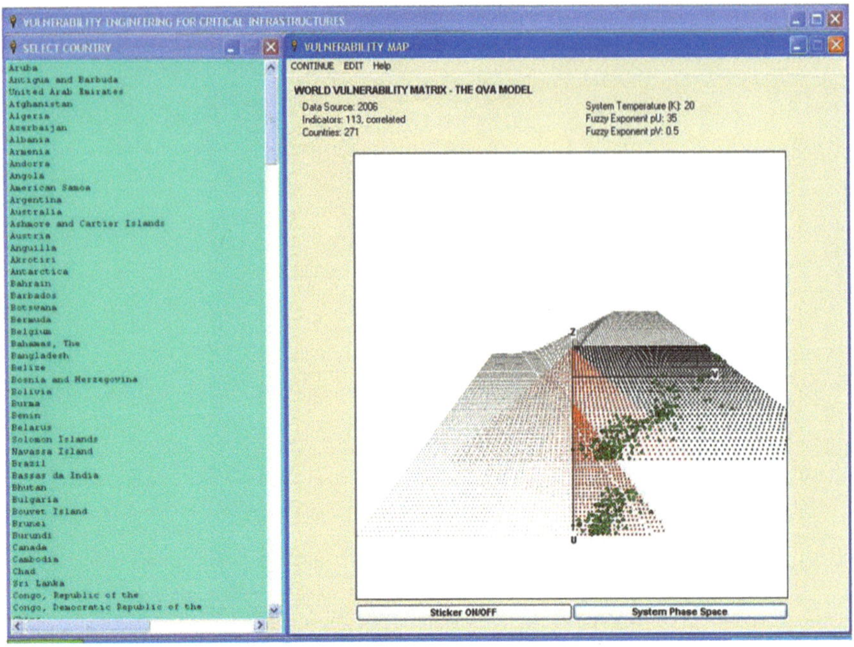

Fig. 4.4 World's SoS in a 3D phase space using U, V, and ζ with view at a 60° slant view, adapted from Gheorghe and Vamanu (2008)

4.1.2 Research Context

The moments in time to correlate with vulnerability maps are as follows:

- 2000—*Business as usual: some better off than others*
- 2001—*The 9/11 Event*
- 2002—*From groggy to angry*
- 2003—*Iraq War: the Invasion*
- 2005—*Iraq War: the Transition to Stability*
- 2006—*Present time*[2]: the World—a better place, at a cost?

[2]Researchers are, of course, aware of the many other events that have taken place since 2006, for instance, the assassination of Benazir Bhutto in 2007, election of Dmitry Medvedev in 2008, inauguration of Barack Obama in 2009, and the likes. The emphasis is present research is not on events, rather the capability of the developed model.

4.1.3 Research Findings

4.1.3.1 The World SOS, 2000–2006: Relaxed Versus Strained Patterns

Authors suggest that there was a relaxed, natural distribution of systems in the years preceding the Iraq War. However, this relaxed and natural distribution of systems transitioned into a strained, rippled distribution featuring coalition/accretion patterns during the years of heavy war as exhibited by international political activities and turmoil around the issue. Another transition of the systems took place, relatively, into a relaxing with the onset of the transition of authority and the gradual, if painful, instatement of normality in Iraq.

The notion of relaxed distribution of systems is captured in the developed model and illustrated in Fig. 4.5.

Using Fig. 4.5 as a baseline, it is now possible to develop a simulation for vulnerability assessment for critical infrastructures following selected moments in time events. The following sequence of graphical representations (Figs. 4.6, 4.7, 4.8, 4.9, 4.10 and 4.11), depicted on a yearly basis, are developed with this view in mind.

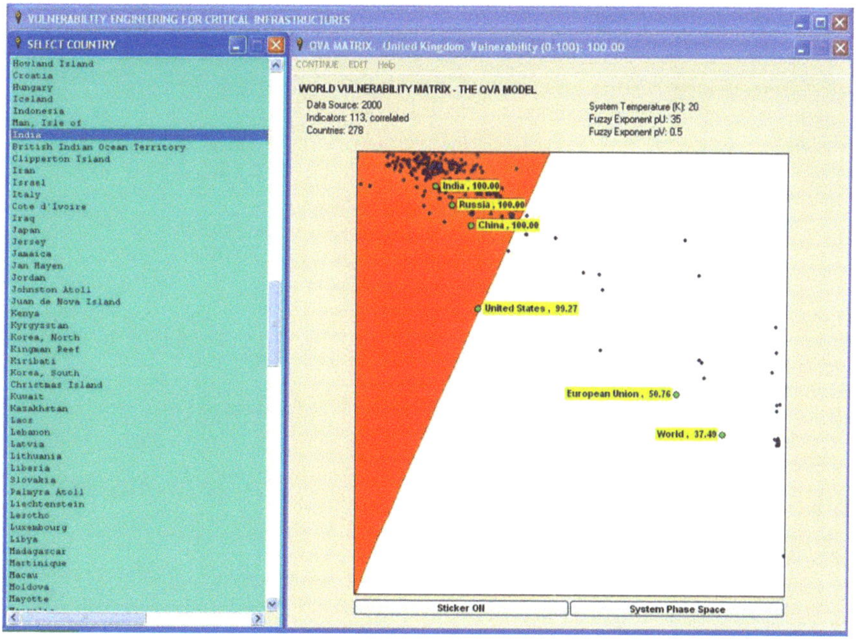

Fig. 4.5 The world SoS in year 2000. *Prior* to 9/11 event, adapted from Gheorghe and Vamanu (2008)

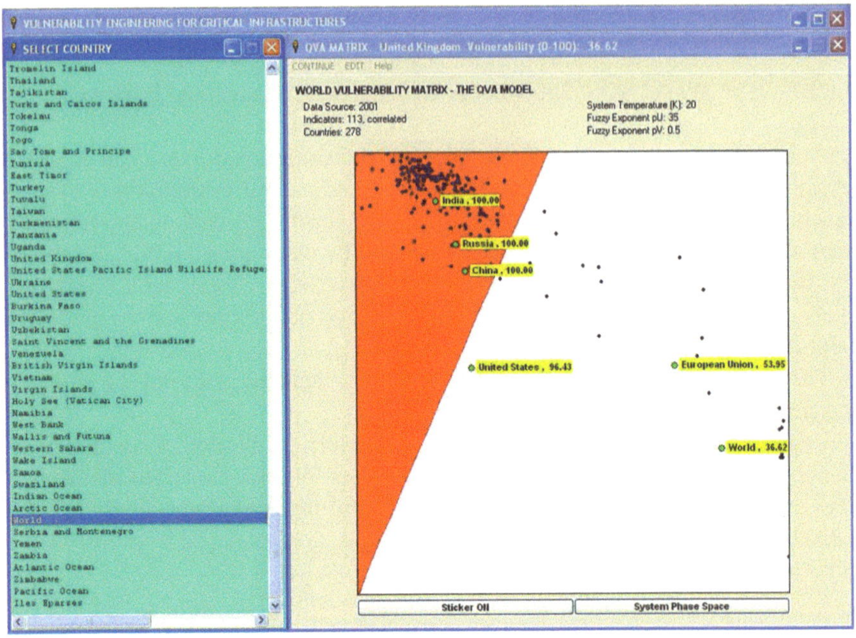

Fig. 4.6 The world SoS in year 2001, adapted from Gheorghe and Vamanu (2008)

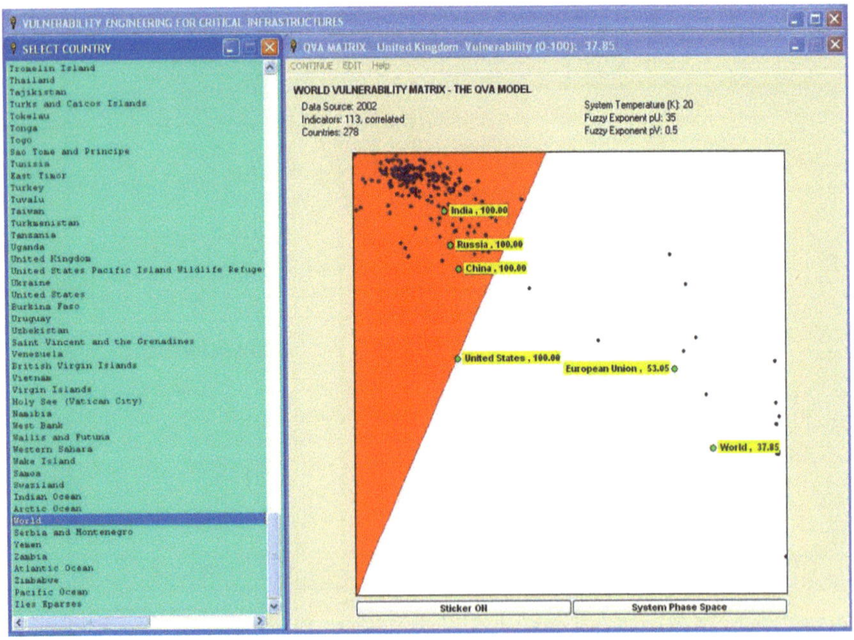

Fig. 4.7 The world SoS in year 2002, adapted from Gheorghe and Vamanu (2008)

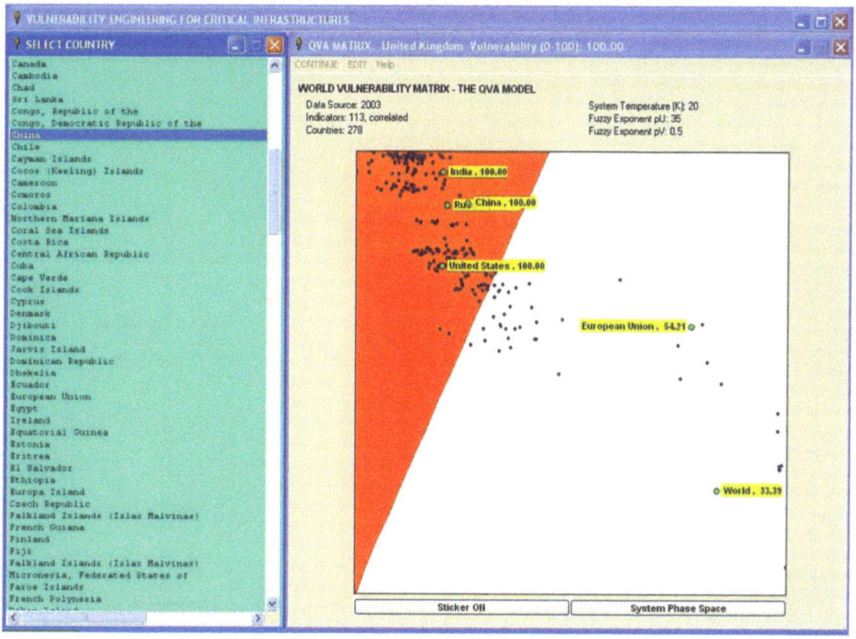

Fig. 4.8 The world SoS in year 2003, adapted from Gheorghe and Vamanu (2008)

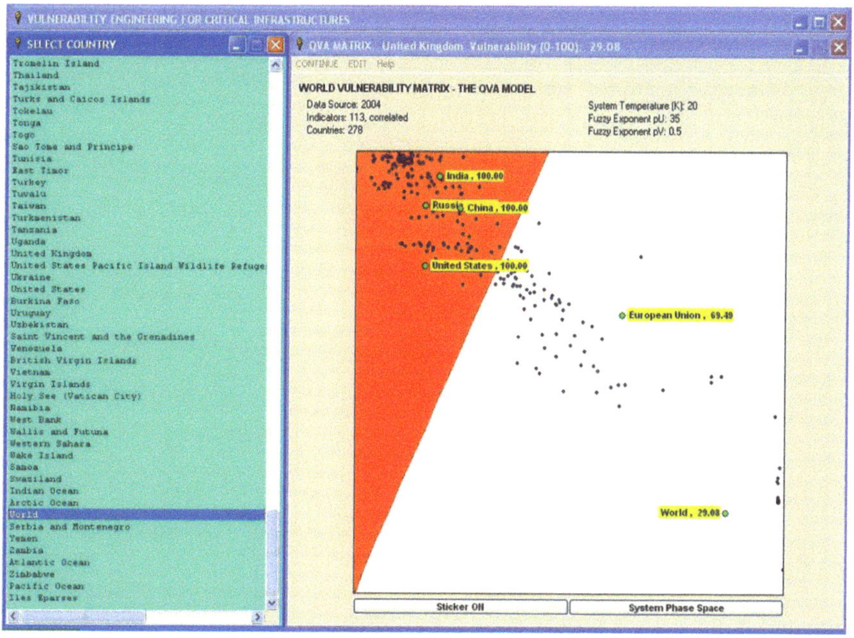

Fig. 4.9 The world SoS in year 2004, adapted from Gheorghe and Vamanu (2008)

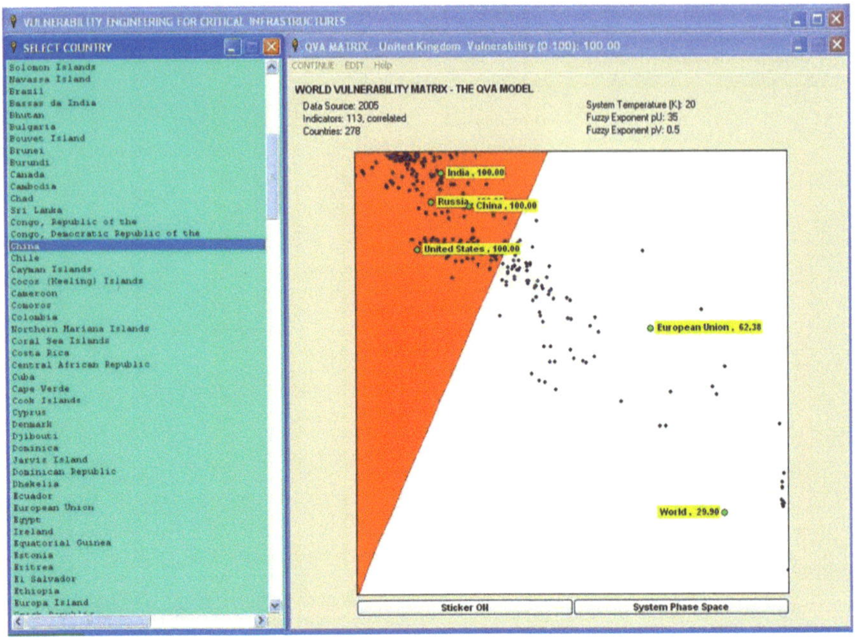

Fig. 4.10 The world SoS in year 2005, adapted from Gheorghe and Vamanu (2008)

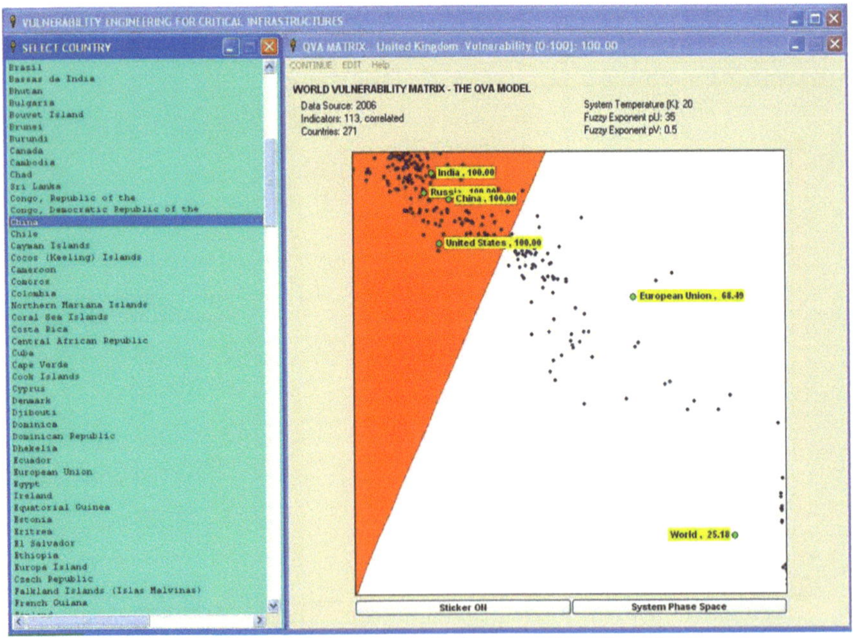

Fig. 4.11 The world SoS in year 2006, adapted from Gheorghe and Vamanu (2008)

4.1.3.2 Selective Comparisons

Interestingly, countries fuzzily perceived as 'akin' tended to share comparable vulnerability behavior. Notice that how the coalition and nations traditionally thought of as depending, as opposed to 'dependent' on the USA, tend to follow in the footsteps of the USA. As previously suggested, notice that the model also suggests that the world eventually became a better place. Of course, recall that present analysis looked at years 2000–2006. It is logical to ask, what of the years that followed, what can analyses by the presented model tell us? What of the years to come?

In the following sequence of figures, results of simulation are presented following world evolving vulnerability performance (i.e., year 2000–2006). Emphasis is placed on European Union (Fig. 4.12), Canada (Fig. 4.13), China (Fig. 4.14), Germany (Fig. 4.15), France (Fig. 4.16), Japan (Fig. 4.17), Russia (Fig. 4.18), UK (Fig. 4.19), and USA (Fig. 4.20).

The same approach was applied to other nations: India (Fig. 4.21), Iran (Fig. 4.22), Iraq (Fig. 4.23), Israel (Fig. 4.24), Italy (Fig. 4.25), North Korea (Fig. 4.26), and the world (Fig. 4.27).

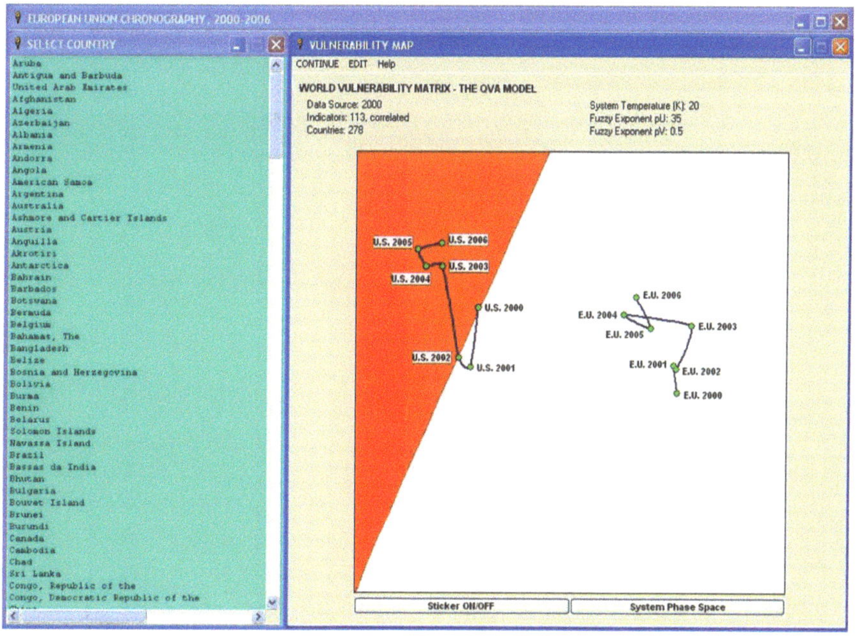

Fig. 4.12 Vulnerability of the European Union in QVA, 2000–2006, adapted from Gheorghe and Vamanu (2008)

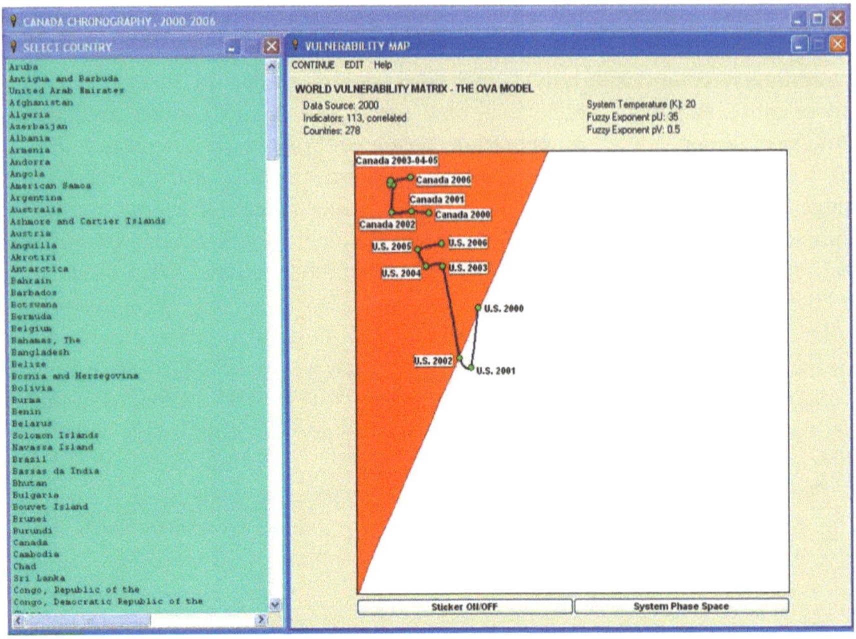

Fig. 4.13 Vulnerability of Canada in QVA, 2000–2006, adapted from Gheorghe and Vamanu (2008)

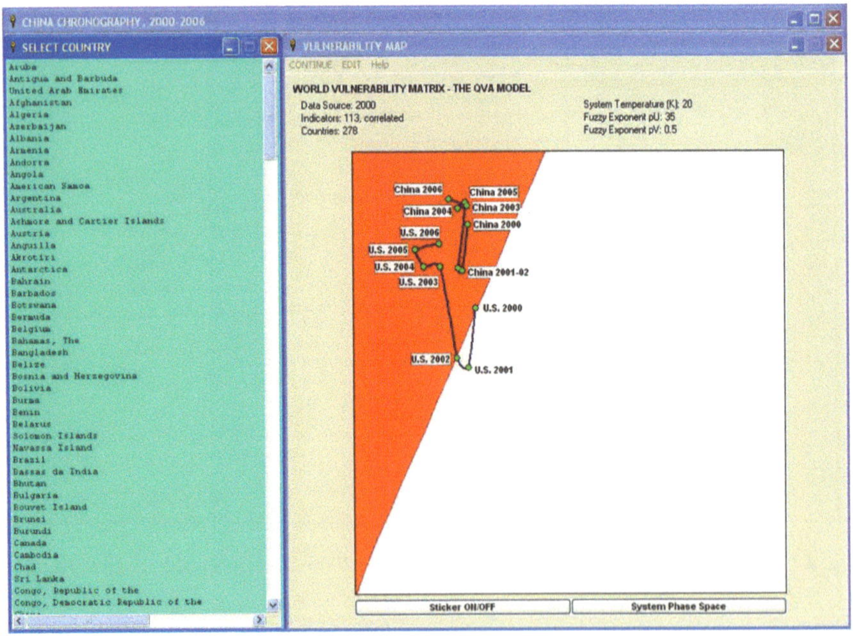

Fig. 4.14 Vulnerability of China in QVA, 2000–2006, adapted from Gheorghe and Vamanu (2008)

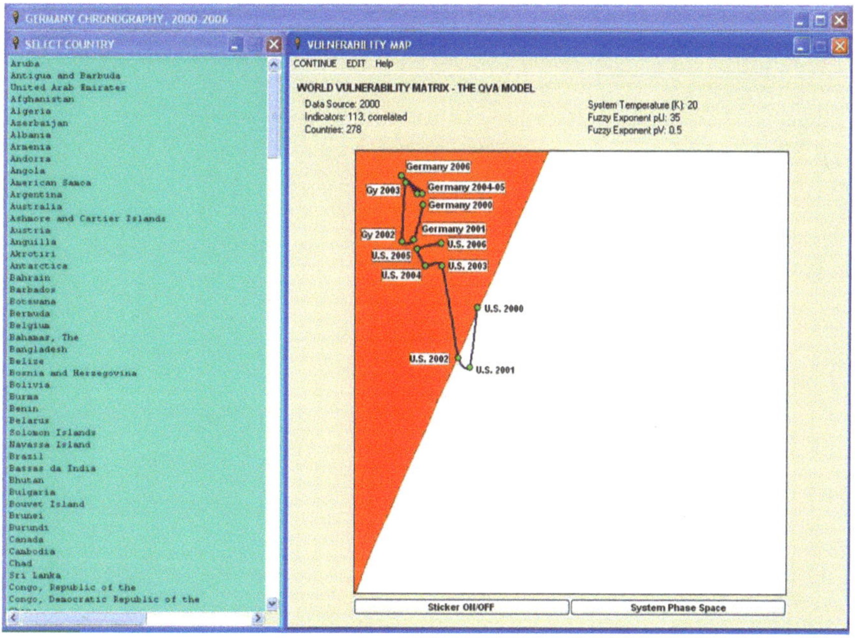

Fig. 4.15 Vulnerability of Germany in QVA, 2000–2006, adapted from Gheorghe and Vamanu (2008)

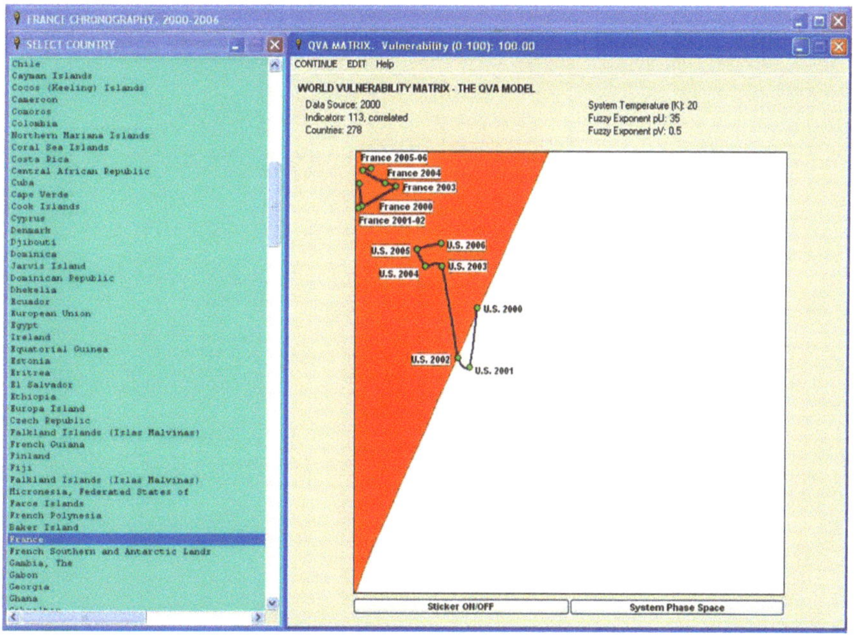

Fig. 4.16 Vulnerability of France in QVA, 2000–2006, adapted from Gheorghe and Vamanu (2008)

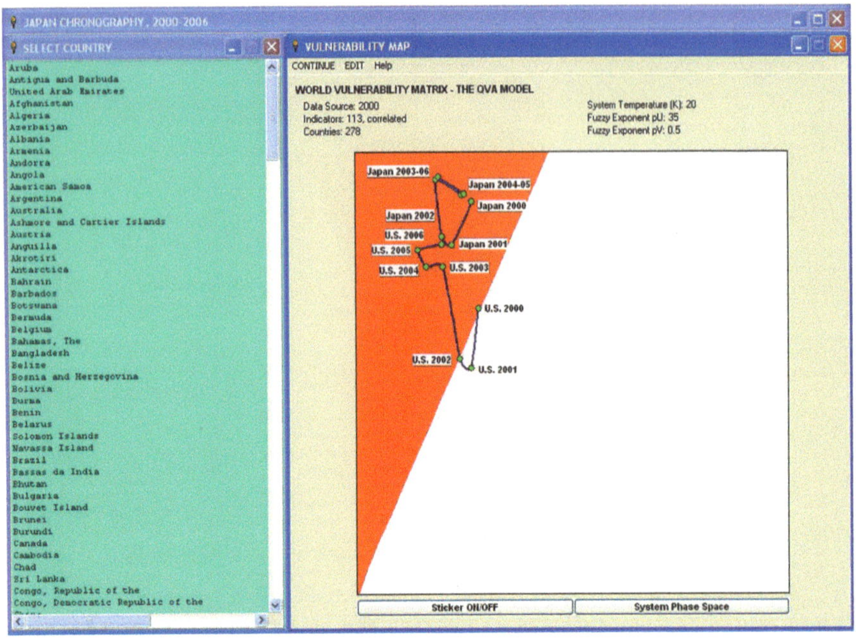

Fig. 4.17 Vulnerability of Japan in QVA, 2000–2006, adapted from Gheorghe and Vamanu (2008)

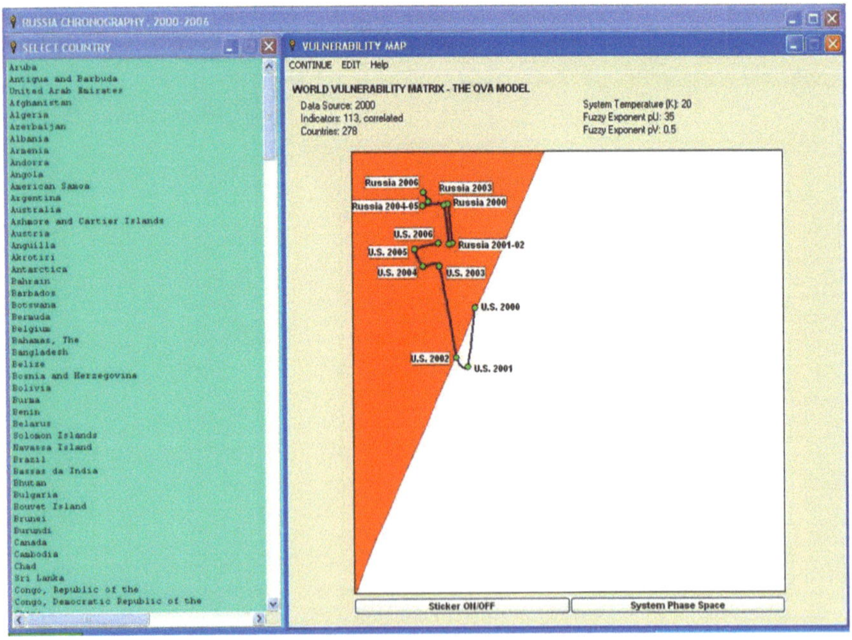

Fig. 4.18 Vulnerability of Russia in QVA, 2000–2006, adapted from Gheorghe and Vamanu (2008)

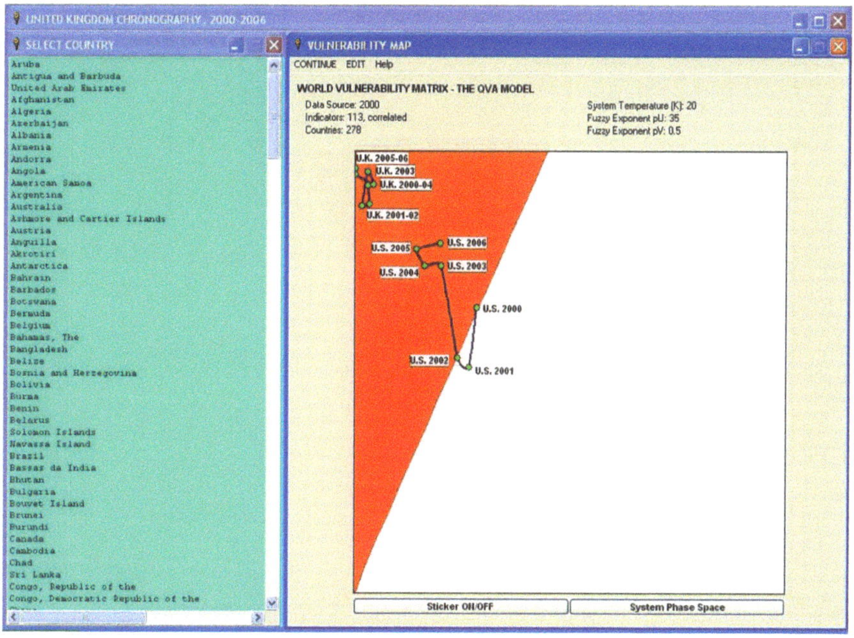

Fig. 4.19 Vulnerability of UK in QVA, 2000–2006, adapted from Gheorghe and Vamanu (2008)

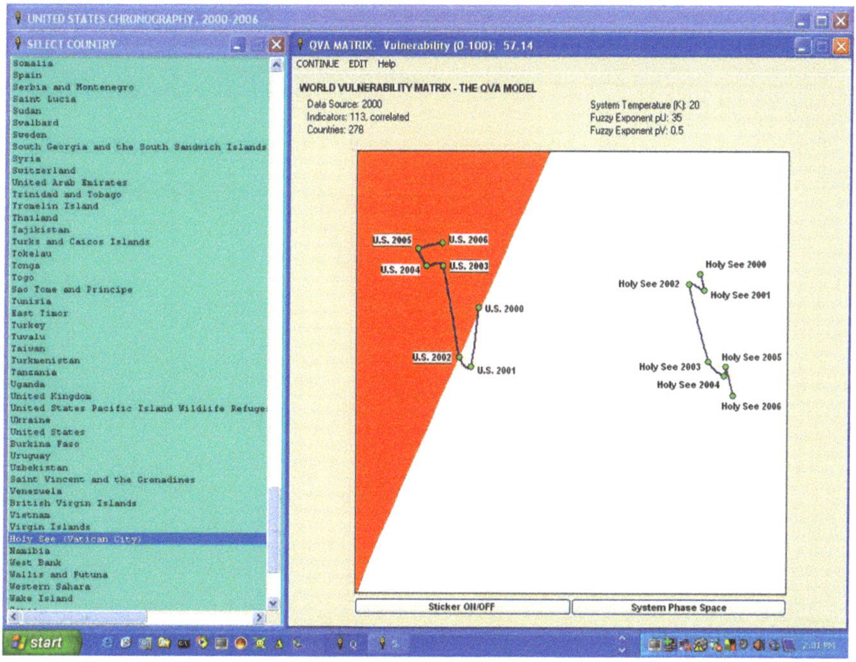

Fig. 4.20 Vulnerability of USA in QVA, 2000–2006, adapted from Gheorghe and Vamanu (2008)

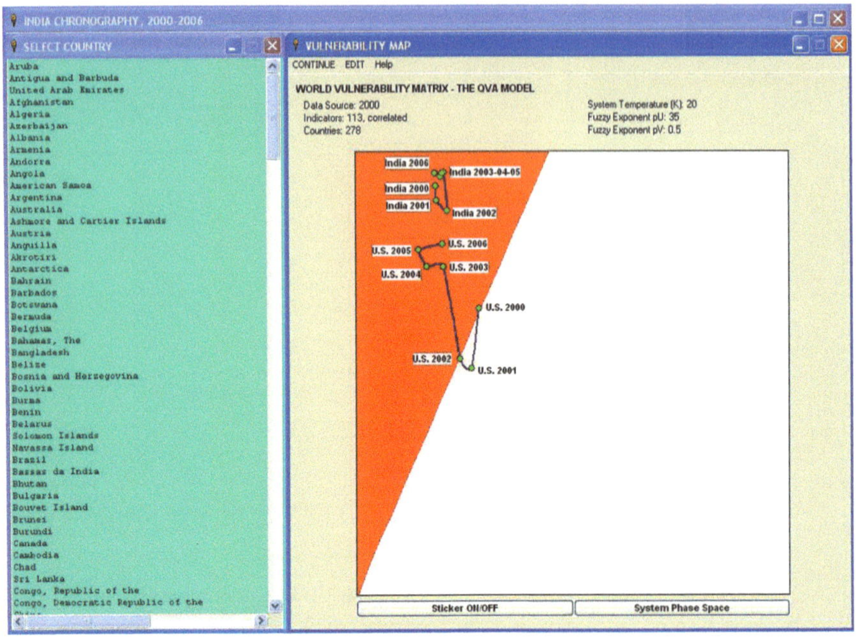

Fig. 4.21 Vulnerability of India in QVA, 2000–2006, adapted from Gheorghe and Vamanu (2008)

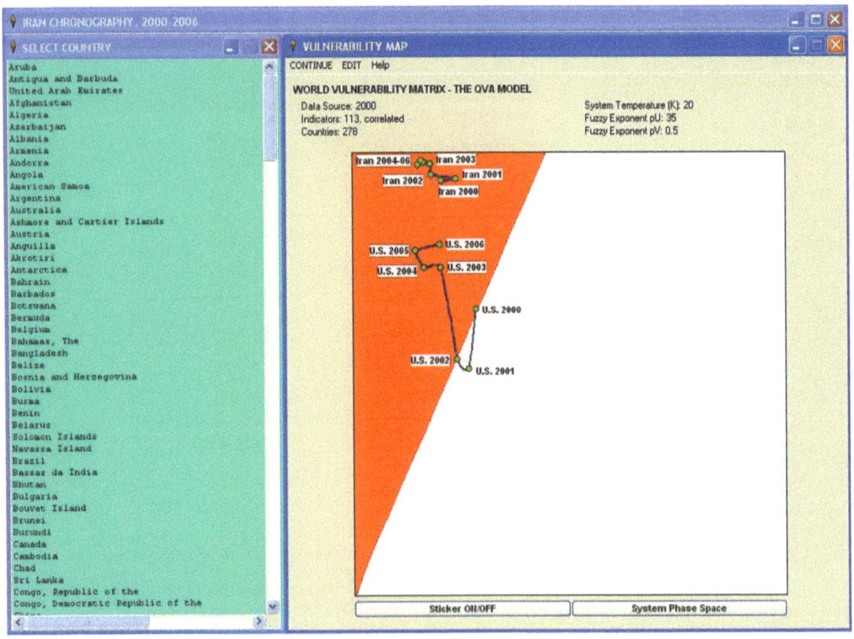

Fig. 4.22 Vulnerability of Iran in QVA, 2000–2006, adapted from Gheorghe and Vamanu (2008)

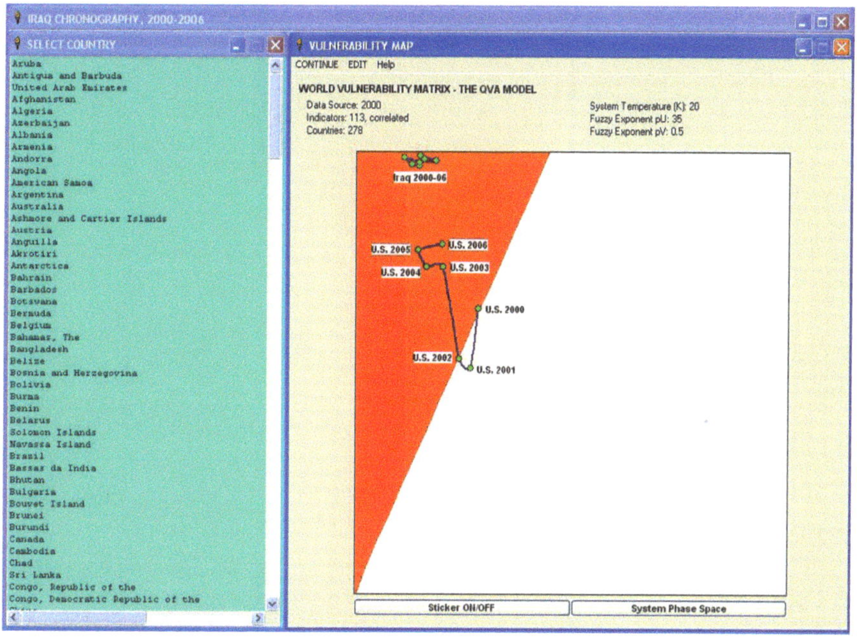

Fig. 4.23 Vulnerability of Iraq in QVA, 2000–2006, adapted from Gheorghe and Vamanu (2008)

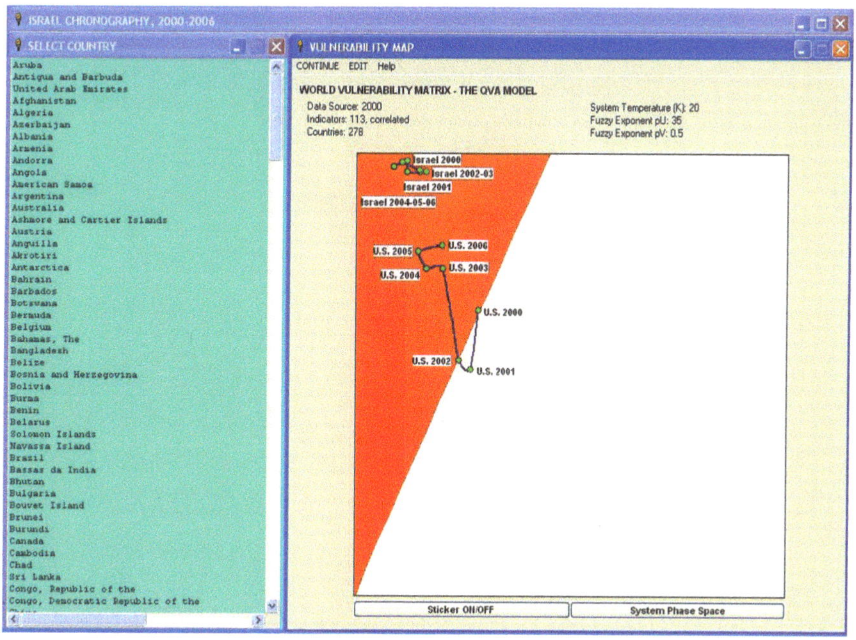

Fig. 4.24 Vulnerability of Israel in QVA, 2000–2006, adapted from Gheorghe and Vamanu (2008)

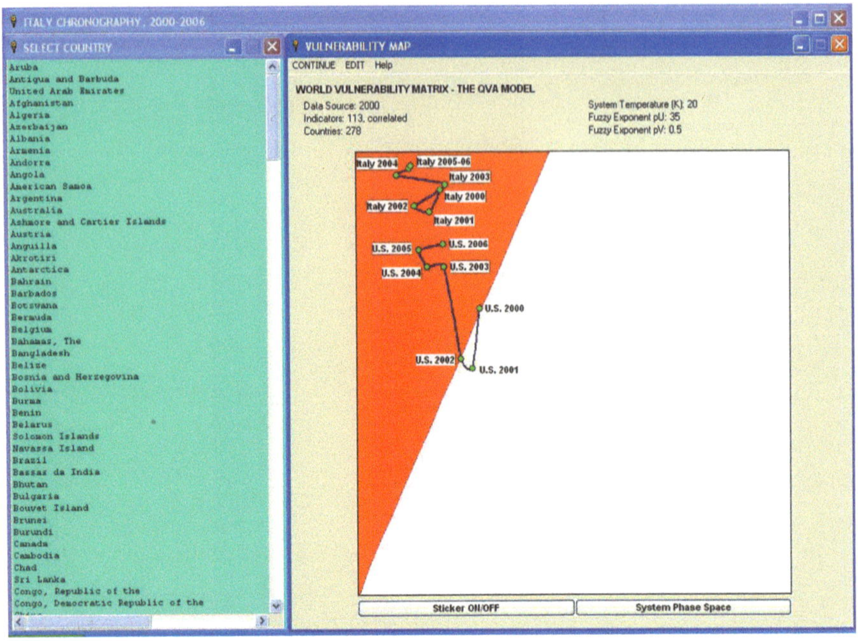

Fig. 4.25 Vulnerability of Italy in QVA, 2000–2006, adapted from Gheorghe and Vamanu (2008)

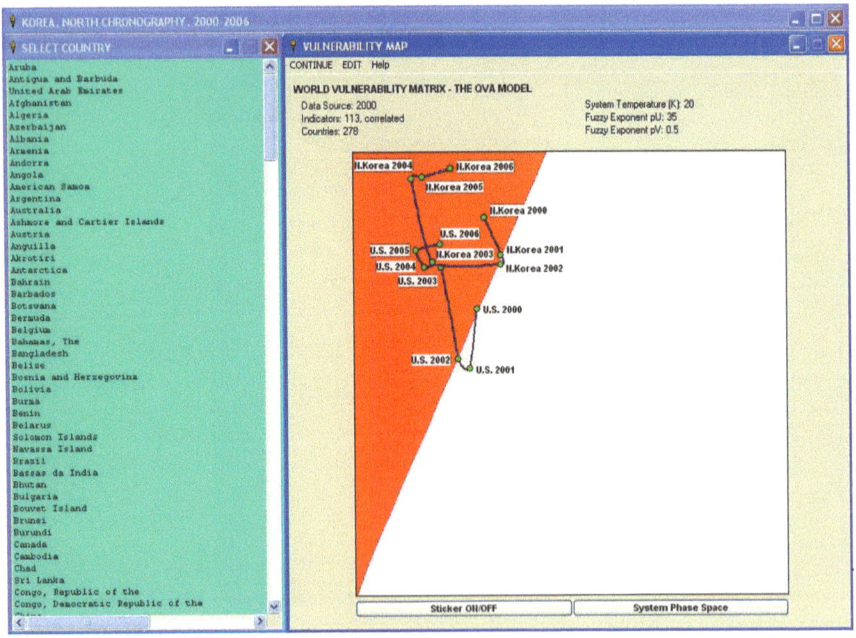

Fig. 4.26 Vulnerability of North Korea in QVA, 2000–2006, adapted from Gheorghe and Vamanu (2008)

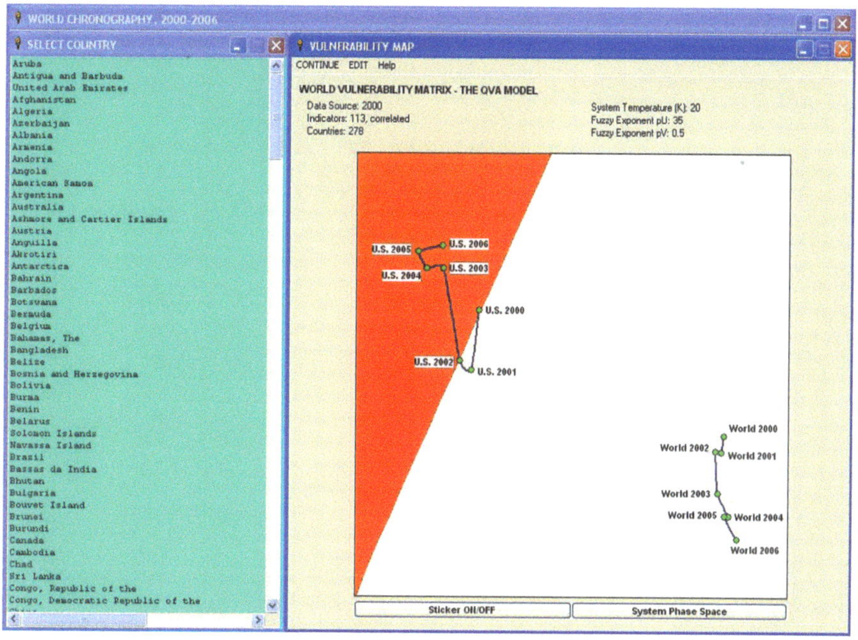

Fig. 4.27 Vulnerability of world in QVA, 2000–2006, adapted from Gheorghe and Vamanu (2008)

4.1.3.3 Synoptic: The World at Large

The developed model offered interesting insights: (i) many great countries of the world, and particularly disfavored countries, *got less vulnerable* as a result of the evolutions relating to the Iraq War, and (ii) contrastingly, and perhaps at the cost of US coalition countries, and countries traditionally depending on the USA *getting more vulnerable*, especially during the 2003–2005 time frame.

Arguably, taking a course back to a lesser vulnerability only begins with 2006. In fact, one could suggest that therein lies a principle for complex systems: *the mighty get the heat, as the weaklings get better off*. However, this 'principle' similar to the 'cold fusion' thing needs a thorough scrutiny. In the next set of figures, a synoptic manner of the evolution of the vulnerability indicators at the world level is presented at different time intervals: 2000–2001 (Fig. 4.28), 2000–2002 (Fig. 4.29), 2000–2003 (Fig. 4.30), 2000–2004 (Fig. 4.31), 2000–2005 (Fig. 4.32), and 2000–2006 (Fig. 4.33).

4.1.3.4 Temperature Effect: Geopolitical Climate and Vulnerability

It is common knowledge that a model may help, and yet, a model may also mislead: here's a trivia in the trade! Notice that different 'temperatures' place nations in

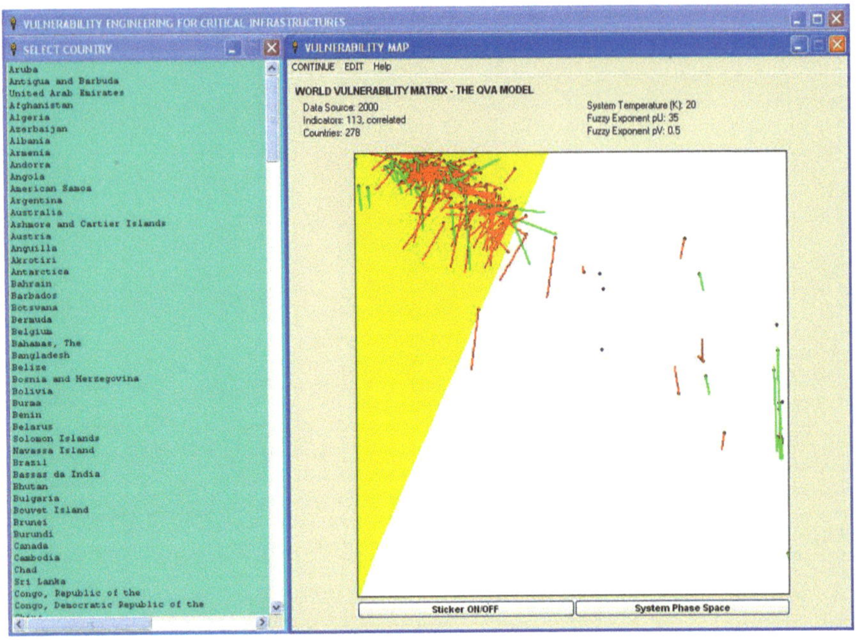

Fig. 4.28 QVA SoS vulnerability flow, 2000–2001, adapted from Gheorghe and Vamanu (2008)

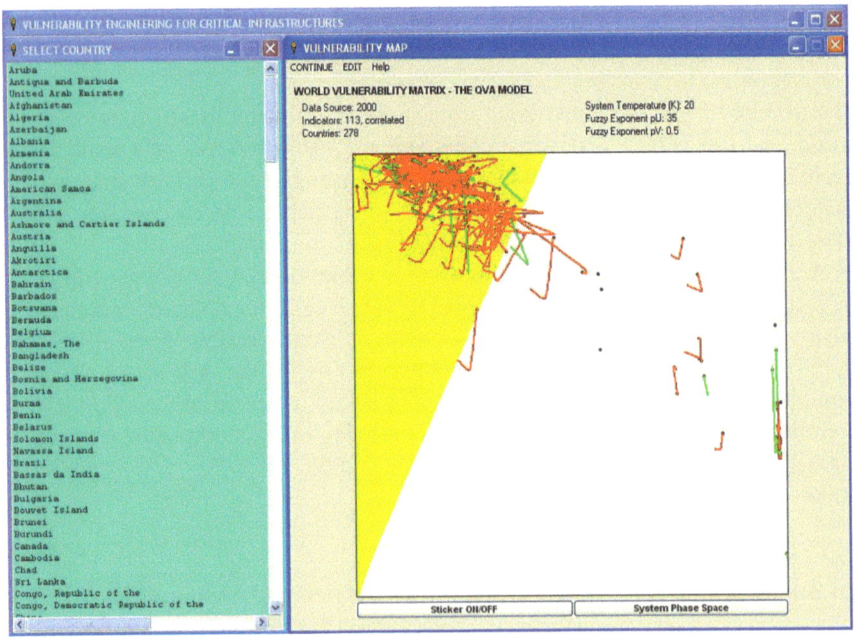

Fig. 4.29 QVA world SoS vulnerability, 2000–2002, adapted from Gheorghe and Vamanu (2008)

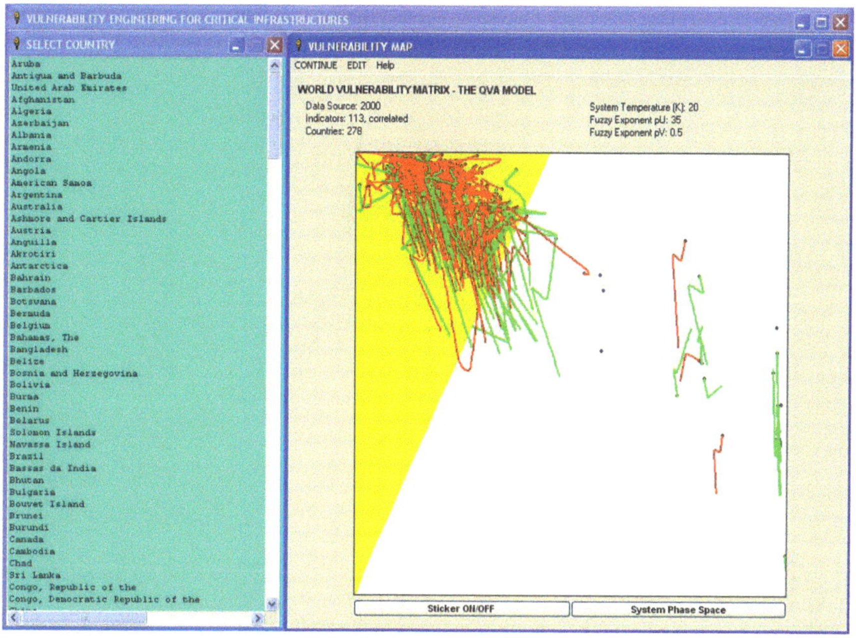

Fig. 4.30 QVA world SoS vulnerability, 2000–2003, adapted from Gheorghe and Vamanu (2008)

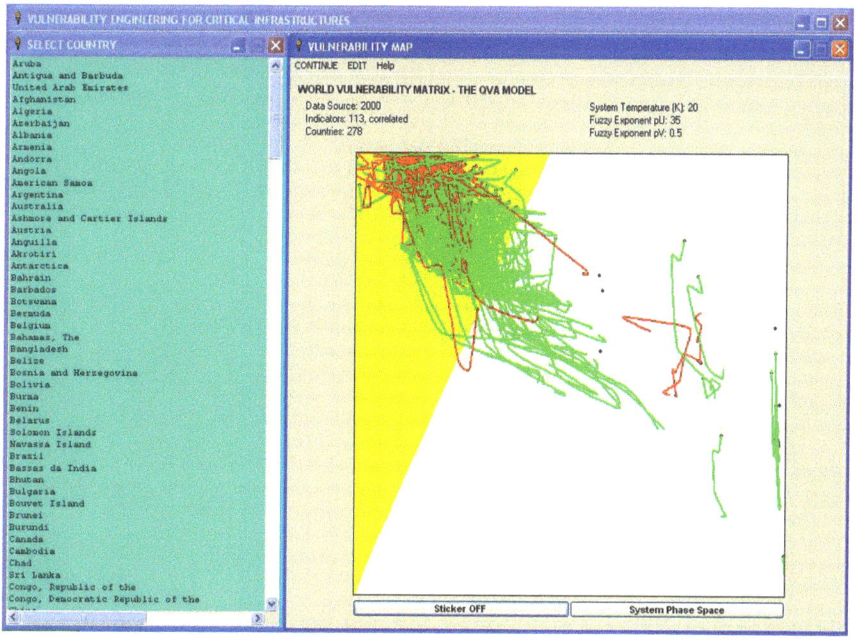

Fig. 4.31 QVA world SoS vulnerability, 2000–2004, adapted from Gheorghe and Vamanu (2008)

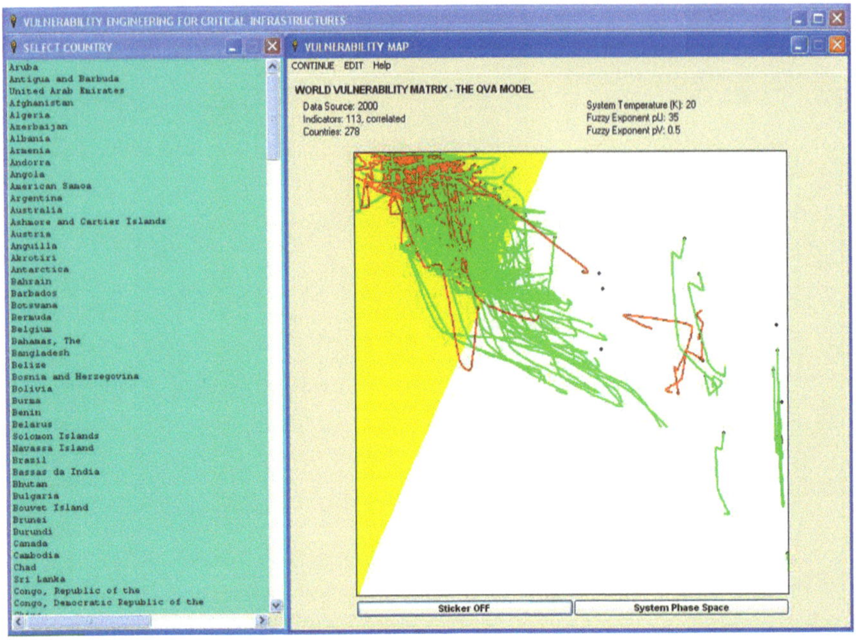

Fig. 4.32 QVA world SoS vulnerability, 2000–2005, adapted from Gheorghe and Vamanu (2008)

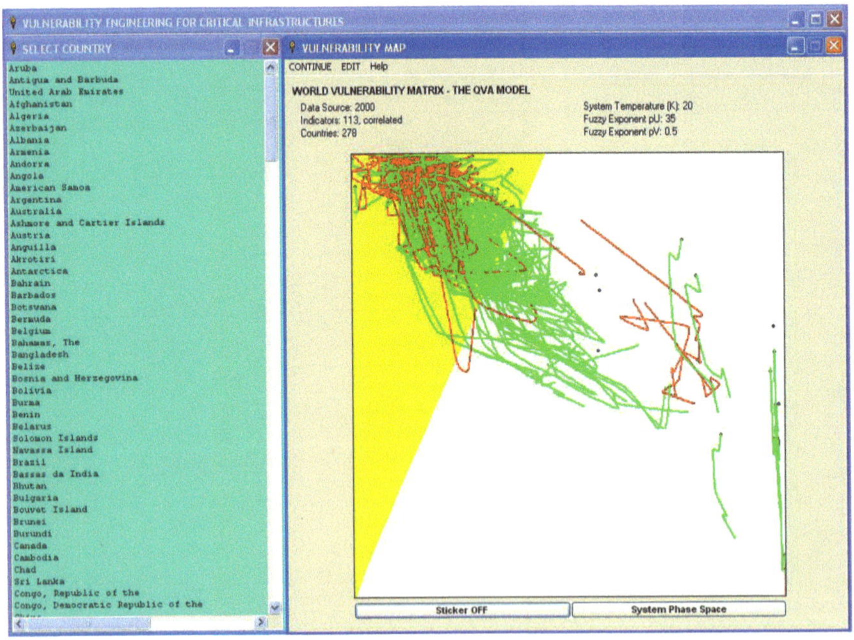

Fig. 4.33 QVA world SoS vulnerability, 2000–2006, adapted from Gheorghe and Vamanu (2008)

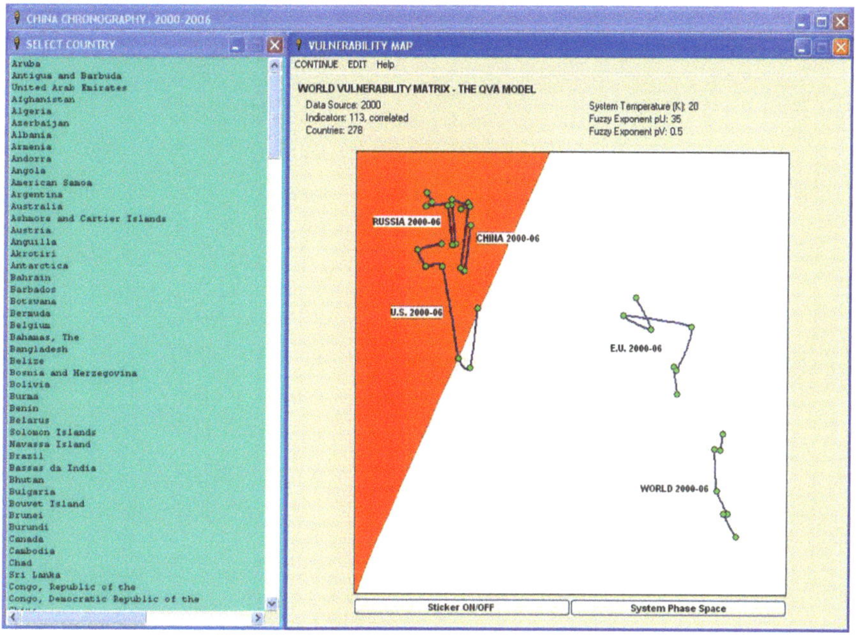

Fig. 4.34 A 'Cold' world: major actors, 2000–2006—at a temperature of 20 K, adapted from Gheorghe and Vamanu (2008)

states: safe or unsafe. In different models, oil prices have been used; they failed to completely satisfy; perhaps, it is time to use another measure such as the Stock Exchange. At this point, the issue of temperature remains a matter of the consensus of analysts. Fixing it along with the fuzzy exponents, p_U and p_V, *calibrates the model.*

Following the model given before, by playing with the parameter, T, called the temperature effect, the simulation indicates effects on the degree of vulnerability 'performance' of some of the countries investigated during present research. Temperature change is taken at 20 (Fig. 4.34), 273 (Fig. 4.35), and 500 (Fig. 4.36) on Kelvin scale.

4.2 Remarks

This chapter pronounces a model for seeing the world through system of systems lens using the world as a platform with a number of selected indicators. Each country is seen as a system. Each system has subsystems. Each subsystem has subsections with a number of indicators. The underlying assumption: The most appropriate indicator for vulnerability assessment is sought in relation to intelligence sources was applied. This leads to using readily and publicly available data

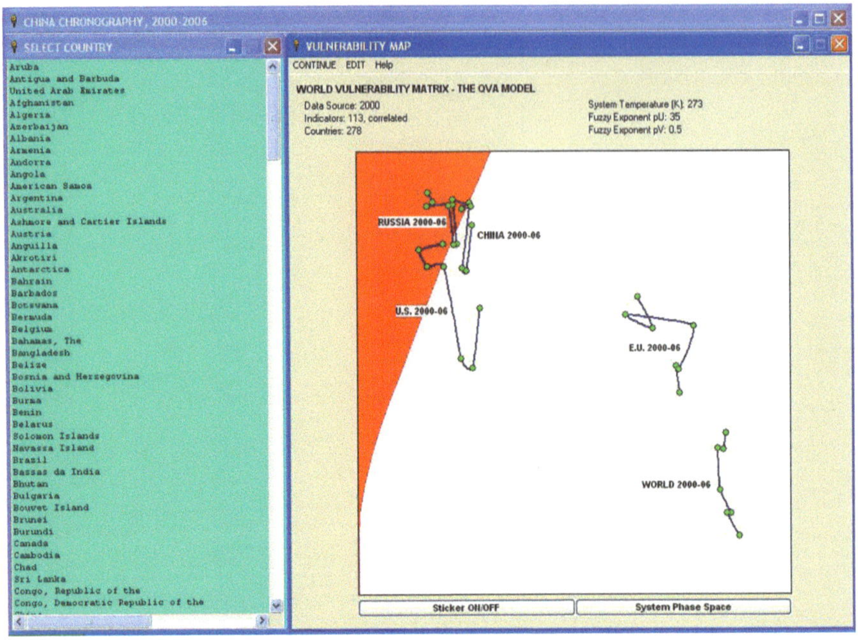

Fig. 4.35 A 'Normal' world: major actors, 2000–2006—at a temperature of 273 K, adapted from Gheorghe and Vamanu (2008)

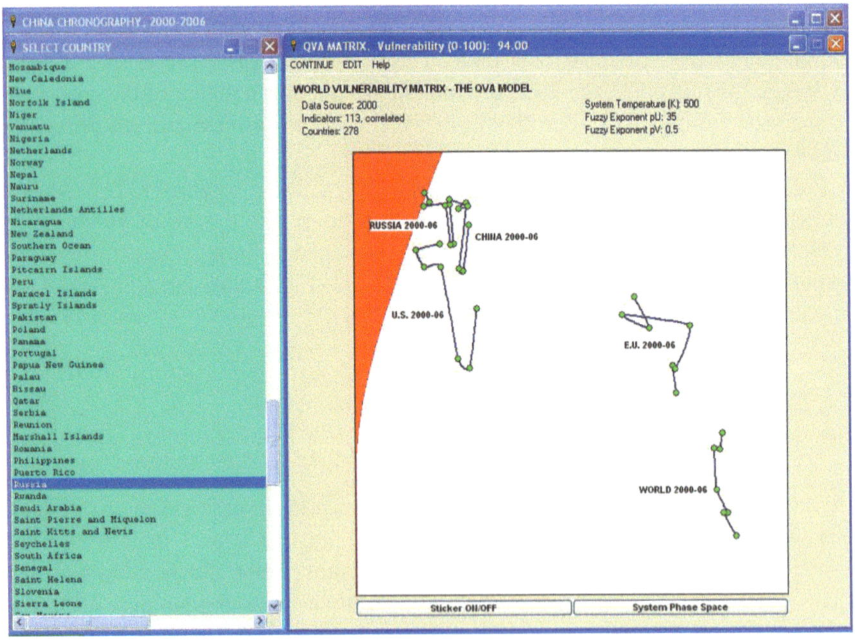

Fig. 4.36 A 'Warm' world: major actors, 2000–2006—at a temperature of 500 K, adapted from Gheorghe and Vamanu (2008)

from the *CIA's World Factbook*. Simulations were developed, and the results suggest a novel approach for understanding the world as a complex system with dynamic behaviors that could be assessed from a vulnerability viewpoint. The results given in this chapter should be taken as an exercise in intelligence data mining for assessing vulnerability for large and complex System of systems. However, there remains a need to improve the presented models, using updated datasets and discerning the meaning associated with such models. Moreover, the presented exercise could be the basis for suggesting a need for data calibration. Certainly, *the truth can be out there*. However, the data must have a high level of fidelity—hence the need for calibration.

References

Calida, B. Y. (2013). *System governance analysis of complex systems* (Ph.D.). Old Dominion University, United States—Virginia.

Calida, B. Y. (2016). Complex system governance: Moving diverse theory to practice. *International Journal of System of Systems Engineering*, *7*(1/2/3), 22–42. https://doi.org/10.1504/IJSSE.2016.076125.

Carter, B. (2015). A metasystem perspective and implications for governance. *International Journal of System of Systems Engineering, 6*(1/2), 90–100. doi:10.1504/IJSSE.2015.068807.

DeLaurentis, D. (2005). Understanding transportation as a system-of-systems design problem. In *43rd AIAA Aerospace Sciences Meeting*. Reno, NV: American Institute of Aeronautics and Astronautics. Retrieved from http://arc.aiaa.org/doi/pdfplus/10.2514/6.2005-123.

Dhar, S. B. (1979). Power system long-range decision analysis under fuzzy environment. *IEEE Transactions on Power Apparatus and Systems, PAS-98*(2), 585–596.

Djavanshir, G. R., Khorramshahgol, R., & Novitzki, J. (2009). Critical characteristics of metasystems: Toward defining metasystems' governance mechanism. *IT Professional, 11*(3), 46–49.

Gheorghe, A. V., & Masera, M. (2014). Infranomics: A discipline-of-disciplines for the XXIst century. In A. V. Gheorghe, M. Masera, & P. F. Katina (Eds.), *Infranomics* (pp. 1–7). New York, NY: Springer.

Gheorghe, A. V., & Vamanu, D. V. (2004). Towards QVA—Quantitative vulnerability assessment: A generic practical model. *Journal of Risk Research, 7*(6), 613–628. doi:10.1080/1366987042000192219.

Gheorghe, A. V., & Vamanu, D. V. (2008). Mining intelligence data in the benefit of critical infrastructures security: Vulnerability modelling, simulation and assessment, system of systems engineering. *International Journal of System of Systems Engineering, 1*(1), 189–221.

Gilmore, R. (1981). *Catastrophe theory for scientists and engineers*. New York: Dover Publications.

Huang, K. (1963). *Statistical mechanics*. New York, NY: Wiley.

Katina, P. F. (2015). *Systems theory-based construct for identifying metasystem pathologies for complex system governance* (Ph.D.). Old Dominion University, United States—Virginia.

Katina, P. F., Despotou, G., Calida, B. Y., Kholodkov, T., & Keating, C. B. (2014). Sustainability of systems of systems. *International Journal of System of Systems Engineering*, *5*(2), 93–113. https://doi.org/10.1504/IJSSE.2014.064833

Katina, P. F., & Keating, C. B. (2015). Critical infrastructures: A perspective from systems of systems. *International Journal of Critical Infrastructures, 11*(4), 316–344. doi:10.1504/IJCIS.2015.073840.

Katina, P. F., & Unal, R. (2015). Application of fuzzy sets in decision analysis for prioritising critical energy infrastructures. *International Journal of Decision Sciences, Risk and Management, 6*(1), 1–15.

Keating, C. B., Rogers, R., Unal, R., Dryer, D., Sousa-Poza, A. A., Safford, R., …, & Rabadi, G. (2003). System of systems engineering. *Engineering Management Journal, 15*(3), 35–44.

Maier, M. W. (1996). Architecting principles for systems-of-systems. In *6th Annual INCOSE Symposium* (pp. 567–574). Boston, MA: INCOSE.

Metropolis, N., Rosenbluth, A. W., Rosenbluth, M. N., Teller, A. H., & Teller, E. (1953). Equation of state calculations by fast computing machines. *The Journal of Chemical Physics, 21*(6), 1087–1092. doi:10.1063/1.1699114.

Pederson, P., Dudenhoeffer, D., Hartley, S., & Permann, M. (2006). *Critical infrastructure interdependency modeling: A survey of US and international research* (No. INL/EXT-06-11464) (p. 125). Idaho Falls, ID: Idaho National Laboratory. Retrieved from http://www.inl.gov/technicalpublications/Documents/3489532.pdf.

Schneider, V., & Bauer, J. M. (2007). Governance: Prospects of complexity theory in revisiting systems theory. In *Annual Meeting of the Midwest Political Science Association* (pp. 1–36). Chicago, IL. Retrieved from https://www.msu.edu/~bauerj/complexity/schneider.pdf.

Sousa-Poza, A. A., Kovacic, S., & Keating, C. B. (2008). System of systems engineering: An emerging multidiscipline. *International Journal of System of Systems Engineering, 1*(1/2), 1–17.

Sprott, J. C. (1993). *Strange attractors: Creating patterns in chaos*. New York: M&T Books.

USAF SAB. (2005). *System of systems engineering for Air Force capability development: Executive summary* (No. SAB-TR-05-04). Washington, DC: US Air Force Scientific Advisory Board. Retrieved from http://www.dtic.mil/get-tr-doc/pdf?AD=ADA442612.

Vamanu, D. V., Gheorghe, A. V., & Vamanu, B. I. (2003). On a generic model in quantitative vulnerability assessment. *Romanian Journal of Physics: Supplement, 48,* 229–237.

Zadeh, L. A. (1965). Fuzzy sets. *Information and Control, 8*(1), 338–353.

Zeeman, E. C. (1977). *Catastrophe theory: Selected papers*. London: Addison-Wesley.

Chapter 5
Use of Cellular Automata in Assessment of Risk and Vulnerability

Abstract The purpose of this chapter is to suggest the use of cellular automaton as a basis for risk and vulnerability assessment (RVA). A cellular automaton (CA) is a collection of 'colored' cells on a grid of specified shape that evolves through a number of discrete time steps according to a set of rules based on the states of neighboring cells. The rules are then applied iteratively for as many time steps as desired. von Neumann was one of the first people to consider such a model, and incorporated a cellular model into his 'universal constructor.' In present research, a model elaborating on CA application in assessing risk and vulnerability with an emphasis on forest, fire, and smoke is provided. Specifically, this research posits a model treating dispersion cloud as collection of 'nanomachines' (model's 'air blobs', or 'particles'), all advected with the wind, and each moving in the advected mass of air according to a simple rule expressing one dominant component of the Navier–Stokes equation for momentum conservation.

5.1 Introduction to CA for RVA

Cellular automata (CA) were studied in the early 1950s as a possible model for biological systems (Wolfram 2002, p. 48). Comprehensive studies of cellular automata have been performed by Stephen Wolfram starting in the 1980s, and Wolfram's fundamental research in the field culminated in the publication of his book *A New Kind of Science* (Wolfram 2002) in which Wolfram presents a gigantic collection of results concerning automata, among which are a number of ground-breaking new discoveries.

Cellular automata[1], sometimes referred to as cellular spaces, tessellation automata, homogeneous structures, cellular structures, tessellation structures, and iterative arrays (Wolfram 1983), consists of a regular grid of cells, each in one of a finite number of states, such as on and off—as opposed to a coupled map lattice. In present research, a model treating *dispersion cloud as collection of 'nanomachines'*

---

[1]The topic of cellular automaton and more toward that application in assessing vulnerability is visited in Chap. 15.

A.V. Gheorghe et al., *Critical Infrastructures, Key Resources, Key Assets*, Topics in Safety, Risk, Reliability and Quality 34,
https://doi.org/10.1007/978-3-319-69224-1_5

(model's '*air blobs*', or '*particles*') is presented. In this model, the dispersion is supported by wind moving in the advected mass of air conferring to a simple rule expression of the Navier–Stokes equation for momentum conservation:

$$\text{Always go where the density of your simile is lower} : (a = -\text{grad}(p)) \tag{5.1}$$

In the above expression, an 'air blob' must avoid two things: (1) other air blob agglomerations, and (2) obstacles in the terrain. In previous work (e.g., CARVA-3D), these tasks are accomplished by an operation that *mimics a neighborhood inspection*: at each time step one or several points in the neighborhood of every blob are selected at random, and tested for the presence of other blobs, or obstacles. Ideally, the testing should be conducted in a 3-D digitized 'world'—a purely numerical, pixilated volume held in a binary file.

The described process can be compute-intensive on a personal computer (PC). To gain computational speed on a PC, the code version employs a hybrid solution, consisting of testing the 2-D display projection of the 3-D world using fast graphics functions (e.g., 'Point', API 'GetPixel') and then filtering out those candidate positions that would 'fall within the volume' of an object, by numerical heights (i.e., the 3rd dimension) comparison. This would of course introduce a certain, spurious 'cylindrical symmetry' in the solution, that can, however, be compensated —within limits—by model parameters calibration in respect with standard dispersion models and coefficients such as the Karlsruhe-Julich approach. Suffice to say that better rules and implementations should be contemplated.

It should be apparent, from the fore mentioned mechanics, that from the code's standpoint, *a 'terrain' is a collection of elevations at computer display's resolution* (i.e., one elevation by pixel). The two obvious manners of obtaining such collections from image maps are: (a) marking, interactively, the polygonal contour of each and every significant object (buildings, etc.); input a height featuring the targeted object; and devise a way for the code to place on record the given height, for all the pixels that make the interior of the polygonal contour, in an ordered 2-D matrix of points and (b) processing the image map so that objects of the same height would uniformly bear the same color coding that height; then input the list of heights—and the respective colors; and finally, instruct the code to automatically scan the view field, pick up heights according to colors, and place the heights in the 2-D matrix of points referred to in the manner above [i.e., (a)].

The following remarks are made regarding manners for obtaining collections form image maps: (i) both procedures eventually lead to the same result—an implicitly 3-D digital representation, or 'world', of the terrain—and thereby would equally make the model functional, (ii) procedure (a) is intensely interactive. However, it holds a decisive advantage and in that implicitly offers, along with the 3-D 'world' file, a map of *object shapes*, in a format that can be rendered, with MS Windows' *Direct 3D* technology, as fast-animated 3-D terrains—a built-in facility that the code prods itself, and (iii) procedure (b), though requiring some initial interactivity, unfolds automatically once the color codes are input—and so may appear more convenient in comparison. However, it appears that at this stage, it cannot provide an object shapes map, so that only static, rough 3-D representations are available with this method.

In summary, CARVA-3D handles the following types of 'maps': **RAW MAPS** (*.bmp): These are inherently 2-D bitmaps (.bmp files) featuring terrain shapes (buildings, etc.) of visible contours and known, or inferable, elevations. The raw maps are the primeval stuff in building up the code's terrain library, **and DIGITAL MAPS** (*.ter): These are indirectly 3-D binary files consisting in 2-D matrices holding object heights on a pixel-by-pixel basis.

The digital maps are the input stuff for the CA-wise dispersion model in CARVA-3D.

SHAPE MAPS (*.sps): These are, in essence, numerical files in string expression holding vertex specifications for *Direct 3D* objects, from which a smooth, and animated, visual 'world' can be obtained. Let it be emphasized that the *shape maps* not only serve user's orientation in the terrain: *They are also meant to carry pictures of the dispersion clouds interspersed with the buildings*—a single map being able to link *any* cloud configuration on the record provided the cloud has been generated on that particular terrain.

However, due to limitations in handling too many objects under *Direct 3D* at a reasonable speed, what the animated output worlds offers, are *snapshots* of the cloud at the time the cloud was captured (by user command), and not the entire time integral of the cloud dispersion—which on the other hand is always obtained by the code as a standard output, and characterizes the environmental contamination in 'dose'- relating terms, as well as by volumes/areas of various degree of exposure intensity.

For a given terrain, the variety of maps above share part of the file names, for easy recognition. An ancillary file with the extension**.TED** is also in the package, holding scaling, and other data. Such *permanent* terrain files are stored in subfolder **MAPS3D**. 'Raw maps' are held in subfolder **MAPSTORE**. Users are advised to place their primary maps in that folder. Files mixing terrain info with *output dispersion cloud data* make 'worlds' in themselves. Such files are held in subfolder **CASE3D**, along with name-sharing ancillary files supporting their use. The code offers an interface to terrain creation with user's participation. While not yet optimized and perhaps tedious at first attempts, it is functional and instrumental in generating the working examples in the installation package. In this developmental stage, any contribution to bug-hunting in this code department is always an ongoing project.

The code is currently being experimented to assess risk and vulnerability of facilities supporting hazardous substances with distribution in complex terrain with emphasis on residential areas with buildings of various heights and spatial distribution. The computational results offer insights into computability of risks and vulnerability to the public living near such facilities.

5.2 Forest Fire Essentials: A Cellular Automaton-Wise, Percolation-Oriented Model

5.2.1 The Forest Model

In the sense of the model and confirming to CA, a *forest* is a two-dimensional space consisting of atomic (indivisible) squared *cells*—the computer display pixels.[2] Every cell in the forest (pixel in display within the forest's boundaries) denotes a *tree pack.* A tree pack is assumed to consist of any number of trees that would behave in a solitary manner during a fire (Allgöwer and Schöning 2006; Hargrove et al., 2000): (i) would include similar trees (either coniferous, or deciduous), (ii) would catch fire simultaneously (share the same *ignition probability*), and (iii) would get extinguished simultaneously after a, species-dependent, *fire duration*.

A tree pack may consist of a single tree. Trees may belong to two 'species'—in fact, categories: *coniferous*; or *deciduous*. The affiliation affects the *ignition probability,* the *fire duration*, and thereby the extinction time for each tree pack. A forest features:

- a certain *geometry*, involving an extension and shape, and implicitly, boundaries identifiable on a work map
- an underlying *terrain*, expressed as elevations derived from appropriate GIS resources
- a certain *density*, in terms of number of tree packs per hectare, which, in conjunction with the knowledge of the area, coming from forest's geometry (see above), would give the total, initial number of tree packs in the forest
- an average *mix* (i.e., a proportion of coniferous, and deciduous, tree packs in total; specifically, the mix) is characterized by the fraction of coniferous tree packs in total

Consistent with the above, a specifically designed algorithm would: (i) allow an interactive marking of forest boundaries, (ii) clear the interior of the forest space thus defined, and (iii) randomly populate the space with tree packs distributed such that both the *forest density* and the *forest mix* are observed, within tolerable error margins. In addition, the code also makes an allowance on *assimilating buildings and artifacts nearing the forest boundaries, with either 'coniferous', or 'deciduous' tree packs*, depending on their presumed fireproofing, so that the fire be enabled to extend itself to such structures outside the forest confines, through the *firebrand mechanism* (projection of chunks of ignited material at some distance downwind from the fire confines, and possibly from the forest boundaries themselves). A summary of the forest fire model is presented in Fig. 5.1 based on VBS Desktop Assistant of Switzerland's Civil Defense Office (2006).

[2]A 1024 × 768 display resolution served as a working reference for the DSS code's current version.

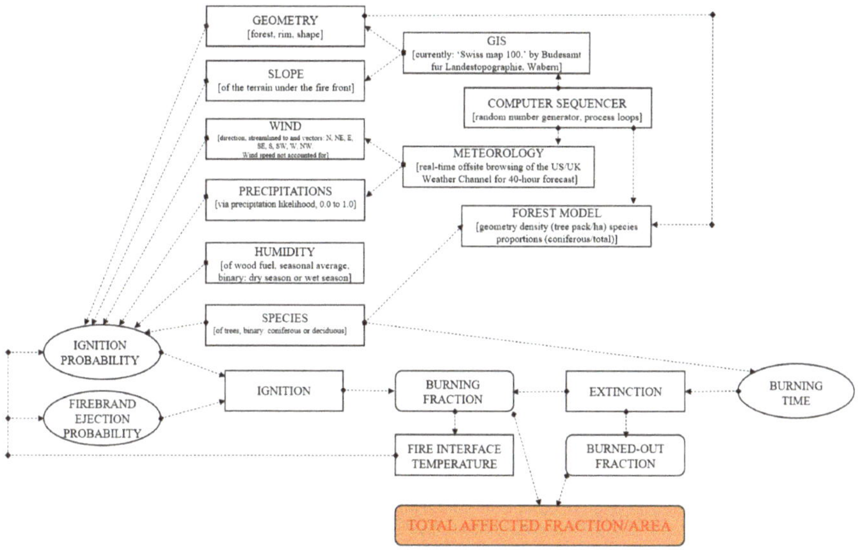

Fig. 5.1 Forest model, modified from Gheorghe and Vamanu (2008)

5.2.2 *The CA Fire Model*

The fire model, depicted in Fig. 5.1, implements the following factual description. A tree or several tree packs could catch fire. This might happen via a natural event (e.g., a thunder strike), and accident (e.g., a road/rail tanker capsizing, followed by loss of containment and spark-initiated mass fire, a plane crash), or by malicious intent. In the computer program, a mouse click would initiate a single-source fire, while a click-and-drag maneuver would result in an instantaneous fire trail, or arbitrary polygon.

The subsequent fire unfolding will be the result of a competition between: (i) a *probabilistic process,* consisting of other pack tree ignition acts, (ii) a *deterministic process,* consisting of burning tree packs getting naturally extinguished, and (iii) by fuel exhaustion, after appointed burning times associated with tree species categories (coniferous, deciduous). One might consider two types of ignition mechanisms: (a) a close-range ignition by, essentially, thermal radiation of flames, covering a limited neighborhood of any ignited tree pack; the squared cell pattern of the forest space would induce the choice for a Moore (8-neighbor enveloping, in the 1st order) neighborhood, and (b) a long-range ignition by firebrands

The likelihood of both processes increases with the *fire interface temperature*—a notion of critical importance for the overall self-organization of the fire process. The fire interface temperature T(K) is the absolute temperature of the layer of air separating the radiating (essentially, visible) flames from their environment.

There is an essential assumption that should not be ignored: *T level is determined not only by the tree pack it refers to, but also by the concurring effect of the entire fraction of burning tree packs in the forest, at any time*. Since the details of such a collective effect are debatable, a simple time-function is chosen to model it, featuring the ambient temperature at the moment of forest fire ignition, and asymptotically saturating itself at the combustion temperature level. In turn, the *combustion temperature* is assumed to be governed only by the normal stoichiometry of carbon oxidation, thereby being taken as independent on the tree species.

A loop is thus created (See Fig. 5.2) at the code's runtime, with the ring consisting of the *ignition probability* (close- and long-range processes considered), the forest *burning fraction*, and the *fire interface temperature* feed backing the first, and tempered down by the competing process of *tree pack extinguishing* themselves after, invariably according to the model, preset burning times of coniferous/deciduous trees.

Beyond these terms, fire evolvement is further conditioned by environmental factors, which include meteorology of the event and terrain. The *meteorology* of the event is assumed to act *globally* over fire area, through *wind* and *precipitations*. The *terrain* is assumed to act *locally*, at tree pack ignition level, through *elevation differences* (slope). This suggests that, essentially, close-range ignition probability involves three aspects:

- close-range ignition probability would be higher for tree packs falling downwind from neighbors already in fire.
- close-range ignition probability would be lower for higher precipitations likelihood.
- close-range ignition probability would be higher for tree packs uphill from neighbors already in fire.

As to the probability of firebrand-induced (long-range) ignition, this is even more roughly modeled as a function of only the fire interface temperature, assuming a maximally possible level (0 through 1) of the event. The firebrand range is, in turn, taken as a fraction exactly equal to the firebrand likelihood, from a maximally possible, assumed, range (meters), thus streamlining model parameterization. However, upon further review;

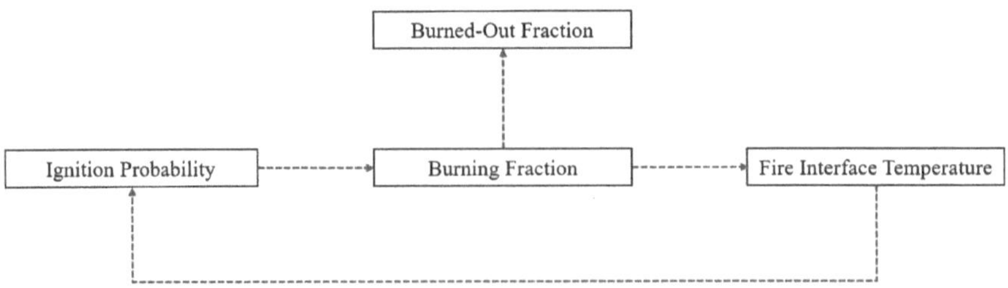

Fig. 5.2 A loop for a fire model

- the wind speed was *not* brought into the model, owing to the intricate and not entirely clear relationship between speed levels and fire intensity, and fire front velocity, essentially governed by air flows dynamics in non-homogeneous, randomly structured, media
- the precipitations were chosen to directly affect the analytic expression of the close-range ignition probability, instead of effecting upon the fire interface temperature—which would have inordinately complicated the functional equation of the latter.

As a general remark covering the modeling style, alternative and more analytical approaches to the various subprocesses described (e.g., the effect of winds, precipitations, and the fire branding) would have involved a series of other, more detailed submodels worth debating in themselves, which would have gone off-limit in respect with the stated scope of the present research and application. A compact briefing of the model parameterization (Table 5.1) based on the *Switzerland Civil Defense Office VBS Desktop Assistant* (Civil Defense Office, 2006) is boxed below.

As with all CA-wise, percolation-oriented models, the *forest density* is a key factor in maintaining, amplifying, and curtailing a fire. Additional feature of *forest geometry* including extension and shape bring into the picture a new factor which is normally ignored in academic exercises. In present research, this has proved to be of dramatic consequences and echoes a known effect in Physics: *decisive influence of boundaries ('finite dimensionality') on cooperative phenomena in many-body systems*. Following from the orientation of the application toward assisting civil defense training, the emphasis on boundary effects would perhaps obscure one key preoccupation of the textbook percolation models—namely the identification of a critical percolation probability threshold, the meaning of which fades away in size and shape-constrained systems.

5.2.3 The Smoke Territory

An indicative visualization of the smoke-covered territory is also provided. This is limited by the cellular automaton-wise nature of the fire model which, in the current version, can only account for wind directions in 8 sectors: N, NE, E, SE, S, SW, W, NW. The reckoning applied considers a Gaussian dispersion (Gheorghe 2005) of the plume governed by: (i) an advection with the 10 m-height above ground wind speed, and (ii) a diffusion driven by time-dependent standard deviations, horizontal and vertical adequate for larger distances from source than the distance-dependent standard deviation (e.g., Karlsruhe-Julich).

Thus, the smoke concentration, $C(r, z, t)$ at a distance r (in meters) from a smoke puff center—the wind-driven image taken at the plume elevation height, H (meters) of the pixel-wise tree pack burning—at a height z (meters) above a flat ground

Table 5.1 Briefing of fire model parameterization

```
MODEL SETTINGS
______________

In the sequel, a 'tree pack' may consist of a single, or of several trees
showing the same ignition likelihood and further behavior.

Feel free to change defaults, if/as appropriate.
'Proceed' from menu.
_______________________________________________

THE FOREST
__________

FOREST DENSITY, FDens (tree packs/ha).
Notice: default based on ca. 50 m2 per tree pack.

FRACTION OF CONIFEROUS TREE PACK IN TOTAL, CFrac.
Notice: The remainder assumed to consist of deciduous trees.

AVERAGE BURNING TIME, CONIFEROUS TREES, tauC (min).
AVERAGE BURNING TIME, DECIDUOUS TREES, tauD (min).

MAXIMAL EXPECTED COMBUSTION TEMPERATURE, Tc (K)
Notice: assumed independent of species.

INITIAL FIRE INTERFACE TEMPERATURE, Te (K.)
Notice: Defined as an average, expected ambient air temperature,
 at the closest (1st order) neighbor-tree-pack distance
 from any tree pack.

SENSITIVITY CONSTANT, of Fire Interface Temperature, alpha.
Notice: A temperature increment is assumed to be induces by mass fire,
 so that the effective temperature at the fire interface be given by:

 T = Te + Tc.{1-exp[(-1).alpha.F]},

 with F - the fraction in total tree packs, of ignited tree packs,
 alpha - a process sensitivity constant,
 chosen to give an increment to near-combustion temperature
 for F=99%.

MAXIMAL FIREBRAND PROBABILITY, Pfbmax.
Notice: the probability of firebrand occurrence, Pfb is assumed to depend on the
 Fire Interface Temperature, T, as follows:

 Pfb = [(T - Te)/Tc].exp(-f/T),

 with f = (Te + Tc).ln(1/Pfbmax)

 ensuring that Pfb = Pfbmax - the maximal, assumed, firebrand probability
 be reached at T = Te + Tc - the maximal, possible, temperature according
 to the temperature equation above, and Pfb be equal to 0 at the
 initial temperature, T = Te.

MAXIMAL FIREBRAND RANGE, Rfbmax (m).
Notice: the firebrand range is assumed to depend on the Fire Interface Temperature, T, as
follows:

 Rfb = Rfbmax.Pfb

 thereby increasing with the Fire Interface Temperature via Pfb (see above).

THE ENVIRONMENT
_______________

SEASON (W(arm) or C(old), season$ (characters string).
Notice: concept set for the Northern Hemisphere, i.e. Warm - Dry, Cold - Wet.

WIND DIRECTION (C for calm), from (N,NE,E,SE,S,SW,W,NW), wDir$.
Notice: For 'real' (geographically identified) forests - inferred from 40-hour meteo
 forecasts obtained from offline browsing the U.S./U.K. Weather Channel
 for the respective location. Assumed uniform in space over the investigated
 forest area, yet varying in time as the forecast tells.
 Remark: since wind speed, and fire intensity and fire front movement
 are correlated in intricate manners, this model version does not account
 for wind speed.
```

```
PRECIPITATIONS (0-1), pre.
Notice:  refers  to  the  'chances  of  precipitations'  entry  of  publicly  available  meteo
forecasts, v.e.g. the U.S./U.K. Weather Channel employed by this code.

TERRAIN SLOPE CONSIDERED (1-yes, 0-no): +sFct)
Notice: model favoring upslope ignition.

THE FIRE
________

The fire is governed by an IGNITION PROBABILITY, P,
indicating the likelihood that a tree pack falling in an n-order Moore neighborhood
of an already ignited tree pack would catch fire.

P depends

 - on the choice of a REFERENCE CONSTANT, a, having, necessarily,
 temperature dimensions (K);
 - on the TREE SPECIES CATEGORY, i.e. 'coniferous', or 'deciduous'; b1
 - on the SEASON, entailing an average humidity of fuel, i.e., 'warm' (dry) b2
 or 'cold' (wet);
 - on the PRECIPITATIONS likelihood, working as an adjustment to the b2p
 season parameter b2, i.e. b2 becoming b2 + b2p;
 - on the WIND, favoring tree packs downwind from the ignited pack; b3
 - on the SLOPE, favoring tree packs uphill from the ignited pack; b4
 - on the FIRE INTERFACE TEMPERATURE, seen as a dynamic forest T (K)
 feature determined by (i) the original (pre-fire) ambient temperature
 and (ii) the temperature increment induced by the fire in proportion
 with the fraction of ignited tree packs;

The model equation is:

P = exp{-a.[ln(1/b1) + (ln(1/b2)+b2p) + ln(1/b3) + ln(1/b4)]/T]}

Model Control Parameters:
Recommended defaults may be changed.

THE MAXIMALLY ATTAINABLE MOORE NEIGHBORHOOD ORDER, nMax.
Notice: the neighborhood order, n, is assumed to depend on the
 Fire Interface Temperature, T,as follows:

 n = 1 + (T - Te).(nMax - 1)/Tc,

 ensuring that n = 1 at the initial Fire Interface Temperature, T = Te
 and n = nMax at the maximally attainable temperature, T = Te + Tc
 (see notations above).
```

(terrain's actual topography ignored), and at a time t (in seconds) into the fire is given the following formula:

$$C(r,z,t) = [1/((2\pi)^{3/2}\sigma_h^2(t)\,\sigma_z(t))]\exp(r^2/(2\sigma_h^2(t)))\exp((z-H)^2/(2\sigma_z^2(t))),$$

with $\sigma_h(t)$, and $\sigma_z(t)$ the horizontal and vertical standard deviations, respectively, computed as

$$\sigma_h(t) = (A_h t)^{K_h}$$
$$\sigma_z(t) = (A_z t)^{K_z}$$

and A, K—textbook-tabulated constants on 5-time intervals.

Upon these, if C_0 is assumed to be the smoke plume front concentration at any given time t, then the radius of a puff-surrounding area that is bordered at the same concentration can be obtained as being:

$$R(t) = \sigma_h(t).\left\{2\ln\left[1/[1/((2\pi)^{3/2}\sigma_h^2(t)\sigma_z(t)C_0)]\right]\right\}^{1/2} \tag{5.2}$$

Upon realization of a fraction of this radius as a '*radius of smoke visibility,*' the code will map the wind-driven projections of all puff centers, depending on the wind speed and direction, as an overall image of the smoke plume, in the terminal phase of the fire.

5.3 An Illustrative Application: Engadin, Zernez, Switzerland

An application of the model was implemented on Zernez, Switzerland. Zernez is a municipality in the district of Inn in the Swiss canton of Graubünden. Since performing present research, Zernez has gone through a transformation as former municipality of Lavin and Susch merged into the municipality of Zernez in January 2015.

This research used Swiss map system and a digital terrain elevation data library originated by the *Bundesamt fur Landestopographie* in Wabern, Switzerland. However, the code of the model still missed several subtleties that would be in order, before qualifying the model as a true fire forecasting and/or response instrument. Nonetheless, code users can consult the 'Help' section of the documentation to adjust orientation issues such as 'road arrows' orientation and placements of qualified institutions and products. The following series of figures (Fig. 5.3 through Fig. 5.15) present telling examples of excerpts from the application. These were developed as part of an *ad hoc* for a decision support system, addressing the ***fire evolution under a wind from Souteasth.*** Present researchers make the following remarks: (i) watch the time counter, lower-right of figures, (ii) red pixels: ignited coniferous packs, (iii) yellow pixels: ignited deciduous packs, (iv) blue pixels:extinguished tree packs, (v) observe effects of forest geometry—temporary fire isolation by roads— and (vi) firebrand ejection depending on fire interface temperature breaking the isolation and leading the fire on, to adjacent forest parcels. (Figures 5.4, 5.5, 5.6, 5.7, 5.8, 5.9, 5.10, 5.11, 5.12, 5.13, 5.14 and 5.15)

5.4 Remarks

Remaining true to the issue of 'governance' in which problem systems are drivers for type of governance, this chapter offers a novel approach for steering our understanding of fore fires using a very familiar approach: Cellular automata. A model of cellular automata for assessing risk and vulnerability is developed with different parameters, uncommon to academic settings. The developed model and the

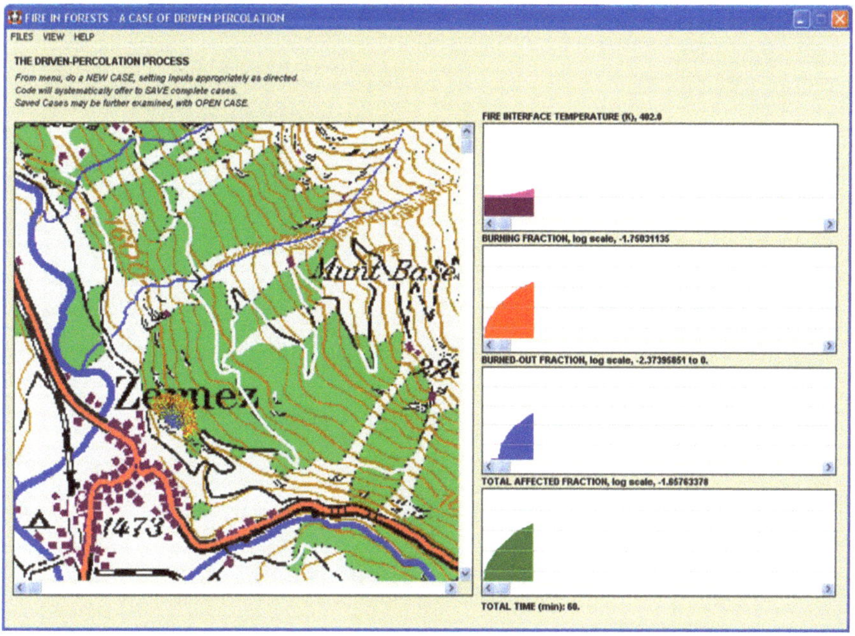

Fig. 5.3 Fire model after 60 min, wind from Southeast, adapted from Gheorghe and Vamanu (2008)

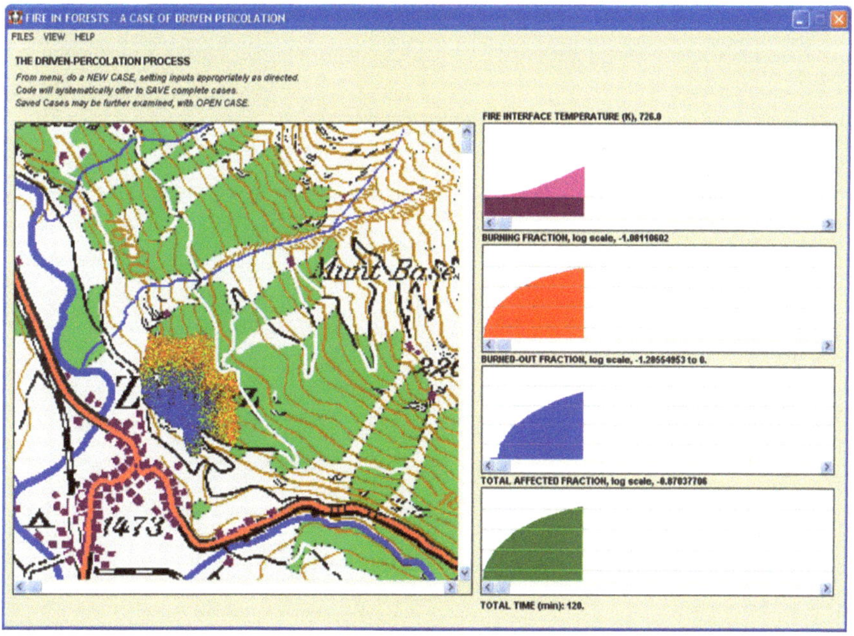

Fig. 5.4 Fire model after 120 min, wind from Southeast, adapted from Gheorghe and Vamanu (2008)

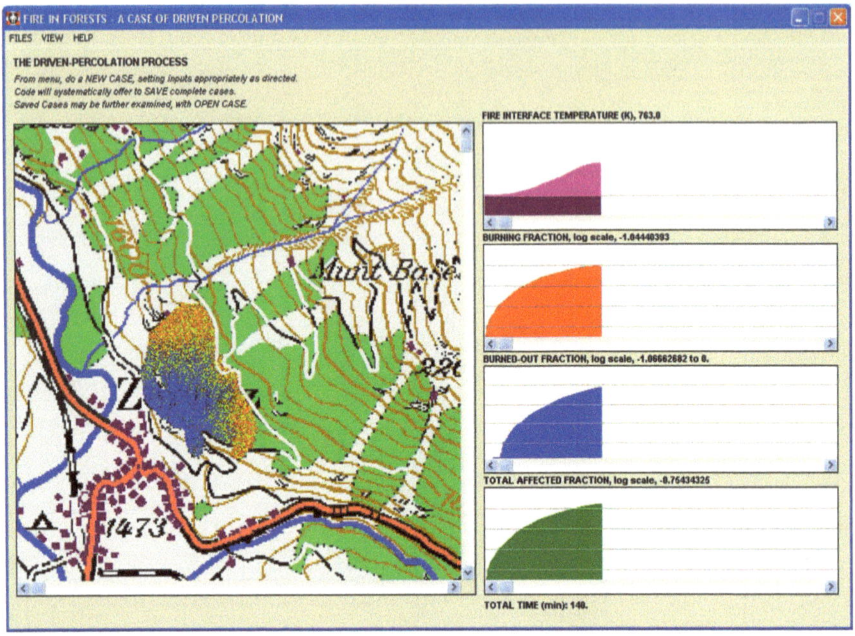

Fig. 5.5 Fire model after 140 min, wind from Southeast, adapted from Gheorghe and Vamanu (2008)

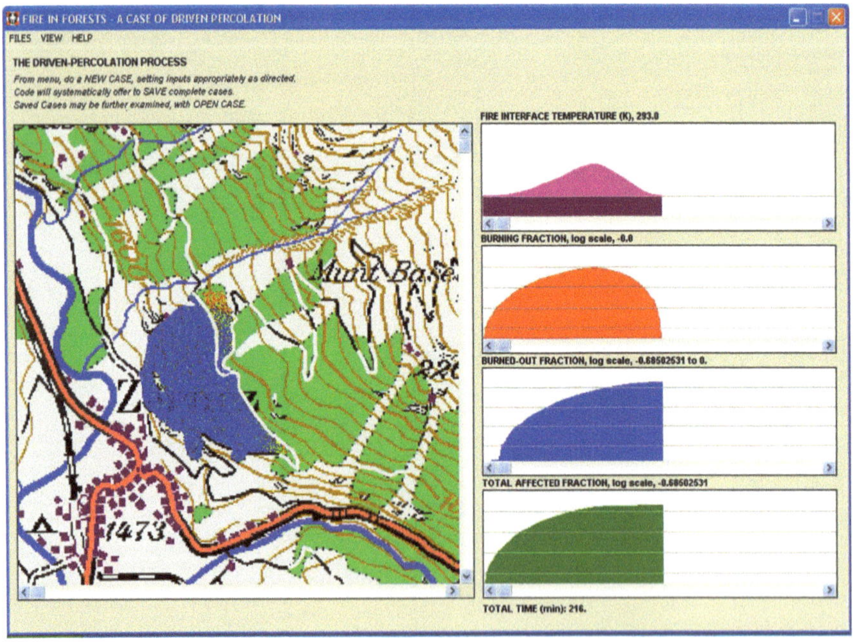

Fig. 5.6 Fire model after 216 min, wind from Southwest, adapted from Gheorghe and Vamanu (2008)

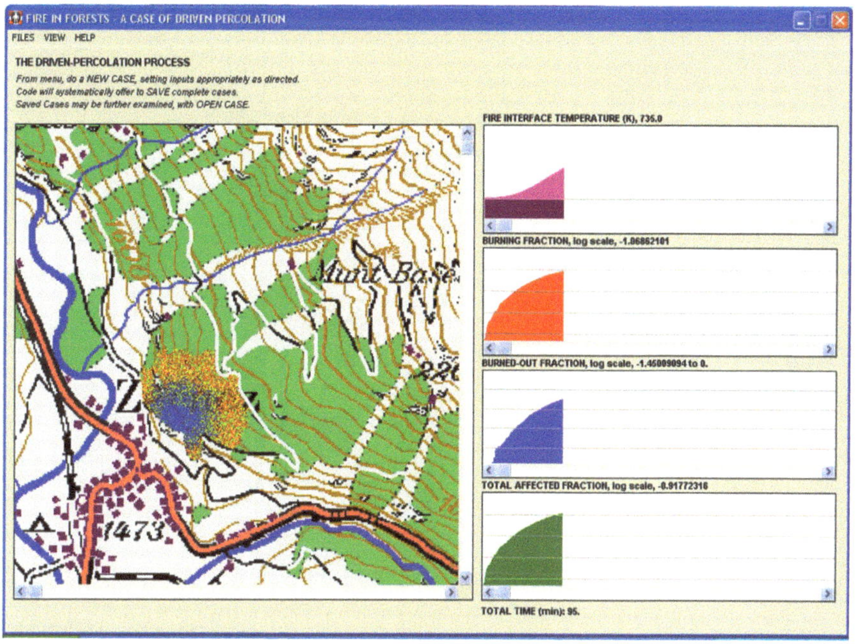

Fig. 5.7 Fire model after 95 min at 735 K, adapted from Gheorghe and Vamanu (2008)

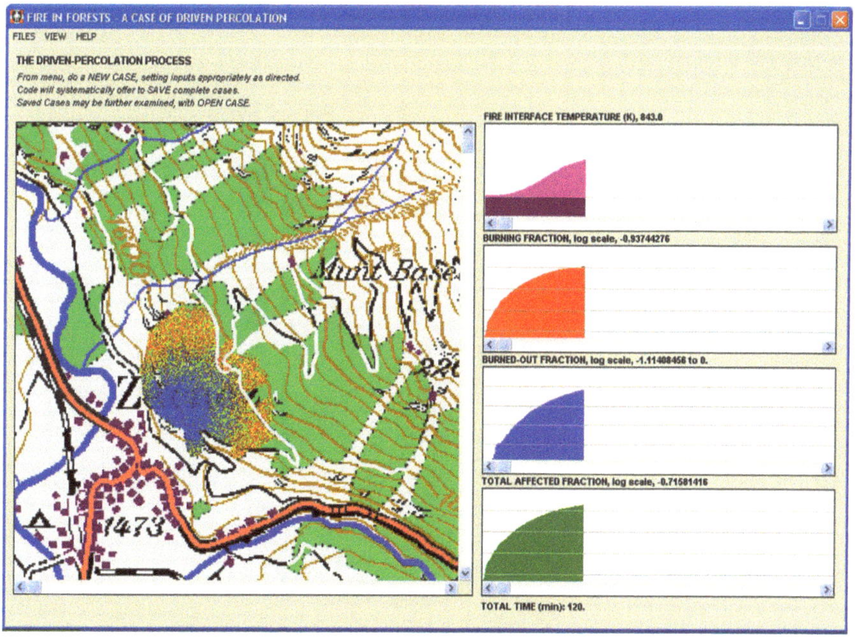

Fig. 5.8 Fire model after 120 min at 843 K, adapted from Gheorghe and Vamanu (2008)

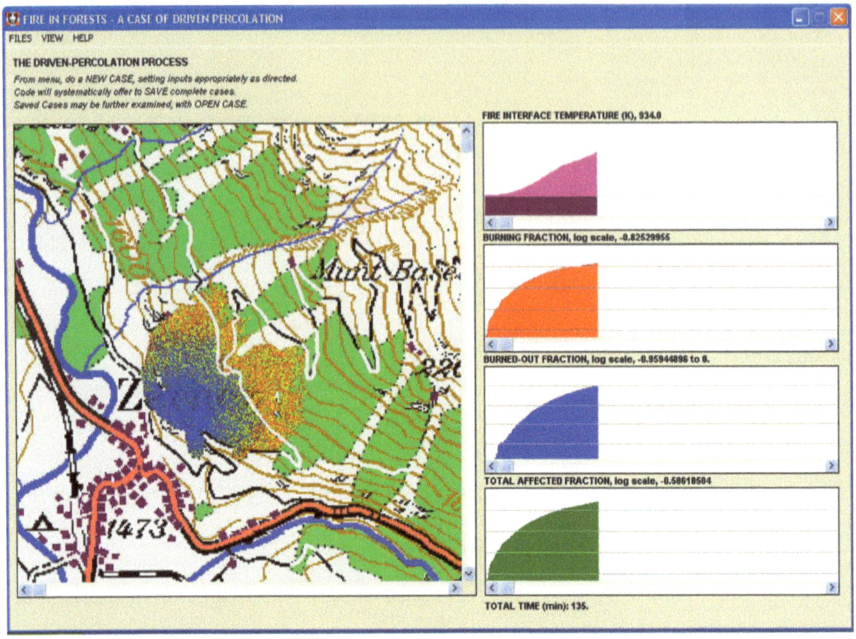

Fig. 5.9 Fire model after 135 min at 934 K, adapted from Gheorghe and Vamanu (2008)

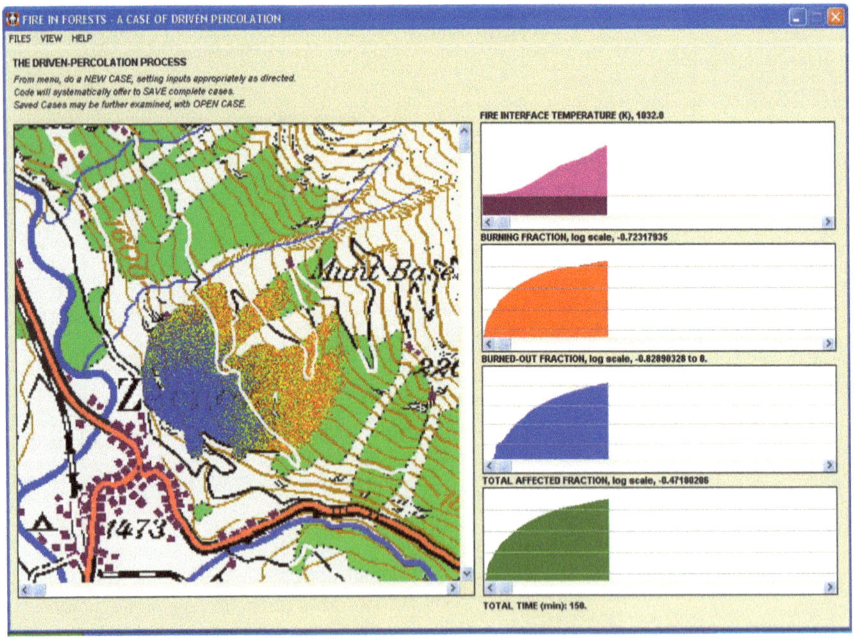

Fig. 5.10 Fire model after 150 min at 1032 K, adapted from Gheorghe and Vamanu (2008)

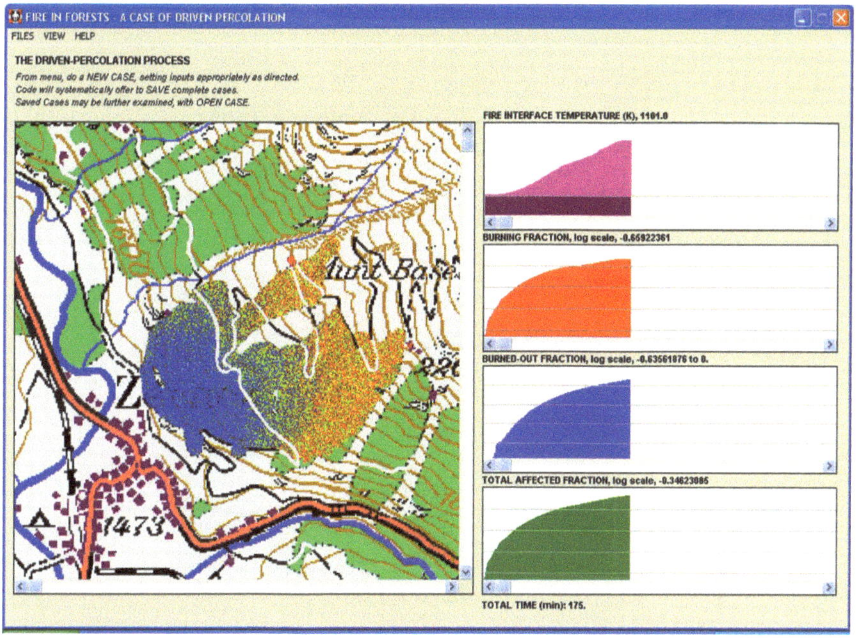

Fig. 5.11 Fire model after 175 min at 1101 K, adapted from Gheorghe and Vamanu (2008)

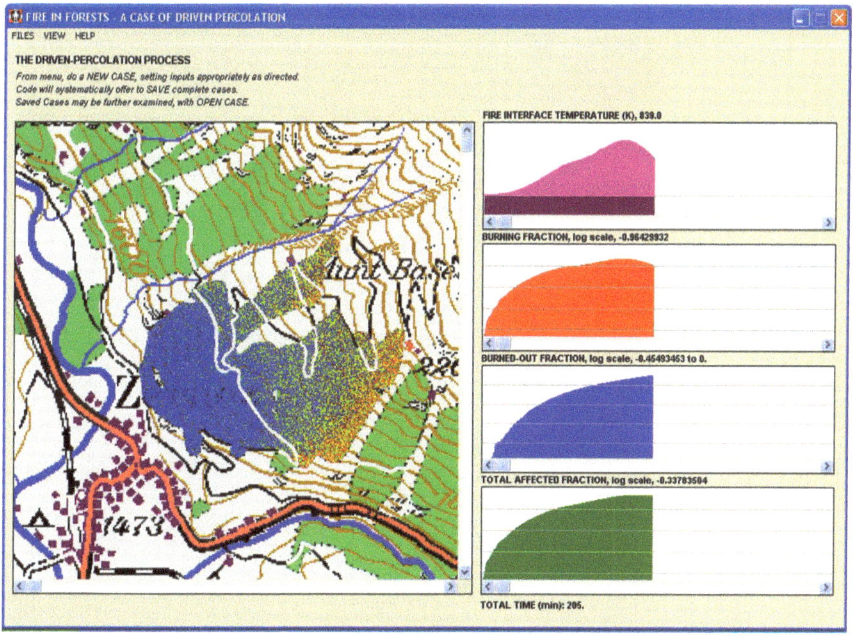

Fig. 5.12 Fire model after 205 min at 839 K, adapted from Gheorghe and Vamanu (2008)

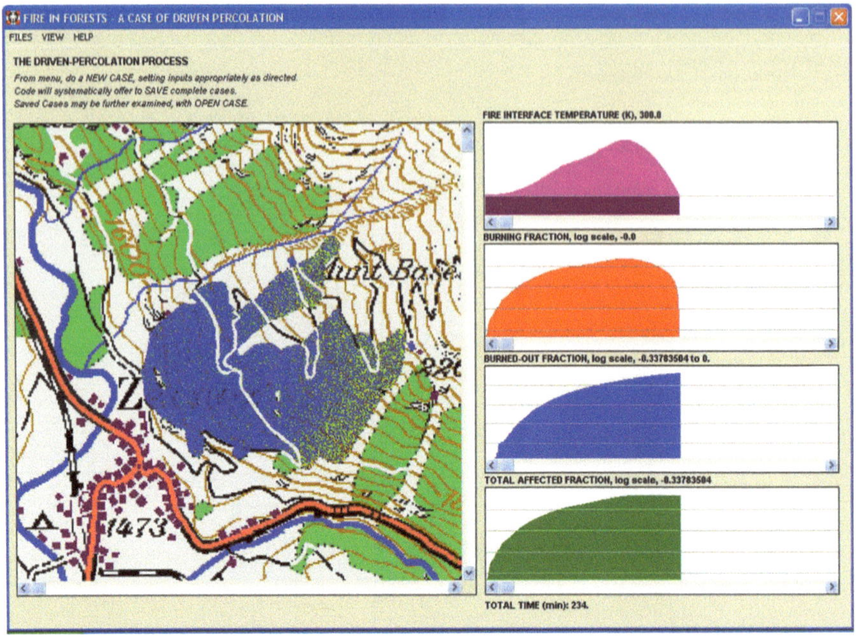

Fig. 5.13 Fire model after 234 min at 300 K, adapted from Gheorghe and Vamanu (2008)

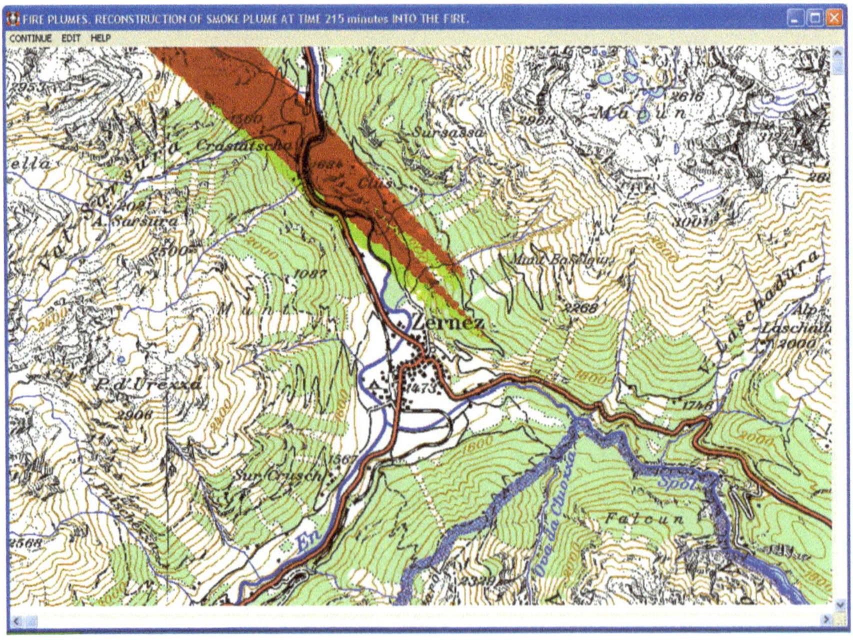

Fig. 5.14 Fire plumes at 215 min into the fire, visibility radii: 0.1, adapted from Gheorghe and Vamanu (2008)

Fig. 5.15 Fire plumes at 215 min into the fire, visibility radii: 1, adapted from Gheorghe and Vamanu (2008)

applications indicate the utility of the research. More importantly, the developed model is generic and can be adopted to different contexts. Moreover, there is a need for the improvement of the presented model by considering a robust listing of factors, beyond environmental, as well as development of industry and regional-specific models. Nonetheless, these limitations should not undermine the utility of the developed models.

A rich area of research CA is the attempt to classify patterns (Ilachinski 2001). These patterns are associated with simple computational models and divided into four classes based on their behavior. In present research, one might be interested in the different models and classes and the implications on engineered systems.

There are cases in which a cellular automaton is *reversible*. The premise for reversibility is that for every current configuration of the cellular automaton, there is exactly one past configuration (pre-image). If this is the case, its time-reversed behavior can also be described as a cellular automaton as suggested Curtis–Hedlund–Lyndon theorem (Gutowitz 1991; Margenstern 2007; Richardson 1972). Again, there is implication for the presented model especially in recreating different scenarios.

References

Allgöwer, B., & Schöning, R. (2006). *Forest fire modeling with GIS in the Swiss National Park.* Santa Fe, CA: National Center for Geographic Information and Analysis. Retrieved from http://www.ncgia.ucsb.edu/conf/SANTA_FE_CDOM/sf_papers/allgower_britta/allgower.html.

Civil Defense Office. (2006). *VBS desktop assistant.* Bern, Switzerland: Department of Defense.

Gheorghe, A. V. (Ed.). (2005). *Integrated risk and vulnerability management assisted by decision support systems: Relevance and impact on governance* (Vol. 8). Dordrecht, The Netherlands: Springer.

Gheorghe, A. V., & Vamanu, D. V. (2008). Forest fire essentials: A cellular automaton-wise, percolation-oriented model. *International Journal of Critical Infrastructures, 4*(4), 430–444.

Gutowitz, H. (Ed.). (1991). *Cellular automata: Theory and experiment* (1st ed.). Cambridge, Mass: The MIT Press.

Hargrove, W. W., Gardner, R. H., Turner, M. G., Romme, W. H., & Despain, D. G. (2000). Simulating fire patterns in heterogeneous landscapes. *Ecological Modelling, 135*(2–3), 243–263. doi:10.1016/S0304-3800(00)00368-9.

Ilachinski, A. (2001). *Cellular automata: A discrete universe* (Reprint ed.). Singapore: World Scientific Publishing Company.

Margenstern, M. (2007). *Cellular automata in hyperbolic spaces: Theory* (Vol. 1). Paris: Gordon Breach.

Richardson, D. (1972). Tessellations with local transformations. *Journal of Computer and System Sciences, 6*(5), 373–388. doi:10.1016/S0022-0000(72)80009-6.

Wolfram, S. (1983). Statistical mechanics of cellular automata. *Reviews of Modern Physics, 55*(3), 601–644.

Wolfram, S. (2002). *A new kind of science.* Champaign, IL: Wolfram Media.

Chapter 6
Nuclear Reactors Vulnerability Assessment—A Generic Model

Abstract It should be evident that present authors have attempted to indicate the various applicability of the general QVA approach. A general consensus is that system's monitored parameters may be aggregated such that the control variables, *U* and *V* of the cuspidal stability model for vulnerability analysis be obtained. In addition, one ought to submit that *U* and *V* are membership functions of a fuzzy theory set of impact indicators. In this chapter, the emphasis is placed on the installation of a nuclear reactor.

6.1 Introduction: QVA in Different Systems

When a QVA concept is tempted in relation to a specific system—such as a coherently functioning industrial installation, the *U* and *V* parameters of the QVA theory should naturally emerge from the physical laws governing the processes that the system hosts. The difference between the 'fast variables' and the 'slow variables' is maintained. However, the actual *U* and *V* parameters of the equation of state may depend on all input variables (which may differ from system to system), whether fast or slow, in intricate fashions.

The most general criterion of the feasibility of a QVA approach to a specific (as opposed to 'general') system may read:

> A system may become a subject of a QVA - in the sense adopted in this work - if its process equations are amenable to an equation of State $Z = Z(u, v)$ having as space of stationary solutions a cuspidal foil featuring stability and instability regions.

Since the process equations may wildly vary from system to system, it becomes virtually impossible to design a universal interface for a QVA code. On the other hand, however, the general framework observed with the 'generic' and 'territorial' vulnerability machines still stand valid. The physical indicator values, Y_i, are again controlled at the interface via their relative correspondents, X_i being obtained from Y_i as:

A.V. Gheorghe et al., *Critical Infrastructures, Key Resources, Key Assets*, Topics in Safety, Risk, Reliability and Quality 34,
https://doi.org/10.1007/978-3-319-69224-1_6

Table 6.1 Reactor QVA parameters

A. INDICATOR MODEL PARAMETERS:		
X_{i1} = 0.2 X_{i2} = 0.6		
B. REFERENCE INDICATOR VALUES:		
INDICATORS	Y_i for	
	X_{i1}	X_{i2}
Fast Variables		
Baseline Reactivity (1/s)	.002	−.002 <
Coolant Flow Rate (kg/s)	10800	108 <
Heat Exchange Coefficient (W/(m2.K))	200	600 <
Slow Variables		
Fuel Temp.Reactivity Coeff.#1 (1/(K.s))	−.00002	.00002 < <
Fuel Temp.Reactivity Coeff.#2 (1/(K.s))	.000002	−.000001 < <
Coolant Temp.Reactivity Coeff.#1 (1/(K.s))	−.00002	00002 < <
Coolant Temp.Reactivity Coeff.#2 (1/(K.s))	.000001	−.000001 < <
Coolant Specific Heat (J/(kg.K))	4182	1000 < <
Heat Exchange Area (m2)	60	20 < <
Fuel Volume (m3)	2	6 < <

$$X_i = AY_i + B; \quad i = 1, 2, \ldots, n$$

The constants A and B are, in turn, derived from the assumed knowledge of pairs of values, Y_{i1}, Y_{i2} and X_{i1}, X_{i2}, for the physical (Y) and normalized (X) indicators, respectively:

$$AY_{i1} + B = X_{i1}; \quad BY_{i2} + B = X_{i2}$$

in which,

$$A = (X_{i2} - X_{i1})/(Y_{i2} - Y_{i1})$$
$$B = (X_{i1}.Y_{i2} - X_{i2}.Y_{i1}))/(Y_{i2} - Y_{i1})$$

A summary of the model with model indicators, indicator values, and variables is presented in Table 6.1.

6.2 Basics of the Model

Notice that in this model, the system is a nuclear reactor with two temperature zones and with delayed neutrons neglected. The present model has the following equations (for an in-depth description of this model with an emphasis on resilience see Gheorghe and Vamanu 2005):

$$dQ/dt = (R_c/L)Q \tag{6.1}$$

$$\left(R_oF * VF * C_pF\right) * dTF/dt = VF * Q - kS * S * (TF - TC) \tag{6.2}$$

$$(R_oC * VC * C_pC) * dTC/dt = kS * S * (TF - TC) - G * C_pC * TC * phi \tag{6.3}$$

where:

Q (W/m^3)	is the core power density;
TF (K)	is the temperature in the zone 1 (fuel + clad);
TC (K)	is the temperature in the zone 2 (primary coolant);
VF (m^3)	is the fuel volume;
VC (m^3)	is the primary coolant volume;
C_pF (J/(kg. K))	is the specific heat of fuel;
C_pC (J/(kg. K))	is the specific heat of primary coolant;
R_oF (kg/m^3)	is the fuel density;
R_oC (kg/m^3))	is the primary coolant density;
kS (W/(m^2. K))	is the global coefficient of heat transfer;
S (m^2)	is the surface of heat transfer between zones 1 and 2;
G (kg/s)	is the coolant massic flow;
R_c (1.0e-5 *dk/k*)	is the total reactivity;
L	is the order of 1.0e-4 s, for thermal neutrons;
phi	is a scaling constant.

With the adiabatic approximation, one has:

$$dTF/dt = 0 \tag{6.4}$$

$$dTC/dt = 0,$$

which makes Eqs. (6.2), (6.3) read:

$$VF * Q - kS * S * (TF - TC) = 0 \tag{6.5}$$

$$kSS(TF - TC) - G * C_pCT * C_{phi} = 0.$$

Solving the system in (6.5), one obtains:

$$TF = AQ \tag{6.6}$$

$$TC = BQ,$$

with

$$A = VF(1/(kSS) + 1/W) \tag{6.7}$$

$$B = VF(1/W)$$

$$W = G * C_p C_{phi}.$$

The total reactivity, Rc, is composed of the reactivity from the control rods, Rc_0, which includes the effect of the fuel burnup, and two temperature-feedback terms relating to fuel and primary coolant, respectively:

$$R_c = R_{c0} + alphaFTF + alphaCTC \tag{6.8}$$

In turn, it is generally admitted that

$$\begin{aligned} alphaC &= a + b * TC \\ alphaF &= c + d * TF \end{aligned} \tag{6.9}$$

Introducing Eq. (6.10) and Eqs. (6.5, 6.6) into Eq. (6.8), and then the resulting Eq. (6.8) into Eq. (6.1), one obtains Eq. (6.11):

$$LdQ/dt = R_{c0}Q + nQ^2 + mQ^3 \tag{6.10}$$

with

$$n = c.A + a.B \tag{6.11}$$

$$m = d.A^2 + b.B^2$$

The change of variable

$$q = Q - epsilon \tag{6.12}$$

transforms Eqs. (6.11, 6.12) into:

$$L.dq/dt = q^3 + U.q + V \tag{6.13}$$

where

$$\begin{aligned} U &= R_{c0}/m - n^3/(3m^2) \\ V &= 2(n/(3m))^3 - n.R_{c0}/(3m^2) \end{aligned} \tag{6.14}$$

and

$$epsilon = -n/(3m). \tag{6.15}$$

The stationary solution is obtained taking in Eq. (6.14)

$$dq/dt = 0,$$

that is:

$$q^3 + U.q + V = 0 \tag{6.16}$$

Equation (6.17) is the 'equation of state' of the system, in the sense of the model, providing a space of states (also known as system's 'characteristic') in the form of a cuspidal foil,

$$q = q(U, V). \tag{6.17}$$

The following series of figures (Figs. 6.1, 6.2, 6.3, and 6.4) present telling examples of excerpts from the QVA application in assessing a reactor.

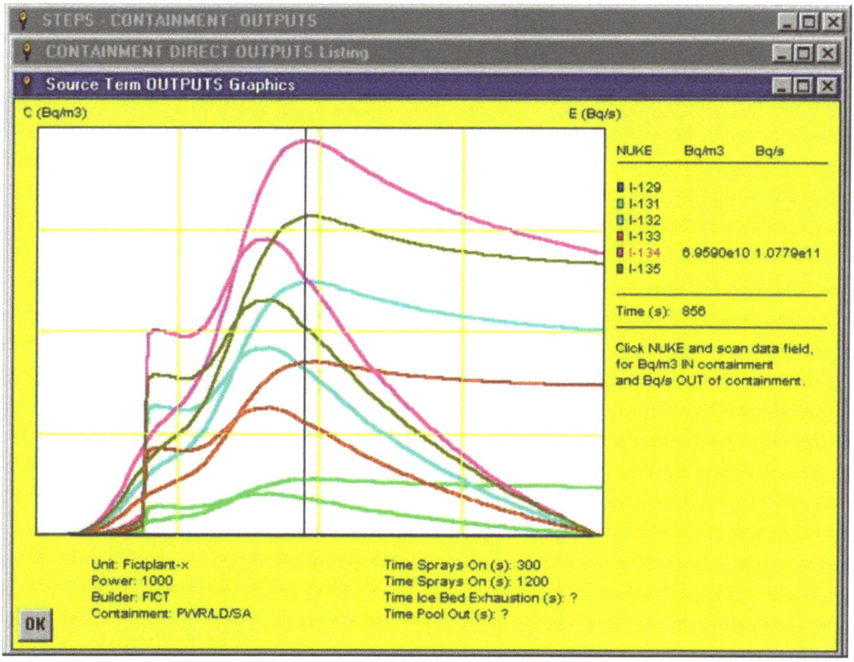

Fig. 6.1 Depiction of different outputs for a reactor

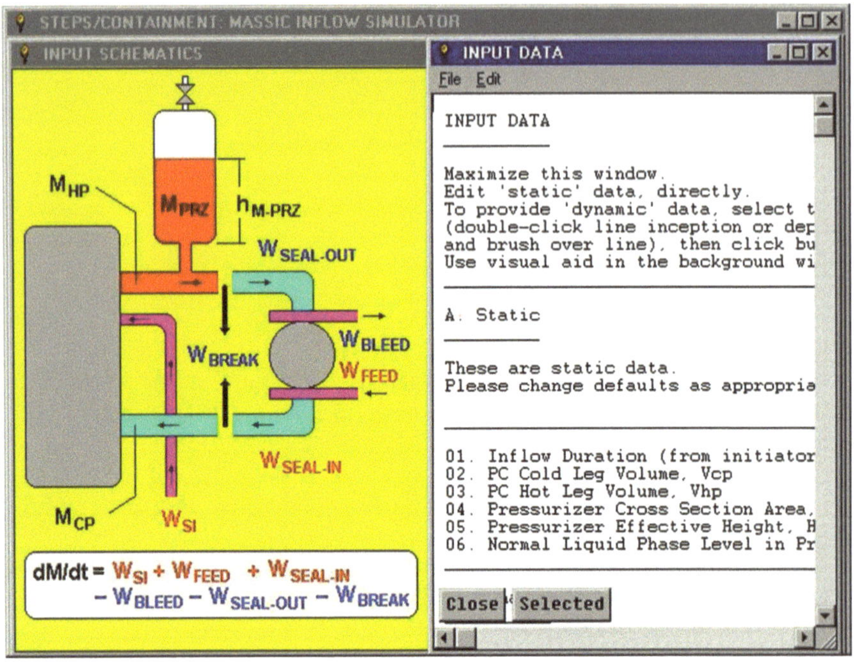

Fig. 6.2 Reactor model with input data, partial list shown

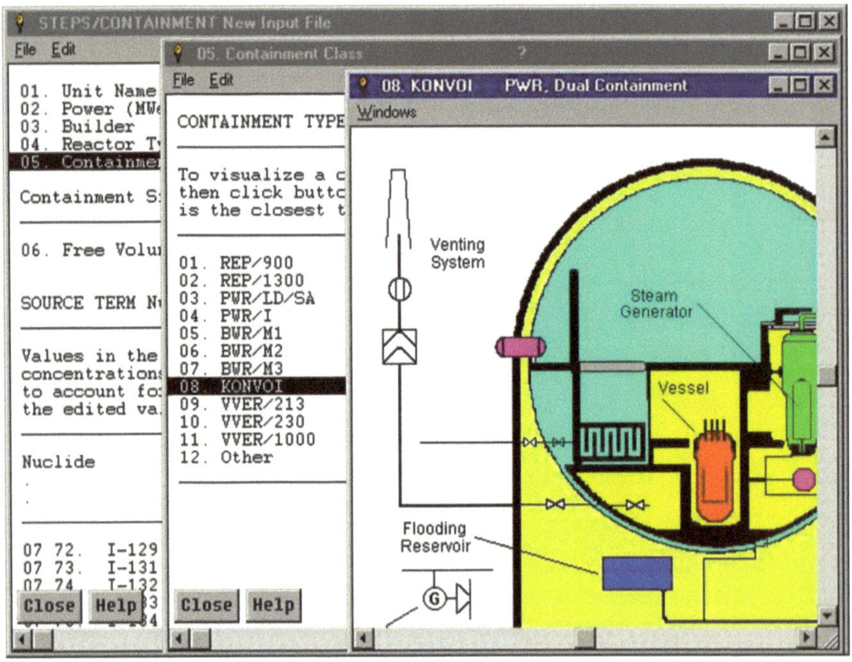

Fig. 6.3 Model for power dual containment

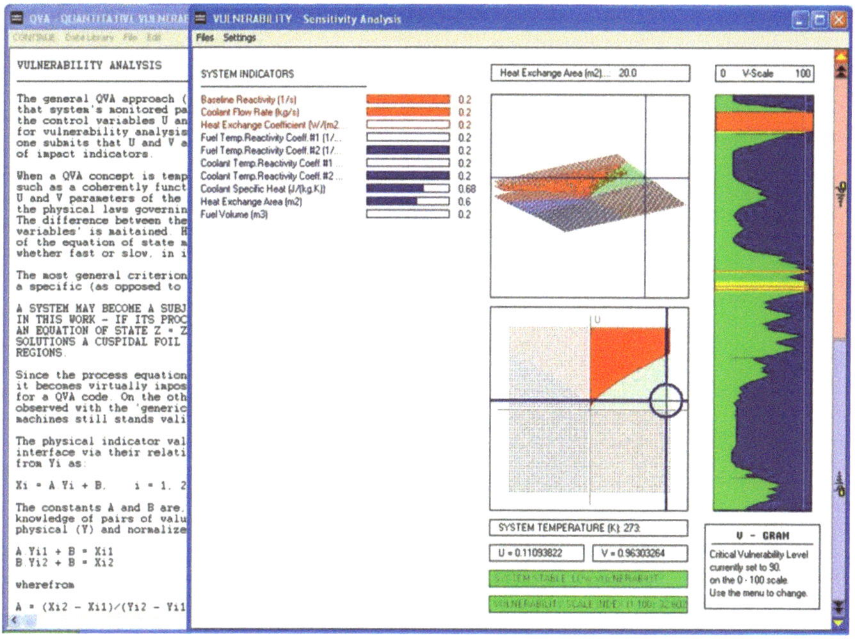

Fig. 6.4 Vulnerability mapping for system indicators

6.3 Remarks

The model of QVA and more succulently its application in 'reactor system' shows generalizability of the generic QVA model. Again, a consideration of fast- and slow-moving variables, especially the inputs, is a key in the application. In this chapter, authors place focus on a specific number of indicators to develop a model that could be used for vulnerability assessment using several indicators commonly associated with nuclear reactors. Notice that the selected indicators are somehow technical in nature. It should not be inferred that non-technical issues do not affect reactors. Quite the contrary. In fact, it has been suggested that culture of the organization, specifically, the safety culture, is a key factor in managing the performance of such systems (Warren 2015). We can conclude that the present model could certainly be modified to include other factors.

References

Gheorghe, A. V., & Vamanu, D. V. (2005). Reading vulnerability in phase portraits: An exercise in probabilistic resilience assessment. *International Journal of Critical Infrastructures, 1*(4), 312–329. https://doi.org/10.1504/IJCIS.2005.006678.

Warren, J. H. (2015). *Safety culture monitoring: A management approach for assessing nuclear safety culture health performance utilizing multiple-criteria decision analysis* (Ph.D.). Old Dominion University, USA, Virginia.

Chapter 7
Emerging Space Treats and Satellites

Abstract For over 20 years Martin Gardner wrote "Mathematical Games and Recreations," a monthly column for Scientific American magazine in which he inspired hundreds of thousands of readers to delve into the large world of mathematics. In one of his columns (January, 2007), he glossed over the issue of anti-satellite weaponry. In a few basic lines of Physics and a computer simulation, one is able to 'illustrate' the feeling that the space, while critical in the sense of critical infrastructure, is readily vulnerable. At this point in this book, the reader is expected to know our evolving issue of critical infrastructures, key resources, and key assets (CIKRKA) complemented by the open bazaar of critical infrastructures of space; undersea; and belowground as suggested in Chap. 1. This chapter is purposefully designed to illustrate a gaming approach, using readily available resources, to hinder operability and even destroy satellite systems. [A variation of this research appears in the *International Journal of Critical Infrastructures* (see Gheorghe and Vamanu in Int J Crit Infrastruct 3:457–470, 2007).]

7.1 Introduction

The term, 'anti-satellite weapons,' is used to describe space weapons designed to incapacitate or destroy satellites for strategic military purposes. Nothing is new here. In fact, since 1950s, world powers (e.g., the USA and the Union of Soviet Socialist Republics, USSR) have been engaged in ground-launched missiles. For example, do a simple search for '*WS-199C High Virgo*' or '*Istrebitel Sputnik.*' Moreover, since then, more ambitious proposals have come to light. For instance, the USA destroyed USA-193 in 2008. USA-193 was a US military reconnaissance satellite (radar imaging) which had been launched in 2006 (USDoD 2008). Despite these advances in technology, there remain questions, perhaps, beyond the realm of technology: *space conflicts with ground consequences*. Think of a situation in which one nation's satellite is intercepted and hacked by another state or nation. Well, how could we get there?

A.V. Gheorghe et al., *Critical Infrastructures, Key Resources, Key Assets*, Topics in Safety, Risk, Reliability and Quality 34,
https://doi.org/10.1007/978-3-319-69224-1_7

For starters, on September 11, 2001, the World Trade Center towers in South Manhattan (New York, USA) were leveled by a terror strike, turning the inconceivable into a shattering reality for thousands and plunging the prime-reference country of the planet into an unparalleled and painful defensive posture.

On November 25, 2002, the *Homeland Security Act* (USDHS 2002) created one of the most significant (re)shufflings of the US Administration, since the creation of the Department of Energy, while leaving the USA and the worldwide open to unprecedented and controversial demands, and experiencing an anxiety unknown since the times of the last Great War.

On October 26, 2006, President Bush signed the *Secure Fence Act of 2006*. That legislation 'authorized the construction of hundreds of miles of additional fencing along our Southern border and gave the Department of Homeland Security to increase the use of advanced technology like cameras, satellites, and unmanned aerial vehicles to reinforce our infrastructure at the border' (US Congress 2006).

On January 11, 2007, China deliberately destroyed one of its weather satellites—a 'test' in more than one way, thought by cool analysts as having the potential to revive a techno-political race believed to be defunct since the 1980s, and by hot diplomats (e.g., see European Union's statement as recorded in *Reuters*) as being 'inconsistent with international efforts to avert an arms race in outer space and undermining the security in outer space.' (Buckley 2007).

As previously suggested, a seriously concerned citizen will undoubtedly be absorbed by an effort to fathom whether or not these events and developments relate to each other, and if so—how? (Gheorghe and Vamanu 2007, p. 459). Instead, a liberal mind will voice the question concerning how much of the political turmoil and emotional increment brought about by any step upwards on the strike-counterstrike ladder is warranted. An informed reader may as well point to technical and budgetary stunts involved in both equipping the space with effective and enhanced Homeland Security capabilities including shooting satellites.

As for present researchers, believers in the relevance of space as a critical infrastructure and the world worth living in and perhaps because of the availability of time for gaming—which might involve a much simpler issue: intellectual capability required to interfere in space systems. The question at hand is 'what is the *intellectual* feasibility of a person willing to engage in menacing space critical infrastructures?'

Looking into the past several decades can offer insights into this question. Having a higher level of education might not even be a factor. In fact, understanding *how feasible* is for a moderately bright mind, and to keep it simple and politically correct, different cultural background must be considered. For instance, consider individual aspiring to produce nerve gas from fertilizer ingredients or enriching uranium while allegedly seeking only mere access to nuclear power. These seem to present valid inception points for a commonsense debate.

It is under this guise that the notes in the sequel endeavor to 'document the Chinese occurrence.' By the end of what is intended as a mere 'game' and pleasant reading, one might hardly escape the conclusion: *while it is delicate and costly in practice, shutting down a space system, critical or otherwise, in theory is a piece of cake*. We write 'in theory' since the assumption is that the operation is 'simply a game.' However, and on a more seriously note, space might not be as vulnerable as your backyard. Nonetheless, it is still vulnerable. If that is the case, then it is best to contemplate space system vulnerabilities and the close-related concepts of, among others, space system risks, space system resilience, space system fragility.

7.2 The Asat Backdrop

Literature on ASAT is substantial, spanning from less technically elaborated newspaper pamphlets to weighty Physics and Engineering, through several legal references and standards. A handy starting point may as well as be a topical entry on Wikipedia (2016) using a simple call-like 'anti-satellite weaponry.' In a more substantial web search, one is flooded with dozens of provocative titles including *The Physics of Space Security—A Reference Manual* (Wright et al. 2005) and *Space-based Weapons—A bibliography* (Rollins 2003). A number of other selective bibliographies mentioning key US Government documents along with informative topical books can be found at http://www.stimson.org/. This listing is just that, a listing. However, relevance to the present topic is the insights that one should glean from this listing:

- the certainty that ASAT is feasible, and in many guises, from direct kinetic hits to near-sci-fi high-power lasers in orbit, via nuclear blasts in an as wide as 1000 km proximity of the targets.
- the evident fact that the technology was demonstrated, if with mixed technical and political results, before being placed on the shelf back in the 1980s, as an asset of possible recourse.
- the equally evident fact that the Great Two of the Cold War era have developed a certain, if cautious, idiosyncrasy about pursuing the development, and the eventual use, of ASAT.
- that is, before the Third Great flexed its muscles this January, which may well reset the entire game.
- It is now possible to design space systems that can 'highjack' other space systems. Once a space system is highjacked it can be used for any other purpose including data theft and ransom demands.

7.3 A Game of Space Systems

Let us first consider the bare facts (AFP 2007; Wikipedia 2016):

> At 5:28 p.m. EST January 11, 2007, the People's Republic of China successfully destroyed a defunct Chinese weather satellite, FY-1C. The destruction was carried out by a modified

medium-range ballistic missile with kinetic ASAT warhead. FY-1C was a weather satellite orbiting Earth in polar orbit at an altitude of about 537 miles (865 km), with a mass of about 750 kg. Launched in 1999, it was the fourth satellite in the Feng Yun series.

7.3.1 The Challenge

In this case, the challenge is an attempt to use basic college knowledge and use a trial and error approach—more or less, try and see if such and such is possible. One might use 'back-of-the-envelope calculation' involving basic Physics putting a ballistic missile into orbit in such a manner as to arrange an 'engagement' with an already in-space orbiting body (i.e., system). But at this point, why not just settle at the idea of getting near-enough to a given target? One once knows the know-how of getting close, they might opt for more ambitious ideas of, for instance, mounting some iterations to the space system, directing it elsewhere, to truly impact the system.

7.3.2 The Requisites

An obvious *pre-requisite* is availability of tools and their selection for the purpose of locating satellites. Notice the framing of the issue at hand: The concern is not necessarily *whether or not* resources are available, but rather picking the right resources—this is on off-the-shelf issue. While the present researchers are 'amateurs', they are also aware of the professional champions—the guys who have for decades, successfully put things into orbit. Sergey Kudryavtsev, of the Sternberg Astronomical Institute of the Moscow State University, concedes that researchers at his institute are developing top-notch analytical methods for calculating satellite orbital perturbations based on the Poincaré method. These approaches would access all perturbations proportional up to, and including, the 5th-order of small parameters. For instance, Kudryavtsev (2002, p. 301) notes that:

> ...The method can precisely calculate the effects of all geodynamical forces on satellite motion given by the most up-to-date IAU [International Astronomical Union] and IERS [International Earth Rotation Service convention] models, such as non-central Earth gravity potential, precession and nutation of the geoequator, polar motion and irregularities in the Earth's rotation, effect of ocean and solid Earth tides, pole tide, and secular variations of gravity coefficients.
>
> Numerical tests prove the method's accuracy to be equivalent to 1-2 cm when calculating positions of high altitude geodetic satellites (like ETALON), and/or of GLONASS navigational spacecraft. The accuracy is stable over 1 year at least and comparable to that of the best tracking measurements of satellites.

In the present research, present authors have themselves at a respectful distance from this league of professional champions and settled for more 'popular' solutions.

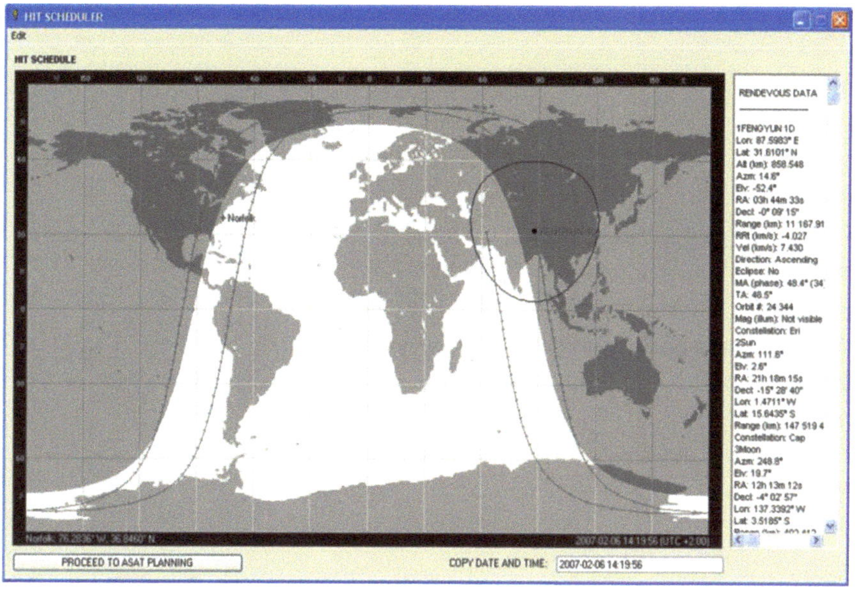

Fig. 7.1 An *ad hoc* ASAT simulator interface to Stoff's 'Orbitron,' adapted from Gheorghe and Vamanu (2007)

In terms of selection of tools, our pick is Sebastian Stoff's *Orbitron* (Stoff 2016). Orbitron is a satellite tracking system for radio amateur and observing purposes with a variety of clients ranging from weather professionals to regular satellite communication users. This free cardware application provides information on the position of satellites at any given moment—in real or simulated time.

In creating a simulation, Google Earth is used to supply maps and geographical information. Prior to I/O interface for ASAT simulator, one needs to create an ad hoc interface to Stoff's 'Orbitron' which enables one to select a target to be hit—or again, get close to the target as possible. Figure 7.1 is an interface developed for this purpose. Figure 7.2 depicts an input-output interface for an ASAT simulator that used in for hitting a specified target. In the present illustration, the target is *FENGYUN 1D*. Figure 7.3 depicts a marked trajectory, target, its vertical, and the visibility circle.

7.3.3 The Solution

Sticking to the 'keep-it-simple' rule, the following action design is adopted:

(a) One's favorite computer, again readily available, is raised to the rank of a Flight Control Center. First, run Sebastian Stoff's *Orbitron* to select a satellite as a target, and predict its position, considering its latitude, longitude, and altitude at

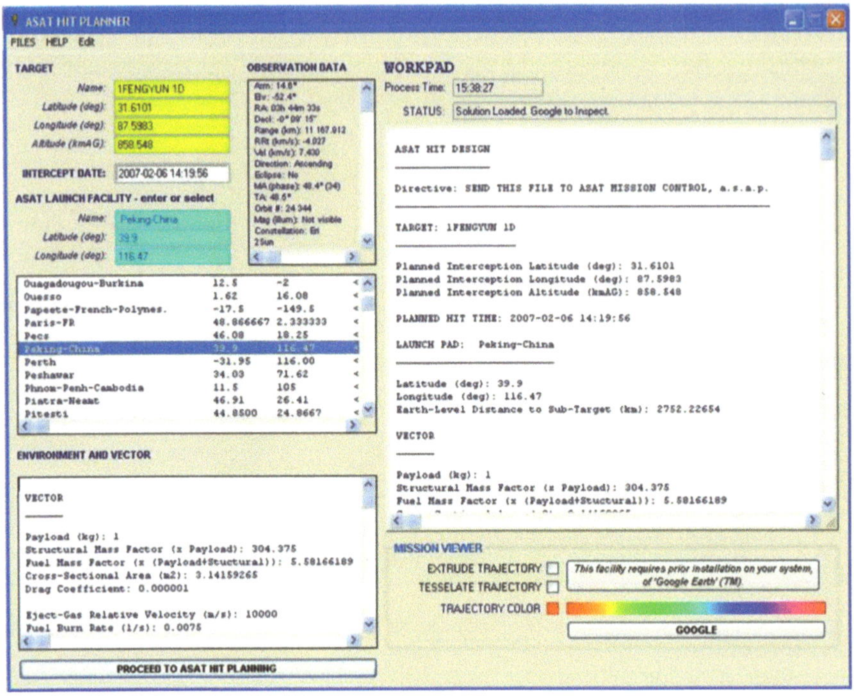

Fig. 7.2 Synoptic I/O interface of ASAT simulator, adapted from Gheorghe and Vamanu (2007)

a comfortable time—say a few tens of minutes ahead. Feed data into your own application, as the 'engagement' or *the rendezvous site*.

(b) One sets a ground-based or a sea-based *launch site*. This is based on the selected latitude and longitude.

(c) Fetch a *one-stage missile* with a warhead of the conventional, high-explosive warhead—*the payload*, of known mass (kg). Infer the quantity of *fuel* indicatively required to take the payload to an Earth orbit, as well as the rocket *structural mass*, and *total mass*, using rules-of-thumb popular everywhere in the literature since decades now (see, e.g., Stoenescu 1962).

(d) Compute a *solution* (i.e., a flight program in soldier's tongue) in the canonical fashion: a clear-pad vertical *liftoff phase*; a controlled orbit *insertion phase*; and a *ballistic fall* onto the target. Assume the least demanding flight control manner—a constant rate (deg/sec.) of imposed increment of the tilt angle during the actively controlled flight phase—with the resulting gravitational fall considered, and a final boost (for simplicity—assumed instantaneously effective) consistent with the tilt angle at engine cutoff time, to reach the last-computed velocity required to ballistic ally rendezvous with the target.

In the preceding, one needs to consider the effect of *air drag* in the ascension phase, using, for example, a common squared-speed law and the 1976 Standard

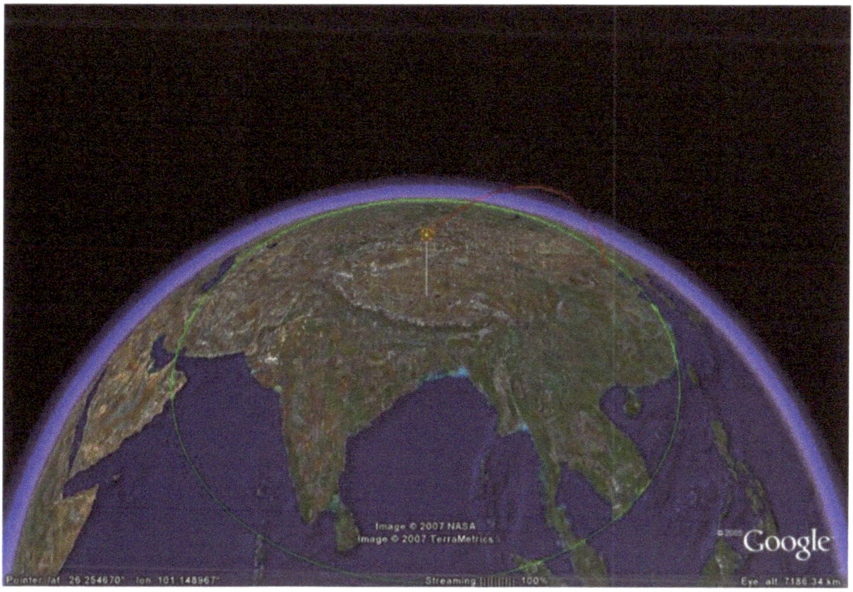

Fig. 7.3 A simulated shooting of FENGYUN ID, adapted from Gheorghe and Vamanu (2007)

Atmosphere model of air density variation with altitude. This process requires: (i) a step-by-step comparison of the current tilt angle imposed at a constant rate during the controlled flight phase, to the ideal angle required at the respective altitude, and velocity, for the rocket to ballistic ally reach target on a momentary *safety ellipse*, (ii) a (virtual) engine cutoff at the precise time when the imposed and gravity-affected tilt angle equals, within an accepted error, the ideal tilt angle referred to above, and (iii) a prompt computation of the velocity at engine cutoff time, required for the rocket to reach target—a velocity to be achieved by an appropriate final boost (a common practice reported in space shuttle flights).

Finally, a 'solution' is retained in terms of required (i) *launch time*, (ii) trajectory *tilt angle* at the ballistic orbit insertion point, and (iii) rocket *velocity* at the ballistic orbit insertion point (at engine cutoff time). Then one proceeds to virtually program rocket's onboard computer(s), starting the countdown, in consideration of the appropriate launch time as determined, and turning the keys to fire away at the appointed time. Figures 7.4 and 7.5 depict the articulated simulation.

7.3.4 The Basic Laws

Every system is governed by certain laws (Keating and Katina 2015). If one is to pull off this kind of a stunt, there are laws that one might want to know and abide

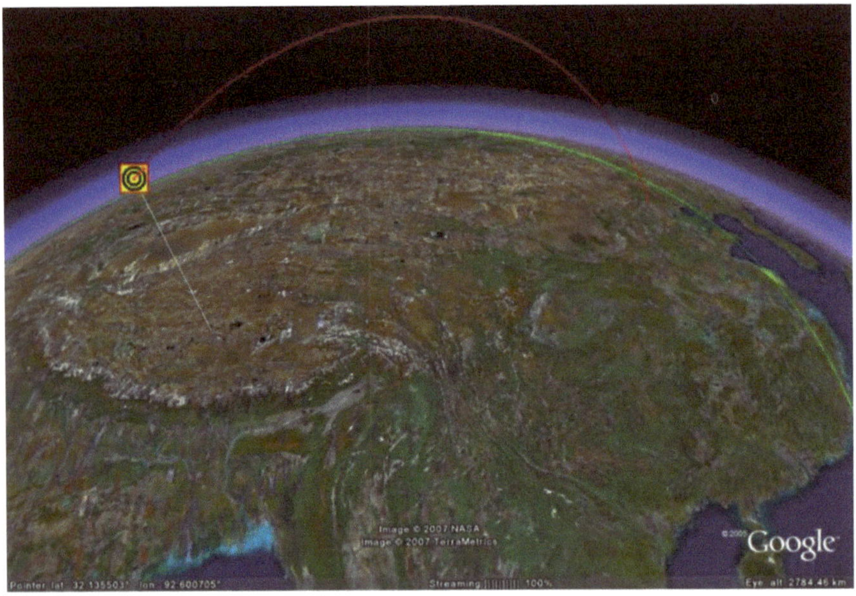

Fig. 7.4 A close-up of a simulated approach to shooting down FENGYUN ID, adapted from Gheorghe and Vamanu (2007)

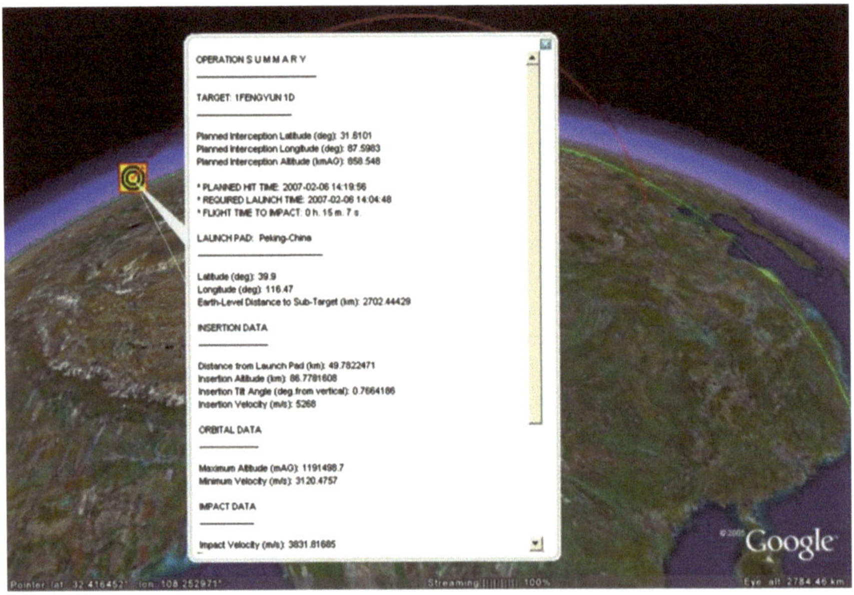

Fig. 7.5 Detailed information regarding the operation, adapted from Gheorghe and Vamanu (2007)

by. In the context of present stunt, it is useful to (re)collect the following textbook lines:

Familiarity with an equation for unmanned flight conics that carry artifacts in space might be mandatory, namely:

$$r = p/(1 + e\cos(\theta - \theta_0)) \tag{7.1}$$

may conveniently be rephrased as

$$r = p/(1 + (p/r_0 - 1)\cos\theta - p\sin\theta/(r_0\,\mathrm{tg}\,\alpha)) \tag{7.2}$$

where

$$p = r_0^2 v_0^2\,\mathrm{tg}^2\alpha/(KM_E(1 + \mathrm{tg}^2\alpha)) \tag{7.3}$$

In the equations above, one has:

r_0 (m)	initial distance of the object from the reference focus of the conic
r (m)	current distance of the object from the reference focus of the conic
θ_0 (m)	initial polar angle of the object from the reference focus of the conic
θ (m)	current polar angle of the object from the reference focus of the conic
v_0 (m/s)	initial velocity imposed upon the object, at the origin of the ballistic flight
α (rad. from the local vertical)	initial tilt angle, at the origin of the ballistic flight
M_E (kg) = 5.9798e^{24}	Earth mass
K (N.m^2/kg^2) = 6.673e^{-11}	universal gravitational (Newton) constant

When applied to the case in point, Eqs. (7.1 and 7.2) describe ASAT weapon's trajectory, r_0 and θ_0 are the polar coordinates of the launch site, and (r, θ) are the current coordinates of the vehicle (rocket), extending down to the weapon-to-target rendezvous point.

Assuming that the rendezvous point—shared by target and vehicle—features the polar coordinates r_1, θ_1 (obtained by processing *Orbitron*'s output—see above). For these coordinates to obey trajectory Eq. (7.3), the following algebraic condition for the tilt angle and launch velocity must be fulfilled:

$$\begin{aligned}&\mathrm{tg}^2\alpha(KM_E r_1(1 - \cos\theta_1) + r_0 r_1 v_0^2\cos(\theta_1 - r_0^2 v_0^2)\\&\quad - \mathrm{tg}\alpha\, r_0 r_1 v_0^2\sin\theta_1 + KM_E r_1(1 - \cos\theta_1)) = 0\end{aligned} \tag{7.4}$$

Thus, in theory, Eq. (7.4) gives the 'solution': for every momentary altitude, velocity, and tilt angle during the controlled flight phase, use Eq. (7.4) to determine *the ideal tilt* that would be required to reach the rendezvous point—considering the ever diminishing distance to the target measured on the Earth's circumference (a perfectly spherical Earth assumed). Keep engines running at a constant regime, and the steering steady, *until the actual tilt equals the ideal tilt* within an acceptable error (e.g., 1.0e-3° of arc). A Launchpad and the launch site for present simulation, and a closer up of the launch site are depicted in Figs. 7.6, 7.7, and 7.8, respectively.

At that moment, one must: (i) compute the difference between the actual speed and the ideal speed required to hit target, (ii) boost the rocket to get to the ideal speed as fast as feasible, and (iii) cut off the engines. At this point, one is able to sit back and watch the simulation, or perhaps try again. In particular, to monitor the velocity and trajectory, tilt angle during the free (ballistic) flight phase use, for every polar radius r_{T} at a moment T:

$$\begin{aligned} v_{\mathrm{T}} &= \left(v_0^2 + 2KM_{\mathrm{E}}(1/r_{\mathrm{T}} - 1/r_0)\right)^{1/2} \\ \sin\alpha_{\mathrm{T}} &= r_0 v_0 \sin\alpha/(r v_{\mathrm{T}}) \end{aligned} \tag{7.5}$$

As to the air drag effects in the forced ascension phase, this may be accounted for in the equations of motion reflecting Newton's Second Law, as follow:

$$\mathrm{d}v/\mathrm{d}t = -g\,\sin\beta - R/m - (v_{\mathrm{r}}/m)\,(\mathrm{d}m/\mathrm{d}t) \tag{7.6}$$

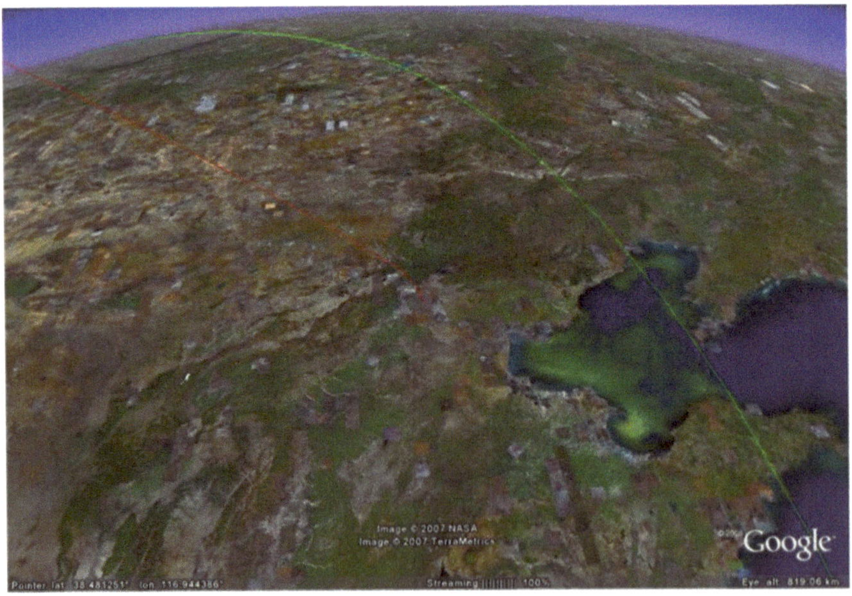

Fig. 7.6 The simulation; closer to the Launchpad, adapted from Gheorghe and Vamanu (2007)

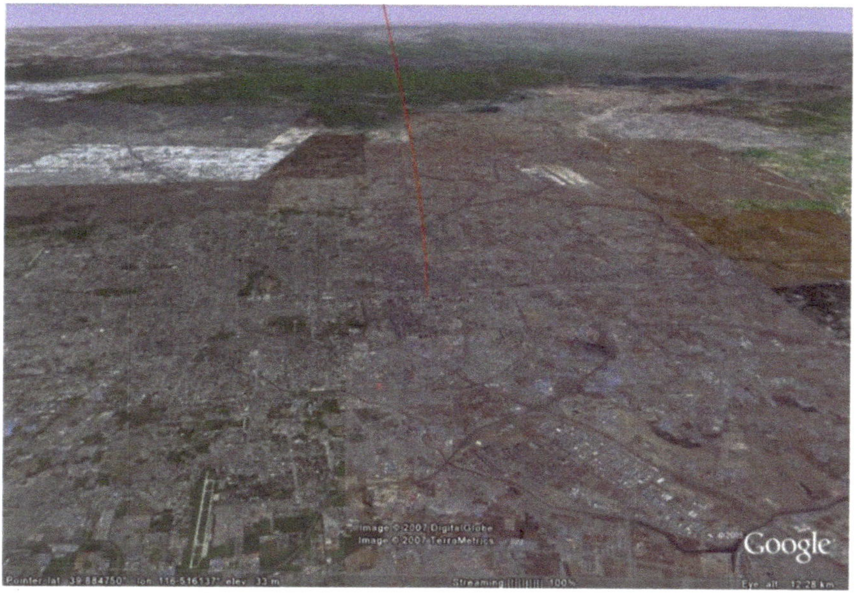

Fig. 7.7 The simulation; the lunch site—the aerial view, adapted from Gheorghe and Vamanu (2007)

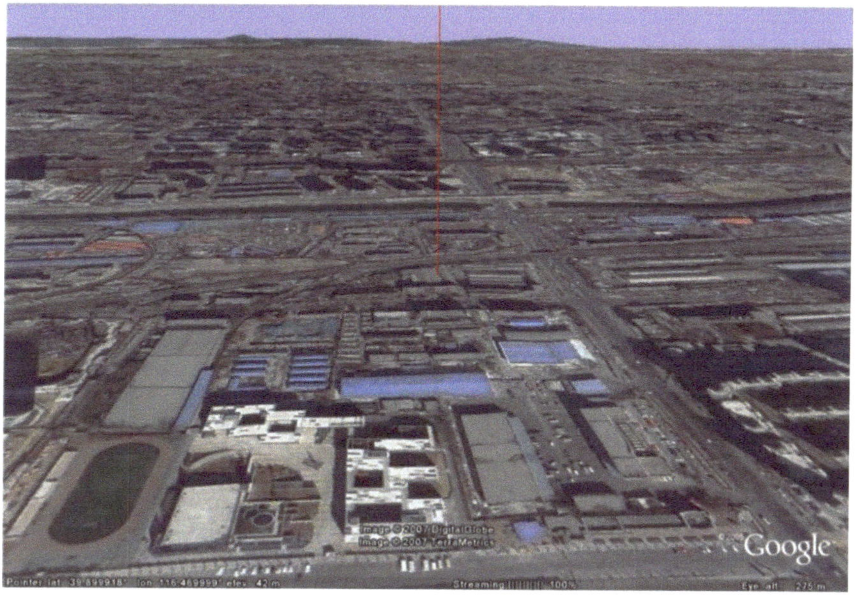

Fig. 7.8 The simulation; Closer to the launch pad—a deliberately absurd location, adapted from Gheorghe and Vamanu (2007)

$$d\beta/dt = -(g/v)\cos\beta \quad (7.7)$$

with the air drag force R (N) given by:

$$R = CA\rho_a v^2/2 \quad (7.8)$$

In these equations, d/dt indicates the 1st derivative, v (m/s) is the vehicle velocity, β (rad. from local horizontal) is the complement of the tilt angle α, v_r (m/s) is rocket engines' ejected gas velocity, C is the aerodynamic shape factor of the vehicle, and A (m^2) is vehicle's effective cross-sectional area.

The total vehicle mass m (kg) is (textbook-routinely) assumed to vary with time T (s) as:

$$m = m_0(1 - \delta T) \quad (7.9)$$

with δ (1/s)—a constant burnout rate.

The gravity constant g (m/s^2) depends on altitude via the polar radius r, as:

$$g = KM_E/r^2 \quad (7.10)$$

where the polar radius can obviously be written as:

$$r = R_E + z \quad (7.11)$$

with the Earth radius taken as R_E = 6.37e6 m, and z (m)—the geometric altitude.

Essentially depending on altitude is also the air density, ρ_a (kg/m^3), described in this exercise as suggested by Richard Shelquist in his scholarly-informative web pages (Shelquist 2006):

$$\rho_a = \left(\frac{a_1 - H}{a_2}\right)^{\frac{1}{a_3}} \quad (7.12)$$

Here, H (km) is the geopotential altitude in kilometers, relating to the common, geometric altitude, z, as:

$$H = zR_E/(z + R_E), \quad (7.13)$$

with, this time, the Earth radius R_E is expressed in kilometers. The coefficients in Eq. (7.11) are: a_1 = 44.3308, a_2 = 42.2665, and a_3 = 0.234969.

The algorithm that can be inferred from the description above is represented in Fig. 7.9.

Paving the way to it are Figs. 7.4, 7.5, 7.6, 7.7 and 7.8, illustrating the workings of an *ad hoc* applet that has been developed as a live-support simulator of ASAT missions. Numerical experiments with the latter have shown that, depending on the finesse of the computational time step assumed one may reach the sought

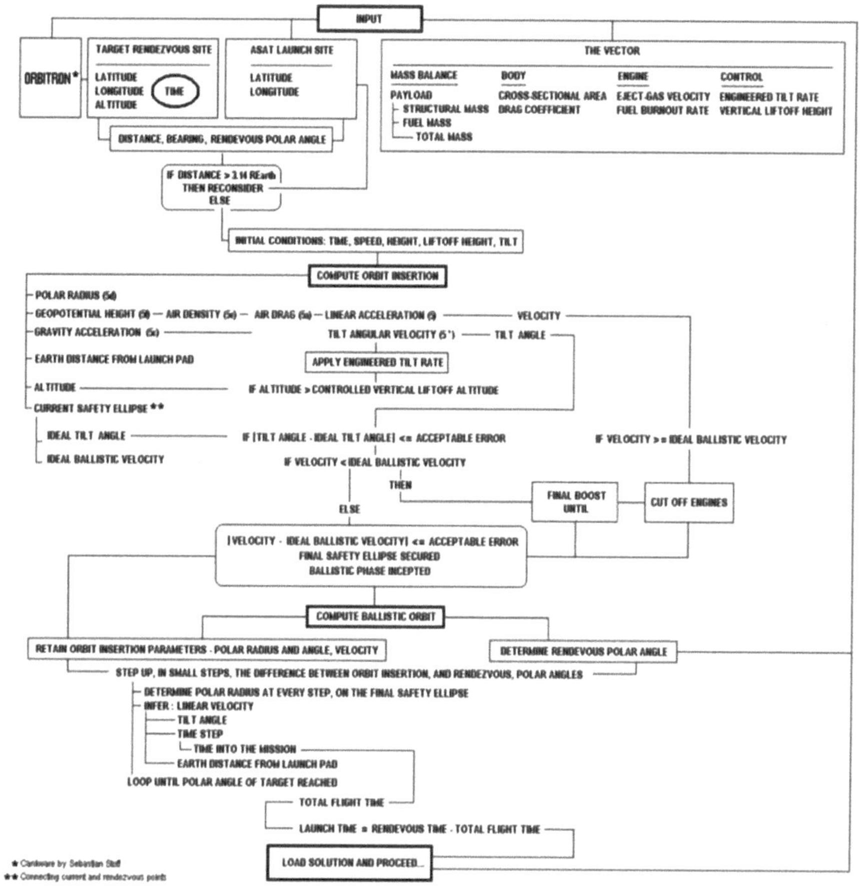

Fig. 7.9 A naïve way of programming an ASAT using ad hoc simulator's algorithm, adapted from Gheorghe and Vamanu (2007)

rendezvous points within a few meters. Table 7.1 below is a sequel excerpt of a few relevant lines from an ASAT mission documentation, as provided by the simulator.

Present authors submit that this simulation 'game' involves a trade-off between accuracy and computational time. In fact, one might consider optimizing the onboard fuel mass so as to minimize, among other factors, dead weight, safety, and loss of control. However, one might with 'increased' knowledge apply 'serious gaming-based approaches' to enable even better understanding of complex system interactions while offering a platform for experimentation of various strategies and scenarios involving space systems (Ancel 2011).

Table 7.1 A partial excerpt of relevant lines from an ASAT mission documentation

ASAT HIT DESIGN
Directive: SEND THIS FILE TO ASAT MISSION CONTROL, a.s.a.p.
MISSION SUMMARY
TARGET: 'FENGYUN 1D'
Planned Interception Latitude (deg): 31.6101 Planned Interception Longitude (deg): 87.5983 Planned Interception Altitude (kmAG): 858.548 *** PLANNED HIT TIME: 2007-02-06 14:19:56** *** REQUIRED LAUNCH TIME: 2007-02-06 14:04:48** *** FLIGHT TIME TO IMPACT: 0 h. 15 m. 7 s.**
LAUNCH PAD: Peking-China
Latitude (deg): 39.9 Longitude (deg): 116.47 Earth-Level Distance to Sub-Target (km): 2702.44429 **INSERTION DATA** Distance from Launch Pad (km): 58.7903947 Insertion Altitude (km): 86.7781608 Insertion Tilt Angle (deg. from vertical): 43.91255 Insertion Velocity (m/s): 5268 **ORBITAL DATA** Maximum Altitude (mAG): 1191498.7 Minimum Velocity (m/s): 3120.4757 **IMPACT DATA** Impact Velocity (m/s): 3831.81685 Impact Tilt Angle (deg. from vertical): 58.4910387 Flight Time to Impact: 0 h. 15 m. 7 s. Visibility Circle Radius (km): 3128.21451 Distance Error (%): 0.65460801 Altitude Error (%): −0.7262643 **VECTOR** Payload (kg): 1 Structural Mass Factor (x Payload): 304.375 Fuel Mass Factor (x (Payload + Stuctural)): 5.58166189 Cross-Sectional Area (m^2): 3.14159265 Drag Coefficient: 0.000001 Eject-Gas Relative Velocity (m/s): 10,000 Fuel Burn Rate (1/s): 0.0075 Controlled Tilt (deg/s): 2 Vertical Liftoff Height (m): 100

(continued)

Table 7.1 (continued)

ASAT HIT DESIGN

ENVIRONMENT: ISA—the 1976 International Standard Atmosphere
Sea level standard pressure (Pa): 10,1325
Sea level standard temperature (K): 288.15
Temperature lapse rate (K/km): 6.5
Gas constant (J/(mol.K)): 8.31432
Molecular weight of dry air (g/mol): 28.9644
Air density law coefficient a1: 44.3308
Air density law coefficient a2: 42.2665
Air density law coefficient a3: 0.234969

CONSTANTS
Number pi: 3.14159265
Earth Radius (m): 6378388
Earth Mass (kg): 5.9798e24
Gravity Constant (N.m^2/kg^2): 0.6673e-10

Insertion Phase Time Step (s): 0.5

Flight Time Distance Altitude Tilt Tilt Target Speed Speed Target Vector Mass
(s) (m from pad) (mAG) (deg) (deg) (m/s) (m/s) (kg)

CONTROLLED FLIGHT TO INSERTION
0.5 0.99792983e-15 16.2979692 0 43.197666 32.5959384 5408 2002.33797
1 0.29981117e-14 48.964497 0 43.3171403 65.3330556 5408 1994.80094
1.5 0.60049012e-14 98.0707182 0 43.4064731 98.2124425 5408 1987.2639
2 0.10022688e-13 163.688319 0 43.4908328 131.235202 5408 1979.72688

....

50 46770.776 83666.8951 43.3802 44.188849 4245.05207 5274 1256.17188
50.5 48260.4179 85218.302 43.8365 44.069641 4301.57943 5271 1248.63484

BALLISTIC FLIGHT TO TARGET
52.7332 58790.3947 86778.1608 43.91255 NA 5268 NA 1241.09781
54.4747 67798.5423 96252.7052 44.01205 NA 5250.8275 NA 1241.09781
56.2244 76806.69 105708.041 44.11247 NA 5233.68393 NA 1241.09781

....

904.965 2761234.69 858548.0 58.7151134 NA 3819.38797 NA 1241.09781
907.631 2770242.84 852312.672 58.4910387 NA 3831.81685 NA 1241.09781

NA—not applicable

7.4 Remarks

In this chapter, yet another approach describing a relevant issue, with emphasis on space critical infrastructure, is suggested. In 'theory,' ASAT ballistic hit looks like a doable 'stunt.' The process is almost insensitive to the choice of the launch point, and relatively simple flight control programs: a matter of constant tilt rates plus final velocity-correcting boosts that may provide sufficient accuracies for conventional brand-tipped explosions at close range from targets be effective in taking these out. It is safe to assume that *any* country in possession of intermediate range vectors can mount an ASAT adventure—even if grotesquely unsophisticated, given a decent engineering capability, a moderate budget, enough determination, and a commensurate political shortsightedness to blind its anticipation of the potentially devastating retaliation to expect on behalf of the target operators. The conclusion: *Space is indeed vulnerable*. And some might argue it is a *wild wild space*. Then again, and as suggested in the present chapter, there are those that might threaten this *status quo*. Perhaps, it is time to craft the governing laws for space as a critical system, similar to those involving critical infrastructures, key resources, and key assets. These issues certainly involve addressing the tension between bilateral and intergovernmental approaches to space governance.

Certainly, and within the quest for methods, tools, techniques, and concepts for dealing with topics at hand (i.e., risk, vulnerability, etc.), one might undertake a probabilistic, deterministic, or a mixed approaches. A *probabilistic* approach enables variation and uncertainty to be quantified, mainly by using distributions instead of fixed values in assessment of risk or any other phenomenon. A distribution describes the range of possible values (e.g., for toxicity) and shows which values within the range are most likely. The result of a probabilistic risk assessment (see, e.g., Gheorghe and Vamanu 2005; Tokgoz 2012) can also be shown as a distribution, showing the range of environmental impacts that are possible, and which impacts within that range are most likely. In this case, such an approach could be used to enable better decision-making regarding, for example pesticide risks, since a full range of possible outcomes is accounted for. On the other hand, a *deterministic* approach tends to treat different factors (e.g., toxicity of pesticides) as if they are fixed and known precisely. Then again, we know that in the real world, factors such as treat and toxicity are not fixed and tend to be emerging and imperfectly known (Gheorghe and Vamanu 2008). Just ask 'experts' in any domain. In fact, in many cases, scientist often extrapolate from 'small' instances to estimate factors (e.g., threat and toxicity) as it is not feasible to measure 'everything' in our natural world. Within this range, there remains a need for developing and applying *hybrid* approaches that combine both, the probabilistic and the deterministic approaches to phenomena. And of course, it goes without saying that the treat does not have to be the traditional sense of atomic, biological, and chemical. Rather, threats of interest can be portrayed as an alphabetical lexicography as portrayed in **Appendix F**.

References

AFP. (2007, February 10). China silent on satellite killer. Retrieved December 31, 2016, from https://web.archive.org/web/20070210230411/http://news.yahoo.com/s/afp/20070119/ts_afp/chinausspacemilitary.

Ancel, E. (2011). A systemic approach to next generation infrastructure data elicitation and planning using serious gaming methods (Ph.D.). Old Dominion University, United States—Virginia.

Buckley, C. (2007, January 23). China confirms satellite test, says no threat. *The Washington Post*. China. Retrieved from http://www.washingtonpost.com/wp-dyn/content/article/2007/01/23/AR2007012300588.html.

Gheorghe, A. V., & Vamanu, D. V. (2005). Reading vulnerability in phase portraits: An exercise in probabilistic resilience assessment. *International Journal of Critical Infrastructures, 1*(4), 312–329. doi:10.1504/IJCIS.2005.006678.

Gheorghe, A. V., & Vamanu, D. V. (2007). Risk and vulnerability games. The anti-satellite weaponry (ASAT). *International Journal of Critical Infrastructures, 3*(3/4), 457–470. doi:10.1504/IJCIS.2007.014120.

Gheorghe, A. V., & Vamanu, D. V. (2008). Mining intelligence data in the benefit of critical infrastructures security: Vulnerability modelling, simulation and assessment, system of systems engineering. *International Journal of System of Systems Engineering, 1*(1), 189–221.

Keating, C. B., & Katina, P. F. (2015). Editorial: Foundational perspectives for the emerging complex system governance field. *International Journal of System of Systems Engineering, 6* (1/2), 1–14.

Kudryavtsev, S. M. (2002). Precision analytical calculation of geodynamical effects on satellite motion. *Celestial Mechanics & Dynamical Astronomy, 82*(4), 301–316. doi:10.1023/A:1015282110947.

Rollins, S. (2003). *Space-based weapons; Maxwell AFB*. Montgomery, AL: Maxwell AFB, AL, Air University Library, August 2003. Retrieved from http://www.au.af.mil/au/aul/bibs/spaceb/space.htm.

Shelquist, R. (2006). Air density and density altitude calculations. Retrieved January 1, 2017, from https://wahiduddin.net/calc/density_altitude.htm.

Stoenescu, A. (1962). *Elemente de cosmonautica (Elements of space engineering)*. Bucureşti: Editura Tehnica.

Stoff, S. (2016). Orbitron: Satellite tracking system, Version 3.71. Gateway. Retrieved from http://www.stoff.pl/.

Tokgoz, B. E. (2012). *Probabilistic resilience quantification and visualization building performance to hurricane wind speeds* (Ph.D.). Old Dominion University, United States—Virginia.

US Congress. (2006). *Secure Fence Act of 2006* (H.R. 6061). Washington, D.C.: Research Library. Retrieved from http://www.aila.org/infonet/text-of-the-secure-fence-act-of-2006-hr-6061.

USDHS. (2002). *Homeland Security Act of 2002* (No. Public Law 107–296) (p. 116 Stat. 2135–2321). Washington, DC: 107th Congress. Retrieved from https://www.dhs.gov/homeland-security-act-2002.

USDoD. (2008). Navy succeeds in intercepting non-functioning satellite. Retrieved December 31, 2016, from http://www.navy.mil/submit/display.asp?story_id=35114.

Wikipedia. (2016, December 22). Anti-satellite weapon. In *Wikipedia*. St. Petersburg, FL: Wikimedia Foundation. Retrieved from https://en.wikipedia.org/w/index.php?title=Anti-satellite_weapon&oldid=756194265.

Wright, D., Grego, L., & Gronlund, L. (2005). *The physics of space security: A reference manual*. Cambridge, MA: American Academy of Arts and Sciences.

Chapter 8
Managerial Vulnerability Assessment Models

Abstract As indicated throughout this book, managers, political pundits, policy makers, scientists, and even hackers, often decisions made without full knowledge of situations. Clearly, this leaves us with changes of enjoying or suffering the outcomes good or bad, respectively. At the same time, one might contend that that are means of increasing changes that the decision being make is 'good.' In this line of thinking, it is generally accepted that making decision based on single criterion is a catalyst for disastrous outcomes. This is especially if such a decision involves multiple conflicting value systems, which could be approached using a multi-criteria decision approach. In this chapter, we provide a literature review on the use of multi-criteria decision analysis. Multiple-criteria decision-making (MCDM) or multiple-criteria decision analysis (MCDA) is a sub-discipline of operations research that explicitly evaluates multiple conflicting criteria in decision making (both in daily life or in professional settings). The review provided therein is to enable those taking the lead with sufficient information regarding the utility of different MCDA-related methods and tools. Emphasis is placed on applications and how such methods have paved the way for decision support systems in complex decision making from a scholarly perspective.

8.1 A Review of Multi-criteria Decision Analysis

Often referred to as Multi-criteria Decision-Making, Multiple Criteria Decision Analysis, and Multi-criteria Decision Aiding; Multi-criteria Decision Analysis (MCDA) is a decision analysis approach that uses problem identification, problem structuring, model building, challenging of thinking using model and information, and then developing of action plan on problems identified to improve the decision-making capacity of a decision maker. Arguably, MCDA is essential for dealing with real decision making that involves qualitative and quantitative evaluation of complex decisions, situations, and scenarios that involve various alternatives. This approach grew mainly due to insufficiencies of single criteria decision analysis. A basic process of MCDA is depicted in Fig. 8.1.

A.V. Gheorghe et al., *Critical Infrastructures, Key Resources, Key Assets*, Topics in Safety, Risk, Reliability and Quality 34,
https://doi.org/10.1007/978-3-319-69224-1_8

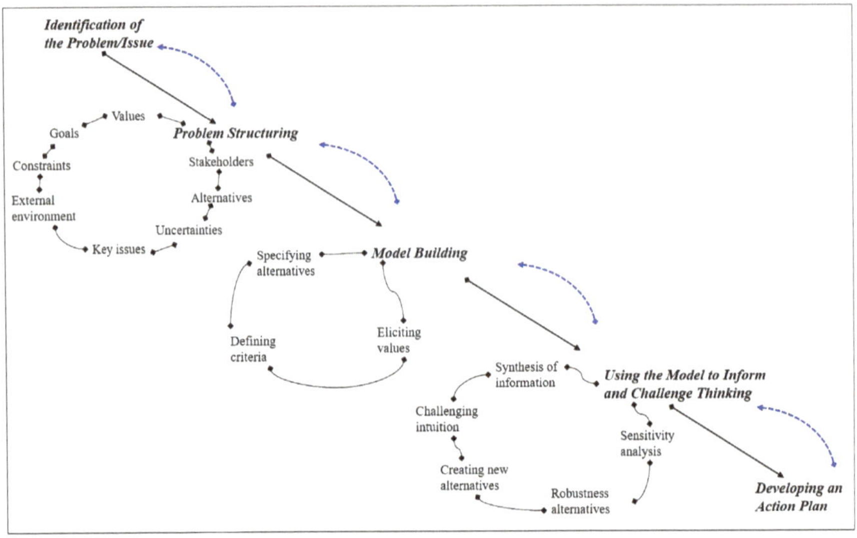

Fig. 8.1 MCDA's basic process, modified from Belton and Stewart (2002)

To further distinguish this methodology from other decision-making approaches, it is necessary to look at the meaning of the term 'criteria.' Merriam-Webster[1] submits, criteria entails *a standard on which a judgment or decision may be based.* At this, we suggest that MCDA uses more than one criterion to bring a decision maker closer to certainty by eliminating as much uncertainty as possible. Notice that the aim of MCDA is not to give the decision make the 'correct' answer, rather and according to Belton and Stewart (2002, p. 3):

- Integrate objective measurements with value judgment;
- Make explicit and manage subjectivity in decision-making process of complex decisions.

Belton and Stewart also stipulate that by aiding decision makers, the major aim of this methodology then becomes 'to facilitate decision makers' learning about and understanding of the problem faced, about their own, other parties' and organizational priorities, values and objectives and through exploring these in the context of the problem to guide them in identifying a preferred course of action' (Belton and Stewart 2002, p. 3). Similar sentiments are echoed by protuberant scholars in decision analysis:

- The major role of formal analysis is to promote good decision analysis. Formal analysis is meant to serve an aid to the decision maker, not as a substitute for him. As a process, it is intended to force hard thinking about the problem area: generation of alternatives, anticipation of future contingencies, examination of

[1]http://www.merriam-webster.com/dictionary/criteria.

dynamic secondary effects, and so forth. Furthermore, a good analysis should illuminate controversy—to find out where basic differences exist in values and uncertainties, to facilitate compromise…to promote good decision making (Keeney and Raiffa 1972), as cited by Belton and Stewart 2002, p. 3)

- …I believe that decision analysis is very delicate, subtle tool that helps decision makers explore and come to understand the beliefs and preferences in the context of a particular problem that faces them. Moreover, the language and formalism of decision analysis facilitate communication between decision makers. Through their greater understanding of the problem and of each other's view of the problem, the decision makers are able to make a better-informed choice (French 1989, p. 1, as cited by Belton and Stewart 2002, p. 4)
- …the theories, methodologies, and models that the analysts may call upon…are designed to help think through the possible changes that a decision process may facilitate so as to make it more consistent with the objectives and value system of the one for whom, or in the name of whom, the decision aiding is being practiced…especially when there are conflicting viewpoints (Roy 1996, p. 1, as cited by Belton and Stewart 2002, p. 4).

Furthermore, it has been suggested that making important decisions often requires treating major uncertainty, longtime horizons, and more complex values issues and since uncertainty is at the heart of most significant decision problems, the process of decision making requires specifying the amount of uncertainty that exits given available information (Howard and Matheson 1983). This is where Multi-criteria Decision Analysis (MCDA) becomes a very powerful tool.

These tools, too numerous for a detailed discussion, include, among others: Aggregated Indices Randomization Method (AIRM), Analytic hierarchy process (AHP), Analytic network process (ANP), Data envelopment analysis, Dominance-based rough set approach (DRSA), ELECTRE (Outranking), PROMETHEE (Outranking), The evidential reasoning approach (ER), Goal programming, Grey relational analysis (GRA), Inner product of vectors (IPV), Measuring Attractiveness by a categorical Based Evaluation Technique (MACBETH), Multi-Attribute Global Inference of Quality (MAGIQ), Multi-attribute utility theory (MAUT), Multi-attribute value theory (MAVT), New Approach to Appraisal (NATA), Nonstructural Fuzzy Decision Support System (NSFDSS), Potentially All Pairwise Rankings of all possible Alternatives (PAPRIKA), Superiority and inferiority ranking method (SIR method), Value analysis (VA), Value engineering and analysis (VE&A), Weighted product model (WPM), and Weighted sum model (WSM). A detailed account of these can be found in Holzgrefe (2015) and Holzgrefe and Hester (2016).

Actual applications in MCDA are vast as illustrated by a large number of literature on the subject. These include and certainly not limited to analysis of integrated planning for transport and land use (Sharifi et al. 2006; Munday et al. 2010), environmental decision making (Steele et al. 2009; Kiker et al. 2005), managing uncertainties in energy savings (Haeri et al. 2007), ranking and prioritization of winners in games (Saaty 2010), financial and banking problems (Doumpos and Zopounidis 2002), addressing mindsets, rationality and emotion (Wenstøp 2005), evaluation of software (Paschetta and Tsoukiàs 2000), enhancing communication and improving emergency management resource allocation (Levy et al. 2007), human resources management (Ensslin et al. 2000), hiring process with regard to anti-discrimination laws (Gardiner and Armstrong-Wright 2000), advancing democratic forms of governments (Bollinger and Pictet 2003), geographic information systems (GIS), and in modeling spatial multiproblems (Murray 2002).

In recent times, this methodology has found its way into democratic processes of some nations. In a direct democracy (e.g., Switzerland), MCDA can be used to ensure that individual perspectives shape national issues including laws by the parliament (Bollinger and Pictet 2003). Steele et al. (2009)'s research also suggests the utility of Analytic Hierarchy Process (AHP) in complex decision making related to the environment. Their research illustrates the effects of weighting final decision (i.e., ranking of options) based 'on the choice of performance scoring scales for the criteria when the criteria weights are held constant' (Steele et al. 2009, p. 26) and how 'sensitivity' of weights assigned influences the decision maker choices.

Therefore, it can be said that MCDA and its methods are useful in many aspects of human decision-making processes particularly in the domain of public sector with intertwined and conflicting objectives of multiple stakeholders. Dyer et al. (1992) noted that there are many documented applications of this methodology and stipulated that in Finland, 'public sector problems involve multiple conflicting objectives' can be examined under this scope of thinking; be it 'in public healthcare systems, environmental policy. Water resources, energy, [and/] or macroeconomic planning, the opportunities for MCDM applications are unlimited' (Dyer et al. 1992, p. 651) as shown above. However, within each method, there are strengths and weakness. Table 8.1 addresses these differences for a select few.

Current Research

Quantitatively speaking, MCDA research encompasses complexity that is inherent within the methodology itself as well as the complexity of the problems it attempts to solve. Adding to this complexity are stakeholders with various views on the problems of decision making as well as question about the methods, tools, and techniques used within MCDA. A general consensus is the lack of 'sufficient integration of systemic social-political context' in decision support systems (Banville et al. 1998, pp. 15–16). Hence, more research is warranted in group decision-making areas including viewing stakeholders from different perspectives of 'group' and within quantitative and qualitative research. Suggestions for using MCDA with group environments are provided below (Banville et al. 1998):

- Compiling all potential actions
- Setting up criteria and attributes to consider
- Evaluation of performances
- Aggregation of performances, however, since this phase within decision structuring is very subjective, it requires refinement from the stakeholders. This could be done through:
- Identification and classification of stakeholders
- Determining stakeholders' participation and roles

Table 8.1 Selected MCDA methods and comparisons, modified from Hudnell (2008)

Method	Important elements	Strengths	Weaknesses
Multi-attribute utility theory	– Expression of overall performance of alternatives in a single, non-monetary number representing the utility of that alternative – Criteria weights often obtained by directly surveying stakeholders	– Easier to compare alternatives whose overall scores are expressed as single numbers – Choice of alternative can be transparent if highest scoring alternative is chosen – Theoretically sound—based on utilitarian philosophy – Many people prefer to express net utility in non-monetary terms	– Maximization of utility may not be important to decision makers – Criteria weights obtained through less rigorous stakeholder surveys may not accurately reflect true stakeholders' preferences – Rigorous stakeholder preference elicitations are expensive
Analytical hierarchy process	– Criteria weights and scores are based on pairwise comparisons of criteria and alternatives, respectively	– Surveying pairwise comparisons is easy to implement	– The weights obtained from pairwise comparison are strongly criticized for not reflecting people's true preferences – Mathematical procedures can yield illogical results. e.g., Ranking developed through AHP can sometimes not be transitive

Outranking	– One option outranks another; If;
	  – It outperforms others on enough criteria
	  – It is outperformed by other in the sense of recording inferior performance on any other criterion
	– Allows options to be classified as 'incomparable'
		– Does not require the reduction of all criteria to single unit
		– Explicit consideration of possibility that very poor performance on a single criterion may eliminate an alternative from consideration, even if that criterion's performance is compensated for by a very good performance on other criteria
			– Does not always take into account whether over-performance on one criterion can make up for under-performance on another
			– The algorithms used in outranking are often relatively complex and not well understood by the decision makers

Table 8.2 Examples of software packages related to MCDA, adapted from Buede (1992)

AHP	MAUA	MCDM	O	PC	%
Criterium	ASA S/W	AIM	ELECTRE	Expert87	Best Choice
Expert Choice	Decision Map	RADIAL	Gaia	Policy PC	P/G%
HIPRE3+	Decision Pad	RID	IDEAS		
Priorities	HIVIEW	Triple C	Pragma		
	ISMAUT	VIG	Promothe		
	Lightyear	VIMDA			
	Logical Decision				
	MAUD				
	PREFCAL				
	SMARTEDGE				
	Treeval				
	VISA				

- Finding all actions, attributes, criteria, and evaluation matrix. This ensures that above 'phases' are aligned with the stakeholders' views
- Verifying the solution's legitimacy by considering whether the most likely decision is a reflection of the group and the individual stakeholders.

On the fronts of *Decision Support Systems* (DSS), Matsatsinis and Samaras (2001) stipulate that making decisions on complex interconnected systems is critical. Something like an MCDA methodology may have to be employed to ensure that all decision makers' opinions (and issues) are well captured especially when members of a system in question have irreconcilable worldviews on the same issue. In this case, the role of a DSS is to *assist* in the organization of knowledge about an ill-structured or unstructured problem. Such a support system could be embedded in computer programs and support technological and managerial decision making by providing complementary knowledge for a more rational decision making (Turban and Aronson 1997). In any case, it is important to recognize that systems, be it space, undersea, and belowground, operate as complex systems in an environment full of ambiguity from fuzzy 'weights' and 'utilities.' However, such 'weights' and 'utilities' could be aggregated using evidential reasoning (EV) of MCDA. Zhou et al. (2010) show that subjective, qualitative information can be modeled despite its incompleteness tendencies using the Dempster-Shafer's theory of evidence. There is no shortage of software related to MCDA for DSS. A sample of such software packages are listed in Table 8.2 (Buede 1992, p. 60).

8.2 An Application: Analytic Hierarchy Process Versus Multi-attribute Utility Theory

8.2.1 Background

At this point, it is common knowledge that some systems are critical since their disruption can have significant effects on public well-being. In terms of decision making, one might wish to engage is an excise of selecting viable investment alternative. In other words, given all that could go wrong, what alternative should you invest in? Regardless of the discomfort with the terms, we can all rest assured that humans do not have the ability to predict or control future events and to some extent, consequences so such events. Nonetheless, choices and actions still must to be made. For instance, consider the energy sector. The alternatives include, among others, natural gas, petroleum, and hydroelectric power. And yet again, several factors might affect such a sector. These include, and certainly among others:

- Physical attributes (*system induced*)—development of consequences, vulnerabilities, and protective strategies on human life and physical well-being relating to fatalities and injuries
- Cyber Attributes (*cyber threat*)—attacks on energy systems by terrorists, criminal organizations, and hackers can cause blackouts as it has been reported before
- Volumetric or throughput attributes (*over dependence*)—if energy production is reduced in half, the consequences will be tremendous
- Temporal/load profile attributes (*temporal aspects*)—energy sector has a strong temporal or time-dependent dimension affected by the season of the year and/or time of day
- Human attributes (*man induced*)—the availability of skilled and experienced technical talent is a concern in the energy sector
- Dependency on other networks (*interdependency*)—energy systems are connected to other systems.

Given such information, it is possible to start analysis for better decision making. This can be done using DSS software. In this section, *Expert Choice* (v. 11.5) is used because of its relation to Analytic Hierarchy Process (AHP), and *Logical Decisions for Windows* (LDW) is employed because it uses Multi-Attribute Utility Theory (MAUT). Notice that in the given situation, issues of System induced (S), Cyber threats (C), Overdependence (O), Temporal aspects (T), Manmade (M), and Interdependence (I) represent a reality that could happen.

8.2.2 Assumptions

Assuming that the three options are the most feasible (i.e., this is done for convenience; these could also change based on stakeholders). Furthermore, assume that;

- AHP and MAUT methods are acceptable methodologies that can be used in complex decision making, and that the results will be accepted and implemented as such
- The results are just results, and that the decision still remains in the hands of the decision makers
- Information used is as current as it can be; meaning that in light of new information, alternatives, possible outcomes, and threats may need to be reassessed.

8.2.3 *Analytic Hierarchy Process*

This section provides information on analytic hierarchy process (AHP) as a decision-making methodology and displays the current problem using Expert Choice. Expert Choice is decision-making software that is based on multi-criteria decision making. Expert Choice implements AHP.

AHP was developed in the 1970s by Thomas L. Saaty. Since then, it has received many refinements. It is used in decision-making problem scenarios involving evaluation of multiple related or unrelated elements, goal searching, and evaluation of alternatives based on stakeholders' possible solutions. It has been suggested that AHP was developed after Saaty's observation of the fact that some of the best scientist and lawyers had no ways of communicating ideas to each other. He was therefore 'motivated to attempt to develop a simple way to help ordinary people make complex decisions' (Forman and Gass 2001, p. 470) and communicate them in a powerful but simple manner. Simplicity within this method has led to widespread usage and acceptance in the USA and across the world for scientists and non-scientists alike. AHP and a method are now used by national governments and leading technology firms around the world (Forman and Gass 2001). For instance, the American Society for Testing and Materials 'adopted AHP as a standard practice for multi-attribute decision analysis of investments related to buildings and building systems…[and is] "extensively in organizations such as the Central Intelligence Agency that have carefully investigated AHP's theoretical underpinnings' (Forman and Gass 2001, p. 470). Steps associated with this method revolve around (Drake 1998; Ragsdale 2001):

1. Deciding upon the criteria for selection
2. Rating the relative importance of these criteria using pairwise comparisons
3. Rating each potential choice relative to each other choice on the basis of each selection criterion—this is achieved by performing pairwise comparisons of the choices
4. Combing the ratings derived in steps 2 and 3 to obtain an overall relative rating for each potential choice.

8.2.4 Analytic Hierarchy Process Using Expert Choice Software

In 1983, Dr. Saaty joined Dr. Ernest Forman, a professor of management science at George Washington University and this partnership led to co-founding of Expert Choice (EC). EC software engages the decision maker and helps in structuring a decision into smaller parts, goal, objectives, and even sub-objectives generation as well as alternative courses of action. This enables the decision maker to make a simple pairwise comparison for alternatives.

In the following example, we illustrate an application of EC in AHP. Starting with Table 8.3, notice the goal, options, and listing of possible questions.

From this information, one is able to start the analysis. First, all information is put into EC software as depicted in Fig. 8.2. The remainder of the figures (Figs. 8.3, 8.4, 8.5, 8.6, and 8.7) offers insights regarding the problem at hand using EC DSS in implementing AHP.

This analysis suggests that investing in option: 'Natural gas' is a viable option for investment. This option remains viable even after the dynamic sensitivity for 'Reduction of time-dependence dimension on systems' is increased from 10.7 to 59.1%. In this specific case, the reader is reminded that the results have to do with the cost associated with investing in the not-so energy option, given the possible occurrence of 'criterion selection.' One might not be able to control the occurrence of such events, but one still has the option of investing in an option with the least negative consequences/impact.

Table 8.3 Basic starting point for a problem situation

Goal and consequence	Relevant description
Goal: *To select an energy alternative for investment in Region X—a fictitious region*	Available options: natural gas; hydroelectric; petroleum
Possible consequences: *The aim is to reduce, control, and eliminate threats or monetary burden of such systems* • The list on the right becomes a de facto criterion for selection in 'step' one above	Criteria for selection: • Physical system-induced threats • Cyberspace threats—minimize threats from cyberspace • Over-dependence • Load profile attributes (temporal aspects) • Human-induced threats—maintaining sector reliability, safety, and security from man-induced risks; skills level • Interdependency among systems—reduction of cascading effects

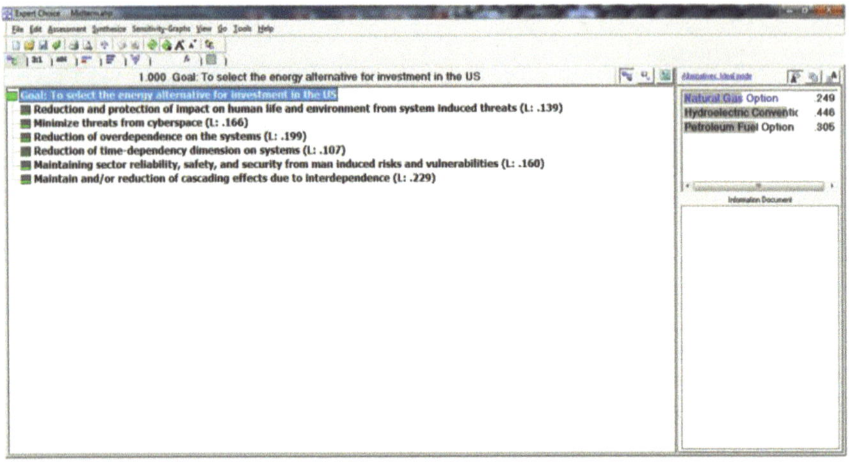

Fig. 8.2 AHP problem hierarchy, alternatives, and goal in EC version 11.5

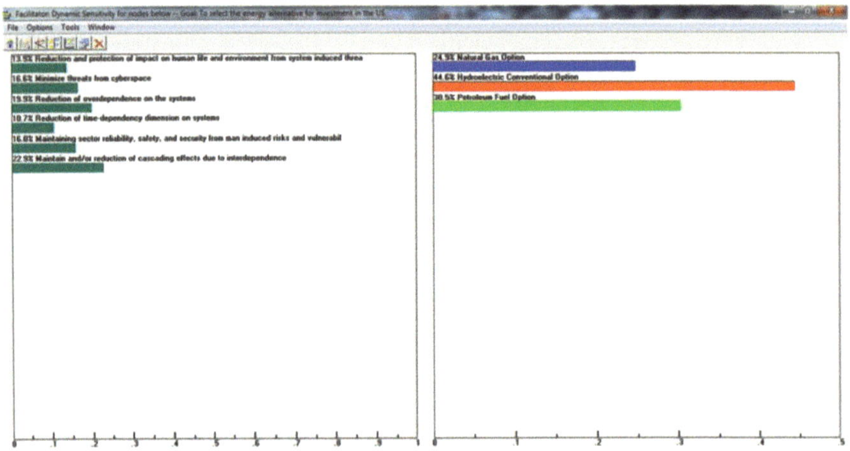

Fig. 8.3 AHP results and the dynamic sensitivity view

8.2.5 *Multi-attribute Utility Theory Approach*

This section offers information on Multi-Attribute Utility Theory (MAUT) as a decision-making methodology. MAUT is intended for use when risk and uncertainties are significant. Dyer et al. (1992) note that MAUT 'focuses on the structure of multicriteria or multiattribute alternatives, usually in the presence of risk or uncertainty, and on methodologies for assessing individuals' values and subjective probabilities. MAUT embraces both a large body of mathematical theory for utility models and a wide range of practical assessment techniques that pay attention to

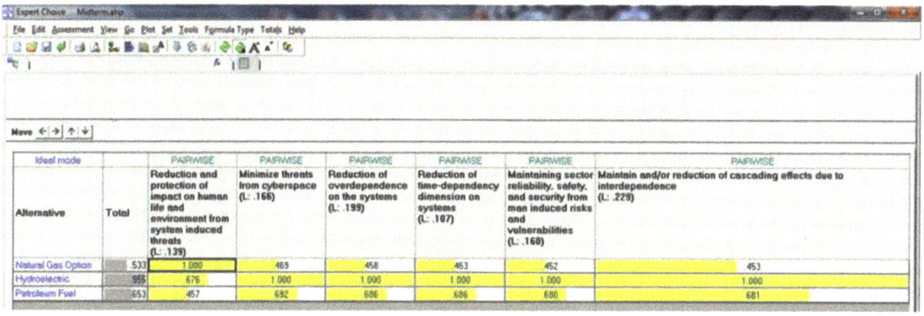

Fig. 8.4 Data grid for the situation at hand

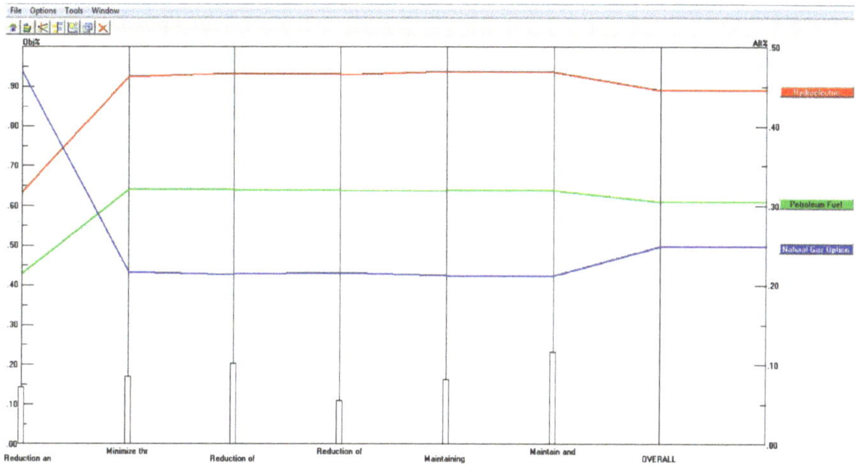

Fig. 8.5 AHP performance sensitivity for 2D goals

limited abilities of assessors. Information obtained from assessment usually feeds into the parent problem to rank alternatives, make a choice, or otherwise clarify a situation for the decision maker. Sensitivity analysis is often involved in the assessment and choice processes' (Dyer et al. 1992, p. 647).

It is worth mentioning that the use of MAUT requires knowing value functions, explicitly. Phases associated with MAUT are provided below (Bellamy 2004; Garvey 2009):

- Identification of the attributes, which collectively describe the overall utility of all relevant decision options
- Identification of the set of actions, projects/programs being evaluated
- Weighting the attributes in terms of their importance
- Transforming the attribute scores, measured in different units, into commensurate units

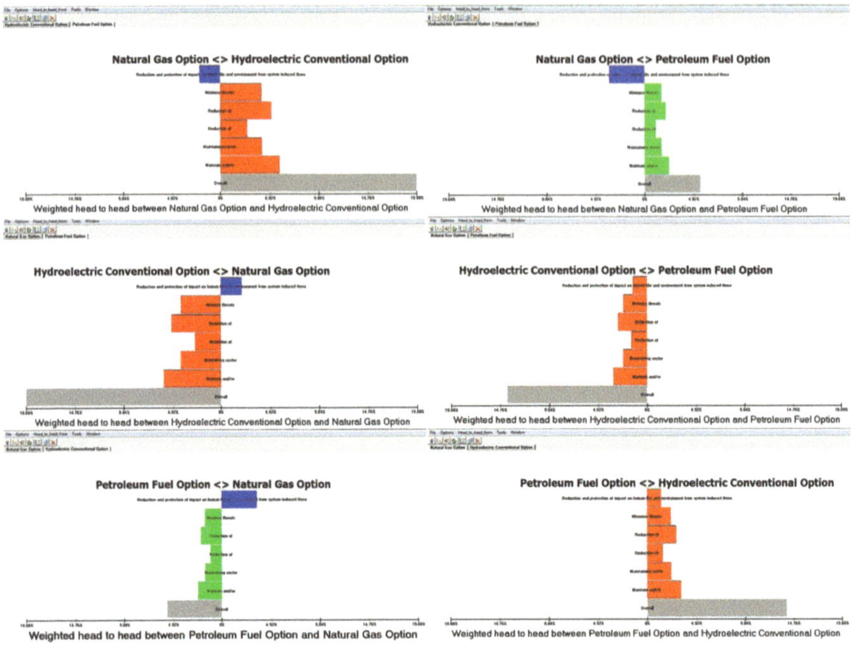

Fig. 8.6 Six different weighted comparison between different options

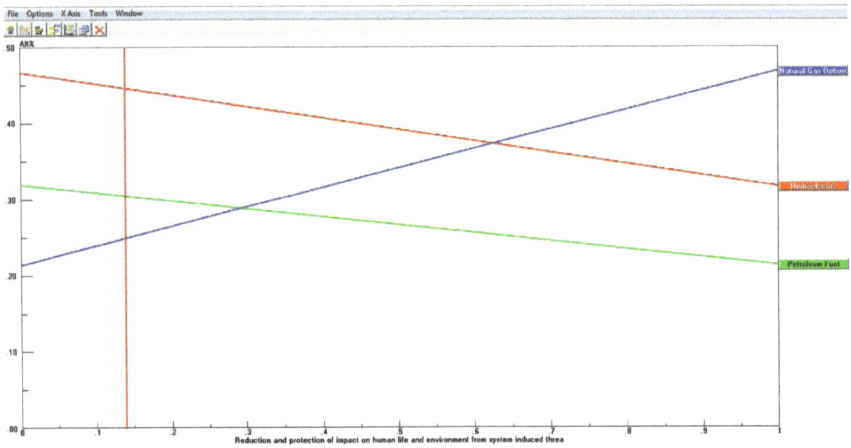

Fig. 8.7 AHP sensitivity analysis for one of selection criteria

- Defining an aggregate utility function, which combines the transformed scores and weights to measure the overall utility of each option
- Conducting a sensitivity analysis on the weights, attribute scores, transformation methods, and types of utility function.

8.2.6 Multi-attribute Utility Theory Approach Using LDW

MAUT process can be implemented in Logical Decisions for Windows (LDW). LDW lets one evaluate choices by considering many variables at once, separating facts from value judgments, and explaining your choice to others. When using LDW (version. 6.2), the following steps are necessary:

- Defining a set of alternatives to be ranked—these are options that we have control over in terms of decision making; these alternatives have to be named and assigned measure levels
- Defining measures to describe the alternatives—LDW uses our measures to describe and rank alternatives. These are the characteristics of alternatives. For current problem, we are using attributes that we think and/or know could happen in the energy sector. These attributes influence selection of alternative
- Entering the level for each measure for each alternative—the above measures are then given a numerical score; translated from a nominal scale (i.e., 'high,' 'Medium,' or 'Low')
- Reviewing preferences so measure level can be combined
- Ranking of alternatives and choosing the best alternative.

The following figures (Figs. 8.8, 8.9, 8.10, 8.11, 8.12, and 8.13) were developed for the problem at hand by repeating what has been done under *Expert Choice* (v. 11.5) with slight modification for Logical Decisions for Windows (LDW). Recall that main goal is still the same and that the second goals show the alternatives. The measures are the criteria or the consequences of which each alternative is evaluated upon.

One of the most important features of LDW is the ability to capture dynamic changes of the decision maker. This can be done using a sensitivity action. This is a dynamic feature of the software that allows for adjustments, at any time, and reflects current changes in the system and operating landscape. An overall comparison for different utilities is also possible. This is illustrated in Fig. 8.14 using different measures of each alternative. The remainder of the figures (Figs. 8.15, 8.16, and 8.17) depicts a comparison between the different energy options.

LDW's other capabilities include allowing the analyst to show some of the elements of AHP (Fig. 8.18). The following figure shows the interdependence of alternatives especially as it relates to weights of each goal. Figure 8.19 depicts the concept of trade-off and weight for analysis of different measures of the energy alternatives.

Authors offer the following remarks regarding the analysis of the present problem (i.e., the issue of investing in an energy alternative). Notice that AHP Expert Choice and MAUT Logical Decision for Windows suggest that the Natural gas option is a viable investment choice. This is only for illustrating purposes. Certainly, several factors would influence the results, including expert and analyst opinions.

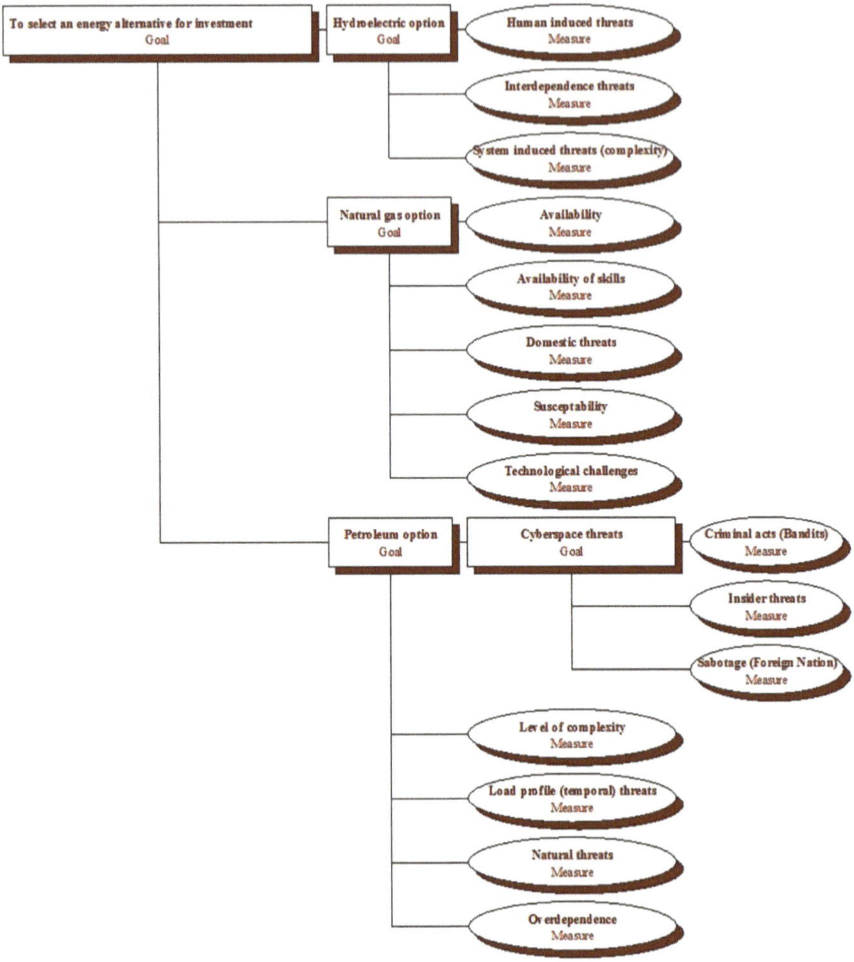

Fig. 8.8 Main goal, alternatives, and measures as captured by LDW

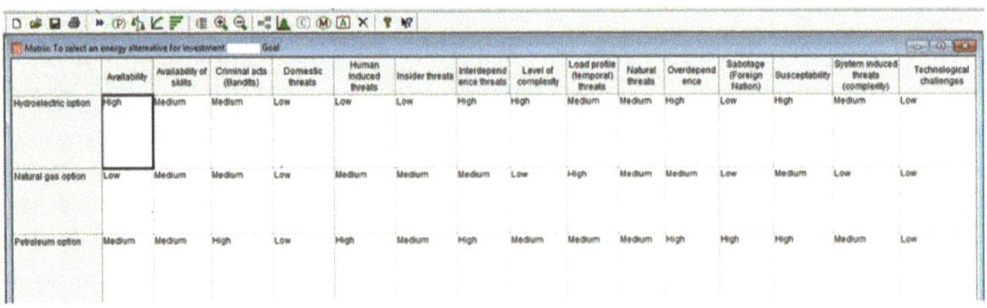

Matrix: To select an energy alternative for investment Goal

	Availability	Availability of skills	Criminal acts (Bandits)	Domestic threats	Human induced threats	Insider threats	Interdependence threats	Level of complexity	Load profile (temporal) threats	Natural threats	Overdependence	Sabotage (Foreign Nation)	Susceptability	System induced threats (complexity)	Technological challenges
Hydroelectric option	High	Medium	Medium	Low	Low	Low	High	High	Medium	Medium	High	Low	High	Medium	Low
Natural gas option	Low	Medium	Medium	Low	Medium	Medium	Medium	Low	High	Medium	Medium	Low	Medium	Low	Low
Petroleum option	Medium	Medium	High	Low	High	Medium	High	Medium	Medium	Medium	High	High	High	Medium	Low

Fig. 8.9 LDW matrix view showing alternatives and evaluations

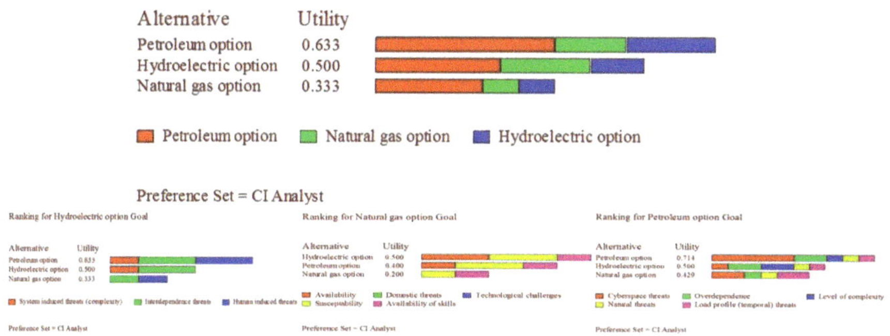

Fig. 8.10 Ranking to select an energy alternative; bottom figures illustrate ranking-based different attributes

Measure	Percentage Weight	Effective Weight
Overdependence	11.1	10.067
Sabotage (Foreign Nation)	11.1	20.134
Criminal acts (Bandits)	11.1	10.067
Insider threats	11.1	10.067
Level of complexity	11.1	20.134
Natural threats	11.1	0.000
Load profile (temporal) threats	11.1	10.067
System induced threats (complexity)	3.7	3.356
Interdependence threats	3.7	3.356
Human induced threats	3.7	6.711
Availability	2.2	4.027
Domestic threats	2.2	0.000
Technological challenges	2.2	0.000
Susceptability	2.2	2.013
Availability of skills	2.2	0.000

Fig. 8.11 Percentage of preferences as noted by an analyst

8.3 Remarks

MCDA methods like AHP, MAUT, and EV are used not to make the decision for the decision maker, but to add a perspective. It is still the job decision maker to make the decision. Such methods and tools might limitations, especially when decision-makers consider issues are difficult to measure such as morals and ethics. Nonetheless, the utility of such methods cannot be disputed. An extension of such

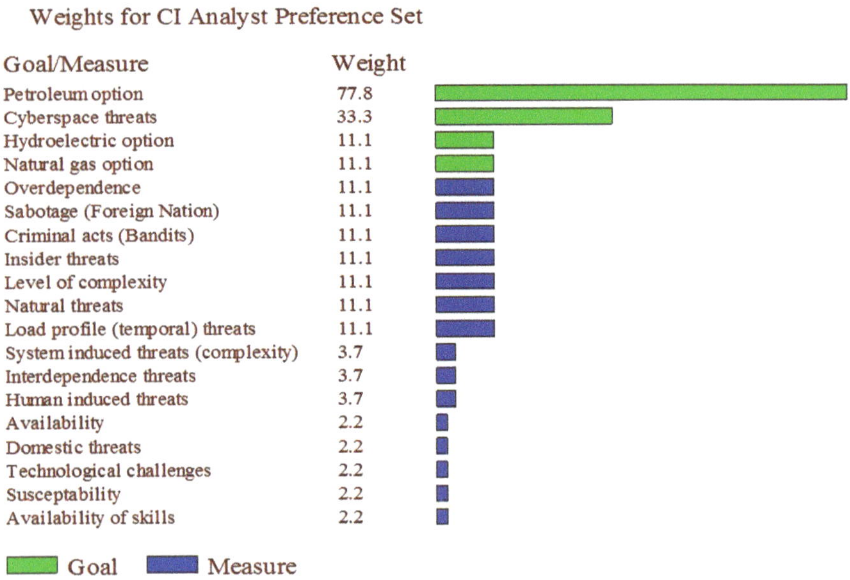

Fig. 8.12 LDW goal/measure and weights for alternatives and attributes

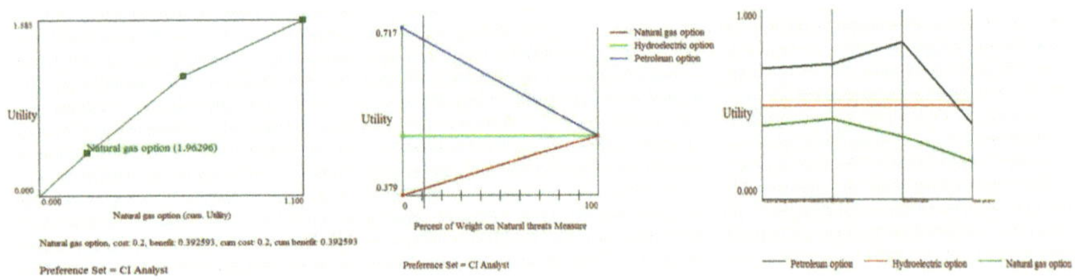

Fig. 8.13 Natural gas option utility to overall cumulative utility; sensitivity analysis for natural threats three alternatives; and LDW results for ranking graph

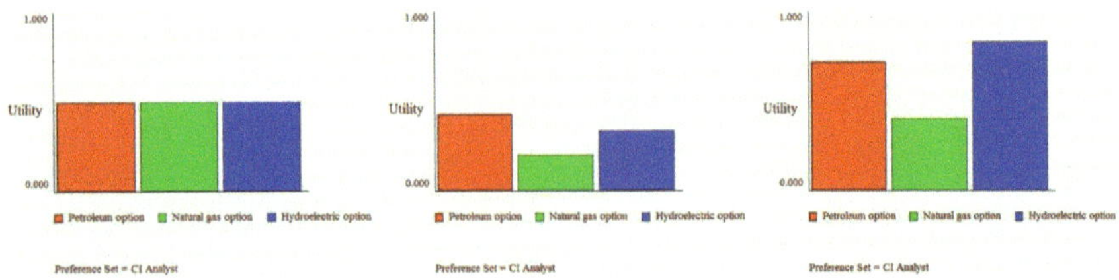

Fig. 8.14 Utilities under options of hydroelectricity; natural gas, and petroleum

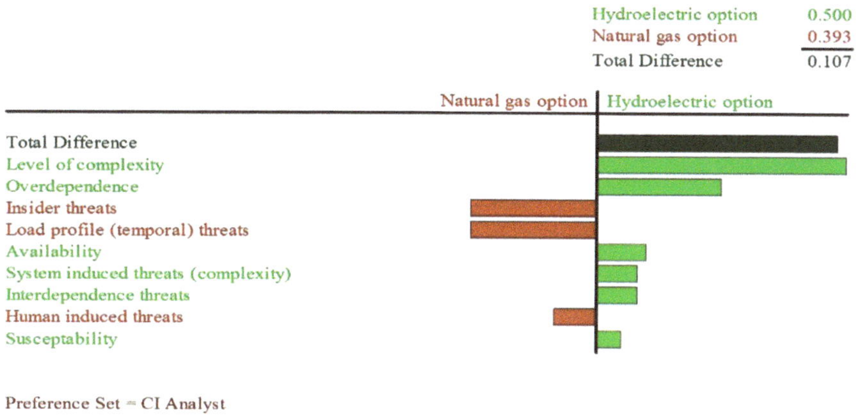

Fig. 8.15 Comparison of natural gas and hydroelectricity options

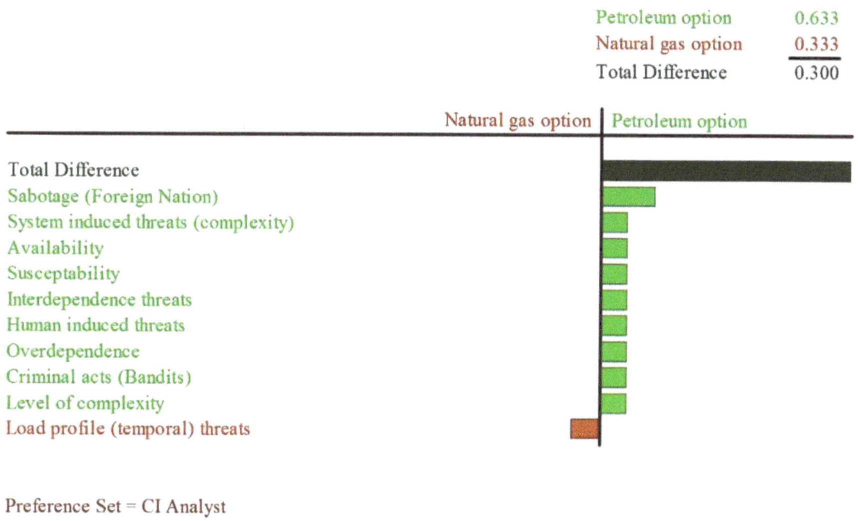

Fig. 8.16 Comparison of natural gas and petroleum options

methods, given the topic of present research a consideration of multiple systems with activities. In such a case, we can assume:

- *n* activities are being considered by a group of interested people having as goals to provide judgments on the relative importance of the activities; and to make sure that the judgments are quantified to an extent, which also permits a quantitative interpretation of the judgments among all activities.

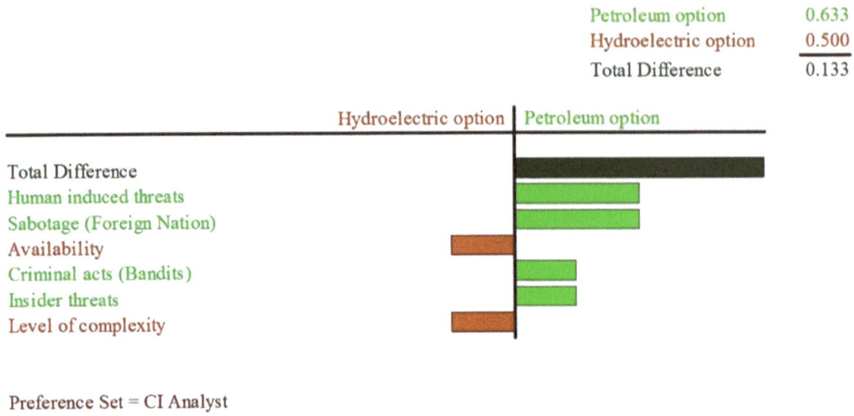

Fig. 8.17 Comparison of hydroelectric and petroleum options

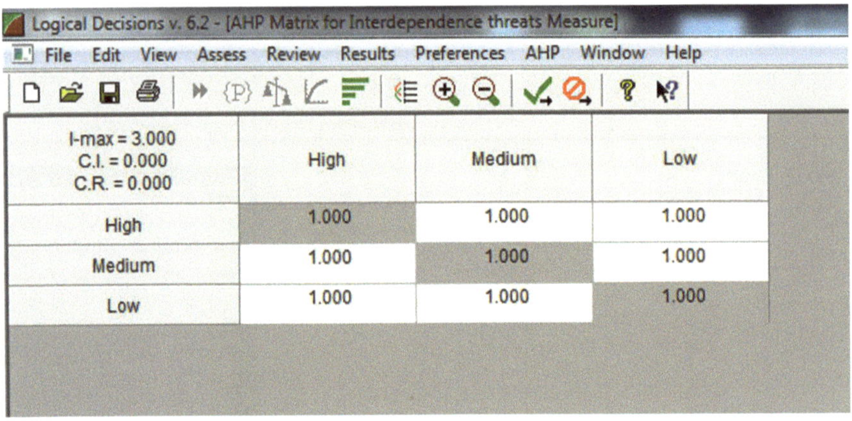

Fig. 8.18 LDW matrix similar to AHP goals

- In formal terms, we could let $C_1, C_2, \ldots, C_n$ be the set of activities. The quantified judgments on the pairs of activities $\{C_i, C_j\}$, $i, j = 1, 2, \ldots, n$ are synoptically represented by an $n \times n$ matrix.

  $A = (a_{ij})$, $i, j = 1, 2, \ldots, n$; The entries a_{ij} are defined by the following rules:

- Rule 1: If $a_{ij} = a$, then $a_{ji} = 1/a$, $a <> 0$.
- Rule 2: If C_i is judged to be of equal relative importance as C_j, then $a_{ij} = 1$; in particular, $a_{ii} = 1$ for all $_i$.

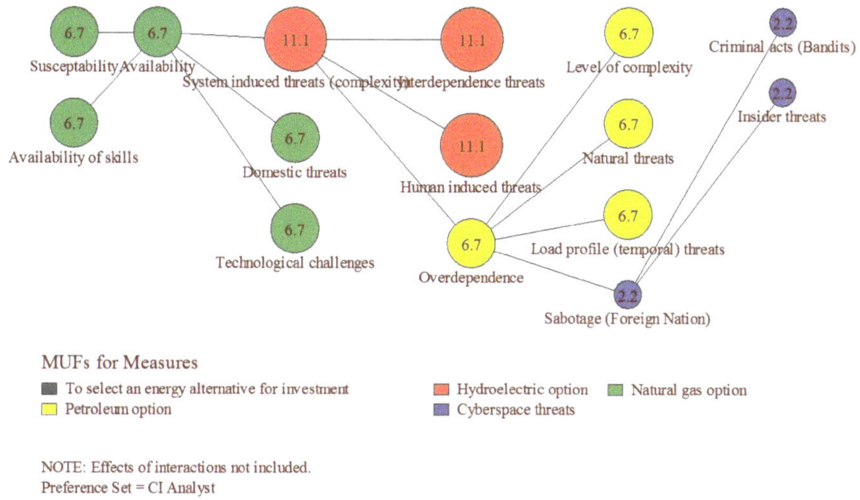

Fig. 8.19 LDW trade-off and weight analysis different measures of alternatives

Upon these, the matrix A has the form:

$$A = \begin{array}{ccccc} 1 & a_{12} & a_{13} & \ldots & a_{1n} \\ a_{21} & 1 & a_{23} & \ldots & a_{2n} \\ a_{33} & a_{32} & 1 & \ldots & a_{3n} \\ \ldots & \ldots & \ldots & \ldots & \ldots \\ a_{1n} & a_{2n} & a_{3n} & \ldots & 1 \end{array} \tag{8.1}$$

A matrix observing such a configuration is known as a 'reciprocal matrix.' Once the reciprocal, evaluation matrix is built up; the problem is to obtain from it a set of numerical weights, w_1, w_2, w_3, …, w_n that would feature the priority to be assigned to the contingencies C_1, C_2,…, C_n. Based on an ample intuitive justification and a sound mathematical reasoning, Saaty's research on AHP (Forman and Gass 2001) proves that: The Priority Vector (w_1, w_2, w_3, …, w_n) sought is the eigenvector of the Maximum Eigenvalue of matrix A; the closer the Maximum Eigenvalue is to n, the more 'consistent' the AHP is considered.

A complex hierarchical structure can be described as a super-set of C_{1k}, C_{2k},…, C_{nk} contingency sets, organized on a number of levels, k = 1, 2,…, m, and interconnected to a certain degree. To prioritize the components in the bottom-level of the hierarchy, one:

(a) Gets the priority vectors of type (a) for every component in a set k, performing the pairwise comparison from the standpoint of every criterion (component) in the preceding, k–1, level of the hierarchy with which logical/functional connections exist; in this, take the overall objective of the hierarchy as level-0, and qualify as 0 the non-connected items, for a mathematically uniform description;

obtain, matrices consisting of priority eigenvectors as columns, and featuring every hierarchy level

(b) Multiplying the obtained priority matrices in the reverse order i.e., from the bottom hierarchy level up; obtain, a unique vector that features the aggregated, multi-criteria evaluation of the priority to be assigned to the items in the bottom (objective-oriented) level of the hierarchy

(c) Computing the Maximum Eigenvalue, λ, associating the Priority Vector; use this to compute the Consistency Index (CI), qualifying the quality of the priority evaluation, as $\mathrm{CI} = (\lambda - n)/(n - 1)$; use, in turn, CI to compute the Consistency Ratio (CR), as the ratio of CI to a 'Random Index' (RI) of reference, CR = CI/RI, explained and computed; a CR lower than 0.1 characterizes the evaluation as 'satisfactorily consistent.'

In practice, the sequence of (a), (b), and (c) makes up the AHP process. A number of methods, of various degree of accuracy, to obtain the priority eigenvector, the corresponding maximum eigenvalue, and indices of consistency of the evaluation, are in use. For the Priority Eigenvector:

1. multiply the n elements in each row of the user-entered Evaluation Matrix;
2. take then nth root of the product;
3. normalize the resulting numbers so that the sum total be 1.

For the Maximum Eigenvalue:

- multiply the Evaluation Matrix on the right by the Priority Eigenvector;
- divide every element of the column-vector obtained, by the original components of the Priority Vector;
- sum all components thus obtained, then divide the sum by the number of components, n;
- the result is an approximation of the Maximum Eigenvalue.

A working example could serve as prototype for developing a code for implementing the described processes. In examining these, note that same code is configurable in different fashions, to accommodate different systems and objectives.

References

Banville, C., Landry, M., Martel, J.-M., & Boulaire, C. (1998). A stakeholder approach to MCDA. *Systems Research and Behavioral Science*, *15*(1), 15–32. https://doi.org/10.1002/(SICI)1099-1743(199801/02)15:1<15::AID-SRES179>3.0.CO;2-B.

Bellamy, J. (Ed.). (2004). *Regional natural resource management planning: The challenges of evaluation as seen through different lenses*. Brisbane, Australia: Science Communications Unit, Natural Resource Sciences, Department of Natural Resources and Mines, QLD. Retrieved from http://research.usc.edu.au/vital/access/manager/Repository/usc:722.

Belton, V., & Stewart, T. (2002). *Multiple criteria decision analysis: An integrated approach*. Norwell, MA: Kluwer Academic Publishers.

Bollinger, D., & Pictet, J. (2003). Potential use of e-democracy in MCDA processes. Analysis on the basis of a Swiss case. *Journal of Multi-criteria Decision Analysis, 12*(2–3), 65–76.
Buede, D. M. (1992). Software review. Overview of the MCDA software market. *Journal of Multi-criteria Decision Analysis, 1*(1), 59–61. https://doi.org/10.1002/mcda.4020010107.
Doumpos, M., & Zopounidis, C. (2002). Multi–criteria classification methods in financial and banking decisions. *International Transactions in Operational Research, 9*(5), 567–581.
Drake, P. R. (1998). Using the analytic hierarchy process in engineering education. *International Journal of Engineering Education, 14*(3), 191–196.
Dyer, J. S., Fishburn, P. C., Steuer, R. E., Wallenius, J., & Zionts, S. (1992). Multiple criteria decision making, multiattribute utility theory: The next ten years. *Management Science, 38*(5), 645–654. https://doi.org/10.1287/mnsc.38.5.645.
Ensslin, L., Dutra, A., & Ensslin, S. R. (2000). MCDA: A constructivist approach to the management of human resources at a governmental agency. *International Transactions in Operational Research, 7*(1), 79–100.
Forman, E. H., & Gass, S. I. (2001). The analytic hierarchy process—An exposition. *Operations Research, 49*(4), 469–486. https://doi.org/10.1287/opre.49.4.469.11231.
French, S. (1989). *Readings in decision analysis*. London, UK: Chapman and Hall.
Gardiner, L. R., & Armstrong-Wright, D. (2000). Employee selection under anti-discrimination law: Implications for multi-criteria group decision support. *Journal of Multi-criteria Decision Analysis, 9*(1–3), 99–109.
Garvey, P. R. (2009). *Analytical methods for risk management: A systems engineering perspective*. Boca Raton: Chapman & Hall/CRC.
Haeri, H., Henriques, D., & Sulyma, I. (2007). Multi-criteria decision analysis: Managing uncertainties in program energy savings cost-effectively. In *2007 Energy Program Evaluation Conference*. Chicago, IL. Retrieved from http://www.cadmusgroup.com/articles/multi-criteria-decision-analysis-managing-uncertainties-in-program-energy-savings-cost-effectively/.
Holzgrefe, J. P. L. (2015). *A framework to simplify the choice of alternative analysis and selection methods* (Ph.D.). Old Dominion University, United States—Virginia.
Holzgrefe, J. P. L., & Hester, P. T. (2016). A framework to choose alternative analysis and selection methods. *International Journal of System of Systems Engineering, 7*(4), 277.
Howard, R. A., & Matheson, J. E. (1983). *Readings on the principles and applications of decision analysis. Vol. 1, General collection., Vol. 2, Professional collection*. Menlo Park, CA: Strategic Decisions Group.
Hudnell, H. K. (Ed.). (2008). Chapter 35 Appendix A: Multi-criteria decision analysis. In *Cyanobacterial harmful algal blooms: State of the science and research needs* (pp. 815–829). New York: Springer.
Keeney, R. L., & Raiffa, H. (1972). A critique of formal analysis in public sector decision making. In A. W. Drake, R. L. Keeney, & P. M. Morse (Eds.), *Analysis of public systems* (pp. 64–75). Cambridge, MA: MIT Press.
Kiker, G. A., Bridges, T. S., Varghese, A., Seager, T. P., & Linkov, I. (2005). Application of multicriteria decision analysis in environmental decision making. *Integrated Environmental Assessment and Management, 1*(2), 95–108.
Levy, J. K., Hartmann, J., Li, K. W., An, Y., & Asgary, A. (2007). Multi-criteria decision support systems for flood hazard mitigation and emergency response in urban watersheds. *JAWRA Journal of the American Water Resources Association, 43*(2), 346–358.
Matsatsinis, N. F., & Samaras, A. P. (2001). MCDA and preference disaggregation in group decision support systems. *European Journal of Operational Research, 130*(2), 414–429.
Munday, P., Jones, A. P., & Lovett, A. A. (2010). Utilising scenarios to facilitate multi-objective land use modelling for broadland, UK, to 2100. *Transactions in GIS, 14*(3), 241–263.
Murray, A. T. (2002). GIS and multicriteria decision analysis, by Jacek Malczewski, 1999. *Geographical Analysis, 34*(1), 91–92. https://doi.org/10.1111/j.1538-4632.2002.tb01077.x.
Paschetta, E., & Tsoukiàs, A. (2000). A real-world MCDA application: Evaluating software. *Journal of Multi-criteria Decision Analysis, 9*(5), 205–226.

Ragsdale, C. T. (2001). *Spreadsheet modeling and decision analysis*. Mason, OH: South-Western College Pub.

Roy, B. (1996). *Multicriteria methodology for decision aiding*. Dordrecht, The Netherlands: Kluwer Academic Publishers.

Saaty, T. L. (2010). Who won the Winter 2010 Olympics? A quest into priorities and rankings. *Journal of Multi-criteria Decision Analysis, 17*(1–2), 25–36.

Sharifi, M. A., Boerboom, L., Shamsudin, K. B., & Veeramuthu, L. (2006). Spatial multiple criteria decision analysis in integrated planning for public transportation and land use development study in Klang Valley, Malaysia (pp. 85–91). Presented at the ISPRS Technical Commission II Symposium, Vienna.

Steele, K., Carmel, Y., Cross, J., & Wilcox, C. (2009). Uses and misuses of multicriteria decision analysis (MCDA) in environmental decision making. *Risk Analysis, 29*(1), 26–33.

Turban, E., & Aronson, J. (1997). *Decision support systems and intelligent systems* (5th ed.). Upper Saddle River, NJ: Prentice Hall PTR.

Wenstøp, F. (2005). Mindsets, rationality and emotion in Multi-criteria Decision Analysis. *Journal of Multi-criteria Decision Analysis, 13*(4), 161–172. https://doi.org/10.1002/mcda.384.

Zhou, M., Liu, X.-B., & Yang, J.-B. (2010). Evidential reasoning-based nonlinear programming model for MCDA under fuzzy weights and utilities. *International Journal of Intelligent Systems, 25*(1), 31–58.

Chapter 9
Airborne Emissions and Territorial Vulnerability Assessment

Abstract This chapter offers insights into a model for assessing vulnerability of territorial kind due to emissions. The procedural agenda for the model are discussed along with break points for chemical and radioactive release. Then the atmospheric dispersion model and its equations are discussed. The main capability of the model, as a decision support system (DSS) code, is to tell consequences of single and multiple releases of widely different durations and time profiles. The model uses a number of user-selected sports in a targeted territory. Results and implications are discussed.

9.1 Procedural Outline

One might be inspired to look for hazards, typically external to the system, affecting the system. On the other hand, one might look at system vulnerabilities. In the present case, emphasis is placed on territorial vulnerability. Specifically, we focus on territorial vulnerability due to airborne emissions. It stands to reason that one requires a good understanding the territory in question (Gheorghe and Vamanu 2005; Oppio and Corsi 2017; Renard and Soto 2014). Along the idea of having requisite knowledge, one might also consider the following fundamentals in the context of airborne emissions and territorial vulnerability:

- A target a *territory* that holds a number of possible effective sources of discharge to the atmosphere—hazardous pollutants (e.g., toxic, radioactive gases, and aerosols).
- Given a *reference point* in a territory (i.e., map), one can establish a *radius of interest* such that an 'order' can be established (e.g., 1 to 1000 km).
- Determine all *sites holding relevant sets of forecast meteorological data*. Meteorological data can include wind direction and speed, cloud cover, and precipitations. These can be established along a radius of interest on the map.
- Use an off-site browser facility to screen the Web source of data and get the model-required meteorological information via appropriate string parsing and physical interpretation procedures, thus building up a meteo forecast file. Each

A.V. Gheorghe et al., *Critical Infrastructures, Key Resources, Key Assets*, Topics in Safety, Risk, Reliability and Quality 34,
https://doi.org/10.1007/978-3-319-69224-1_9

field in the file holds information on: **hour:minute, wind direction, wind speed, cloud fraction, Pasquill stability, Precipitation Intensity (hh: mm/deg. N by E/mph/eights/A–F/0- user given mm/h)**

- In the present case, https://weather.com/en-GB was used for consideration of a typical 8-hour forecast. Then, one marks an investigation for any finite number of *release sources*, by setting their geographic position (WGS84 latitude, longitude), and height above ground—all sources assumed point-wise.
- Using the forecast file, one interpolates, in space and time, the specified data to determine and place on record the *trajectories* of puff releases starting from each and every source.[1] One needs to make the interpolation in time linear, while conducting the spatial interpolation between nodal sites in the source (e.g., https://weather.com/en-GB), list via weighted averages of values provided, at each time, by any number of sites that come closest to the current puff center position. This code version uses as weights the reciprocals of some powers of the distances of sites to the puff center.

There are a couple of caveats to keep in mind: First, the current version of the code employs the meteo information publicly offered by the popular *the Weather Channel* (https://weather.com/), alternative primary sources can be utilized. Of course, when one selects a radius of interest, one restricts the number of data-reporting sites from the thousands in the code's data library, starting with a single station for short radii in the order of a few kilometers.

In the last bullet-point process, one must also consider the *airborne pollutant concentration at ground level*, corresponding to the puff center. This requires accounting for atmospheric stability (Pasquill classes), determined by standard correlations of class to wind speed, cloud cover, daytime, and even the season—fall, winter, spring, or summer. Furthermore, consider ground and inversion lid reflection. The inversion is associated with Pasquill stability in a user-determined fashion. Finally, one must have winds corrected for vertical shear in velocity magnitude. There is spatial expansion since use time-dependent dispersion coefficients, in consideration of their covering wider time spans.

- Break point: select between a case for *chemical releases*, and a case for *radioactive releases*.

9.1.1 The Case for Chemical Release

Define a *source term* to be attached to the release file. The definition requires:

- the specification of the substance that is released;

[1]Note that the release file that is built up in a manner of Sect. 9.1 is independent on other details on the source term(s) and can be used with equal effectiveness for, typically, single-chemical releases and radioactive releases featuring complex mixes of nuclides, of different radio-dosimetric impact.

- the extraction from the code's resident data library of the features of interest in the assessment of the consequences of a loss-of-containment event; these include the Immediately Dangerous for Life and Health (IDLH) limit; the Threshold Limit Value (TLV); the Short-Term Exposure Limit (STEL); the Emergency Response Planning Guidelines (ERPG 1, 2, 3); the *Lethality Probit Function* coefficients and exponent (see Committee for the Prevention of Disasters 1992; Gheorghe and Vamanu 1996; Vamanu et al. 2016).
- the specification of the source strength (kg/puff) for every released puff—with the assignment of an average, uniformly assigned, value as the simplest case; the time profile of the puff loads may be designed to mimic intermittent releases.

Use the release file obtained as described at Sect. 9.1. To determine, (i) chemical doses, (ii) lethality probit values, (iii) lethality percentages, and (iv) the maximal airborne concentrations reached at user-selected locations in the targeted territory. The collection of these files is provided at the decision support *output*. At this point, we find it necessary to inform the reader that, and in accordance with frequent practice: chemical releases the decay and that the dry deposition and wet deposition (i.e., washout) were though of a being of secondary importance and were therefore discarded in the present code version.

9.1.2 The Case for Radioactive Release

One needs to define a *source term* to be attached to the release file. The definition involves:

- the specification of the isotopic mix that is released. One can consider up to 155 species.
- the extraction from the code's resident data library of the features of interest in the assessment of the consequences of a loss-of-containment event. These include the half-lives of the nuclides, activity-to-dose-conversion factors (mrem/(uCi/m3 airborne), mrem/(uCi/m2 deposition), mrem/(uCi ingested)) while considering pathways. These are related to ingestion, (re)suspended deposition, ingestion of contaminated food and a variety of model-related factors such as land productivity, retention factors on crops, and even diet—all can interactively be trimmed in the user interface.
- the specification of the source strength (Ci/puff) for every released puff—with the assignment of an average, uniformly assigned, value as the simplest case; again, the time profile of the puff loads may be designed to mimic intermittent releases.

Use the release file obtained as described at (Sect. 9.1) to obtain the following:

```
ALERT INFO:
___________

- Maximal Airborne Activity Concentration (uCi/m3)
 reached at Time (s. into release)

RADIOMETRY:
___________

- Time-Integrated Concentration (uCi.s/m3)
- Ground Dry Deposition (uCi/m2)
- Ground Wet Deposition (uCi/m2)
- Ground Total Deposition (uCi/m2)

EARLY PHASE DOSES
___________

- Air Immersion External Effective Dose Equivalent (mrem)
- Deposition External Effective Dose Equivalent (mrem)
- Inhalation Committed (50 y) Effective Dose Equivalent (mrem)
- Inhalation Acute Bone Dose (mrem)
- Inhalation Acute Lung Dose (mrem)
- Inhalation Committed Dose Equivalent to Thyroid Dose (mrem)
- Deposition External Exposure Rate (mR/h)
- Total Acute Bone Dose (TABD, mrem)
- Total Acute Lung Dose (TALD, mrem)
- Total Effective Dose Equivalent (TEDE, mrem)
- Deposition 4-day Dose, External&Inhalation of Resuspension, Non-Arid
Land (mrem in 4 d)
- Deposition 4-day Dose, External&Inhalation of Resuspension, Arid Land
(mrem in 4 d)

INTERMEDIATE PHASE DOSES
___________

- 1st-year Dose from Deposition,
External&Inhalation of Resuspension, Non-Arid Land (mrem in 1st year)
- 1st-year Dose from Deposition,
External & Inhalation of Resuspension, Arid Land (mrem in 1st year)
- 2nd-year Dose from Deposition,
External & Inhalation of Resuspension, Non-Arid Land (mrem in 2 year)
- 50-year Dose from Deposition, External & Inhalation of Resuspension,
Non-Arid Land (mrem in 50 year)
```

```
- 1st-year Dose from Deposition, Inhalation of Resuspension,
Non-Arid Land (mrem in 1st year)
- 1st-year Skin Dose from Deposition (mrem in 1st year)

INGESTION PHASE DOSES
___________

Dose from Milk, Cream, Cheese, Ice Cream (mrem)
Dose from Milk, Infant 1 yr-old (mrem)
Dose from Water, Adult or Child (mrem)
Dose from Fats, Oils (mrem)
Dose from Flour, Cereal (mrem)
Dose from Backery (mrem)
Dose from Meat (mrem)
Dose from Poultry (mrem)
Dose from Fish, Shellfish (mrem)
Dose from Eggs (kg/d) (mrem)
Dose from Sugar, Syrup, Honey, Molasses (mrem)
Potatoes, Sweet Potatoes (mrem)
Dose from Fresh Vegetables (mrem)
Dose from Fresh Fruit (mrem)
Dose from Canned Vegetables (mrem)
Dose from Vegetable Juice (single strength) (mrem)
Dose from Canned, Frozen Fruit (mrem)
Dose from Fruit Juice (mrem)
Dose from Other Beverages (coffee etc.) (mrem)
Dose from Soup, Gravies (mrem)
Dose from Nuts, Peanut Butter (mrem)
Dose from Total Diet (mrem)
```

See the collection of these as the decision support output. Note that for radioactive releases the decay, as well as the dry deposition and wet deposition (washout), is of fundamental consequence.

9.2 Model for Atmospheric Dispersion

9.2.1 Model Equations

The atmospheric transport model is described in fundamentals (Sect. 9.1). It draws upon the *forecast files* introduced at fundamental in the same section to generate kinematic quantities—stepwise space displacements of the released puff centers at the input-height above ground. The substantive quantities in the records of the forecast files, namely: **Wind direction, wind speed, cloud fraction, Pasquill stability, precipitation intensity** are subject to a linear time-interpolation between successive values spaced at 1-hour in the site of interest (e.g., https://weather.com/), time sequence, and to a weighted interpolation between the values coming from the nearest *n* data reporting sites to the current position of the puff center—*n* being set by user.

In other words, if $Q(t_1)$ and $Q(t_1 + 3600)$ are two successive, 1-hour-spaced values of any of the quantities in the record above, the value of Q at time t seconds ($t \geq t_1$ and $t < t_1 + 3600$) is:

$$Q(t) = Q(t_1) + (t - t_1)(Q(t_1 + 3600) - Q(t_1))/3600 \tag{9.1}$$

with 3600 s in 1 h considered.

On the other hand, if $Q_i\,(t), i = 1, 2, \ldots, n$ are the values reported at time t by the n closest stations to the current position of a puff center, and d_i are the distances of the reporting sites i to the puff center, then the value assumed by $Q(t)$ in consideration of these is:

$$Q_i(t) = \sum \left(\frac{Q_i}{d_i^p}\right) / \sum \left(\frac{1}{d_i^p}\right) \tag{9.2}$$

with the exponent p defaulted to a value of 2, and left at user's discretion at the interface. To account for the atmospheric dispersion of released pollutants 'SNIFFER' employs a standard puff sequencer. The following briefs on the essential model equations.

The airborne concentration at an arbitrary point in the terrain, at a height of z meters above ground, and at a time t seconds from the puff launch time, from a trail of T puffs, $j = 1, 2, \ldots, T$, launched at 1 arbitrary (user-given) time interval from each other is:

$$C(z;t) = \sum\nolimits_{j=i}^{T} C(r_j, z; t, j) \tag{9.3}$$

where, in the right-hand side, the contribution from the j-th puff is

$$\mathrm{C}(r_j, z; t, j) = \mathrm{C}(0, z; t, j)\exp(-r_j^2/(2.\sigma_h^2(t-j))) \tag{9.4}$$

with $\mathrm{C}(0, z; t, j)$ the concentration at the j-th puff center, r_j [m] the linear distance from the j-th puff center to the observation spot, and $\sigma_h(t-j)$—the horizontal Gaussian standard deviation as a function of the puff 'age', $t-j$ (remember that time j is puff's j launch time).

In turn, puff center concentration is given by:

$$C(0,z;t,j) = Q_j f_{decay}(t,j;\lambda) f_{dry}(t,j;V_g) f_{wet}(t,j;\Lambda)\Big\{\exp\Big[-(z-\ (H-v_s(t-j)))^2\Big]$$
$$+\ \exp\Big[-(z+\ (H_j-v_s(t-j)))^2$$
$$+\ \exp\Big[-(z-\ (2H_{inv}-H_j-v_s(t-j)))^2\Big]\Big\}/\ \Big[2\Pi^{3/2}\sigma_h^2(t-j)\sigma_v(t-j)\Big] \tag{9.5}$$

Here,

Q_j	is the load [kg, C_i] of the puff j;
$\sigma_h(t-j)$[m]	is the vertical Gaussian standard deviation as a function of the puff 'age', $t-j$;
H_j [m]	is the average height above ground, of the j-th puff's center;
H_{inv} [m]	is the Pasquill stability-dependent inversion lid height;
v_s [m/s]	is a settling velocity of pollutant particles—if appropriate; for all intend and purpose, to avoid ambiguities between the deposition velocity and settling velocity, this model version ignores gravitational settling as such and relies only on the deposition velocity V_g.
$f_{decay}(t, j; \lambda)$	is the decay factor depending on the decay constant λ [1/s], the latter relating to the nuclide half-life, $T_{1/2}$ [s] as

$$\lambda = ln(2)/T_{1/2}, \tag{9.6}$$

With $ln(2) \cong 0.693$—the natural log of 2. The equation for f_{decay} is:

$$f_{decay}(t,j;\lambda)\ = \exp(-\lambda(t-j)) \tag{9.7}$$

f_{dry} $(t, j; V_g)$ is the dry depletion factor depending on a dry deposition velocity, V_g [m/s]. The equation for f_{dry} is:

$$f_{dry}(t,j;V_g)\ =\ \exp\{-V_g\,(2/\Pi)^{1/2}\int_j^t d\tau\ \exp[-(H_j-v_s(\tau-j))^2/(2\sigma_v^2(t-j))]\ /\sigma_v(\tau-j)\} \tag{9.8}$$

$f_{wet}(t, j; \Lambda)$ is the wet depletion factor depending on a washout constant, $\Lambda[1/s]$, the latter relating to the precipitation intensity, I_{rain} [mm/h] as

$$\Lambda = \Lambda_o(\text{nuclide})I_{\text{rain}}^{\text{E(nuclide)}}. \tag{9.9}$$

Here, Λ_o and E are a nuclide-sensitive coefficient and an exponent, respectively. The equation for f_{wet} is:

$$f_{wet}(t,j;\Lambda) \;= \exp[-\Lambda(T_{endrain}(r,0)\; - T_{startrain}(r,0)] \tag{9.10}$$

where $T_{start\ rain}$ and $T_{end\ rain}$ indicate the times the rain starts and ends, respectively, at the observation spot situated at r metres on the horizontal, form the puff center, and at ground level.

The total dry deposition from the trail of puffs, at the observation spot and time t, $D_{dry}(t)$, is given by:

$$D^T_{dry\,j}(t) = \sum D_{dry}(r_j;t,j) \tag{9.11}$$

where the contribution from the j-th puff is:

$$D_{dry}(r_j;t,j) \;= V_g C(r_j,0;t,j) \tag{9.12}$$

with the second, concentration factor in the right-hand side of Eq. (9.12) obtained from Eq. (9.4) for

$$z = 0\,\text{m (ground level)}.$$

Similarly, the total wet deposition from the trail of puffs is:

$$D^T_{wet\,j}(t) = \sum D_{wet}(r_j;t,j) \tag{9.13}$$

where the contribution from the j-th puff is:

$$D_{wet}(r_j;t,j) \;= \Lambda\, f_{wet}(t,j;\Lambda) Q_j \exp[-r^2/\;(2.s_h^2(t-j))]\;/\;(2p.s_h^2(t-j)) \tag{9.14}$$

with f_{wet} obtained from Eq. (9.10).

9.2.2 The System of Dispersion

The following is a synthetic description of the dispersion system employed with the code, as it appears at the user interface.

```
TIME-DEPENDENT DISPERSION LAW (Doury)
_______________________________________

The Time Law: SIGMAx = (Ah x t)^Kh
SIGMAy = (Ah x t)^Kh
SIGMAz = (Av x t)^Kv
```

```
Feel free to update data.
You may also wish to SAVE the updated system,
or to OPEN an archived file.

Current settings will become effective on CLOSING.
__________________________________________________

Time Ah Kh Av Kv
__________________________________________________

Atmospheric Stability Class 1 (strong diffusion)
__________________________________________________

0 4.05e-1 .859 .42 .814 < 0
2.40e2 1.35e-1 1.130 1.00 .685 < 1
3.28e3 1.35e-1 1.130 20.00 .500 < 2
9.70e4 4.63e-1 1.000 20.00 .500 < 3
5.08e5 6.50e0 .824 20.00 .500 < 4
1.30e6 2.00e5 .500 20.00 .500 < 5

Atmospheric Stability Class 2 (weak diffusion)
__________________________________________________

0 4.05e-1 .859 .20 .500 < 6
2.40e2 1.35e-1 1.130 .20 .500 < 7
3.28e3 1.35e-1 1.130 .20 .500 < 8
9.70e4 4.63e-1 1.000 .20 .500 < 9
5.08e5 6.50e0 .824 .20 .500 < 10
1.30e6 2.00e5 .500 .20 .500 < 11
__________________________________________________

WIND POWER LAW EXPONENTS, pw
____________________________

Strong Diffusion Weak Diffusion
------------------------------ ------------------
A B C D E F
------------------------------ ------------------
0.07 0.13 0.21 0.34 0.44 0.44 < 12
__________________________________________________
```

```
RECOMMENDED INVERSION HEIGHTS
________________________________

A B C D E F
------------------------------ ------------------
1800 1400 1000 800 125 80  < 13
______________________________________________________
```

9.2.3 Additional Input Conventions

The use of a public source of meteorological data confronts the code developer with the need to interpret some verbal information targeting the layman in quantitative terms. The following is an example of such correlation set(s):

```
CONVENTIONS
___________________________________________________________

The defaults apply to U.S. Weather Channel/UKWeather.com
public info, that is the primary source of meteo data
for which this code version is developed.
One may therefore change the values, yet NOT the keywords.

An option for an alternative primary source would require
negociated code adjustments (see menu's 'About', 'Contact'.

Current settings will become effective on CLOSING.
____________________________________________________________
Contextual Keyword Cloud Fraction Rain Intensity
. (eights) (mm/h)
____________________________________________________________

Rain: 8 1.0
Snow: 8 1.0
Light: 7 0.25
Showers: 6 0.5
Scattered: 5 0.1
Cloudy: 8 0.0
Mostly_cloudy: 6 0.0
Late: 5 0.1
Storm: 7 0.75
Partly: 4 0.0
Fog: 7 0.0
Flurries: 6 0.2
```

```
Sunny: 0 0.0
Mostly_sunny: 2 0.0
Fair: 3 0.0
Clear: 0 0.0
```

Also at the interface level is a capability of changing the key parameter of the meteo sites interpolation scheme:

```
SITE INTERPOLATION SCHEME
_________________________

Site interpolation is performed via a weighted averaging
of the meteo data coming from the closest n sites to the
current position of the puff center, with a power, p, of
the reciprocal site-distances to puff center, as weights.

Default p is 2
_________________________________________________________

Adopted p: 2
_________________________________________________________
```

9.3 Remarks

Vulnerability assessment in a form of a decision support system can offer several benefits to the use. In the present chapter, a chief capability is present in the form of assessment for consequences of single or multiple releases of toxic emissions. Such assessments must be equipped with durations and time profiles for user-selected spots in the targeted territory.

A representative map (see e.g., Fig. 9.1) could be developed from such a model. This working model is consistent with the running time constraints on hardware (i.e., PCs). As such relatively complex models may take time to perform, especially when working with prolonged emissions from numerous sources, and when the pollutant consists of a rich mix of components (e.g., radionuclides).

A complementary approach is a N-WATCHDOG approach. N-WATCHDOG is an experimental software under development at *Universitatea din București* (Bucharest, Romania). The software can deliver a variety of user features, customized interactive and/or services analytical analysis and alerting of nuclear-related risks and vulnerabilities (Vamanu 2014). The software includes advanced capabilities for simulation and visualization as in 'serious gaming' that could be used to increase society awareness of risk, threats, vulnerability as well as

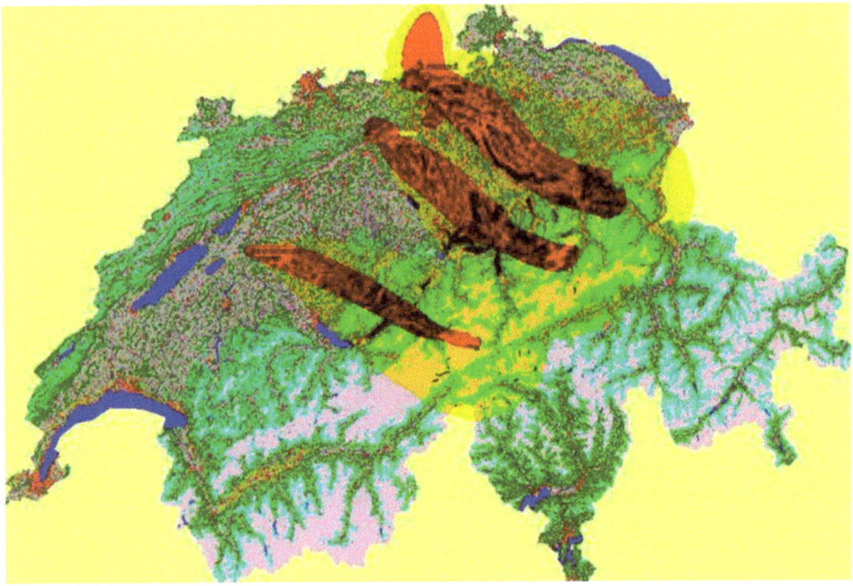

Fig. 9.1 Territorial vulnerability evaluation due to airborne emissions

create a more aligned public perception of risks and merits inherent in nuclear energy, especially in the post-Fukushima Daiichi nuclear disaster.

For convenience, the code should also offer *a synoptic working mode*. Under this mode, the user is called to define a certain, limited area of interest, that is subsequently gridded by the code and calculations are automatically conducted in the grid's knots. Areas in which the characteristic values of time-integrated concentration, doses, lethalities, etc. are found to be large, are then graphically rendered on maps. However, the user is advised that the synoptic mode can be time-consuming.

Noteworthy is the fact that all outputs are placed in immediate comparison *with critical exposure levels*, relating to either (a) health effects or (b) required countermeasures. A variety of nationally enforced or internationally recommended levels could conveniently be provided, online. While the dose (chemical, radiation) and dose-effect working modes are of essence, a more preliminary information may prove particularly attractive to some user categories including the field response operatives and top-ranking officials. These groups might have different objectives associated with information. The information classification, such as *most exposed (alert) areas* in the terrain and *the attention areas*, might be of relevance different decision-making scenarios. In this example, the most exposed areas are obtained by cloud trajectory reconstruction on maps taken at appropriate scales. The attention areas are computed and overlaid on maps, as the stripes of a time-evolving with of a user-give multiple of the dispersion's horizontal standard deviation, for instance $3\sigma_h$, or, more conservatively, $6\sigma_h$.

Therefore, the discussed code architecture has multiple utilities including (i) aiding in crisis as part of management decision support, (ii) emergency preparedness as part of hands-on training and drill tool, and (iii) desktop or portable tool (mobile platform) to keep track of the pollution situation in heavily exposed environments (e.g., region-wide oil fields, large industrial parks, or mega-city agglomerations). A starting place for such research might involve VULPET—a software platform for assessing vulnerability in petrochemical industry developed for *Swiss Re*. More information on VULPET is provided in Appendix G.

References

Committee for the Prevention of Disasters. (1992). *Methods for the determination of possible damage*. The Hague, The Netherlands: Sdu Uitgevers, Den Haag.

Gheorghe, A. V., & Vamanu, D. V. (1996). *Emergency planning knowledge*. Zurich, Switzerland: vdf Hochschulverlag AG an der ETH Zurich.

Gheorghe, A. V., & Vamanu, D. V. (2005). Daily regional vulnerability of infrastructures to obnoxious agents—How vulnerable are you today. In *Proceedings of the Annual IIASA-DPRI Meeting on Integrated Disaster Risk Management*. Beijing, China.

Oppio, A., & Corsi, S. (2017). Territorial vulnerability and local conflicts perspectives for waste disposals siting. A case study in Lombardy region (Italy). *Journal of Cleaner Production*, *141*, 1528–1538. https://doi.org/10.1016/j.jclepro.2016.09.203.

Renard, F., & Soto, D. (2014). Measuring territorial vulnerability? An attempt of qualification and quantification. In B. Murgante, S. Misra, A. M. A. C. Rocha, C. Torre, J. G. Rocha, M. I. Falcão, … O. Gervasi (Eds.), *Computational science and its applications—ICCSA 2014* (Vol. 8582, pp. 331–343). Cham: Springer International Publishing. Retrieved from http://link.springer.com/10.1007/978-3-319-09147-1_24.

Vamanu, B. I., Gheorghe, A. V., & Katina, P. F. (2016). *Critical infrastructures: Risk and vulnerability assessment in transportation of dangerous goods—Transportation by road and rail* (Vol. 31). Cham, Switzerland: Springer.

Vamanu, D. V. (2014). *Sistem de alertare timpurie si asistare computerizata a deciziilor, bazat pe evaluarea anticipativa a dinamicii rapide a vulnerabilitatilor induse in teritoriu de obiectivele nucleare* (No. PN-II-PT-PCCA-2013-4-0262). Bucureşti, Romania: Institutul Naţional de Fizică şi Inginerie Nucleară "Horia Hulubei".

Chapter 10
System Resilience Governance

Abstract Globalization, Digitalization, Forth industry revolution (interconnected cyber-physical systems), Internet of 'Everything' all require new method, concepts, and solutions to document, understand, analyze, operate, control, and transform critical infrastructure as a whole or in parts. It is not new that critical infrastructures are highly interconnected and collaborative and therefore susceptible to domino effects supported by their systemic dependencies. Yet, it is still often a surprise when something does not work. The reason for this is deeply embedded in the very issue governance: the often-unknown system purpose of parts of a comprehensive system of systems landscape, the dynamic driven by a volatile environment, the ongoing change, cyber events, and the impact of politic or social media. A matter of fact is in-transparency and a missing or not properly maintained dependency does not help to manage a normal as well as a complex situation—a system of system landscape under special circumstances.

10.1 Related Terms

10.1.1 Governance

At this point in time, one can only hope that, if we have learned anything, in regard to critical infrastructures and management, is that critical infrastructures are complex interconnected system of systems, very difficult to govern, and require slightly different rules and regulations. These statements are made in the context of responsibility, accountability, ownership, concerned and informed parties. Governance might have to be transparent, traceable, and maintained.

Governance transparency is not only a legal issue but also an operational one. The Governance is characterized based on a well-know concept called RACI: Responsible, Accountable, Concerned, and Involved (Meredith et al. 2016; Smith and Erwin 2005). This concept is enhanced by ownership and implemented as a model-driven multi-attribute system description called RACIO (Dickstein 2008). The RACIO concept is additionally applied to the organization system definition

A.V. Gheorghe et al., *Critical Infrastructures, Key Resources, Key Assets*, Topics in Safety, Risk, Reliability and Quality 34,
https://doi.org/10.1007/978-3-319-69224-1_10

where each entity in a model is modeled by reusing this concept and to document their governance. Governance is the option to documenting organizational readiness with facts and figures. With additional attributes, a description about organizational distributed intelligence is important information to understand organizational actions and reactions without focusing individuals. To understand decentralized intelligence is not only a technical documentation but also an organizational advantage. Historically, centralized organizations are mostly slow, sluggish, inefficient, and not very effective because local opportunities are not recognized. Therefore, decentralized organization with a transparent governance is more agile and effective. To document governance as a multi-attribute model, different important triples are in place, e.g., between asset, processes, person, role, board, organization, legal entities or groups, cost center as well as a job description regarding managing competences and individual skills.

To carry out a task with the applied competences and responsibilities, a role with defined skills is required. This role has a relation to a person who has skills and an intersection with the required knowledge. The multi-level and model-driven-documented system of systems governance based on the RACIO concept allows a comprehensive organizational description and readiness check. The option to apply distributed intelligence concepts to improve productivity based on efficiency (time) and effectiveness (quality, maturity, and trust) is a further value.

Additionally, a well-documented governance description based on the RATIO concept is supporting organizational transparency and tractability.

10.1.2 The Era of Living System of Systems

A living system of systems description[1] consists of a number of specific interconnected entities (called artifacts) from a generic predefined classification system. The model-driven approach provides a possible quality and maintained virtual picture of a real system of systems landscape. More than one source can be discovered and mapped (i.e., discover, map, visualize, and improve is a systemic, systematic and standardized approach to digitalize a simple or complex system of systems landscape with all definitions, rules, and dependencies as a model) to get a redundancy-free, maintained, and interconnected collaboration map available as a multi-attribute holistic model in best quality and maturity. After discovering and mapping (digitalization) the model, content can be automatically generated as stakeholder-oriented visualizations based on incorporated rules. Different visualizations depending on the selection and content can be provided as tables, pictures, text, dashboard items, or in any required combination. The holistic model has more

---

[1]In an economy, there are many 'System of systems' including individual enterprise, financial service organization, nuclear plant, public service, political system, and cross-boarder telecommunication provider.

than one implicit structure, depending on the entry point and applied filters; therefore, many different structures can be shown. Depending on the rules, essentials of a system of systems can be aggregated, collapsed, expanded, visualized, or in details documented.

A living system of system description consists of a certain amount of generic or special artifacts (entities) out of a classification system. Arguably, the most important artifact of a system of systems description (enterprise) is called *ValueChain*. This description covers an end-to-end view (e.g., order to cash). This cross-department description digitalized as a 'ValueChain' covers besides all conditions, rule and attributes also indicators and deviations as well as possible incidents or change requests. The artifact 'Processes' with all individual incorporated 'Activities' have a vertical view with a focus to document a control flow orientation. 'Processes' are supported by different type of resource artifacts (e.g., 'Application,' 'Organization,' 'People,' 'Tools,' and 'Logistics'). In a stack below are 'Infrastructure' and 'Facilities' to get a vertical picture of the system of system landscape (hierarchy and tower). The artifacts 'Influence,' 'Business Rule,' and 'Control' are available to document, measure, and report the outbound conformance. For inbound conformance, the principles can be reused. The artifact 'Dependency' is used to document, measure, and control interconnectivity and systemic relationship of a higher order. A multi-attribute system of systems description is very complex because it is a digitalized virtual representation of a real system of systems landscape.

A Living system of systems model exists mostly in two normal situations (in operation {as-is} and in transformation {to-be}) and sometimes also in a complex situation (exceptional circumstances). The difference between a normal and a complex situation is that the complex situation is unknown (not always a black swan), unexpected, and stressor-driven. Most of the time from outside, influencing established system boundaries or across multiple systems, and usually there is not enough information available to understand all kind of risks and consequences, its impact or to make sustainable decisions. Energy fallouts, system faults, terror attacks, cyber threats, and many others are typical stressors that can produce a complex situation.

During the development, operation, assessment, or transformation of living system of systems, many stakeholder groups require information. Therefore, it is important to have a consistent and trusted system of systems description available at any time. Here, all the required capabilities must be documented and available to deal with the system of systems dynamic, to understand performability, conformability, changeability, and riskability.[2] With this unique, redundancy-free, and

---

[2]Performability, conformability, changeability, and riskability are an evidence, entrepreneurial and economically driven approach (concept and pattern) to document and represent generic required capabilities to manage system of systems traceable and sustainable. Performability focuses on product, market, revenue, cost, profit, and solvability. Conformability covers how all legal requirements, commitments, and liabilities can be met, how promises are managed and agreed contracts and SLAs fulfilled. Changeability manages all change requests, internal and external

model-driven system of systems description, economic and entrepreneurial decisions can be supported, and possible consequences analyzed, visualized or simulated.

10.1.3 System Context

To manage a complex situation, usually only a part of the system of systems is involved. There are many reasons for this. Independent of availability and quality, a 'system context' can be built at any time and can cover the entire system or just a part, perhaps also with an overlap or intersection with another system of systems or system context. Each System context consists of interconnected artifacts and is aggregated on different layers, depending on the structure of the underlying living system of systems. System context can be built dynamically. It does not always represent a consistent system landscape but often a logical entity.

A system context is essential to get a manageable and understandable abstract of the system of systems and is visualized and represented on, for instance, a system resilience governance profile[3] as a bubble. The dependencies between different system contexts are visualized in an aggregated form. So, if a system has a dependency on some level, it will be automatically visualized on the highest level. The dependencies always have a direction, strength, and an impact. In some cases, content (work product) is also involved. This moves from one context to the other in a specific scenario. System context can exist on different levels and must not be free of overlap. The involved artifacts in a system context are always unique and free of redundancy. They are maintained by the interconnected living system. If this is not possible, quality and maturity depend on the manual maintenance cycle. The system context is the essential foundation to manage complex situations with dynamic capabilities.

10.1.4 Dynamic Capabilities

In the system of systems engineering and management, it is important to have a sustainable, traceable, and managed system of systems description in best possible

(Footnote 2 continued)

demands, lifecycle and innovation GAPs, incidents and maintains issues. Riskability is the balance or the intersection between the four topics with a special focus. This capability is difficult to manage, prevent, forecast, and predict because it is fuzzy and often depends on people's behavior, attitudes, and current circumstances.

[3]System resilience governance profile is a comprehensive representation of a specified, validated, and assessed certain amount, part or entire system of systems landscape.

quality every time and everywhere available. The system resilience governance architecture consists of distinct instruments is developed to meet this request. All these instruments provide (partly or fully) the required information about system of systems to get dynamic capabilities to manage (normal) complex situations across the entire life cycle manageable documented. Dynamic capabilities provide a holistic view and description about all involved artifacts in a requested situation, with all their details (e.g., attributes, rules, exceptions, and dependencies). The system context summarizes also all artifacts of involved or concerned with other system of systems and their dependencies.

All dynamic capability artifacts are from type relational higher order and all involved artifacts in a specific scenario, along with their dependencies, are summarized under a dynamic capability. A capability can have its own governance, applied activity, individual risk, and risk dependencies, system and system dependencies, and additional profile information. The homeostatic behavior and dependencies are managed by applied model rules.[4] The dynamic capability is a comprehensive way to document what is required and shows all artifacts and dependencies. A capability can have its own individual configurable visualizations and indicators.[5]

Many different types of capabilities are applicable. A risk capability specifies what is required to manage or mitigate a specific risk [gross] to a target value/position = risk [net], and what consequences are acceptable (appetite) or the managed/controlled transfer. A performance capability describes what is required (value chain, process, application, organization, information, etc.) to offer a product under local regulations to a customer. An organization capability specifies what is required to run a lean, efficient, and effective organization that is profitable and that conforms (organizational readiness and distributed intelligence). A resilience capability shows what is required to manage and control a system resilience governance profile. A transformation capability specifies what kind of resources are required, to transform a current situation, to make it transparent and traceable for the future under observance of quality, functionality, time, budget, and so on.

[4]The artifact capability in a specific system of system context (e.g., enterprise) collects and covers dependencies, and shows what is used to offer a product (what) at a specific location (where) under the valid regulations (what has to be done). The process indicates how a capability is performed, while the organization (who) performs a capability. The value chain (why) offers the value (asset, product, and service) at a place under local conditions. The application with the applied infrastructure (whereby) supports the capability. The information shows which data is used or required to perform consistently and at high quality, with maturity and trusted capability.

[5]Capabilities can be documented and visualized in different ways (e.g., capability risk matrix, capability visualization (i.e., heat map, sensitivity, benchmark, quality–maturity–trust, and time series), and capability control matrix) as a check to ensure that everything running as required.

10.1.5 Complex Situation

A complex situation is, in fact, a normal situation at a given moment or in a given circumstance.[6] The difference is perhaps just the time, the unbelievable place, the incredible brutal act, or the surprise where not all information about a system of systems are available, specific information are missing or unknown at this time to understand a situation or to make a decision.

The model-driven multi-attribute model-driven system of system description context is a virtual picture of a real situation and can be at any time systematically reduced or enlarged. Therefore, a valid digitalized virtual image about a real or imaginary system is available as a model. To manage a complex situation, additional steps and techniques are helpful and sometimes essential. To manage a complex situation properly, additional information is perhaps required. This information can be incorporated real time on the meta-, physical- as well as information layer. Additional system- and risk assessments are required to find the most effective and efficient ways to manage that specific complex situation. The most important and critical factors in managing a complex situation are time, knowledge, patients, overview, and ownership. Therefore, a model-driven multi-attribute system of systems description, available as a maintained virtual picture, always offers a value for comprehensive analyses, simulations, and action planning.

10.1.6 Consequences

One of the biggest issues today in the discipline risk management, especially in the process steps risk assessment, and risk mitigation, is the constantly underestimated or ignored consequences. This matter of fact is often the main reason for additional risks, additional damage and the initializer for domino effects across systemic dependencies. Consequences are sometimes not clearly visible or not until an unpredictable time shift. The reason for this difficulty is different systems are involved or other governance.

Consequences, in a multi-attribute model-driven system of system description, are either modeled as dependency, as risk shift scenario of an existing risk,[7] as additional attribute or as additional risk. All those options are dynamically

[6]A common known complex situation is 9/11. The terror attack was unknown, unexpected before time of the event. The involved and concerned system of systems where unknown, had a complex governance, information where not available about a certain time and consequences where not all seen and managed over a period of time.

[7]Consequences characterized as impact or risk shift are on all systemic-related (dependent) risks. A risk shift on a systemic related risk is visible shift depends on dependency strength, -impact {vector shift} or attribute modification.

applicable to manage possible consequences. The identification of possible consequences is still difficult and requires skills, knowledge, and special expertise.

10.2 Meta-Model

To manage critical infrastructure under this dynamic environment requires more flexibility, transparency, and traceability in administration, operation, technology, business, and legal. Under the current fourth industry revolution and the increasing legal pressures, a paradigm shift is required and new instruments are necessary. Additionally, and in reference to Albert Einstein's call for different approaches, a System Resilience Governance Profile has been developed to manage complex situations in present times.

The system resilience governance profile is a standardized comprehensive profile with focus of living system of systems engineering and management and sustainable development. We make the following observations:

- Incorporated fragility rules and predetermined breaking points to protect a system for total damage driven by domino effects
- A comprehensive collaboration and system dependency management across different types (legal, technical, organizational, political, and administration)
- A focus not only on system of system level across the entire life cycle but also on specific events called complex situations[8]
- A documentation option and a rule-based multi-attribute assessment approach for complex governance- and overlapping ownership structure with diametric interest
- Possibilities to document and cover organizational issues regarding, efficiency, effectiveness, feasibility, and manageability topics
- A choice to measure and show dynamics across the strength, impact, and effects of systemic dependencies
- An opportunity to validate, assess, and compare automatically the quality, maturity by the living system of systems maintained information and their trust

This paradigm shift is what we call *system resilience governance,* and the visualization is supported by a comprehensive *System Resilience Governance Profile*. The profile covers and abstracts all aspects of a complex living system of system landscape divided in manageable system context.

All interdependencies between the individual system contexts are aggregated across different levels, covered, and visualized. All legal, regulation, policy as well

[8]To differentiate between a 'normal' and a 'complex' situation, think in terms of 'time.' In complex situations, time for actions is short and the rate of change of the environment in which the system is embedded is dynamic. Thus, there is no enough time to collect additional data or to start talking, actions are required.

as directives are incorporated by the multi-attribute aspect called conformability and attached to the involved or concerned system context. The performance, profit, and value aspects are integrated and covered under the aspect performability. The balance between conformability and performability with additional comprehensive assessments, events and vulnerability profiles are measured, aggregated and visible under the topic called riskability. Additionally, to the three mentioned abilities, all aspect of change and transformation of a system context are summarized and managed by changeability. Also, this aspect will influence the Riskability and possible normal situations as well as exceptional complex situations of the living system of systems.

To develop such a profile, the meta-model shown inFig. 10.1 is an essential building block but is not always required. The system resilience governance profile covers all dynamic capabilities to understand, manage, and transform a living system of systems landscape in a normal as well as in a complex situation on a comprehensive high management level as well as on a detail level. All entities in this meta-model can be decomposed and for analyses and report aggregated based on predefined entity rules.

A *Complex Situation* shows a living system of systems landscape and there involved and concerned entities driven by a specific stressor (event) under specific circumstances. Each component is documented as a model with all their capabilities, attributes, and dependencies.

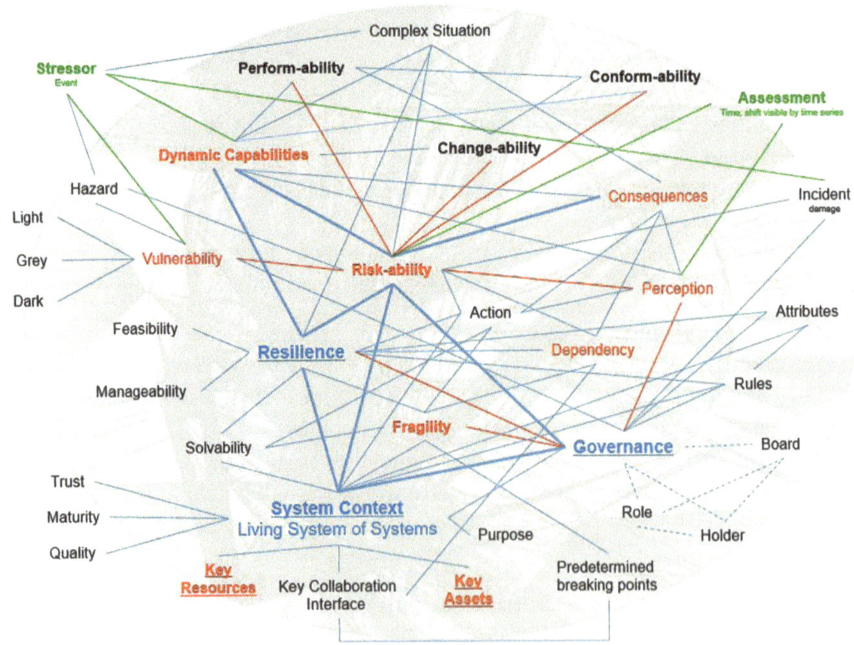

Fig. 10.1 System resilience governance meta-model

To avoid useless complexity instead of a comprehensive system of system description, a system of system landscape can be documented with interconnected *system context* (well-defined disjunctive or overlapping number of artifacts). The purpose of each individual system context or the related system of systems with all their standardized entities (artifacts) and dependencies is to outlined and characterized all required attributes, rules, and exceptions. A special aggregation of elements (cluster of artifacts) in key resources, key assents as well as their key collaborations and interfaces exists to follow the expanded standard of homeland security.

Depending on the maintenance cycle or the degree of interconnection to the real existing system landscape in operation, quality, maturity as well as the trust of information can be measured or calculated based on a set of predefined rules. For each artifact, system, system of systems, or system context, the related governance with the individual mapped responsibility, accountability, concerned or involved parties can be mapped, characterized, measured, and assessed. The governance self is divided between person (Holder), role, job, board, and the related organizational definition. By the fact that each artifact covers attributes (responsible, accountable, concerned, and involved) as link to person and role, the entire governance can be sustainable documented.

The system of systems resilience description depends on each individual system context, the related dependency, the specified governance, and the applied *risk-ability* (capability to identify, assess, mitigate, control, and monitor risks with their dependencies), the applied and validated type of actions (manage, control, and mitigate risk) as well as individual rules and attributes. Each context is assessed and depends on the outcome of feasibility and manageability assessment positioned on a four-quadrant map where also the fragility and solvability of a system context can be shown.

Related to the stakeholder dependency risk *perception* and *consequences* can be identified, documented, calculated, reviewed, implemented, and visualized related to their governance. Dependent of all individual available attributes, rules, facts and figures and indicators the performance, conformance and change-related description can be developed, documented, and visualized. If all required dynamic capabilities are outlined the system ability to manage and improve performance (*perform-ability*), to fulfill internal and external legal requirements, internal policies and rules (*conform-ability*), to close required modifications, requests for change or planned improvements (*change-ability*) as well as to understand, mitigate and monitor identified risks (*risk-ability*) a comprehensive and sustainable system of system documentation is characterized and available.

All the abilities are summarized to an aggravated behavior of a system of systems landscape call dynamic capabilities. One ought to ask: are dynamic capabilities required? or are present capabilities capable of managing and transforming a system of systems landscape? At any time, all or a required part of dynamic capabilities with all the related entities can be clustered to support an assessment. In this case assessment is driven to support a comprehensive *vulnerability*[9] analysis. To protect a system of systems landscape for total damage based on dynamic capabilities, the resilience profile, and the fragility attributes and governance rules and definitions, predetermined breaking points can be defined and incorporated to prevent or stop domino effects belonging to systemic dependencies. For the sake of completeness, there are incidents (to document claims) as well as hazard (to characterize and describe possible risk) visible and incorporated in the meta-model.

Based on this model, all essential assets, resources, and dependencies of critical infrastructure can be digitalized, characterized, and documented. This documentation can be used for development, operation, transformation, and any kind of assessments. Additionally, all those models can be used to compare the entire or part of the living system of systems landscape. All this work can be supported because all elements of a critical infrastructure are highly standardized. In the century of the forth industry revolution, there is no excuse not to use such efficient and effective concepts to understand, monitor, and transform critical infrastructure.

10.3 System Resilience Governance Architecture

The unstoppable advance of globalization, the shift of the individual markets, the resulting increase in networking and complexity, the incredible amount of reactive government-released regulations, and many other trends alter the risk landscape and the hazard potential of the entire system of systems world. To manage the market dynamics, the complexity, the sustainability, and the simultaneously and constantly changing threats driven by corruption and terror, new thinking models, concepts, system abilities, skills, talents, and solutions are required. This might be especially the case with critical infrastructure especially the new comer of space critical infrastructure. The potential of misuse appears growing and there is a lack of transparency and governance.

The system resilience governance architecture is a solution framework to deal with the paradigm shifts in the system of systems engineering and risk dependency management. With the integrated system resilience governance instruments, a normal or complex situation can be completely and systematically documented as a multi-attribute system of systems description. This digitalized virtual picture of a real living system of systems landscape or a specific system context can be used for advanced analyses, operation, transformation, assessment, and benchmarking tasks. Therefore, all of the required dynamic capabilities to manage a normal- or complex

[9]Vulnerability analyses of a living system of systems landscape can be distinguished in bright-field, dark-field or gray-field analyses. In case of all required information to support a comprehensive vulnerability analyses (attributes, function, rules, relations) are available in detail a bright-field analyses can be supported. If Information is only available on meta- or principle level or just structure are known a dark-field analyses can be developed. For all analyses where not enough or only partially information is available a grey-field can be done or is suggested.

situation are documented and available in a standardized form. The available instruments of the system resilience governance architecture are summarized (clustered) in four areas and based on the specified documented in Fig. 10.1 above.

The system of systems management cluster supports the systematic documentation of a living system. To document and manage a complex situation, additional instruments are available under complex situation management, and with the instruments under systems resilience governance evaluation, assessment and profiling of a (normal or) complex situation is supported. The system resilience governance framework is in fact a methodology and approach that contains all the methods and techniques required to develop a consistent and high-quality description of a real system context.

System resilience governance architecture is a model-driven system design solution for how dynamic capabilities are modeled and implemented. It covers the entire solution. The system resilience governance framework is a methodology for the multi-attribute model-driven system of systems operation and transformation description. This description is documented as a cookbook, where concepts, the approach, and the methodology for standardization, digitalization, collaboration, visualization, and industrialization are summarized.

Multi-attribute system of systems description represents a real system that is digitalized, documented, and visualized as a model. Every element of a system, system of systems, or system context is represented as an artifact according to the definitions specified in a classification system. In the classification system, artifacts are modeled as a combination or assembly of predefined objects with a standardized set of functionalities applied. There is more focus on dependency, collaboration, and managed interfaces. One way to manage relations on a higher order (in a context-sensitive manner) is to offer stakeholders options to look around the corner. Multi-attribute governance management instruments and visualization techniques offer solutions to deal with fear- and circumstance-driven perceptions, and the consequent risk assessments.

Multi-attribute decision support helps manage more dynamics and market volatility. Outlined multi-conformance management concept to manage complexity in legislation interpretation, GAP analyses and implementation. Multi-attribute risk assessment and risk dependency management for individual assessments and evaluations.

10.4 System Resilience Governance Architecture Instruments

With all the outlined instruments, the system resilience governance architecture can be measured, managed, and transformed in a holistic way with the maximum stakeholder empowerment and value. The system resilience governance architecture consists of four clusters of specific instruments. These instruments represent the dynamic capabilities to manage complex (or normal) situations. Each instrument follows definitions and rules, and has a visualization focus.

The target of all instruments is that real systems, after digitalizing, are available as a virtual picture, where they are represented as unique model-driven multi-attribute system descriptions. A model (multi-attribute system description) is based on artifacts instantiated from a standardized classification system where artifacts are available as node, edge, or interconnected to each other. With this kind of model, every normal and complex situation can be documented. The holistic view of a digitalized real system landscape can be measured, controlled by the multi-attribute indicators. Based on holistic assessments, evaluation, and ratings, information about the model can be visualized in comprehensive and simple diagrams or highly collaborative pictures and documents, and reused by other instruments. Moreover, the governance can be comprehensively documented and transparently visualized. All specific actions to manage the living system of systems, individual systems, a subset of systems called system context or individual artifacts and their dependencies are incorporated and, therefore, a part of the entire model. Every system context is documented in a comprehensive way as a standardized model (virtual picture).

Each dynamic capability is implicitly documented and can be individually measured. The entire system governance is visible and can be measured and expanded. Through integration and the multi-attribute aspect, measures like the solvability of a system can be visualized. With predefined instruments, a normal or complex situation can be compared and benchmarked. The rules and conditions can be modified and applied at any time. Dependencies and relations are managed like all other elements of a system. New instruments with analytics and comprehensive calculations can be additionally configured and integrated.

In the following list, all instruments are described in brief along with their focus. All instruments are documented, visualized, represented, and applied as a multi-attribute model description based on the favorite object-oriented concepts (encapsulation, inheritance, class object, and message).

- System of systems engineering principles cover all concepts, definitions, templates, visualization techniques, quality definitions, and rules to document a system of systems in general, a living system as a digitalized virtual picture of real systems or a specific system context
- A living system of systems is the digitalized virtual picture of a real system representation in all possible dimensions. This picture is available as a multi-attribute system of systems description. It is mainly maintained by discover, map, and visualize utilities
- Risk evaluation is a comprehensive assessment and visualization of all risk types of a living system or system context. There is an incorporated process for defining, assessing, measuring, mitigating, and monitoring risks as well as the required definitions and rules for risk-control frameworks, incidents, and action management. A risk evaluation can be enriched by a risk dependency map
- Based on the dashboard, all facts and figures, indicators, and measurements of a living system of systems with the formulas and equations are represented. There are many different visualization options available

- The scenario allows and supports a rule-based multi-assessment of a living system of systems regarding vulnerability, performance, conformance, risk, and change or against other configurable aspects
- With the assessment, a specific system context of a living system can be validated against standardized topics specified by rules. This assessment can also be used for a system context involved in a complex situation
- The indicator is a fuzzy set-calculated comprehensive number of a set of standardized assessments. This number shows the temperature of a system context—like measuring a fever—within a normal and a complex situation. There is also an option to produce an indicator on a higher aggregation like a system landscape
- With a scorecard, every single assessment can be detailed and supplemented by the timeline (past, current, and target) to visualize changes and sensitivity shown by specific thresholds. The scorecard also has benchmark functionality per assessment topic
- The road map covers all actions (task, measure, and activities) applied while using instruments of the system resilience architecture in operation and transformation, supplemented by resources and other attributes. With this instrument, actions can be structured, aggregated, prioritized, and set in an order
- The profile represents various system context as a comprehensive multi-attribute decision support for visualization. Not only the context but also risks, dependencies, and technical and resource restrictions, with advanced ratings, measures, and qualifications are incorporated. The profile is the highest possible aggregation about a system context in a complex or normal situation in this architecture. It is also a visualization and representation of dynamic capabilities as a specific context-sensitive and holistic system landscape with encapsulation of all relevant essentials
- A complex situation is usually initialized by a special event or call stressor and shows a living system or a system of systems under special circumstances. Therefore, in a complex situation, a special context is in focus and not the entire system. The key features of a complex situation are often time pressure, availability of information, lack of responsibility, and quality and maturity of relevant and essential information. Most of the time, the event was unknown and unexpected for all involved and concerned stakeholders
- The risk dependency map is a special instrument for managing, analyzing, simulating, and visualizing the dynamics, dependencies, influence, and flow of risks and their causality. This instrument supports the analyses of cause-effect and impact. With this map, even loops and mappings can be identified as well as problem and possible solution risk
- Under special circumstances with risk development, many produced risk evaluation results, produced over time or in a special simulation scenario can be shown in a special kind of timeline diagram
- With the predetermined breaking points analyses, there is an option to analyze and visualize possible damage of risk as well as to apply breaks in order to avoid or protect a system from total damage. This instrument can also be used to separate two systems, like in building construction

The system resilience governance architecture uses 14 instruments (articulated above) to document, visualize, and control dynamic capabilities in order to manage complex situations. The corresponding model is depicted in Fig. 10.2: *System Resilience Governance Architecture Meta Model.*

With this architecture, the target of dynamic capabilities to manage complex situations can be met. The architecture solution and procedures support understanding, encapsulating, and visualizing the complexity of a complex situation and the systemic dependencies, identifying the information demand in a dynamic environment, forecasting risk and possible impact, simulating resilience, and implementing action patterns to control and handle continuous change. Especially with the available instruments in this architecture, complex situations can by managed by modeling and measuring system resilience governance.

With this model-driven system of systems description approach, discovered, mapped, and visualized content about system of systems, a living system or a specific system context can be aggregated, stored in a content- and context-sensitive way, and a multi-attribute decision support can be used. Regarding the used and applied standards, language, and taxonomy in the classification system, people can by empowered and a common understanding can be supported. Information available as a multi-attribute model description is powerful. It allows one to develop, control, manage, and transform a complex situation in a dynamic environment.

In a standard approach, system resilience governance can be documented and therefore offer an approach to manage. In a number of steps, a system can

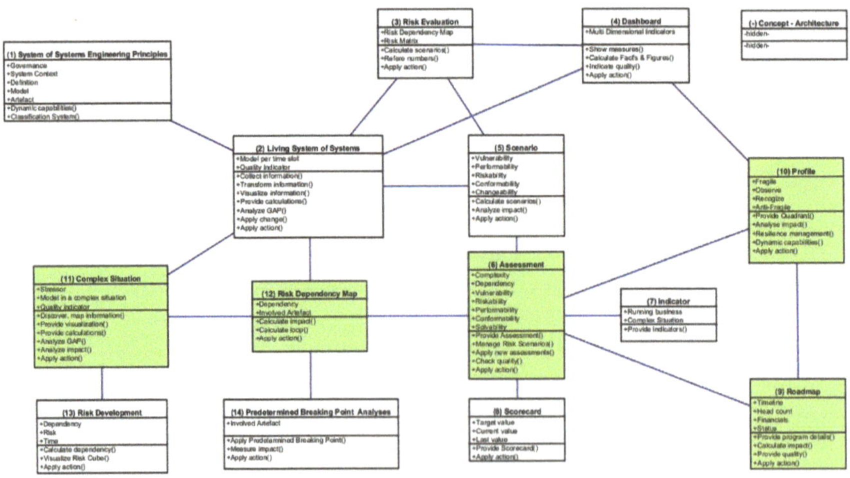

Fig. 10.2 System resilience governance architecture meta-model: this model is elaborated in Chap. 11

systematically digitalized, visualized, and then the results can be used as a basis for system development. These results could then be used as a basis for management, empowerment, and system comparability.

The multi-attribute decision support is a comprehensive balance between performance, conformance, risk, change, and the applied actions implemented and available as a model-attribute model-driven system of system description. Arguably, the presented methodology could be instrumental in thinking and doing something regarding some of the current pressing issues as documented in, for instance, Rosato et al. (2008), Taleb (2010), Vamanu et al. (2016), World Economic Forum (2012, 2016). Additional business and leadership value can be generated as well.

10.5 System Resilience Governance Profile

The most important and comprehensive instrument to document and manage a system of system landscape under special circumstances (complex situation) is the system resilience governance profile shown in Fig. 10.3. These dynamic capability visualizations provide additional information regarding to feasibility and manageability as a combination and aggregation of risk evaluation, risk dependency map, governance, type of fragility, and road map.

The system resilience governance profile has two occurrences and four quadrants. On the horizontal axis (abscissa), the technical, organizational, or legal

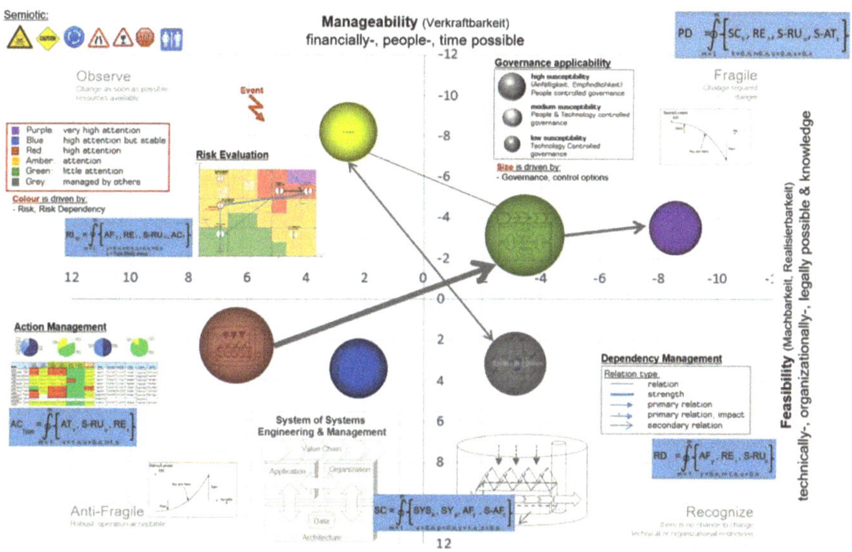

Fig. 10.3 System resilience governance profile

feasibility of the possibility to apply available knowledge is qualified. On the vertical axis (ordinate), manageability depends on resource, time, and money, whereby the ability to execute is qualified. Figure 10.3 shows a system context in a specific status visualized as a bubble with its dependencies. If a system context is on the fragile quadrant, change is required because the system can be harmed. The impact can damage or destroy the system context. Thus, actions are required to the maximum awareness and power. Systems on the anti-fragile corner operate robustly or are on an acceptable operation. System context on an observe quadrant changes as soon as possible or once the required resources become available.

System context on a recognize quadrant has no chance to change right away because there are technical or organizational restrictions. Each system context has, besides its position, also a size and a color with an applied rating. The color is driven by risk evaluation—purple: very high attention; blue: high attention but stable; red: high attention; amber: attention; green: little attention; grey: managed by others. Size is mainly driven by how the governance is managed, while the color represents the risk result. Bubble size shows the level of susceptibility, whether high susceptibility, people-controlled governance, medium susceptibility, people- and technology-controlled governance, low susceptibility, or technology-controlled governance. All dependencies have a direction, strength, and an impact, which are maintained by the model aggregation. All the definitions are specified and implemented based on rules to provide maximum flexibility. The dynamic is based on connectivity, dependency, and rules.

The system resilience governance profile is the most comprehensive and essential multi-attribute decision support instrument consisting of a living system of systems description, risk evaluation, risk dependency map, governance definition, and all types of resources and actions for a value-focused road map.

To execute a resilience profile attributes, rules and conditions, and exceptions have to be defined. Semiotic, empathy, homeostasis and DNA rules have to be specified and applied. Each profile consists of four quadrants (Observe, Fragile, Anti-Fragile, and Recognize), and the related rules are specified in a visualization model. The profile has two important axes to provide a scale to measure system context in a comprehensive manner. The x-axis shows the feasibility on a numeric scale from -12 to $+12$ expressing the technical, organizational, legal possibilities to implement a solution and related availability of technology and knowledge. The y-axis shows the manageability on a numeric scale from -12 to $+12$ expressing the availability of people, financials, other resources and time for a sustainable solution implementation. Each bubble is representing a system context and the aggregated dependencies. The relation types specify the elation type as well as the direction and strength of the relation. The size and color are depending on assessment, quality, rules, conditions, and state. The position of a bubble is expressing the calculated position according to rules and incorporated information in the road map. The relationships between the system contexts (bubbles) are formed either by dependencies or by system context overlap. Size, direction, and impact of each dependency derive from the definitions and rules. The size of each bubble directly depends on the governance and how this multi-attribute and multi-layer governance is applied. High susceptibility means people-controlled governance. Medium susceptibility means people and technology controlled in a defined share governance. Low susceptibility means technology-controlled governance. The profile information is bidirectional interchanged dynamic capability mapped where the entire dependencies are visualized.

Essential for a system resilience governance profile is a digitalized and virtualized real existing system of system landscape. All well-known standard method for system of systems engineering and management are still applicable to manage, maintain, and transform the landscape as a virtualized model. Virtualization via model helps different stakeholders to filter, document, and visualize specific contexts of real existing system of systems with all these inter- and intra-collaborations and dependencies. Each specific system context with all inherent defined dependencies can be individually assessed, managed, transformed, and visualized in a standardized instrument called system resilience governance profile.

Essential for the entire life cycle management of a system context is a well-defined and maintained governance with an integrated susceptibility assessment. Each system context can subsequently be validated in a comprehensive risk evaluation where standardized risk assessment and mitigation method are used for an extensive and target-focused action management. Supported by this standardized procedure and approach, a comprehensive and comparable system resilience governance profile with a comprehensive set of definition, attributes, and rules can be visualized.

10.6 System Resilience Governance Profile Calculations

10.6.1 System Context Calculation

A system context in general is $\subseteq \sum$ system of systems where the system context can be also equal to a system self. The system context is the best possible option to get a focus encapsulated. The overlap to possible other system context can either be ignored or be properly managed by the fact that the system context or system dependencies are managed by a comprehensive relation artifact. Not only system of systems can be aggregated and represented by a system context but also dependency can be shown in a similar way. On an aggregation level, only the attributes are shown from each dependency and the strength, the direction as well as the impact. There is a further visual separation applied to show the dependencies a primary or a secondary in case of an applied scenario.

The system context (SC) is in fact an aggregation of system of systems (SYS), system (SY), artifacts (AF), special artifacts (S-AF) available in a model.

10.6.2 Action Management Calculation

All kind of resources (time, money, people, technology, legal knowledge, legal interpretation, etc.) and their dependencies are calculated by rules. Required knowledge and skills are managed in a comprehensive way, time slots are imported, when a resource is required and what are their dependencies and under what kind preconditions and conditions (e.g., knowledge and skills to implement new technology). The availability at the right time is crucial. Sequence, status, and dependencies are calculated and visualized in a road map together with many other information.

10.6.3 Road map Calculation

Road map is influencing the profile; especially, the position of a context is influenced by the road map calculation. Fragile: a system context gets in this quadrant if it is not feasibly and not manageable to execute all applied actions. Anti-Fragile: a system context gets in this quadrant if it is feasibly and manageable to execute all applied actions. Observe: a system context gets in this quadrant if it is feasibly but not manageable to execute all applied actions. Recognize: a system context gets in this quadrant if it is not feasibly but manageable to execute all applied actions.

Road map or a specific set of action can also be influencing dependencies and similar to the risk evaluation a dependency can reposition a system context in a profile or an impact can be simulated and visualized.

10.6.4 Profile Calculation

A profile is divided into four quadrants. A system context gets in the Fragile quadrant if it is not feasibly and not manageable to execute all applied actions, into the Anti-Fragile quadrant if it is feasibly and manageable to execute all applied actions, into the Observe quadrant if it is feasibly but not manageable to execute all applied actions and into Recognize quadrant if it is not feasibly but manageable to execute all applied actions.

The system context (SC) shown in a profile (PD) is in fact an aggregation of system of systems relation (RE), special rules (S-RU) and special attribute (S-AT) of a specific system context. There is also an option to apply additional rules and filters to manage complexity of a profile. The *color* of the system context depends on the risk and implicit also from the life cycle attribute. The system context *dependencies* are an aggregation of all specified and modeled dependencies of a system context self and from applied risk dependencies. The *size* of a system context shows the complexity and quality of the applied governance. The *position* is influenced by the road map.

There are many other calculations relevant to generate a system resilience governance profile. The essential calculations are mentioned to visualize a system of systems critical infrastructure landscape and to validate them in a standardized approach.

References

Dickstein, D. (2008). *No excuses: A business process approach to managing operational risk.* New York, NY: Wiley.

Meredith, J. R., Mantel, S. J., & Shafer, S. M. (2016). *Project panagement, Binder ready version: A managerial approach* (9th ed.). Hoboken, N.J.: Wiley.

Rosato, V., Issacharoff, L., Tiriticco, F., Meloni, S., De Porcellinis, S., & Setola, R. (2008). Modelling interdependent infrastructures using interacting dynamical models. *International Journal of Critical Infrastructures, 4*(1), 63–79.

Smith, M. L., & Erwin, J. (2005). *Role and responsibility charting (RACI).* Project Management Institute, Inc. Retrieved from https://www.projectmanagement.com/deliverables/234137/RACI-Matrix.

Taleb, N. N. (2010). *The black swan: The impact of the highly improbable.* New York, NY: Random House Trade Paperbacks Edition.

Vamanu, B. I., Gheorghe, A. V., & Katina, P. F. (2016). *Critical infrastructures: Risk and vulnerability assessment in transportation of dangerous goods—Transportation by road and rail* (Vol. 31). Cham, Switzerland: Springer International Publishing.

World Economic Forum. (2012). *Partnering for Cyber Resilience: Risk and responsibility in a hyperconnected world—Principles and guidelines* (No. REF 270912). Geneva, Switzerland: World Economic Forum. Retrieved from http://www3.weforum.org/docs/WEF_IT_PartneringCyberResilience_Guidelines_2012.pdf.

World Economic Forum. (2016). *The global risks report 2016, 11th Edition* (No. REF 080116). Geneva, Switzerland: World Economic Forum. Retrieved from http://www3.weforum.org/docs/GRR/WEF_GRR16.pdf.

Chapter 11
Dynamic Capability Model

Abstract This chapter addresses how understanding current, future, and intermediate situations for critical infrastructure with all relevant components and dependencies is supported by a model-driven approach.

11.1 Model-Driven Approach

A model-driven system resilience governance architecture is divided into three areas: *systems management, system resilience governance evaluation, and complex situation management*. The system of systems management covers all the rules and definitions to document and support business in terms of the status of the operation, transformation, and control. Here, business is usually documented in a normal situation.

In system resilience governance evaluation, a specific context of a system of systems is selected for advanced validation, assessment, and profiling. With all these activities, the sustainable, efficient, and effective system of systems management is guaranteed, and a value of these model-driven comprehensive analyses and documentation approach increases the quality. In addition, all the required dynamic capabilities to manage a complex (or normal) situation are documented and available for analysis, reporting, and management.

The third area of the architecture, called complex situation management, supports analyzing, mapping, and managing a complex situation initialized by a stressor. For the description of a complex situation, the definition and rules used to manage a normal situation can be used as well. Here, some additional instruments to manage a complex situation are available. The difference between normal and complex situations is often only the unknown and unexpected moment, the available time to respond, and the availability and quality of information. System of systems management, system resilience governance evaluation, and complex situation management together provide a comprehensive architecture and the required instruments for dynamic capabilities to manage complex (or normal) situations.

A.V. Gheorghe et al., *Critical Infrastructures, Key Resources, Key Assets*, Topics in Safety, Risk, Reliability and Quality 34,
https://doi.org/10.1007/978-3-319-69224-1_11

With system resilience governance architecture instruments, the paradigm shifts that are mainly driven by globalization and systemic dependencies can be handled properly and in a holistic way. There is no intention to save the world by applying these instruments. Rather, the goal is to give people the opportunity to identify, document, manage, modify, and control a complex system of systems, all its internal elements and assets, as well as its internal and external dependencies in a standard format.

11.2 System Resilience Governance Architecture/Instruments

Figure 11.1 shows the meta-model with the suggested instruments along with their relations. Based on these instruments, a living system of systems can be standardized and digitalized to handle normal and complex situations. The instruments are documented, visualized, and represented as a multi-attribute model description based on the favorite object-oriented concepts (encapsulation, inheritance, class object, and message). The following caveats are noted:

> System of systems engineering principles cover all concepts, definitions, templates, visualization techniques, quality definitions, and rules to document a system of systems in general, a living system as a digitalized virtual picture of real systems or a specific system context. The incorporated model rules allow for documenting system of systems on a specific level of information management (1)

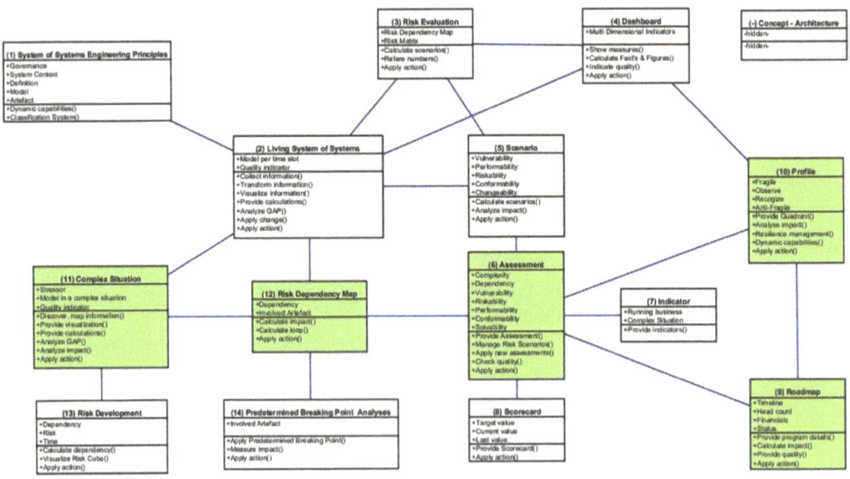

Fig. 11.1 Meta-model for system resilience governance architecture

11.2.1 System Meta Information

For a system description and their connectivity in the system resilience governance architecture, predefined artifacts will be reused from the classification system and incorporated in different types of models. If an element in the real world does not have a representation in the classification system, a specific type can be constructed. The aim of the model is to have after digitalization, for each element in the real world a virtual representative. The element/specific artifact represents the capabilities in the model. Patterns are also sometimes applied, which have a high similarity. For example, processes such as managing finance, lead/guide the company, providing logistics, manage people, and perform market. These patterns also exist for indicators, risks, and specifically for the application of regulations. This standardization and application of such patterns increase the quality, transparency, and comparability of models.

11.2.2 Systemic Resilience Governance Instruments

The system of Systems Engineering Principles:

- *Structured information*: In practice, metadata are available in a defined number called 'instance' and are in a context to each other (interconnected) in accordance with definitions. Based on these findings and definitions, an organization in all details can be uniformly documented and visualized. The governance, for example, can be full- or semi-automatically generated, visualized, and maintained.
- *Business Data:* Available information can be accumulated, aggregated, or improved by rules and patterns in systemic-systematic approach and visualized as any type of indicators. These figures are also based on rules in context and are exemplary in many cases. It should be mentioned, as examples, EBIT, solvability index or the risk capital allocation. Comprehensive standardized indicators are very powerful to support benchmarking as long its clear where they are coming from, how they are constructed, and how the quality and maturity looks like.
- *Instance Data:* To consider dynamic information from operation during life cycle (flow), instance data are in focus. This collected information is called big data. In the system resilience governance architecture, systemic systematic consideration might offer a great advantage through treading scheme related to where, how and what is collected information are provided and available. This scheme will improve the readability of big data and will support possible transformation.

A living system of systems is the digitalized virtual picture of a real system representation in all possible dimensions. This picture is available as a multi-attribute system of systems description. It is mainly maintained by discover, map, and visualize utilities (2)

A living system of systems description or, in this case, an enterprise description consists of a number of specific-interconnected artifacts from the classification system. The model-driven approach provides the best possible quality and a maintained virtual picture of the reality. More than one source can be discovered and mapped to get a redundancy-free, maintained, interconnected collaboration map available as a multi-attribute holistic model. After discovering and mapping the model, content can be automatically visualized as stakeholder-oriented based on applied rules. Different visualizations depending on the selection and content can be provided as tables, pictures, text, dashboard items, or in any required combination. The holistic model has more than one implicit structure, depending on the entry point and applied filters; therefore, many different structures can be shown. Depending on the rules, the essentials of a system of systems can be aggregated and visualized.[1] There is always something under flux; sometimes a transformation is on. The enterprise tower presents a comprehensive picture of the digitalized reality in a multi-layer diagram, where details can be filtered, aggregated, and visualized for different stakeholders with varying requests.

It is not possible to build and maintain this picture manually due to complexity. Clearly a methodology has to be deployed to produce information from data, to move from information to knowledge. This might be supported by technology (software and hardware) that is able to handle the complexity producing visualizations that are able produce insights that can empower people. The world is too complex for its information to be maintained manually, especially in the over-regulated world today, where more interpretation is required, and where implementation and validation are complex and difficult.

In a digital world without solutions, where sources are automatically maintained, it is no longer possible to manage complex situations and possible risks manually. The trust in manually maintained sources and individually rated risks is low because a complex situation requires valid information.

[1]Usually, the most important information from an enterprise is the end-to-end view—the value chain of the offered product and services—and the available performance implemented as a horizontal end-to-end view. Process landscapes and individual activities have a vertical view and a control flow orientation. Processes are supported by application, organization, and logistics. In a stack below are infrastructure and facilities to get the picture. The influence factors document, measure, control, and report the outbound conformance. For inbound conformance, the principles can be reused. The dependencies and interconnectivity are used to measure the systemic relationship. A multi-attribute system of systems description is very complex because it is a digitalized picture of reality. Therefore, some tools (navigators) are required to manage the complexity. Using a navigator, a network of dependencies can be visualized as a structure based on rules and filters. Seeing or understanding all details is not always relevant. Details can be aggregated, clustered, and visualized. A tool is required here because such complexity cannot be managed manually.

The difference between a normal and a complex situation is that the complex situation is unknown, unexpected, stressor-driven most of the time from outside, influencing established system boundaries or across multiple systems, and usually with not enough information available to understand risk and its impact or to make sustainable decisions and understand possible consequences. Energy fallouts, system faults, terror attacks, cyber threats, and so on are stressors that can produce a complex situation.

Risk evaluation is a comprehensive assessment and visualization of all risk types of a living system or system context. There is an incorporated process for defining, assessing, measuring, mitigating, and monitoring risks as well as the required definitions and rules for risk-control frameworks, incidents, and action management. A risk evaluation can be enriched by a risk dependency map (3)

Risks are visualized traditionally in a symmetric matrix in colors of the traffic light. The risk evaluation instrument in the system resilience governance architecture can be asymmetric. The color is mainly driven by the risk evaluation and the risk dependency map. The source of the dependencies is from the system context and the living system of systems connectivity. Its impact and likelihood scale can be numeric or linguistic. In addition, the color can be defined by rules without any limitations. Risks can be positioned in different statuses and conditions, usually a minimum of two (gross and net). Each mitigation strategy (gross to net) is documented as multi-attribute actions, and in the net status with risk appetite and consequences as well as transfer options. Gross and net risk can have dependencies on other risks. Risk dependencies are characterized by strength, direction, and an impact. The impact on the influenced risk is applied as a vector to visualize the dependency effect.

The *risk evaluation matrix*, with the applied shifting scenarios, risk dependencies, and impact analyses, is the main input for the system resilience governance profile (bubble) color.

The actions applied for a shifting of risk (applied mitigation scenario) and detailed in a road map can be visualized on a profile. On this map, a stressor is visualized because the company lost its main customer and found itself in a complex situation because of the unknown, unexpected event with a big impact. Impact appears might appear in the finances as indicated by measures associated with liquidity, profit, and solvability, especially if an organization loses its biggest customers. Immediately, operations will start influencing other systems-like cost-cutting activities launched by finance to get the numbers under control. The impact and the consequent actions change the risk color of both system contexts. The team will be influenced and discouraged, which can affect the efficiency and quality of the delivery. Often, applied actions generate effects that were not planned or expected. The sales force action also initialized by finance has no immediate second effect. Often, during an assessment, the action or reaction can be visualized, analyzed, and modified.

System risk management is not strictly a serial process, where one component affects only the next. It is a multidirectional, iterative process in which almost any component can and does influence another. Contradictory validation can influence either risk shifting scenarios or also override and suspend them, and stakeholder groups could apply different scenarios.

The risk matrix is a common and well-known instrument to visualize risk. In the system resilience architecture, some additional options are individually applicable. To each risk, one to many risk scenarios as well as possible shifting can be applied. Each risk position as well as the specific risk shifting has his governance definition based on RACIO. Per definition, each individual risk can have one to many risk dependencies applied. If risk dependencies are in between of two risks on this matrix, the dependencies can be visualized.

> Based on the dashboard, all facts and figures, indicators, and measurements of a living system of systems with the formulas and equations are represented. There are many different visualization options available (4)

With the know-your-interfaces, all relationships in the system and peripheral systems are documented and measured. The individual types of relationships are evaluated and used to calculate the collaboration complexity. This instrument should also provide clarity on whether a system can operate in a dependent, independent, or interdependent manner. The information about the relationship network is also used in detail in the dependency map and the assessment. The cross-linking always assumes an important role because it is, on the one hand, difficult to handle, and, on the other hand, has a strong effect on the quality and complexity of a system. The instrument Know your Governance is not only corporate governance or organizational responsibility for the system, but also the entire artifact ownership. Implicitly not only the visibility is thus documented, but also any particulate component of the system and all relationships. The present shows again that the data and the system, and specifically the interfaces, are specifically regulated responsibilities or insufficiently transparent. This means that the traceability in compliance is making it difficult to guarantee. With this instrument, the strategic as well as operational governance is regulated holistically.

> The scenario allows and supports a rule-based multi-assessment of a living system of systems regarding vulnerability, performance, conformance, risk, and change or against other configurable aspects (5)

The resilience profile has dependencies to all other visualizations but specifically to risk evaluation. The color is mainly driven by the risk evaluation and the risk dependency map. The source of the dependencies is from the system context and the living system of systems connectivity.

Risks are visualized traditionally in a symmetric matrix in colors of the traffic light. The risk evaluation instrument in the system resilience governance architecture can be asymmetric. Its impact and likelihood scale can be numeric or linguistic. In addition, the color can be defined by rules without any limitations. Risks can be positioned in different statuses and conditions, usually a minimum of two (gross and net). Each mitigation strategy (gross to net) is documented as multi-attribute actions, and in the net status with risk appetite and consequences as well as transfer options. Gross and net risk can have dependencies on other risks. Risk dependencies are characterized by strength, direction, and an impact. The impact on the influenced risk is applied as a vector to visualize the dependency effect.

The risk evaluation matrix, with the applied shifting scenarios, risk dependencies, and impact analyses, is the main input for the resilience profile system context (bubble) color.

The actions applied for a shifting of risk (applied mitigation scenario) and detailed in a road map can be visualized on a profile. On this map, a stressor is visualized because the company lost its main customer and found itself in a complex situation because of the unknown, unexpected event with a big impact. The impact appears to the finance system because liquidity, profit, and solvability are affected due to the company having lost its biggest customer. Immediately, operations will start influencing other systems-like cost-cutting activities launched by finance to get the numbers under control. The impact and the consequent actions change the risk color of both system contexts. The team will be influenced and discouraged, which can affect the efficiency and quality of the delivery. Often, applied actions generate effects that were not planned or expected. The sales force action also initialized by finance has no immediate second effect. Often, during an assessment, the action or reaction can be visualized, analyzed, and modified.

With the assessment, a specific system context of a living system can be validated against standardized topics specified by rules. This assessment can also be used for a system context involved in a complex situation (6)

All the elements of a problem are not always significant for an assessment; at this point, the definition is no carried out. The right viewing space is selected, the influencing factors are determined, a reasonable allocation of the system context is performed (level of detail is set), the dependencies and interfaces are defined, and possible events are listed. The system context is derived from the operating model and evaluated. Not every system has to be validated in the same detail.

With this validation, a living system of systems landscape can be validated respectably how a system context can be built to manage complexity and complex change. All for the specific scenario relevant system of systems has to be selected, and a system context is generated. The related systems to the selected system context have to be selected and documented either. The specific and selected system context and all connected parties are validated against predefined scenarios. Between individual scenarios, an intersection has to be managed separately. The reason of the overlap, which indicates the contractions, arises from the fact that 100% performance by 100% conformance is not possible. Based on the risk or system connectivity, a cause effect or impact simulation can be generated and visualized. In each scenario, the ability will be checked. The question in mind is: *Does the individual system context (the focus of a scenario validation) have the capability to fulfill the performance targets and requirements?*

> The abstract of a living system of systems called system context with all incorporated dependencies, rules, and attributes, is the main input for the system resilience governance profile (bubble) with all dependencies to others.

The indicator is a fuzzy set-calculated comprehensive number of a set of standardized assessments. This number shows the temperature of a system context—like measuring a fever—within a normal and a complex situation. There is also an option to produce an indicator on a higher aggregation like a system landscape (7)

With a scorecard, every single assessment can be detailed and supplemented by the timeline (past, current, target) to visualize changes and sensitivity shown by specific thresholds. The scorecard also has benchmark functionality per assessment topic (8)

Multi-attribute indicators about a system context, system of systems, a single system or an artifact visualize standardized techniques, facts and figures, numbers, time series, and multi-attribute indicators to empower stakeholder groups in system management.

The multi-attribute system description, the scorecard as well as some scenarios are used to provide a set of standardized instruments. Instruments are measurements, facts and figures, and some calculation represented as a 'Dashboard.' All indicators can be used individually or in combination with system management, measure progress, system comparison, or benchmarking. By appropriate links and rules, own instruments can be created and integrated into the framework. All these instruments support systems management empowerment.

The roadmap covers identified actions, documented and applied in each step or visible on all the different instruments. All the applied actions are collected for details and analyses on the roadmap. The visualization of a roadmap can be provided in certain level of details. The roadmap will affect the positioning of a system context on the profile. The roadmap is not just a planning step. It is also a comprehensive examination of all types of resources. After validation, all modifications will influence the source. An identified stressor will increase the dynamic additionally (9)

Moreover, in a complex situation initialized by a stressor, the same steps and diagrams are involved and affected. Depending on the environment and the stakeholder, a complex situation can be modeled, documented, measured, and managed separately or together with a normal situation. The definitions of priority and sequence on the road map depend on many different attributes. Priorities are difficult to manage because stakeholder conflicts and contradictions are not always transparent and simple to answer. The road map covers all kinds of actions. All types of resources are managed in this step and visualized on the map. This includes people, knowledge, skills, all-time dimensions, financials; all other resources, infrastructure, logistics as well as technology availability; and their actions and resources as well as time constraints for setup, development, deployment, and operation. The benefit, quality, and the attitudes of the involved and concerned people are also considered here.

All specified and validate actions applied to a specific system context, rules, and attributes are the main input for the system resilience governance profile (bubble) position in one of the four quadrants.

The profile represents various system context as a comprehensive multi-attribute decision support for visualization. Not only the context but also risks, dependencies, and technical and resource restrictions, with advanced ratings, measures, and qualifications are incorporated. The profile is the highest possible aggregation about a system context in a complex or normal situation in this architecture. It is also a visualization and representation of dynamic capabilities as a specific context-sensitive and holistic system landscape with encapsulation of all relevant essentials. On this map, it is clear that the team management process with the motivation risk attached is critical but also human-driven. People are unpredictable and often driven by money, fear, envy, and so on. Hence, this problem is difficult to measure and manage (10)

This dynamic visualized on a risk dependency map will influence the risk evaluation. The stressor could be visualized, and it would influence a risk evaluation or an applied mitigation scenario. Actions are documented and validated on the road map. Appetite and consequences are visible within the net risk. There is a risk dependency of the finance net risk on the operation gross risk. Each risk dependency has a rule-based influence (strength, direction, and impact) on a target risk. The effect on a specific target risk is applied as a rule-based vector that shows a possible shift on the matrix. Other dependencies are also visible. Not all risks have dependencies, and the effect is not known for all dependencies. The dynamic is driven by all the dependencies. The aggregated risk mapped to a system context is visualized on the profile. It is sometimes better to manage risk on the profile (whole) than to be lost in details regarding individual risks (part).

A complex situation is usually initialized by a special event or call stressor, and shows a living system or a system of systems under special circumstances. Therefore, in a complex situation, a special context is in focus and not the entire system. The key features of a complex situation are often time pressure, availability of information, lack of responsibility, and quality and maturity of relevant and essential information. Most of the time, the event was unknown and unexpected for all involved and concerned stakeholders (11)

The risk dependency map is a special instrument for managing, analyzing, simulating, and visualizing the dynamics, dependencies, influence, and flow of risks and their causality. This instrument supports the analyses of cause-effect and impact. With this map, even loops and mappings can be identified as well as problem and possible solution risk (12)

The dependency map (Instrument 12) is a multi-layer visualization. Aggregation is also visible. With the risk dependency map, the artifacts where an individual risk is applied (in this example, process and their attached risk), and risk dependencies dependent on or independent of artifact relations are visualized. The stressor is visible as well as the dynamic of the impact. Risk dependency and possible involved WorkProducts are visualized as well.

The risk dependency map visualizes risk mapping on a multi-level and the risk dynamic along with the dependencies. It is an essential input for strength and impact as well as for showing additional relations between the system contexts represented by a system resilience governance profile.

Under special circumstances with risk development, many produced risk evaluation results, produced over time or in a special simulation scenario can be shown in a special kind of timeline diagram (13)

The dependency map (Instrument 12) is a multi-layer visualization. Aggregation is also visible. With the risk dependency map, the artifacts where an individual risk is applied (in this example, process and their attached risk), and risk dependencies dependent on or independent of artifact relations are visualized. The stressor is visible as well as the dynamic of the impact. Risk dependency and possible involved WorkProducts are visualized as well.

The risk dependency map visualizes risk mapping on a multi-level and the risk dynamic along with the dependencies. It is an essential input for strength and impact as well as for showing additional relations between the system contexts represented by a system resilience governance profile.

With the predetermined breaking points analyses, there is an option to analyze and visualize possible damage of risk as well as to apply breaks in order to avoid or protect a system from total damage. This instrument can also be used to separate two systems, like in building construction (14)

With all the outlined instruments, the system resilience governance architecture can be measured, managed, and transformed in a holistic way with the maximum degree of stakeholder empowerment and value. The system resilience governance architecture today comprises 14 specific instruments. These instruments represent the dynamic capabilities to manage complex (normal) situations. Each instrument follows their definitions and rules and focuses on visualization.

With this architecture, the target of dynamic capabilities to manage complex situations can be met. The architecture solution and procedures support understanding, encapsulating, and visualizing the complexity of a complex situation and the systemic dependencies, identifying the information demand in a dynamic environment, forecasting risk and possible impact, simulating resilience, and implementing action patterns to control and handle continuous change. Especially with the available instruments in this architecture, complex situations can be managed by modeling and measuring system resilience governance.

With this model-driven system of systems description approach, discovered, mapped, and visualized content about the system of systems, a living system or a specific system context can be aggregated, stored in a content- and context-sensitive way, and a multi-attribute decision support can be used. Regarding the used and applied standards, language, and taxonomy in the classification system, people can be empowered, and a common understanding can be supported. Information available as a multi-attribute model description is powerful. It allows one to develop, control, manage, and transform a complex situation in a dynamic environment. With a standard approach, system resilience governance can be documented and managed. In 14 steps, a system can be systematically digitalized and visualized. Different results can be developed using this procedure. Based on the standard results, management, empowerment, and comparability can be supported.

The multi-attribute decision support is a comprehensive balance between performance, conformance, risk, change, and the applied actions implemented and available as a model-driven multi-attribute system of systems description. Again, with this methodology, it's possible to address some of the most pressing solution requirements, challenges, and issues (Freixas et al. 2015; Hollnagel et al. 2006; OCEG 2015; Vester 2007). Additional business and leadership value can be generated as well.

11.3 Living System of Systems; System Context

A living system of systems description or, in this case, an enterprise description consists of a number of specific-interconnected artifacts from the classification system. The model-driven approach provides the best possible quality and a maintained virtual picture of the reality. More than one source can be discovered and mapped to get a redundancy-free, maintained, interconnected collaboration map available as a multi-attribute holistic model. After discovering and mapping the model, content can be automatically visualized as stakeholder-oriented based on applied rules. Different visualizations depending on the selection and content can be provided as tables, pictures, text, dashboard items, or in any required combination.

It's essential to recall that the difference between normal and complex situation. In a complex situation, the situation itself might not be unknown, it is unexpected, driven by issues outside the control of oneself. Energy fallouts, system faults, terror attacks, cyber threats, and so on are stressors that can produce a complex situation.

During the development, operation, assessment, and transformation of a living enterprise (system of systems), many stakeholder groups require information. If the common artifacts to document an enterprise are used, numerous common information requests about the enterprise, customer, product, report, transaction, and so on are available in standardized visualizations. The big picture provides different information and dependencies relating to many different things. Assets are of high significance or use for the system. The meaning may be an important protective value for the system itself; it has its uniqueness, was expensive to purchase, or is expensive to maintain. Things, for example, can be people, processes, products, locations, or infrastructure, among others. In most cases, an asset is a group of artifacts. The value of a system may be what a system passes on to the customer or partner, but there is also internal value. The performance of a system offering is agreed upon with a customer in service level agreement (SLA) or, in general, with regard to what a system is able to produce. System risk comprises all risks applied to artifacts and summarized, visualized, or aggregated on a system level. Conformance is a representation on the system level of a summary of all legal regulations, policies, and guidelines, and the fulfillment of all internal commitments, policies, and agreements. Productivity is often measured in terms of cost, income, or number of units produced in a specific period. Quality, maturity, and trust indicators relate to fulfillment and availability of information or services used inside the system or provided to customers and partners. Changes are a summary of ideas, demands, requirements, and requests validated and mapped to artifacts to show possible gaps as well as how they are identified and planned to be closed or options to improve or optimize business. Generally, across a system of systems landscape, artifacts are redundancy-free in many different pictures. The handling is always the same. An individual, for example, exists, only at one time. It can be available and used at different places with different focus and attributes, but as it is available only at one time, the engineer has to decide who the real source of an artifact is. There are many technical options for managing redundancy using a computer.

Performability focuses on product, market, revenue, cost, profit, and solvability. Conformability covers how all legal requirements, commitments, and liabilities can be met, how promises are managed and agreed contracts and SLAs fulfilled. Changeability manages all change requests and internal and external demands, closes GAPs, applies innovation, and maintains assets. Riskability is the balance or the intersection between the other four topics. This capability is difficult to manage, prevent, forecast, and predict because it is fuzzy and often depends on people's behavior and attitudes.

With the multi-attribute system resilience governance decision support based on a model-driven multi-attribute system description, the requirements can be met. The holistic view and overlap of performance, conformance, change, and risk can be sustainably managed. A fully digitalized system of systems landscape can be very complex as shown in Fig. 11.2 (below). Documentation of a living system of systems is complex. The quality of documentation is always known, but it depends on the discovered digitalized sources. Some systems are documented like a 'black box,' others like a glasshouse, and the rest in perfect detail. This insecurity and uncertainty are normal in a complex situation because there is no time and no alternative to ensure more quality. In a normal situation, the availability and quality are sometimes better. Based on the artifact and the incorporated concept about encapsulation, this can be controlled. In most cases, the artifact type node is better documented; therefore, it is more relevant than the

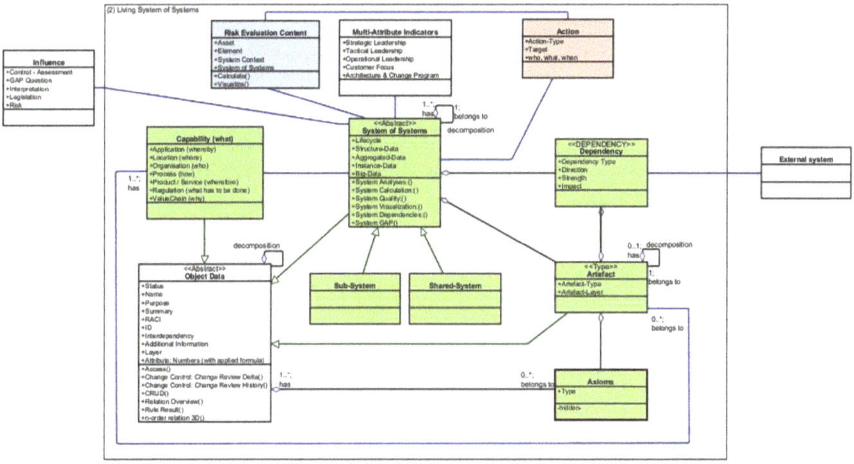

Fig. 11.2 Living system of systems meta-model

artifact type edge, because there are more relations than elements to manage $n * \left(\frac{n-1}{2}\right)$. The dynamics and the unmanageability are often driven by unknown or badly documented and managed dependencies. With the comprehensive and highly standardized model-driven multi-attribute system of systems description, all kinds of living system of systems and system context can be documented to support engineering and management disciplines like development, assessment, operation, and transformation. To understand a living system, a multi-attribute system description is required.

If all the artifacts are also measured and controlled, a sustainable management can be guaranteed. This involves:

- *Standardization*: speak the same language, support measurability, and control
- *Digitalization*: reduce maintenance and improve quality, guarantee traceability
- *Collaboration*: manage the whole and not just the part, improve quality
- *Visualization*: empower people, get a common understanding
- *Automation*: enable, optimize, and improve availability
- *Industrialization*: improve productivity, efficiency, and effectiveness

It is important to focus on the importance of a sustainably managed system of systems description. This is the most important part in the system resilience architecture as shown in Fig. 11.2 depicting living system of systems meta-model. The system resilience governance architecture consists of distinct instruments. All these instruments provide the required information about dynamic capabilities to manage complex situations. Dynamic capabilities provide a holistic view of all involved artifacts in a specific situation, with all their details. The system context summarizes all involved artifacts systems and their dependencies. The resilience profile is a comprehensive view on system context assessments. The governance model shows the entire engagement of a living system, system context, or all involved artifacts. Risk evaluation shows all risks in different statuses and their dependencies as an asymmetric or symmetric risk matrix. The risk dependency map is a visualization of all risks, risk dependencies, and their origin as a map. The road map covers all the activities, along with their dependencies, attributes, status, and rules. All the instruments document the uniqueness of a living system, are available at the best possible quality, and are fully interconnected to provide a homeostatic picture.

In the corresponding model definition, all relevant details are explicitly documented and visualized. The system of systems borderline is essential. This borderline is limitlessly thin and sharp. This is the only guarantee that either a specific artifact is inside or outside a system and only dependencies (virtual = air, Wi-Fi, real = interfaces) can cross this borderline or scope. It is very important to obtain the precise scope of a virtual picture about a real system landscape. The virtual picture is represented as a multi-attribute system of systems description available as a model about reused artifacts of a classification system. There are two specific artifacts to model the outbound of a specific system. With the artifact influence, any kind of outbound representation of a system of systems can be modeled and documented. The very unique characteristic of an influence is unidirectional. Therefore, influences are used to document, for example, environment characteristics or legislation because they have to be accepted, because an interaction or modification is impossible. There is a second outbound artifact called External System (sometimes also External Agent). With an external agent, most of the time, a bidirectional dependency is agreed and often depends on the system context specified by an agreement, for instance, SLA.

A system of systems is divided into a subsystem and shared systems. These two types are important to separate because the characteristics are completely different. To fulfill the target of the system resilience governance architecture to document systemic inbound and outbound dependencies in a systematic way manually, semi or fully automatic, to obtain a fully digitalized picture about a real system landscape, this separation is essential and helpful. A brief example to clarify the definition. In a house, several subsystems are available, like the living room, kitchen, dining room, parent and child rooms, and bedroom. In each room, a specific temperature is expected, in all rooms Wi-Fi is essential, and in some rooms, water is required. Therefore, engineers will design three shared systems: a water system, a Wi-Fi system, and a heating system. The characteristic is, for example, the heating system is in a specific room, and there are interfaces built for each room (e.g., pipe system). Design, development, control, and maintenance from these two types of systems are different and, so, the split is required. To stay quickly by this house example to focus again to inbound and outbound as well as the essentiality to have a well-defined borderline. The weather outside the house is given, and the interfaces are absolutely transparent (window, tor), and by other sometime, the worker has different permits (main current supply line and in-house installation).

The entire multi-attribute system of systems description is made by reusing Artifacts from a classification system. All measurement, indicators, and fact and figures are specified, collected, and visualized by the class multi-attribute indicator. The risk evaluation content class is the content bridge to the risk evaluation instrument. With the action class, a container for any measure is specified. With the entire model, the capabilities of a specific digitized real system landscape are documented as a multi-attribute system description.

11.4 Dynamic Capabilities to Manage Complex Situations

To manage critical infrastructure in a normal or complex situation requires a high-quality documentation. Based on this documentation also complex situations can be properly managed by applied dynamic capabilities. Dynamic capabilities are a relational higher-order artifact type. All involved artifacts in a specific scenario, along with their dependencies, are summarized under a capability. A capability can have its own governance, applied activity, individual risk, risk dependencies, system and system dependencies, and additional profile information. The homeostatic dependencies are managed by applied model rules.

The artifact capability collects and covers dependencies and shows what is used to offer a product (what) at a specific location (where) under the valid regulations (what has to be done). The dynamic capability is a comprehensive way to document what is required and shows the dynamic in all the other artifacts, systems, and their dependencies. What the dependencies look like and what kind of visualization is possible are shown in Fig. 11.3. A capability can have its governance, as well as many individual configurable visualizations and indicators.

The process indicates how a capability is performed, while the organization (who) performs a capability. The value chain (why) offers the value (asset, product, and service) at a place under local conditions. The application with the applied infrastructure (whereby) supports the capability. The information shows which data is used or required to perform consistently and at high quality, with maturity and trusted capability. Capabilities can be documented and visualized in many different ways, as capability risk matrix, capability visualization (heat map, sensitivity, benchmark, quality–maturity–trust, and time series), capability control matrix to check is everything running as required and defined, capability governance diagram, and multi-layer quality check to visualize and validate all the mappings.

Many different types of capabilities are applicable. A risk capability specifies what is required to manage or mitigate a specific risk [gross] to a target value/position = risk [net], and what consequences are acceptable (appetite) or the managed/controlled transfer. A performance capability describes what is required (value chain, process, application, organization, information, etc.) to offer a product under local regulations to a customer. An organization capability specifies what is required to run a lean, efficient, and effective organization that is profitable and that conforms (organizational readiness, distributed intelligence). A resilience capability

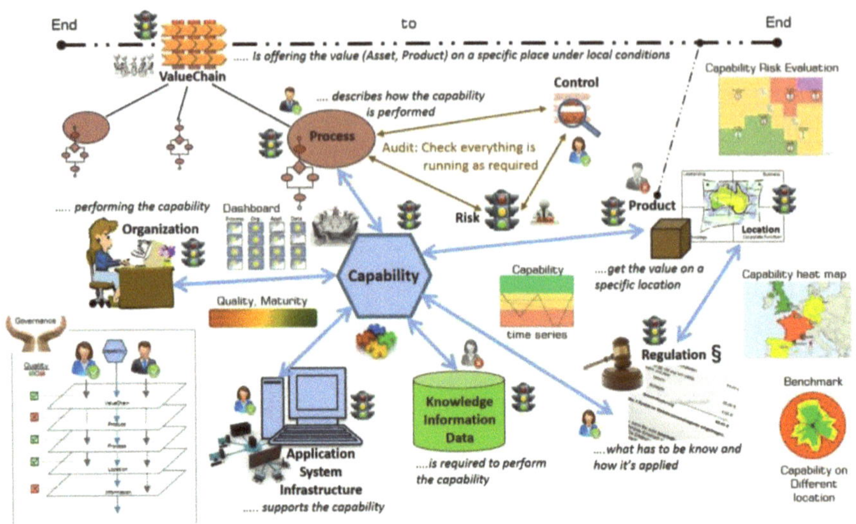

Fig. 11.3 Dynamic capabilities model

shows what is required to manage and control a system resilience governance profile.

A transformation capability specifies what is required (resources) to transform a current situation to make it transparent and traceable for the future under observance of quality, functionality, time, budget, and so on. All dynamic capabilities are artifacts that are highly dependent on a relational higher order.

> The holistic (relation on a higher order, relation object) multi-attribute decision supports to manage complex situations on a system of systems level regarding balanced performance, conformance, risk, and change are called dynamic capabilities.

References

Freixas, X., Peydró, J. -L., & Laeven, L. (2015). *Systemic risk, crises, and macroprudential regulation*. Cambridge: MA: The MIT Press.

Hollnagel, E., Woods, D., & Leveson, N. (Eds.). (2006). *Resilience engineering: Concepts and precepts*. New York: NY: CRC Press.

OCEG. (2015). GRC Capability Model 3.0 (Red Book). Retrieved February 21, 2017, from http://www.oceg.org/resources/red-book-3/.

Vester, F. (2007). *The art of interconnected thinking: Tools and concepts for a new approach to tackling complexity*. München: Mcb Verlag.

Part III
Working Examples

Chapter 12
Processing Switzerland

Abstract This chapter explores an application of QVA model. It presents a model and computer-run reports at the national level using Switzerland as an example. The reader might find it necessary to refer to Chaps. 2 and 3 and Appendix C for explanatory notes.

12.1 The Model

The model includes:

- Indicators Inter-comparison and Correlations
- Vulnerability Model 1: Targeting Strategic Goals through the Index Method
- Vulnerability Model 2: Targeting Governance Robustness through the Matrix Method
- Vulnerability Model 3: Targeting Penetrability of Complex Systems
- Vulnerability Model 4: Targeting Resilience under Operational Stress—a Probabilistic Resilience Analysis Working Case
- Vulnerability Model 5: Targeting Territorial Vulnerabilities.

The illustration/concept demonstration nature of the exercises needs to be clearly emphasized—the true solutions are looming somewhere ahead, to be defined as 'cahiers de travail,' perhaps in the aftermath of this study. A supportive tool in the form of a 'decision support system' was developed to implement QVA. An interface for the QVA DSS is depicted in Fig. 12.1.

12.2 Model Indicators

This section operates on the belief that any analysis targeting risk and vulnerability should heavily depend on the volume and appropriateness of *indicators* featuring the subject in question.

A.V. Gheorghe et al., *Critical Infrastructures, Key Resources, Key Assets*, Topics in Safety, Risk, Reliability and Quality 34,
https://doi.org/10.1007/978-3-319-69224-1_12

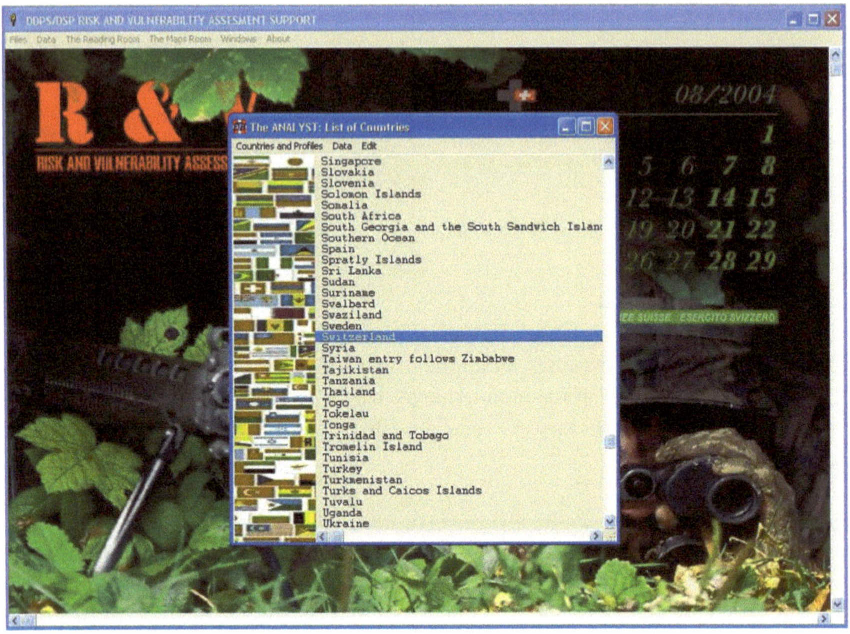

Fig. 12.1 An interface of QVA support tool

The notion of 'indicator' is given in a perhaps more comprehensive than usual meaning: The term lends itself, on the one hand, to *metrical* quantities such as, among others, country area, length of borders, total population, GDP, mortality rate, number of Internet providers, number of ocean fleet oil tankers, and number of males fit for military service. On the other hand, an indicator is also derived based on the descriptive (textual) entries in the primary database of the investigative operation. The supportive software that has been developed for the purpose would seek for key words that were identified, by the code users/developers, as bearing a worthy relevance in risk/vulnerability evaluation. In a way of a few random examples, one may thus think of 'earthquakes,' 'floods,' 'volcanism,' 'draughts,' yet also of 'fundamentalism,' 'riot,' 'revolt,' 'poverty,' 'corruption,' 'litigations,' 'totalitarian,' and the likes. One may thus identify countries that are more at risk than others and, by attributing values on a given scale (e.g., 1–100) to the relevance of the key words, one may also categorize the countries, and perform a multi-attribute analysis on these.

The quantity that has been derived based on key words, and their valuing has been termed, in the context, a *'verbose indicator.'* In this concept demo phase, the elaboration of the aforementioned idea halted at a phase of *country categorizing*, and an incipient analysis of the *statistical correlations* between various indicators, with all the considerable potential of the method still laying ahead.

Much effort has been allocated to the identification of a primary data library that would fit the specific needs and intentionally biased orientation of a risk/vulnerability-oriented analysis. From the first brainstorming on, it became clear that the sought data library would exceed the regular scope of the standard yearly statistical. To find out whether a country is vulnerable, to what, to what extent, one should know what that country resources or otherwise assets are, what its history briefing is, what system—societal, political—is there prevailing, down to who is who and who is doing what. This, in broad lines, would rhyme with an *intelligence* data library. While a fully fledged analysis might have to make recourse to the Swiss intelligence resources of the kind, an interim, valid, substitute was proven to be the USA's *The World Factbook*, which has become the chief online source of primary information for the present drill.

To illustrate, the following is the manner in which the 'Switzerland' entry is featured in the said resource. Following the primary data, listing is the key words retained by the authors of this study and their rating, as well as a restructuring of an indicator selection for the purposes of further analyses (see Chap. 3, Sect. 3.1.1 for an application at the world-level).

```
SWITZERLAND
_____________________________________________________________

1. BACKGROUND: Switzerland's independence and neutrality have long been
honored by
the major European powers, and Switzerland was not involved in
either of the two World Wars. The political and economic integration
of Europe over the past half century, as well as Switzerland's role
in many UN and international organizations, has strengthened
Switzerland's ties with its neighbors. However, the country did not
officially become a UN member until 2002. Switzerland remains active
in many UN and international organizations, but retains a strong
commitment to neutrality.

2. LOCATION Central Europe, east of France, north of Italy

3. GEOGRAPHIC COORDINATES 47 00 N, 8 00 E

4. MAP REFERENCES Europe

5. AREA total: 41,290 sq km
water: 1,520 sq km
land: 39,770 sq km

6. AREA - COMPARATIVE slightly less than twice the size of New Jersey
```

7. LAND BOUNDARIES total: 1,852 km
border countries: Austria 164 km, France 573 km, Italy 740 km, Liechtenstein 41 km, Germany 334 km

8. COASTLINE 0 km (landlocked)

9. MARITIME CLAIMS none (landlocked)

10. CLIMATE temperate, but varies with altitude; cold, cloudy, rainy/snowy winters; cool to warm, cloudy, humid summers with occasional showers

11. TERRAIN mostly mountains (Alps in south, Jura in northwest) with a central plateau of rolling hills, plains, and large lakes

12. ELEVATION EXTREMES lowest point: Lake Maggiore 195 m
highest point: Dufourspitze 4,634 m

13. NATURAL RESOURCES hydropower potential, timber, salt

14. LAND USE arable land: 10.57%
permanent crops: 0.61%
other: 88.82% (1998 est.)

15. IRRIGATED LAND 250 sq km (1998 est.)

16. NATURAL HAZARDS avalanches, landslides, flash floods

17. ENVIRONMENT - CURRENT ISSUES air pollution from vehicle emissions and open-air
burning; acid rain; water pollution from increased use of agricultural fertilizers; loss of biodiversity

18. ENVIRONMENT, INTNTNL.AGREEMENTS: party to: Air Pollution, Air Pollution-Nitrogen
Oxides, Air Pollution-Persistent Organic Pollutants, Air Pollution-Sulphur 85, Air Pollution-Sulphur 94, Air Pollution-Volatile Organic Compounds, Antarctic Treaty, Biodiversity, ClimateChange, Desertification, Endangered Species, Environmental Modification, Hazardous Wastes, Marine Dumping, Marine Life Conservation, Nuclear Test Ban, Ozone Layer Protection, Ship Pollution, Tropical Timber 83, Tropical Timber 94, Wetlands, Whaling,
signed, but not ratified: Antarctic Environmental Protocol, Climate Change-Kyoto Protocol, Law of the Sea.

19. GEOGRAPHY - NOTE landlocked; crossroads of northern and southern Europe; along with southeastern France, northern Italy, and southwestern Austria, has the highest elevations in the Alps

20. POPULATION 7,318,638 (July 2003 est.)

21. AGE STRUCTURE 0-14 years: 16.6% (male 623,428; female 591,709)
15-64 years: 67.8% (male 2,519,302; female 2,439,560)
65 years and over: 15.6% (male 470,257; female 674,382) (2003 est.)

22. MEDIAN AGE total: 40.2 years
male: 39.3 years
female: 41.2 years (2002)

23. POPULATION GROWTH RATE 0.21% (2003 est.)

24. BIRTH RATE 9.59 births/1,000 population (2003 est.)

25. DEATH RATE 8.82 deaths/1,000 population (2003 est.)

26. NET MIGRATION RATE 1.37 migrant(s)/1,000 population (2003 est.)

27. SEX RATIO at birth: 1.05 male(s)/female
under 15 years: 1.05 male(s)/female
15-64 years: 1.03 male(s)/female
65 years and over: 0.7 male(s)/female
total population: 0.98 male(s)/female (2003 est.)

28. INFANT MORTALITY RATE total: 4.36 deaths/1,000 live births
female: 4.25 deaths/1,000 live births (2003 est.)
male: 4.47 deaths/1,000 live births

29. LIFE EXPECTANCY AT BIRTH total population: 79.99 years
male: 77.11 years
female: 83.02 years (2003 est.)

30. TOTAL FERTILITY RATE 1.48 children born/woman (2003 est.)

31. HIV/AIDS - ADULT PREVALENCE RATE 0.5% (2001 est.)

32. HIV/AIDS - PEOPLE LIVING WITH ~ 19,000 (2001 est.)

33. HIV/AIDS - DEATHS less than 100 (2001 est.)

34. NATIONALITY noun: Swiss (singular and plural)
adjective: Swiss

35. ETHNIC GROUPS German 65%, French 18%, Italian 10%, Romansch 1%, other 6%

36. RELIGIONS Roman Catholic 46.1%, Protestant 40%, other 5%, none 8.9% (1990)

37. LANGUAGES German (official) 63.7%, French (official) 19.2%, Italian (official) 7.6%, Romansch (official) 0.6%, other 8.9%

38. LITERACY definition: age 15 and over can read and write
total population: 99% (1980 est.)

39. COUNTRY NAME conventional long form: Swiss Confederation
conventional short form: Switzerland
local short form: Schweiz (German), Suisse (French), Svizzera (Italian)
local long form: Schweizerische Eidgenossenschaft (German), Confederation Suisse (French), Confederazione Svizzera (Italian)

40. GOVERNMENT TYPE federal republic

41. CAPITAL Bern

42. ADMINISTRATIVE DIVISIONS 26 cantons (cantons, singular - canton in French;
cantoni, singular- cantone in Italian; kantone, singular - kanton in German); Aargau, Appenzell Ausser-Rhoden, Appenzell Inner-Rhoden, Basel-Landschaft, Basel-Stadt, Bern, Fribourg, Geneve, Glarus, Graubunden, Jura, Luzern, Neuchatel, Nidwalden, Obwalden, Sankt Gallen, Schaffhausen, Schwyz, Solothurn, Thurgau, Ticino, Uri, Valais, Vaud, Zug, Zurich.

43. INDEPENDENCE 1 August 1291 (Founding of the Swiss Confederation)

44. NATIONAL HOLIDAY Founding of the Swiss Confederation, 1 August (1291)

45. CONSTITUTION 18 December 1998

46. LEGAL SYSTEM civil law system influenced by customary law; judicial review of legislative acts, except with respect to federal decrees of general obligatory character; accepts compulsory ICJ jurisdiction, with reservations.

47. SUFFRAGE 18 years of age; universal

48. EXECUTIVE BRANCH chief of state: President Pascal COUCHEPIN (since 1 January 2003); Vice President Ruth METZLER (since 1 January 2003); note – the president is both the chief of state and head of government. Head of government: President Pascal COUCHEPIN (since 1 January 2003); Vice President Ruth METZLER (since 1 January 2003); note the president is both the chief of state and head of government. Cabinet: Federal Council or Bundesrat (in German), Conseil Federal (in French), Consiglio Federale (in Italian) elected by the Federal Assembly usually from among its own members for a four-year term elections: president and vice president elected by the Federal Assembly from among the members of the Federal Council for one-year terms that run concurrently; election last held NA December 2002 (next to be held NA December 2003). Election results: Pascal COUCHEPIN elected president; percent of Federal Assembly vote - NA%; Ruth METZLER elected vice president; percent of legislative vote - NA%.

49. LEGISLATIVE BRANCH bicameral Federal Assembly or Bundesversammlung (in German), Assemblee Federale (in French), Assemblea Federale (in Italian), consists of the Council of States or Standerat (in German), Conseil des Etats (in French), Consiglio degli Stati (in Italian) (46 seats - members serve four-year terms) and the National Council or Nationalrat (in German), Conseil National (in French), Consiglio Nazionale (in Italian) (200 seats - members are elected by popular vote on the basis of proportional representation to serve four-year terms) elections: Council of States - last held NA 1999 (each canton determines when the next election will be held); National Council - last held 19 October 2003 (next to be held NA October 2007) election results: Council of States - percent of vote by party - NA%; seats by party - FDP 18, CVP 15, SVP 7, SPS 6; National Council - percent of vote by party - SVP 27.7%, SPS 24.2%, FDP 16%, CVP 12.9%, Greens 7.7%, other small parties all under 5%; seats by party - SVP 55, SPS 52, FDP 36, CVP

28, Green Party 13, other small parties 16.

50. JUDICIAL BRANCH Federal Supreme Court (judges elected for six-year terms by the Federal Assembly)

51. POLITICAL PARTIES AND LEADERS Christian Democratic People's Party (Christichdemokratische Volkspartei der Schweiz or CVP, Parti Democrate-Chretien Suisse or Cristiano Popolare Svizzero or PDC, Partida Cristiandemocratica dalla Svizra or PCD) [Philipp STAEHELIN, president]; Green Party (Grune Partei der Schweiz or Grune, Parti Ecologiste Suisse or Les Verts, Partito Ecologista Svizzero or I Verdi, Partida Ecologica Svizra or La Verda) [Ruth GENNER and Patrice MUGNY, co-presidents]; Radical Free Democratic Party (Freisinnig-Demokratische Partei der Schweiz or FDP, Parti Radical-Democratique Suisse or PRD, Partitio Liberal-Radicale Svizzero or PLR) [Christiane LANGENBERGER, president]; Social Democratic Party (Sozialdemokratische Partei der Schweiz or SPS, Parti Socialist Suisse or PSS, Partito Socialista Svizzero or PSS, Partida Socialdemocratica de la Svizra or PSS) [Christiane BRUNNER, president]; Swiss People's Party (Schweizerische Volkspartei or SVP, Union Democratique du Centre or UDC, Unione Democratica de Centro or UDC, Uniun Democratica dal Center or UDC) [Ueli MAURER, president]; and other minor parties.

52. POLITICAL PRESSURE GROUPS AND LEADERS NA

53. INTERNTNL. ORGANIZATION PARTICIPATION ACCT, AfDB, AsDB, Australia Group, BIS, CE, CERN, EAPC, EBRD, ECE, EFTA, ESA, FAO, G-10, IADB, IAEA, IBRD, ICAO, ICC, ICCt, ICFTU, ICRM, IDA, IEA, IFAD, IFC, IFRCS, ILO, IMF, IMO, Interpol, IOC, IOM, ISO, ITU, LAIA (observer), MONUC, NAM (guest), NEA, NSG, OAS, (observer), OECD, OPCW, OSCE, PCA, PFP, UN, UNCTAD, UNESCO, UNHCR, UNIDO, UNITAR, UNMEE, UNMIBH, UNMIK, UNMOP, UNOMIG, UNTSO, UNU, UPU, WCL, WCO, WHO, WIPO, WMO, WToO, WTrO, ZC

54. DIPLOMATIC REPRESENTATION IN THE US chief of mission: Ambassador Christian
BLICKENSTORFER
consulate(s): Boston
consulate(s) general: Atlanta, Chicago, Houston,
Los Angeles, New York, and
San Francisco
FAX: [1] (202) 387-2564
telephone: [1] (202) 745-7900
chancery: 2900 Cathedral Avenue NW, Washington,
DC 20008

55. DIPLOMATIC REPRESENTATION FROM THE US chief of mission: Ambassador Mercer REYNOLDS III
embassy: Jubilaeumsstrasse 93, 3005 Bern
mailing address: use embassy street address
telephone: [41] (031) 357 70 11
FAX: [41] (031) 357 73 44

56. FLAG DESCRIPTION red square with a bold, equilateral white cross in the center that does not extend to the edges of the flag

57. ECONOMY - OVERVIEW Switzerland is a prosperous and stable modern market economy with low unemployment, a highly skilled labor force, and a per capita GDP larger than that of the big western European economies. The Swiss in recent years have brought their economic practices largely into conformity with the EU's to enhance their international competitiveness. Switzerland remains a safe haven for investors, because it has maintained a degree of bank secrecy and has kept up the franc's long-term external value. Reflecting the anemic economic conditions of Europe, GDP growth dropped in 2001 to about 0.8%, to 0.2% in 2002, and to -0.3% in 2003.

58. GDP purchasing power parity - $233.4 billion (2002 est.)

59. GDP - REAL GROWTH RATE 0.1% (2002 est.)

60. GDP - PER CAPITA purchasing power parity - $32,000 (2002 est.)

61. GDP - COMPOSITION BY SECTOR agriculture: 2%
industry: 34%

services: 64% (2002 est.)

62. POPULATION BELOW POVERTY LINE NA%

63. HOUSEHOLD INCOME OR CONSUMPTION lowest 10%: 2.6%
BY PERCENTAGE SHARE highest 10%: 25.2% (1992)

64. DISTRIBUTION OF FAMILY INCOME - GINI INDEX 33.1 (1992)

65. INFLATION RATE (CONSUMER PRICES) 0.5% (2002 est.)

66. LABOR FORCE 4 million (2001)

67. LABOR FORCE - BY OCCUPATION services 69.1%, industry 26.3%, agriculture 4.6%
(1998)

68. UNEMPLOYMENT RATE 1.9% (2002 est.)

69. BUDGET revenues: $30 billion
expenditures: $30 billion, including capital
expenditures of $NA (2001 est.)

70. INDUSTRIES machinery, chemicals, watches, textiles,
precision instruments

71. INDUSTRIAL PRODUCTION GROWTH RATE 3.2% (2001)

72. ELECTRICITY - PRODUCTION 68.68 billion kWh (2001)

73. ELECTRICITY - PRODUCTION BY SOURCE fossil fuel: 1.3%
hydro: 59.5%
other: 2% (2001)
nuclear: 37.1%

74. ELECTRICITY - CONSUMPTION 53.43 billion kWh (2001)

75. ELECTRICITY - EXPORTS 34.54 billion kWh (2001)

76. ELECTRICITY - IMPORTS 24.1 billion kWh (2001)

77. OIL - PRODUCTION 0 bbl/day (2001 est.)

78. OIL - CONSUMPTION 290,400 bbl/day (2001 est.)

79. OIL - EXPORTS 10,420 bbl/day (2001)

80. OIL - IMPORTS 289,500 bbl/day (2001)

81. NATURAL GAS - PRODUCTION 0 cu m (2001 est.)

82. NATURAL GAS - CONSUMPTION 3.093 billion cu m (2001 est.)

83. NATURAL GAS - EXPORTS 0 cu m (2001 est.)

84. NATURAL GAS - IMPORTS 3.093 billion cu m (2001 est.)

85. AGRICULTURE - PRODUCTS grains, fruits, vegetables; meat, eggs

86. EXPORTS $100.3 billion f.o.b. (2002 est.)

87. EXPORTS - COMMODITIES machinery, chemicals, metals, watches, agricultural products

88. EXPORTS - PARTNERS Germany 19.2%, US 10.2%, Italy 9.6%, France 8.9%, UK 7.7% (2002)

89. IMPORTS $94.4 billion f.o.b. (2002 est.)

90. IMPORTS - COMMODITIES machinery, chemicals, vehicles, metals; agricultural products, textiles

91. IMPORTS - PARTNERS Germany 27.4%, France 11.4%, Italy 9.7%, US 8.5%, Russia 5.8%, UK 5.4%, Austria 4.6%, Netherlands 4.1% (2002)

92. DEBT - EXTERNAL $NA

93. ECONOMIC AID - DONOR ODA, $1.1 billion (1995)

94. CURRENCY Swiss franc (CHF)

95. CURRENCY CODE CHF

96. EXCHANGE RATES Swiss francs per US dollar - 1.56 (2002), 1.69 (2001), 1.69 (2000), 1.5 (1999), 1.45 (1998)

97. FISCAL YEAR calendar year

98. TELEPHONES - MAIN LINES IN USE 4.82 million (1998)

99. TELEPHONES - MOBILE CELLULAR 1.967 million (1999)

100. TELEPHONE SYSTEM general assessment: excellent domestic and international services.
domestic: extensive cable and microwave radio relay networks

international: satellite earth stations –
2 Intelsat (Atlantic Ocean
and Indian Ocean)

101. RADIO BROADCAST STATIONS AM 4, FM 113 (plus many low power stations), shortwave 2 (1998)

102. TELEVISION BROADCAST STATIONS 115 (plus 1,919 repeaters) (1995)

103. INTERNET COUNTRY CODE .ch

104. INTERNET SERVICE PROVIDERS (ISPS) 44 (Switzerland and Liechtenstein) (2000)

105. INTERNET USERS 3.85 million (2002)

106. RAILWAYS total: 4,511 km
standard gauge: 3,483 km 1.435-m gauge (3,472 km electrified)
narrow gauge: 982 km 1.000-m gauge (975 km electrified); 46 km
0.800-m gauge (46 km electrified) (2002)

107. HIGHWAYS total: 71,011 km
paved: 71,011 km (including 1,638 of expressways)
unpaved: 0 km (2000)

108. WATERWAYS 65 km
note: The Rhine carries heavy traffic on the
Basel-Rheinfelden and Schaffhausen-Bodensee
stretches; there are also 12 navigable lakes

109. PIPELINES gas 1,831 km; oil 212 km; refined products
7 km (2003)

110. PORTS AND HARBORS Basel

111. MERCHANT MARINE total: 29 ships (1,000 GRT or over)
597,049 GRT/1,051,380 DWT note: includes some
foreign-owned ships registered here as a flag
of convenience: UK 6, US 1 (2002 est.) ships
by type: bulk 16, cargo 6, chemical tanker 2,
container 2, passenger 1, petroleum tanker 1,
specialized tanker 1

112. AIRPORTS 66 (2002)

113. AIRPORTS - WITH PAVED RUNWAYS total: 41
over 3,047 m: 3
2,438 to 3,047 m: 5
914 to 1,523 m: 9
under 914 m: 14 (2002)

1,524 to 2,437 m: 10

114. AIRPORTS - WITH UNPAVED RUNWAYS total: 25
1524 to 2437 m: 1
under 914 m: 24 (2002)

115. HELIPORTS 1 (2002)

116. MILITARY BRANCHES Army, Air Force, Frontier Guards, Fortification Guards

117. MILITARY MANPOWER - MILITARY AGE 20 years of age (2003 est.)

118. MILITARY MANPOWER - AVAILABILITY males age 15-49: 1,834, 638 (2003 est.)

119. MILITARY MANPOWER –
FIT FOR MILITARY SERVICE males age 15-49: 1,552,728 (2003 est.)

120. MILITARY MANPOWER –
REACHING MILITARY AGE ANNUALLY males: 42,761 (2003 est.)

121. MILITARY EXPENDITURES - DOLLAR FIGURE $2.548 billion (FY01)

122. MILITARY EXPENDITURES - PERCENT OF GDP 1% (FY01)

123. DISPUTES - INTERNATIONAL none

124. ILLICIT DRUGS because of more stringent government regulations, used significantly less as a money-laundering center; transit country for and consumer of South American cocaine and Southwest Asian heroin.

VULNERABILITY ANALYSIS (to be developed at a later stage)

Verbose Indicators:

---

pover 0.99000099e-1
unemploy 0.14850015
migrant 0.49500049e-1
drug 0.49500049e-1
petroleum 0.99000099e-1
oil 0.99000099e-1
pollution 0.24750025e-1
flood 0.14850015
HIV/AIDS 0.99000099e-1
pover 0.99000099e-1

```
unemploy 0.14850015
migrant 0.49500049e-1
drug 0.49500049e-1
petroleum 0.99000099e-1
oil 0.99000099e-1
pollution 0.24750025e-1
flood 0.14850015
HIV/AIDS 0.99000099e-1
Russia 0.67314884
polar 1.24223602
cold 0.62111801
gas 0.48335124
petroleum 0.53705693
oil 0.53705693
lead 0.21482277
iron 0.13426423
salt 0.10741139
tin 0.26852846
flood 0.47898455
avalanche 0.59873069e-1
landslide 0.17961921
air pollution 0.14930721
emission 0.14930721
pollution 0.59722886e-1
fertilizer 0.59722886e-1
biodiversity 0.11944577e-1
extensive 0.11944577e-1
loss 0.11944577e-1
acid 0.89584329e-1
spec 0.11944577e-1
landlocked 0.64935065
migrant 2.27963526
dense 1.51975684
republic 0.29012417
federal 0.29012417
federation 0.29012417
boundar 1.28410915
claim 1.44462279
heroin 0.46685341
coca 0.42016807
lab 0.18674136
money-launder 0.11671335
```

```
Total Verbose Index: 16.9647195

Numerical Indicators
```

```
Frame
-----
Area, Total (sq.km): 41290
Area, Land (sq.km): 39770
Area, Water (sq.km): 1520
Boundary Length, Total (km): 1852
Coastline Length (km): 0
Land Use
--------
Land Use, Arable (%): 10.57%
Land Use, Permanent Crops (%): 0.61%
Land Use, Other (%): 88.82%
Irrigated Land (sq.km): 250
Population
----------
Population (person): 7318638
Age Structure, 0-14 y (%): 16.6%
Age Structure, 15-64 y (%): 67.8%
Age Structure, 65y and over (%): 15.6%
Median Age, total (year): 40.2
Median Age, male (year): 39.3
Median Age, female (year): 41.2
Population Growth Rate (%): 0.21%
Birth Rate (births/1000 person): 9.59
Death Rate (deaths/1000 person): 8.82
Net Migration Rate (migrants/1000 person): 1.37
Sex Ratio, at birth (males/females): 1.05
Sex Ratio, under 15 y (males/females): 1.05
Sex Ratio, 15-64 y (males/females): 1.03
Sex Ratio, 65 y and over (males/females): 0.7
Infant Mortality Rate, total (deaths/1000 live births): 4.36
Infant Mortality Rate, male (deaths/1000 live births): 4.25
Infant Mortality Rate, female (deaths/1000 live births): 4.25
Life Expectancy at Birth, total (year): 79.99
Life Expectancy at Birth, male (year): 77.11
Life Expectancy at Birth, female (year): 83.02
Total Fertility Rate (children born/woman): 1.48
HIV/AIDS - Adult Prevalence Rate (%): 0.5%
```

```
HIV/AIDS - People Living with HIV/AIDS (%): 19000
HIV/AIDS - Deaths (%): less
Literacy, total (%): 99%
Literacy, male (%): NA
Literacy, female (%): NA

Economy
-------
GDP (billion US$ purchasing power parity): 233.4
GDP - Real Growth Rate (%): 0.1%
GDP - per capita (US$/person purchasing power parity): 32,000
GDP - Composition by Sector, Agriculture (%): 2%
GDP - Composition by Sector, Industry (%): 34%
GDP - Composition by Sector, Services (%): 64%
Population below poverty line (%): NA%
Household income or consumption by percentage share, lowest 10% (%): 2.6%
Household income or consumption by percentage share, highest 10% (%): 25.2%
Distribution of family income - Gini index: 33.1
Inflation rate (consumer prices, %): 0.5%
Labor Force (million persons): 4
Unemployment rate (%): 1.9%
Budget, Revenues (billion US$): 30
Budget, Expenditures (billion US$): 30
Budget, Capital Expenditures (billion US$): NA
Industrial production growth rate (%): 3.2%
Electricity - production (billion kWh): 68.68
Electricity - production by source, fossil fuel (%): 1.3%
Electricity - production by source, hydro (%): 59.5%
Electricity - production by source, other (%): 2%
Electricity - production by source, nuclear (%): 37.1%
Electricity - consumption (billion kWh): 53.43
Electricity - exports (billion kWh): 34.54
Electricity - imports (billion kWh): 24.1
Oil - production (bbl/day): 0
Oil - consumption (bbl/day): 0.0002904
Oil - exports (bbl/day): 0.1042e-4
Oil - imports (bbl/day): 0.0002895
Natural Gas - production (billion cu m): 0
Natural Gas - consumption (billion cu m): 3.093
Natural Gas - exports (billion cu m): 0
Natural Gas - imports (billion cu m): 3.093
Exports (billion US$): 100.3
Imports (billion US$): 94.4
Debt - external (billion US$): 0
Exchange Rates, year 2002 (to US$): 1.56
```

```
Exchange Rates, year 2001 (to US$): 1.69
Exchange Rates, year 2000 (to US$): 1.69
Exchange Rates, year 1999 (to US$): |
Exchange Rates, year 1998 (to US$): 1.45

Communications
--------------
Telephones - main lines in use (million): 4.82
Telephones - mobile cellular (million): 1.967
Radio broadcast stations, AM: 4,
Radio broadcast stations, FM: 113
Radio broadcast stations, shortwave: 2
Television broadcast stations: 115

IT Ubiquity
-----------
Internet Service Providers (ISPS): 44
Internet users (million): 3.85

Transportation
--------------
Railways, total (km): 4,511
Railways, broad gauge (km): NA
Railways, standard gauge (km): 3,483
Railways, narrow gauge (km): 982
Highways, total (km): 71,011
Highways, paved (km): 71,011
Highways, unpaved (km): NA
Waterways (km): 65
Pipelines, gas (km): 1831
Pipelines, condensate (km): NA
Pipelines, liquid petroleum gas (km): NA
Pipelines, oil (km): 212
Pipelines, refined/oil/petroleum products (km): 7
Pipelines, oil/gas/water (km): NA
Pipelines, water (km): NA
Pipelines, unknown (km): NA
Merchant marine, total: 29
Merchant marine, barge carrier: NA
Merchant marine, bulk: 16,
Merchant marine, cargo: 6,
Merchant marine, liquefied gas tanker: NA
Merchant marine, chemical tanker: 2,
Merchant marine, combination bulk: NA
Merchant marine, combination ore/oil: NA
Merchant marine, container: 2,
Merchant marine, freighter: NA
```

```
Merchant marine, heavy lift carrier: NA
Merchant marine, passenger/cargo: NA
Merchant marine, passenger: 1,
Merchant marine, petroleum tanker: 1,
Merchant marine, refrigerated cargo: NA
Merchant marine, roll on/roll off: NA
Merchant marine, short-sea passenger: NA
Merchant marine, specialized tanker: 1
Merchant marine, vehicle carrier: NA
Merchant marine, large-load carrier: NA
Airports - with paved runways: 41
Airports - with unpaved runways: 25

Defense
-------
Military manpower - availability (males age 15-49): 1834638
Military manpower - fit for military service (males age 15-49): 1552728
Military manpower - reaching military age annually (males): 42761
Military expenditures - dollar figure (billion US$): 2.548
Military expenditures - percent of GDP (%): 1
```

12.3 Targeting Strategic Goals

Several methods could be used. We selected the Index Method and the Matrix Method. The Index Method attempts to illustrate a risk management-oriented approach featuring a strong societal orientation, as prevailing in the methodological attitude of, for instance, Swedish Association of Local Authorities, and described in operative terms in a pre-study by Nilsson and colleagues (2001). Specifically, the module offers one possible implementation of an assess and rank solution to describing a local municipal authority's vulnerability, as captured in the '*Appendix 3*' of Nilsson et al. (2001) and elaborated in an applied to a transport system in 'Chap. 6' of Vamanu et al. (2016). Following these elaborations, Fig. 12.2 was developed to depict assess and rank solution describing Switzerland's municipal authority's vulnerability based on the data suggested by the 'indicators' above.

Similar to the Index Method, the Matrix Method attempts to illustrate a risk management-oriented approach featuring a strong societal orientation. However, the model based on the Matrix Method offers one possible implementation of a solution to describing a local municipal authority's 'robustness' (i.e., capability to manage risks, as a direct opposite of vulnerability). The reader is directed to '*Appendix 4*' of Nilsson et al. (2001) and Chap. 6 of Vamanu et al. (2016) for detailed discussion. Following these detailed discussions and applying the specified data regarding Switzerland, Fig. 12.3 was developed to depict governance robustness and Fig. 12.4 to depict vulnerability acceptability.

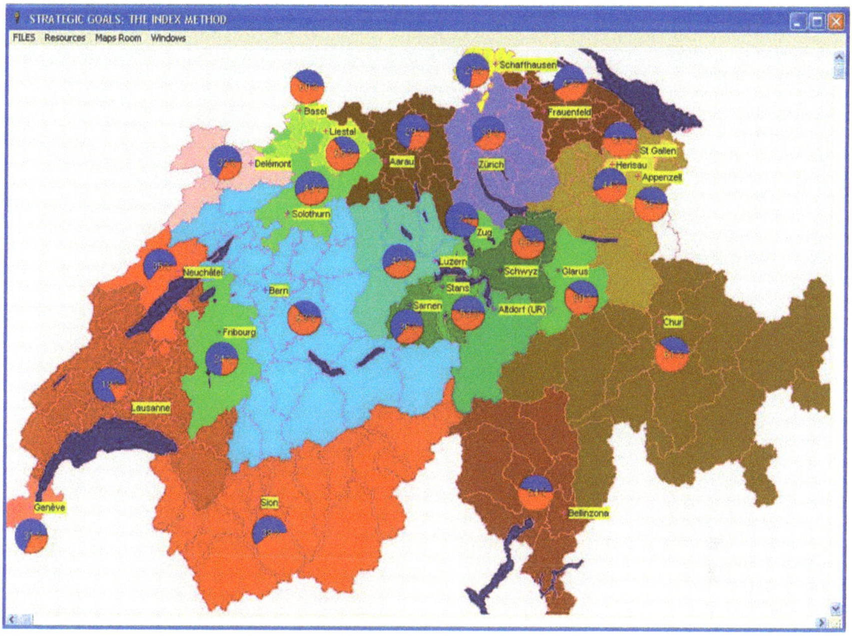

Fig. 12.2 Targeting strategic goals through the index method

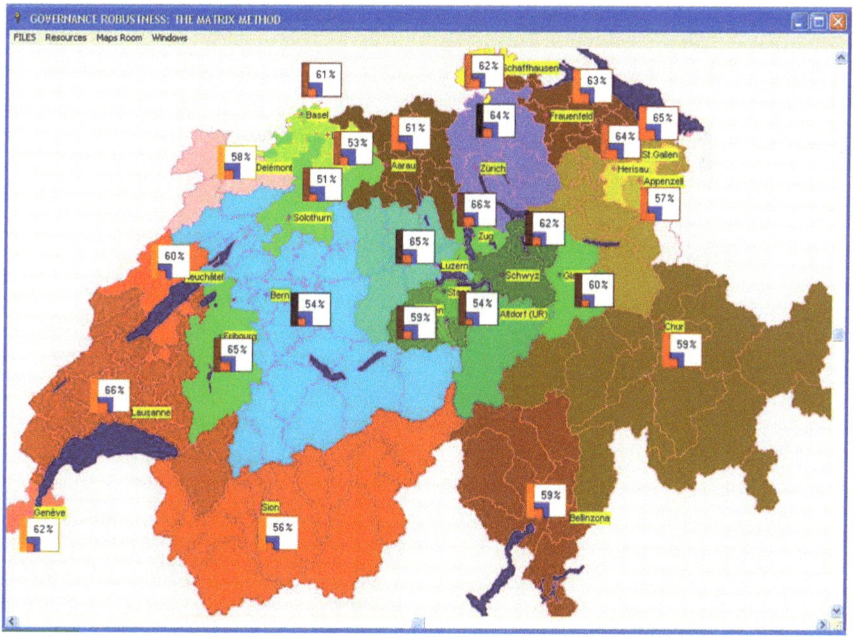

Fig. 12.3 Switzerland's governance robustness through the matrix method

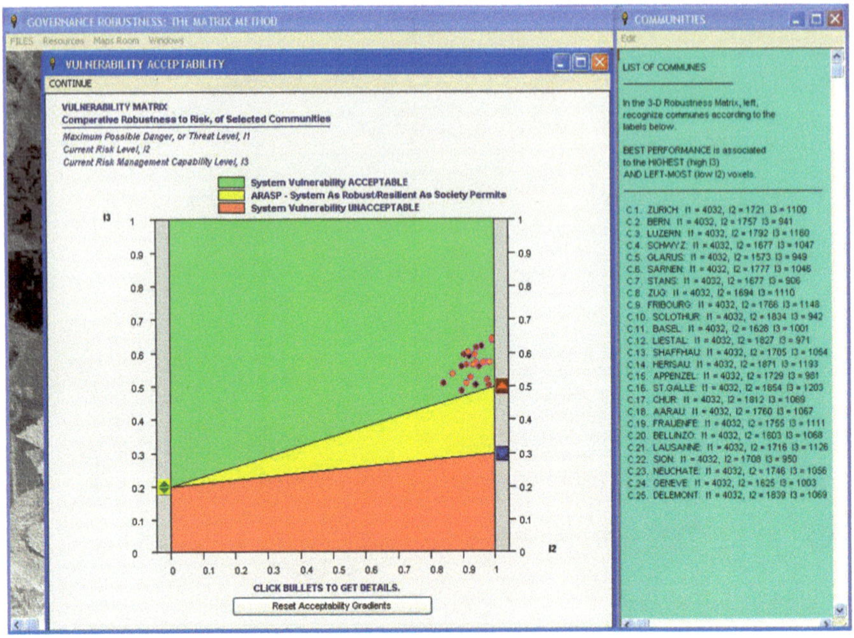

Fig. 12.4 Switzerland's vulnerability acceptability

In both cases, the Index and Matrix method applications are meant to support the notion that the only productive way to approach vulnerability at this early stage into the building up of an awareness on the matter is to exercise due respect for the perception of various actors in the emerging intellectual play, and firmly resist the temptation of academic reductions and model standardization.

12.4 Complexity-Induced Vulnerability

A cross-cutting line of work in a QVA package is the concept of complexity-induced vulnerability (see Gheorghe and Vamanu 2004). Substantiating the notion of a QVA, the models introduced under the preceding menu entries range from physical analogies with the order–disorder phenomena and phase transitions to multi-criteria (matrix-wise) processing of opinion pools. Pervading these all, however, was the notion of cooperative behavior of multi-component (many bodies) systems feeding from their internal connectivity (member interactions). The single key word used to capture both aspects of colloquial language is COMPLEXITY, although some would ostensibly argue that the meaning of 'complexity' may well transcend the ad hoc acceptance here adopted. In Chapter 2,

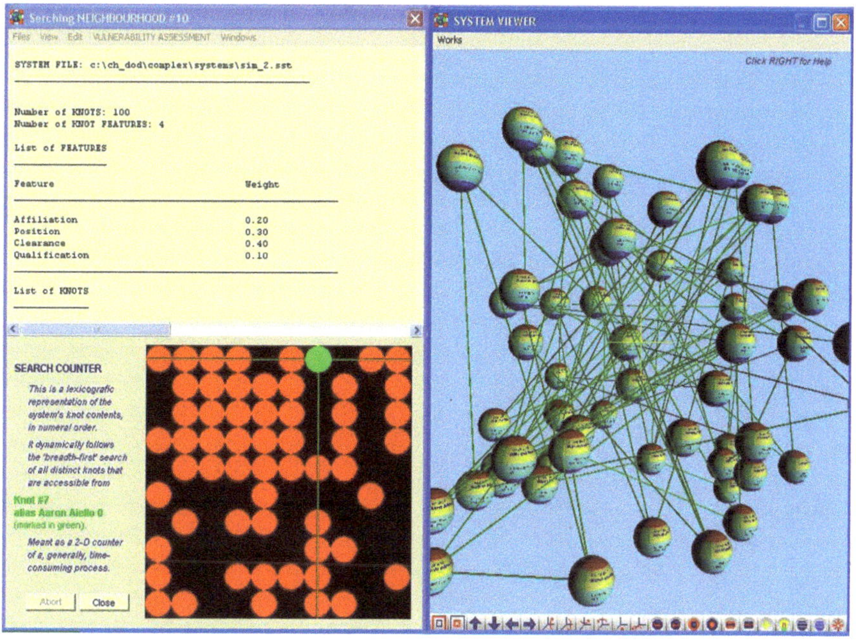

Fig. 12.5 Targeting penetrability of a system (i.e., Switzerland)

Sect. 2.2.4 provides detailed information on complexity and system penetrability. In the present case, Fig. 12.5 was developed.

A special line of thought has eventually emerged from the finding above, setting the task of taking a straightforward approach to complexity as a source of vulnerability. The practical goal is to attach a relevant metrics to the internal connectivity of multi-component systems so that this be turned to account from a QVA-oriented standpoint.

12.5 Probabilistic Resilience Assessment

In the Sect. 2.1.1 (Chap. 2), it was suggested that discipline of *System Dynamic Stability*, especially phase space system topology could be used as a basis for probabilistic resiliency assessment. In QVA, the module offers an interface to evaluate the identity, and structural, fault probability in a complex system subject to

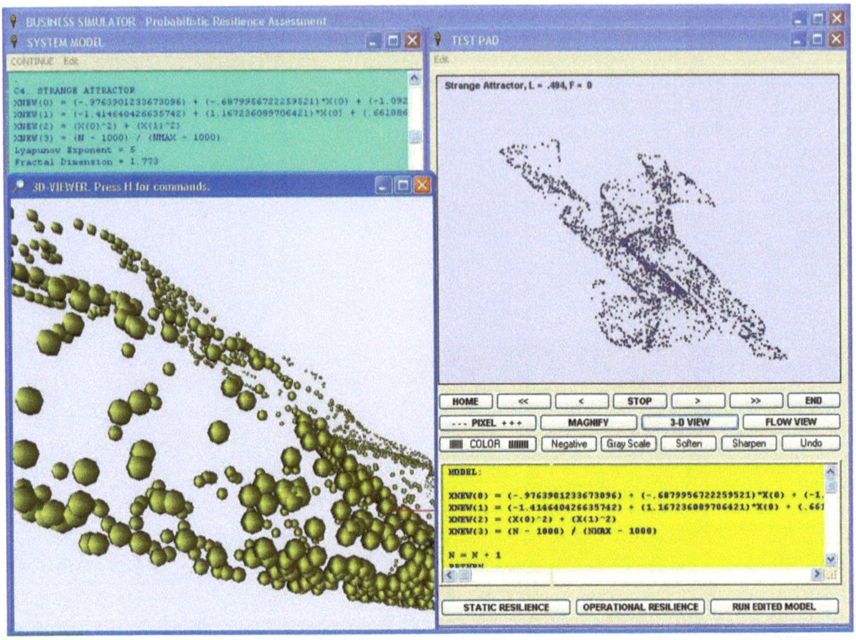

Fig. 12.6 Probabilistic resilience assessment

spontaneous or engineered fluctuations in its control parameters.[1] This was used as the basis for the following graphic (Fig. 12.6).

The application is meant to further support the notion that the only productive way to approach vulnerability at this early stage into the building up of an awareness on the matter is to exercise due respect for the perception of various actors in the emerging intellectual play, and firmly resist the temptation of academic reductions and model standardization.

12.6 Territorial Vulnerability

This aspect of research was done for the canton of Valais, Switzerland, specifically Monthey district and the neighboring communes. The application follows a generic QVA method. To get the physical indicators required, risk classification and prioritization as suggested by van den brand (1996) methodology were employed. To

---

[1]It should be noted that the concept of resilience has numerous applications, ranging from large social systems to small-scale systems. For those interested in quantification and evaluation resilience in buildings, Tokgoz's (2012) research provides a 'dashboard' for resilience acceptability related to desired resilience of buildings corresponding to hurricane categories; complete to numerical estimations and models based on HAZUS–MH: Hazards U.S. Multi-Hazard.

obtain this vulnerability, hazardous substance database, which is linked to the local code's data banks, as well as the shorthand method is used to evaluate effect—distances and areas from fire, explosions, and toxicity following environmental releases of such substances. Data related to the selected area of study is depicted in Fig. 12.7.

From this point on, the local code's GIS is employed to determine people and property (land) affectations within the obtained areas and distances. Effect indicators are combined with indicators reflective of the managerial capability to mitigate risks, in order to fully enable the functioning of the generic QVA procedure.

In the background of this application, the 'stability-related vulnerability' is one cross-cutting line of work within AIDRAM's QVA project (Gheorghe 2004). A generic model is proposed, providing, in essence:

- a two-parameter description and the respective equation of state, for any multicomponent, multi-indicator system featuring two states: *operable* and *inoperable.*
- a division of the two-parameter phase space of the system into *vulnerability basins.*
- a 0–100 *Vulnerability Scale* and the means to measure the respective 'Vulnerability Index,' as an operational expression of a QVA.

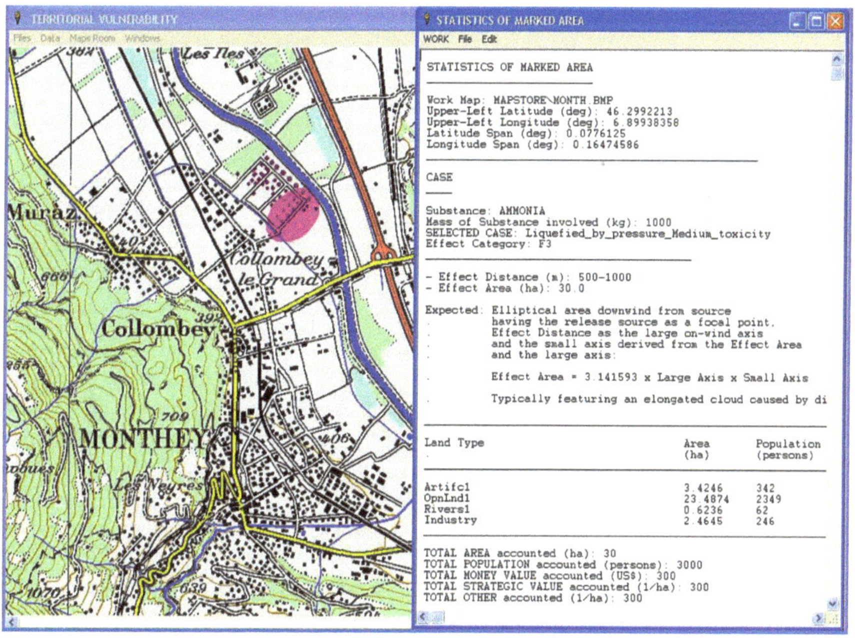

Fig. 12.7 Statistic of the marked area

A method to diagnose the vulnerability of complex systems featuring large numbers of indicators, both internal and external, as well as to dynamically monitor the time-evolvement of the vulnerability as the indicators change, is thus articulated and demonstrated. The method is generic and algorithmic, and is believed to having the potential to accommodate a virtually unlimited variety of applications. A related 'Sensitivity Analysis' for this model is depicted below (Fig. 12.8) indicating system indicators, the basis for the sensitivity.

12.7 Remarks

Clearly, given the present model of QVA, especially its indicators along with indicators of a system, it is possible to perform analysis targeting risk and vulnerability. This suggests that research such as the one highlighted in this chapter could be performed an any given location of interest. In fact, present researchers have conducted similar research in Hampton region of Virginia (USA) (Fig. 12.9) as well as other sites around the world (see Fig. 12.10).

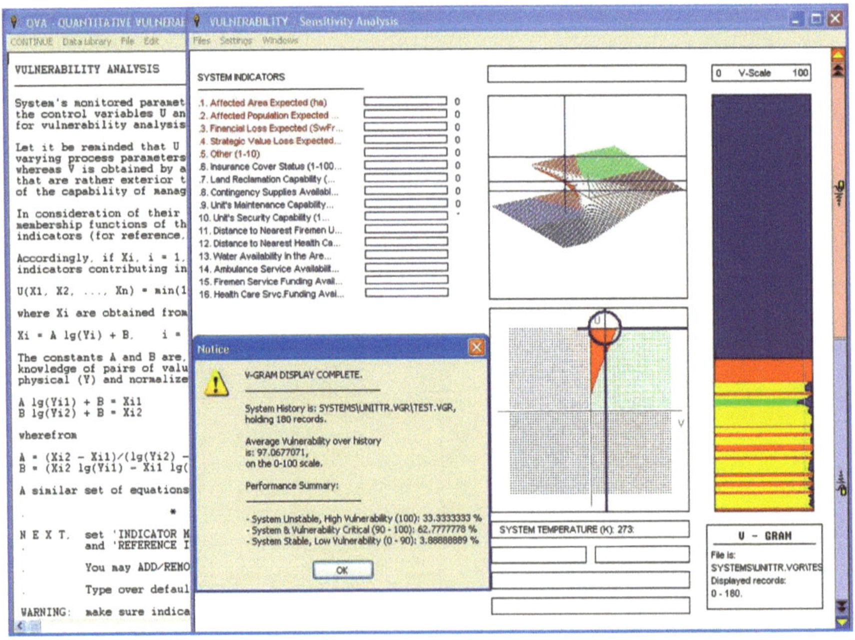

Fig. 12.8 A V-Gram display for sensitivity analysis in QVA

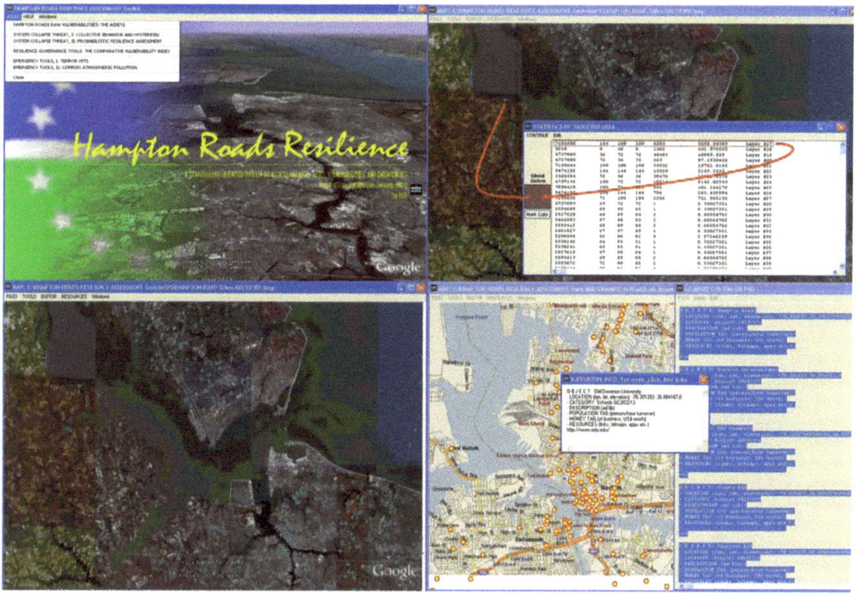

Fig. 12.9 Initializing QVA research in Hampton Roads

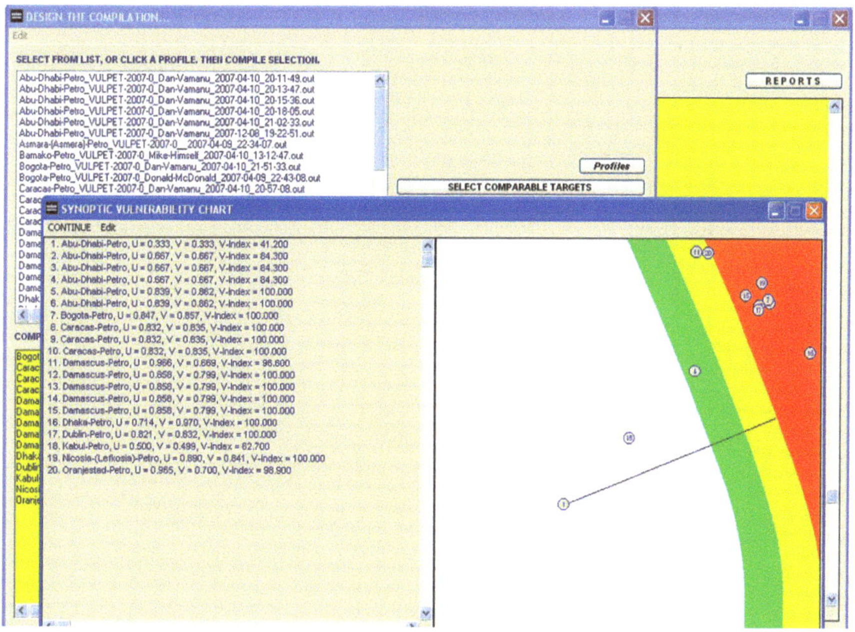

Fig. 12.10 A design compilation of different research sites

References

Gheorghe, A. V. (2004). *The hidden faults: Towards a standing method to assess Switzerland's vulnerabilities*. Zurich, Switzerland: Laboratory of Safety Analysis, ETH Zurich.

Gheorghe, A. V., & Vamanu, D. V. (2004). Complexity induced vulnerability. *International Journal of Critical Infrastructures, 1*(1), 76–84.

Nilsson, J., Magnusson, S., Hallin, P., & Lenntorp, B. (2001). *Models for vulnerability auditing and distribution of governmental economical means at the local authority level*. Lund, Sweden: LUCRAM: Lund University Centre for Risk Analysis and Management.

Tokgoz, B. E. (2012). *Probabilistic resilience quantification and visualization building performance to hurricane wind speeds* (Ph.D.). Old Dominion University, United States—Virginia.

van den Brand, D. (1996). *Manual for the classification and prioritization of risks due to major accidents in process and related industries* (No. IAEA-TECDOC-727 (Rev. 1)). Vienna: International Atomic Energy Agency.

Vamanu, B. I., Gheorghe, A. V., & Katina, P. F. (2016). *Critical infrastructures: Risk and vulnerability assessment in transportation of dangerous goods—Transportation by road and rail* (Vol. 31). Cham, Switzerland: Springer International Publishing.

Chapter 13
Vulnerability Analysis and Swiss Reduction—Building a Framework for Ranking Solutions

Abstract Having processed Switzerland with indicators, strategic goals, vulnerabilities, and looking for means to addressing assessment of probabilistic resilience as well as vulnerabilities (Chap. 12), this chapter offers practical means to classify vulnerability, a nexus, and categorization of metrics for risk and vulnerability prioritization.

13.1 Vulnerability High-Level Classification

The immediate concerns, and complementary to the preceding chapter, involves:

- Methods for analysis of vulnerability at politico—strategic level: a commented list of findings, principles, recommendations stemming out from the review on the 'Agora' debate in the preceding chapter that does apply to the 'Swiss system' case.
- Strategic–operational (from confederation to cantons) issues.
- Interactive IT decision support environment.
- Future aspects: integration of risk and vulnerability analysis into a nexus approach.

Starting from an agreed goal—definition on vulnerability—one can identify a hierarchy of criteria and indicators in order to describe a way how to structure the issue at politico–strategic level or at the tactical operative level. This is a dynamic and interactive process which can be carried out through expert interactions as well as by using literature studies. In general, the approach of using multi-criteria decision analysis is not new and has been used in other problem-solving situations, when a problem has to be appropriate structures in view of integrating criteria, weighting factors for each individual criterion as well as a hierarchy for criteria architecture.

The example introduced in the Fig. 13.1 below is a generic and indicates a series of criteria introduced for assessing vulnerability at the national level. It considers a variety of hazards, namely man-induced, man-made, and natural hazards. Other

A.V. Gheorghe et al., *Critical Infrastructures, Key Resources, Key Assets*, Topics in Safety, Risk, Reliability and Quality 34,
https://doi.org/10.1007/978-3-319-69224-1_13

class of criteria could refer to external or internal potentially induced hazards which could have origins at global, European, or regional levels, down to the cantonal dimension.

In view of assessing politico–strategic as well as tactical–operational vulnerability-related issues, by addressing a systemic approach, we propose to exercise dedicated tools, as well as new ones, capable to handle in an analytic, as well as hybrid (experimental-analytic) formats the new aspects related mainly to the concept of vulnerability.

A simplified schema to design a hierarchy of models and instruments for assessing vulnerability in the specific case of Switzerland is outlined in Fig. 13.2 below. By following the logic, it is possible to understand how the security, as a common good, could be achieved at some given price, and in the end, every citizen is part of the process of assuring a given level at the societal level.

Addressing indicators formation, and their degree of complexity or completeness for solving real problems, is considered as a separate task where specialists and policy makers should be fully involved by using well-known techniques (e.g., brainstorming) or tools (e.g., Think Tools). Such work already has been experimented in Switzerland within work related to risk analysis in Switzerland. The specificity of vulnerability assessment tasks requires, *inter alia*:

- an adequate mind-set to address indicators for vulnerability-related aspects, within the security policy.
- understanding the indicators which differentiate from risk analysis, and which can explain the robustness of various systems, from technical to society levels, to external or internal threats.
- understanding the type of models to be used in order to address specific aspects related to vulnerability evaluation and management.

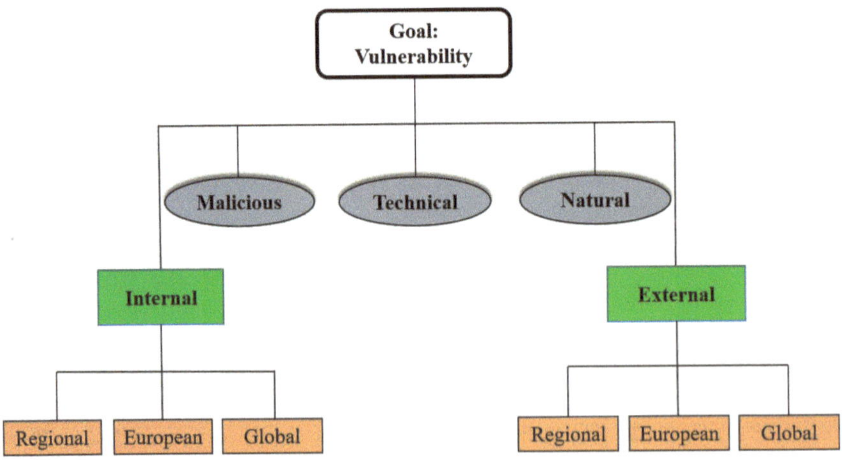

Fig. 13.1 A high-level classification of vulnerability

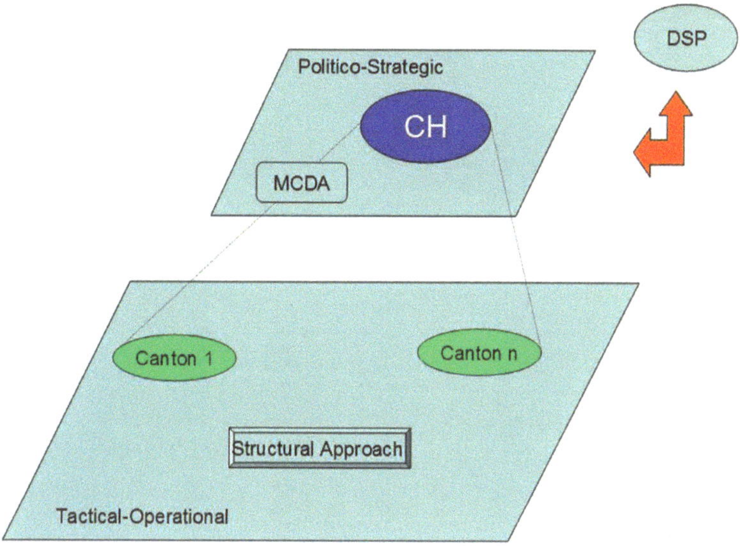

Fig. 13.2 A model relating Canton relationship in the context of the nation

- associating economic-related information and data for assessing what one can define as vulnerability economics, in order to assess the necessary economic efforts to diminish a certain level of vulnerability at local, regional, or national level.
- involving site and location-specific vulnerability assessment by use of GIS technology, and by involving space-related information which can further enhance the knowledge space for making vulnerability informed decisions. For instance, visit: http://money.cnn.com/news/newsfeeds/articles/prnewswire/C1149.htm.

Figure 13.3 indicates the extension of the definition domain of the security concept, by fully integrating the concepts of risk and vulnerability. It becomes more useful to deal distinctively with the above two concepts in order to make further integration into the operative concept of security.

The present study focuses on the operative concept of vulnerability while other intensive work into the past has been done with relevance to risks in Switzerland (see Braun 1998) and related *Riskoprofil Schweiz* publications). The two approaches are complementary; the author of the present study is fully aware of advantages and limitations of either of the two topics which lead to a comprehensive and modern view on security at national or international levels.

By promoting an activity related to living vulnerability assessment and management, Switzerland could be today at the front end of work and activities in this field and can have a competitive advantage in many political and economic situations related to the issue of national security. Options could be identified for work on various types of subjects which could be labeled also as:

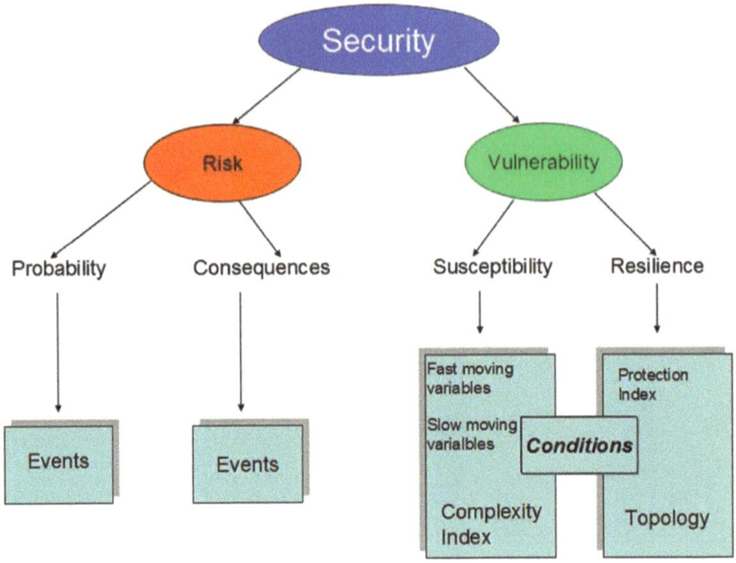

Fig. 13.3 An extension of security relating risk and vulnerability

- Human security
- Environment security
- Technological security
- IT security
- National security

The intelligent blending of political experience, tradition and international status, combined with analytical tools for prediction and pattern identification of vulnerable scenarios is a step into the direction of proper vulnerability governance and management.

The use of AHP (Analytic Hierarchic Process) methodology helps in ranking practical measures addressed to diminish the vulnerability level at a given entity. In the following, Boxes are presented the general, *de minimis*, analytic framework in order to address and lately solve specific problems related to vulnerability assessment and management.

Following the broad understanding of governance (OECD 2003) involving scientific analysis of risks, integration of societal perception and amplification of risk into the risk assessment process, structuring decision-making in a consistent rational and democratic way (with a multitude of 'abstract' societal values involved) to transparent and open communication, an attempt has been made to identify problem fields judged to be of high relevance and for which the DSP–VBS (Directorate for Security Policy—Federal Department of Defence, Civil Protection and Sports) may provide unique solutions. The problem fields concentrate on increasing traditional or emerging large-scale risks that have trans-boundary ramifications. The OECD (2003) has referred to these risks as 'systemic risks.' This term denotes the associated risks to human health and the environment. Systemic risks are at the crossroads between natural events (partially altered and amplified by human action such as the emission of greenhouse gases), economic social and technological developments, and policy-driven actions, both at the domestic and international level.

Systemic risks are characterized by complexity, genuine uncertainty, and ambiguity. Better governance of systemic risks has been pointed out as being necessary and urgent. Following a similar line of thinking, within the context of the present work, the above idea will be extended to the concept of vulnerability ranking with reference to issues of national security. In this framework, general presentation on the ranking method based on the techniques of multi-criteria decision analysis will be made, and made it available for better decisions related to what might be called 'vulnerability governance.'

For setting up the Evaluation Matrix, it has been assumed that political actors/ decision-makers at business and government level need scientific expertise and assistance as a means to identify critical issues, to provide unified knowledge, and to demonstrate options for resolving conflicts and problems. From the multitude of problems, those could be considered, which calls for either a broader international enterprise, or complementary to a sectoral national approach.

In a first step, the problem fields have to be consciously outlined in a broad manner. The Strategy Team or the so-called The Council could continually tackle these possible fields, develop expertise in them, and pick out of them proprietary single domains and aspects for which specific actions and tasks will be defined (…). Although deliverables do present specific differences, they generally go in the direction of:

- Compilation, verification, and 'harmonization' of scientifically sound methods, tools, and data; revelation of remaining disputes and prevailing uncertainties; provision of verified risk information ('white books');
- Consensually formulated fundamental approaches, methodologies to be applied and promising procedures, endorsed best practices ('generic guidelines' on methods for assessment, regulatory principles, process rules);
- Improved efficiency in risk management ('acceptable trade-offs'), better prevention of crisis situations, better early detection and adequate handling of changing risk patterns ('recommendations'), all from a more comprehensive (multi-disciplinary, cross-sectoral) national and international perspective (…).

In the quest for a reliable method of planning, prioritizing, and allocating DSP–VBS resources in manners best responsive to perceived needs for ranking vulnerability-related issues, activities, or projects, the notion of an Evaluation Matrix emerged, together with the necessity to provide for a comprehensive processing of the latter. Eventually targeted for this purpose was Saaty's Analytic Hierarchy Process (Saaty 1980).

Assuming: n activities are being considered by a group of interested people having as goals:

(i) to provide judgments on the relative importance of the activities; and
(ii) to make sure that the judgments are quantified to an extent which also permits a quantitative interpretation of the judgments among all activities.

In formal terms, let C_1, C_2,…, C_n be the set of activities. The quantified judgments on the pairs of activities $\{C_i, C_j\}$, $i, j = 1, 2,..., n$ are synoptically represented by an $n \times n$ matrix:

$$\mathrm{A} = \left(a_{ij}\right), i, j = 1, 2, \ldots, n$$

The entries a_{ij} are defined by the following rules:

Rule 1: If $a_{\mathrm{ij}} = a$, then $a_{\mathrm{ji}} = 1/a$, $a <> 0$.
Rule 2: If $\mathrm{C_i}$ is judged to be of equal relative importance as $\mathrm{C_j}$, then $a_{\mathrm{ij}} = 1$; in particular, $a_{ij} = 1$ for all i.

Upon these, the matrix A has the form;

$$\mathrm{A} = \begin{bmatrix} 1 & a_{12} & a_{13} & \cdots & a_{1n} \\ a_{21} & 1 & a_{23} & \cdots & a_{2n} \\ a_{33} & a_{32} & 1 & \cdots & a_{3n} \\ \cdots & \cdots & \cdots & \cdots & \cdots \\ a_{1n} & a_{2n} & a_{3n} & \cdots & 1 \end{bmatrix}$$

A matrix observing such a configuration is known as a ‘reciprocal matrix.’ Once the reciprocal, Evaluation Matrix is built up, the problem is to obtain from a set of numerical weights, w_1, w_2, w_3,…, w_{n} that would feature the priority to be assigned to the contingencies $\mathrm{C_1}$, $\mathrm{C_2}$,…, $\mathrm{C_n}$. Based on an ample intuitive justification and a sound mathematical reasoning, Saaty (1980). proves that:

a. The Priority Vector (w_1, w_2, w_3,…, w_{n}) sought is the Eigenvector of the Maximum Eigenvalue of matrix A; the closer the Maximum Eigenvalue is to n, the more ‘consistent’ the AHP is considered)
b. A complex hierarchical structure can be described as a super-set of $C_1^k, C_2^k, \ldots, C_n^k$ contingency sets, organized on a number of levels, $k = 1, 2,..., m$, and interconnected to a certain degree. To prioritize the components in the bottom-level of the hierarchy,

   (1) Get the Priority Vectors of type (a) for every component in a set k, performing the pairwise comparison withthe standpoint of every criterion (component) in the preceding, k−1, level of the hierarchy with which logical/functional connections exist; in this, take the overall objective of the hierarchy as level-0, and qualify as 0 the non-connected items, for a mathematically uniform description; obtain, matrices consisting of Priority Eigenvectors as columns, and featuring every hierarchy level.

(2) Multiply the obtained priority matrices in the reverse order, that is, from the bottom hierarchy level up; obtain, a unique vector that features the aggregated, multicriterial evaluation of the priority to be assigned to the items in the bottom (objective-oriented) level of the hierarchy.

(3) Compute the Maximum Eigenvalue, λ, associating the Priority Vector; use this to compute the Consistency Index (CI), qualifying the quality of the priority evaluation, as CI = $(\lambda - n)/(n-1)$; use, in turn, CI to compute the Consistency Ratio, CR, as the ratio of CI to a 'Random Index' (RI) of reference, CR = CI/RI, explained and computed in the aforementioned book; a CR lower than 0.1 characterizes the evaluation as 'satisfactorily consistent.'

Again, in a practical sense, *the sequence b.1-3 makes up the* AHP. A number of methods, of various degree of accuracy, to obtain the Priority Eigenvector, the corresponding maximum Eigenvalue, and indices of consistency of the evaluation, are in use.

For the Priority Eigenvector, (i) multiply the n elements in each row of the user-entered Evaluation Matrix, (ii) take then n th root of the product, and (iii) normalize the resulting numbers so that the sum-total be 1. For the Maximum Eigenvalue, (i) multiply the Evaluation Matrix on the right by the Priority Eigenvector, (ii) divide every element of the column-vector obtained, by the original components of the Priority Vector, and (iii) sum all components thus obtained, then divide the sum by the number of components, n; the result is an approximation of the Maximum Eigenvalue.

The working examples which could be provided with the future code's installation package are meant to illustrate the procedure as described. In examining these, note that the same raw material—in this case, an original DSP–VBS Evaluation Matrix can be configured in different fashions, so as to place on the bottom hierarchy level—the 'level of objectives'—various items of interest.

13.2 The Code Design

First, enable users to fill in, and file for further use, Evaluation Matrix forms. Second, there needs to be a software implementation of Analytic Hierarchy Process (AHP), meant to serve as a platform to interpret evaluation matrices from a multi-criteria assessment perspective, including:

1. An interface for the interactive hierarchic structuring of Evaluation Matrix (or any other AHP-targeted objective, for that matter) criteria, and for conveniently filling-in pairwise comparison matrices.
2. The AHP-algorithm machine itself.
3. Standard output archive/de-archive/export facilities.

4. A *de minimis* on-line documentation.

When interested in assessing the vulnerability from the point of view of various circumstantial situation, and addressing aspects of national security, one might adopt three characterizations for vulnerability assessment: (i) vulnerability during the *peacetime*, (ii) vulnerability in case of *potential war*, and (iii) vulnerability in *war situations*.

This study addresses specific issues in relation to vulnerability assessment during, for example, peacetime up to cases of potential war. In case that the topic of vulnerability becomes an issue of national security in Switzerland, future research has to answer to the questions such as: 'It is possible to develop generic methodologies and models in order to address the above-mentioned vulnerability situations, as part of the general security policy?'

A distinct cycle approach to the issue of vulnerability assessment could be developed by taking into consideration: (i) peace situations, (ii) contingency situations, (iii) crisis situations, and (iv) conflict up to a war situation.

Recent research indicates the possibility to address risk and vulnerability by use of a nexus approach. More details are given in the following notes.

13.3 A Unified Representation for Critical Infrastructure

Consider that by modeling work numerical results are available and they could, in the end, be associated with risk indicators and vulnerability indicators. Understanding the way how solutions were generated the risk indicator is a function of the predictability numbers (which can be associated with the likelihood indicator in a risk-related assessment), generated by solving the sets of specific models as well as the maneuverability number (which can be associated with the significance indicator in a risk-related assessment).

In a similar manner, the stability indicator introduces a derived measure for the systems vulnerability, and the numerical and/or linguistic performance indicator could be easily associated, as a derived figure for vulnerability index in case of high interdependencies of critical infrastructures. By adopting this line of argumentation, one can now develop a risk—vulnerability nexus which will allow giving indications, in a totally coupling manner, on the evolution of a given system of systems in the new nexus space.

Figure 13.4 below represents a new nexus matrix, which would allow indicating, simultaneously, the degree of risk and vulnerability under which complex dynamic interactions are to be addressed and managed.

The current body of formal, academic knowledge still offers vast, and yet poorly exploited resources to a better understanding of the structure and dynamics of modern societal activities. Turning to account such resources would be a profitable exercise, because being low-cost and highly insightful for a wide category of users that do not necessarily require a formal academic training, because the message of

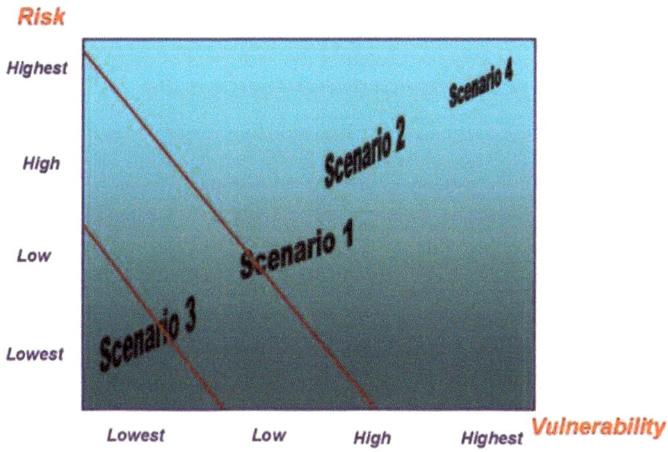

Fig. 13.4 A nexus matrix for risk and vulnerability

the mathematical models involved is expressed: (i) in plain terms and figures such as costs, transactions, and profit; and (ii) in visual, intuitive shapes for which the eye gets self-educated.

13.4 Remarks

By adopting adequate metrics to risk, vulnerability and nexus, one can develop a more elaborated understanding of how critical infrastructures have to be addressed in order to develop safe and stable connections and system evolution. In Fig. 13.5 below, one can see the categorization of the above concepts in a format which integrate all the concepts related to risk and vulnerability of complex systems of systems, namely critical infrastructures, which in turn have a determined influence on vulnerability assessment and national security.

Risk Metric	Vulnerability Metric	Nexus Metric
QRA Quantitative Risk Assessment	QVA Quantitative Vulnerability Assessment	QNA Quantitative Nexus Assessment
CRA Comparative Risk Assessment	CVA Comparative Vulnerability Assessment	CNA Comparative Nexus Assessment
Risk -Criteria	Vulnerability Criteria	Nexus Criteria
Risk Perception	Vulnerability Perception	Nexus Perception

Fig. 13.5 A categorization of metrics for risk and vulnerability, adapted from Gheorghe (2004)

References

Braun, H. (1998). Umfassende Risikoanalyse Schweiz: ein Projekt als Grundlage zur Weiterentwicklung der schweizerischen Sicherheitspolitik. *ASMZ: Sicherheit Schweiz: Allgemeine Schweizerische Militärzeitschrift, 164*. https://doi.org/10.5169/seals-65397.

Gheorghe, A. V. (2004). *The hidden faults: Towards a standing method to assess Switzerland's vulnerabilities*. Zurich, Switzerland: Laboratory of Safety Analysis, ETH Zurich.

OECD. (2003). *Emerging risks in the 21st century: An agenda for action*. Paris, France: OECD Publishing.

Saaty, T. L. (1980). *The analytic hierarchy process*. New York, NY: McGraw-Hill.

Chapter 14
The Case for Sihl Dam

Abstract This chapter presents a case and results that were undertaken at the request of the *Swiss Risk Engineering Company*, regarding the potential consequences of a virtual Sihl Dam break event. Note that the modeling is limited to a 'worst case scenario' describing a full destruction of the dam wall within a short time span and the consequent release of the whole water reservoir volume. To compare results and show the strength of the proposed model and the associated decision support system, a similar consequence assessment published on the Internet by the Polizei department, Zurich was used as a frame of reference.

14.1 An Overview

Commercially available consequence analysis tools regarding dam break are found in the area of emergency management but not for insurance purposes. In the insurance business field, own developed tools are used. Table 14.1 provides a comparative summary.

Clearly, there is no 'winner' yet, in the dam break modeling competition, and chances are that no winner is ever even conceivable: Some models and codes will always be likely to be better than others—for any given specific case—infrastructure, event, and environment considered.

We start with differentiation from other approaches. Let's considering the initial case, originating from the *Emergency Preparedness and Response* unit within the Department of Police of the City of Zurich (see *Stadt Zurich, Polizeidepartement FAQ: Wasseralarm Sihlsee)*, the following aspects were given a special attention for the purpose: The recognition that a major abnormal event can indeed occur, calling for the definition of comprehensive evacuation and warning areas. The worst case concerns the expectation of an 8 m-high flood wave. The flood wave arrival time is 1 h 25 min to the proximal city border down-flow (Leimbach); 1 h 50 min to downtown Zurich; and 2 h 50 min to the distant city border in Altstetten.

A.V. Gheorghe et al., *Critical Infrastructures, Key Resources, Key Assets*, Topics in Safety, Risk, Reliability and Quality 34,
https://doi.org/10.1007/978-3-319-69224-1_14

Table 14.1 Comparative summary of analysis tools for dam break

	Tool for the scope of emergency planning and response	Tool for reinsurance application
Objective	Provide sufficient *and comparable* safety coverage to all subjects at risk in a given area	Provide sufficient *and individualized* safety coverage to each and every subject at risk in a given area
Method	Conservative	Cautious
Approach	Area-oriented assessment and planning of prevention/intervention	Hot Spot-oriented assessment and prevention/intervention

14.2 The Findings

The most severe risk relating to a water reservoir management is the dam breaking. The single other abnormality that may compare to a dam break includes *dam overflow by flash floods of tributaries* into the reservoir.

Taking as chief target indicator the *potential extension of the flooded areas downstream*, the dam break mechanism—piping, instantaneous removal etc. is of lesser consequence than the event itself. Other indicators like the *maximal flood front-wave and the wave velocity* (celerity) may be break-type-sensitive; such aspects are not likely to bear too heavily on the final assessment.

Dam Break Modeling as a department of the more general field of Open Channel Flows makes a highly complex, in mathematical and physical terms, endeavor, also featuring an intense and enduring (decade-long) academic competition.

14.2.1 Tool Description

The ***ad hoc*** **software** package (see structure in Fig. 14.1) integrates:

- intelligent GIS resources
- dedicated modules to implement the screening models
- imported documentary files and software
- an appropriate interface.

The ***analysis*** proceeds on the following lines:

- Selecting theater of action and securing the appropriate map folder.
- The interactive definition of 'open channel'—offensive of the flood wave.
- Running the working (screening) models and comparing results.
- Selecting a reference screening model for final assessment.

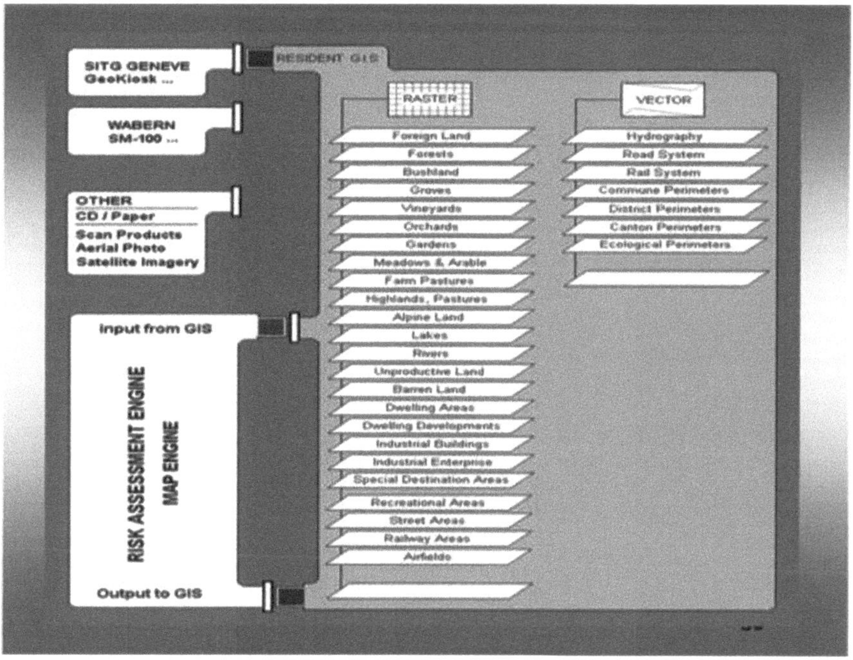

Fig. 14.1 Structure of an intelligent GIS

Dam break-induced flood simulation:

- the flood wave progress along open channel
- the wave front distance from dam (m) versus time (s)
- the wave front maximal height (m) as a function of time (s)
- the wave front velocity (m/s) versus time (s)
- the maximal water-flooded area available at each point along channel, obtained via an up-filling algorithm that reads the maximal wave front height and the terrain elevations around the channel.

The cadastral statistics of the flooded area is based on the GIS data of the flood-filled areas obtained, and is used as an expression of the dam break event consequences.

The map was derived from Centre of Excellence on Risk and Safety Sciences (KOVERS) digital resources. Since the map scale of the overall view is unacceptable for a detailed analysis of the river channel, 6 (six) other maps were prepared at larger scales. In the following Annex, algorithm especially that of the calculation of the wave front velocity is described.

MODEL #1: VISCOUS DAMBREAK FLOW IN INCLINED CHANNELS.

Source: **Bed slope effect on the dam break problem Effet de l'inclinaison du canal sur une rupture de barrage**

BLAISE NSOM, *Université de Savoie. UFR SFA. Campus Scientifique du Technolac, 73376, Le Bourget du Lac, Cedex, France*

KHALED DEBIANE, *Laboratoire de Rhéologie (UJF-CNRS-INPG), BP 53, 38041 Grenoble Cedex, France*

JEAN-MICHEL PIAU, *Laboratoire de Rhéologie (UJF-CNRS-INPG), BP 53, 38041 Grenoble Cedex, France*

Resume: Nous étudions ici, le problème de la rupture de barrage sur un canal pentu. L'écoulement d'une série de fluides newtoniens est généré par la rupture d'un barrage dans un canal entièrement transparent et d'inclinaison variable. L'évolution de la hauteur de fluide en une station donnée ainsi que celle du front d'onde sont établies à l'aide de moyens de mesure ultrasonore et d'Analyse d'images, respectivement. Par ailleurs, les équations de Navier Stokes et de continuité sont adimensionnalisées en régime visqueux et résolues avec l'hypothèse des eaux peu profondes. Enfin, les résultats expérimentaux et théoriques sont comparés avec succès.

The concept demonstration algorithm implements the following time loop:

```
FOR t = 0 to Time_Span STEP Time_Step          'Start looping.
                                                       'Time_Span (s) and Time_Step
                                                       'are user-input (UI)

tx = t*rho*g*cos(alpha)*(H^3)/(12*miu*(L^2))   'tx     ()        the non-dimensional time
                                                       'rho    (kg/m3)   fluid's density (UI)
                                                       'g      (m/s2)    gravity acceleration
                                                       'alpha  (deg)     channel slope angle (UI)

                                                       'H      (m)       dam height (UI)
                                                       'miu    (Pa.s)    fluid's dynamic viscosity
(UI)
                                                       'L      (m)       reservoir length (UI)

IF alpha = 0 THEN                               'The horizntal channel case

        IF tx <= 0.1 THEN                       'Correlational rules for non-dimensional
                xf = 0.969*tx^0.5                      'wave front distance from dam, xf ()
        ELSE                                           'and wave front height, hfx ()

        ELSE                                           'and wave front height, hfx ()
                xf = 1.860*tx^0.2 - 0.902
        END IF
        hfx = 0.640/(xf + 1)

ELSE                                                   'The inclined channel case

        IF tx <= 1.23 THEN                      'Correlational rules for non-dimensional
                xf = 1.071*tx^0.5                      'wave front distance from dam, xf ()
        ELSE                                           'and wave front height, hfx ()
                xf = 2.052*tx^0.31 - 1.0
        END IF
        hfx = 0.553/(xf + 1)^0.8

END IF
Vf = (xf - xf0)*L                                      'Wave front velocity, average over time step, m/s
xf0 = xf

x = xf*L                                               'Physical wave front distance, x (m)
hf = hfx*H                                             'Physical wave front height, hf (m)

Print & plot t, x, hf, V                               'Render as computed

NEXT t                                                 'Loop
```

14.2.2 Parameters and Criteria

While the preceding phases pertain to a *preprocessing* of the requisite data, the computation of *the maximal wave front position* along the channel, *wave height*, and *wave velocity* as a function of time makes a key phase into the assessment. In particular, it is the wave height that determines—within the algorithm designed—the extension of the maximally expected inundated area around every running point along the water flow.

The phase is comprised of two major steps. In a 1st step, the dam break model is run, over a distance and a time in tune with those of interest. In the case in point, the time is set at 10,200 s (i.e., 2 h 50 m) as suggested by the *Wasseralarm* document. As to the distance of interest, it follows automatically from the run of the dam break model. The situation is such that, 'on the natural assumption, the key data in the *Wasseralarm* document, namely the wave arrival times are taken as presenting a satisfactory degree of confidence, having the wave front arriving at the announced checkpoint around the times specified in the reference document of the emergency managers and stands as a valid test for appropriateness of the screening model.'

The numerical experiments performed with the two screening models in attention (see Sect. 14.2) have evidenced the remarkable adequacy of the Nsom et al. (2000; Nsom 2002) model of viscous flow on an inclined channel, from the standpoint just emphasized. The non-dimensional nature of the model, and the fact that two of its key control parameters—the fluid density (kg/m^3) and the dynamic viscosity (Pa.s) are inferred from experiments conducted in similitude conditions on a hydraulic test ground specially designed for the purpose, being also confronted with data in the literature, may be at the origin of the said quality. The bare fact is that:

> adopting a model of the downcoming fluid presenting a density of 1406 kg/m^3 and a dynamic viscosity of 12 Pa.s (muddy water plus solid debris?)—which are exactly the values inferred by Nsom et al. from…the Sihl dam break wave front turns out to reach the monitoring endpoint near Altstetten, at ca. 60 km downstream from the dam, in almost exactly the time prescribed, i.e., 2 h 50 min, or 10,200 seconds.

A full account of the results obtained on this line is given in the sequel, in the context of the flood consequences determination. Let it be said that, a slight trimming operated in the value of the dynamic viscosity would immediately calibrate the model so as to fit the *Wasseralarm* arrival times and checkpoints to any degree of accuracy. However, the authors have not indulged in such an exercise, thought of being futile in consideration of all uncertainties and inaccuracies involved in a screening-type assessment. The emphasis is rather placed on the genuine adequacy of the model, in its original parameters, for a description consistent with the Emergency Manager's appraisal.

A discussion of a full flood wave on second-by-second basis is described in this section. However, the first and last quarter-hour are rendered here, as a sample:

```
CASE: Sihl
```

```
To run these data,
go to menu's 'CONTINUE', 'Run Current Input'.
You may also type in alternative data,
or open an alternative input from menu.
```

```
RESERVOIR LENGTH, L (m): 8000
DAM HEIGHT H, (m): 19
CHANNEL SLOPE ANGLE, alpha (deg): -0.83621468e-2
FLUID DENSITY, rho (kg/m3): 1406
FLUID DYNAMIC VISCOSITY, miu (Pa.s): 12
```

14.2.3 Tool Design and Methodology

In the present case, the 'theater of action' parameters are as follows:

- Land South of Zurich City, including the Sihlsee Dam mouth
- the Sihl River down to Altstetten.

The dam (i.e., Sihl Dam) itself is situated according to the following coordinates:

- Northernmost Latitude (°): 47.411664; Westernmost Longitude (°): 8.45388238
- Southernmost Latitude (°): 47.145; Easternmost Longitude (°): 8.85723
- The dam is 8'000 m long and 19 m high

The channel files offer a description and features of the open channel including:

- the geographic coordinates of the vertices in the polygonal lines of definition (latitude, longitude)
- the local elevation (m ASL)
- the direction cosines of the flow with respect to horizontal *XY* axes (*X*—eastward, *Y*—southward)
- the slope of every polygonal segment, inferred from the elevation database
- the path length, measured from the dam as an origin.

While the preceding phases pertain to a *preprocessing* of the requisite data, the computation of *the maximal wave front position* along the channel, *wave height*, and *wave velocity* as a function of time make a key phase into the assessment. In particular, it is the *wave height that determines—within the algorithm designed—the extension of the maximally expected inundated area* around every running point along the water flow.

- The phase is comprised of two major steps
- The dam break model is run, over a distance and a time in tune with those of interest. The time is set at 10,200 s (i.e., 2 h 50 m), as suggested by the *Wasseralarm* document
- As to the distance of interest, it follows automatically from the run of the dam break model (See Fig. 14.2). The results of the model for current case are depicted in Fig. 14.3.

14.3 The Consequence Assessment

For each and every current point on the channel line, given by its distance to the dam and featuring an elevation *z*0 m ASL:

- Pick the respective (interpolated) maximal wave height *h* from the flow wave file;
- Determine by iteration the polygonal contour made of all points upstream of the current point (within 180°), placed on the terrain at a height higher than *z*0 m, and lower than *z*0 + *h* m

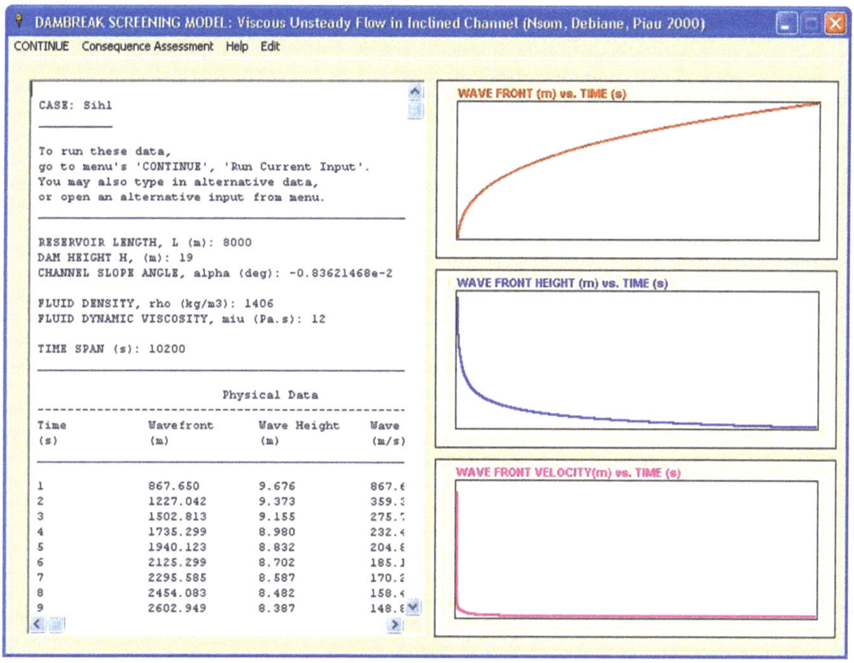

Fig. 14.2 Modeling Viscous unsteady flow in include channel

CASE: Sihl

To run these data,
go to menu's 'CONTINUE', 'Run Current Input'.
You may also type in alternative data,
or open an alternative input from menu.

RESERVOIR LENGTH, L (m): 8000
DAM HEIGHT H, (m): 19
CHANNEL SLOPE ANGLE, alpha (deg): -0.83621468e-2
FLUID DENSITY, rho (kg/m3): 1406
FLUID DYNAMIC VISCOSITY, miu (Pa.s): 12
TIME SPAN (s): 10200

	Physical Data			Non-dimensional Data		
Time (s)	Wavefront (m)	Wave Height (m)	Wave Velocity (m/s)	Time tx()	Wavefront xf()	
1	867.650	9.676	867.650	0.010255	0.108456	<
2	1227.042	9.373	359.392	0.020510	0.153380	<
3	1502.813	9.155	275.771	0.030765	0.187852	<
4	1735.299	8.980	232.486	0.041019	0.216912	<
5	1940.123	8.832	204.824	0.051274	0.242515	<
6	2125.299	8.702	185.175	0.061529	0.265662	<
7	2295.585	8.587	170.286	0.071784	0.286948	<
8	2454.083	8.482	158.499	0.082039	0.306760	<
9	2602.949	8.387	148.865	0.092294	0.325369	<
10	2743.749	8.299	140.800	0.102549	0.342969	<
11	2877.668	8.217	133.919	0.112803	0.359708	<
12	3005.626	8.141	127.958	0.123058	0.375703	<
13	3128.355	8.069	122.729	0.133313	0.391044	<
14	3246.447	8.001	118.092	0.143568	0.405806	<

Fig. 14.3 Results of the model

- Perform an adequate (Bezier) interpolation of the jagged poly-line obtained, in consideration of the fact that jags are rather spurious effects of the 100 m squared cell elevation, population and land use grid, bilinearly interpolated
- Fill the polygon—meaning that you have marked the upstream filled maximal area at the current point
- Take up the next point (entry) in the flow wave file.

The data contained in the DAMBREAK ASSESSMENT FILE:

1 Landscape

- Northernmost Latitude (°): 47.411664; Westernmost Longitude (°): 8.45388238
- Southernmost Latitude (°): 47.145; Easternmost Longitude (°): 8.85723.

2 physics

- RESERVOIR LENGTH, L (m): 8000
- DAM HEIGHT H, (m): 19; CHANNEL SLOPE ANGLE, alpha (°): −0.83621468e-2
- FLUID DENSITY, rho (kg/m^3): 1406; FLUID DYNAMIC VISCOSITY, miu (Pa.s): 12
- TIME SPAN (s): 10,200.

3 Statistics of potentially affected area

Land Type. (ha)	Area Population Money		Strategic Other	
	(persons)	Value (a.u.)	Value (a.u.) a.u.)	
Arable1	262.0166	1048	2620.17	2620.17 2620.17
ForstO1	166.2725	166	1662.72	1662.72 1662.72
Orchard1	14.0298	126	140.3 140.3 140.3	
Rivers	34.3162	412	343.16	343.16 343.16
FarmPast	45.5022	91	455.02 455.02 455.02	
UnPrd	1.3271	3	13.27 13.27 13.27	
Vineyard	2.2751	14	22.75 22.75 22.75	

- TOTAL AREA accounted (ha): 841
- TOTAL POPULATION accounted (persons): 10,105
- TOTAL MONEY VALUE accounted (a.u.): 8410
- TOTAL STRATEGIC VALUE accounted (a.u.): 8410
- TOTAL OTHER accounted (a.u.): 8410

Abbreviations

Arable	arable farm land including meadows
ForstO	Old (Tall) Forest
FarmPast	farm pastures
UnPrd	unproductive land

14.3.1 Flooding Course and Extension Within 60 Km

A general information regarding the watercourse for the Sihl River in case of total dam break is given in the Fig. 14.4. Figure 14.5 shows the flooding area within Zurich covering with 1.9 m height the Bahnhofstrasse in Zürich, with the consequence of flooding the main station.
At this point, we make the following observations:

- The extension of planned evacuation in the emergency management-oriented document is warranted, in the lower Sihl River segment (Fig. 14.6)
- Maintaining a wide-stripe evacuation and attention areas on the entire lower half of Sihl River (conservative), low-liability policy geared to population protection, may be excessive in objective terms
- Even more excessive seems the assumption that the flood wave would reach Zurich City area at a wave height of 8 m—a height plausible—model confirmed, only in the near-dam area
- It appears that the actual situation in case of a dam break-triggered flood wave would come closer to the notion of *hot spots* scattered along the river, rather than to a stripe-type impact area. Such hot spots are apparent along the entire Sihl channel, which otherwise appear as a naturally well-channeled river, even on a low-resolution elevation grid
- *Engineered flood channeling*, as a factor of key consequence in the mitigation of any flash flood, including the dam break-induced waves.

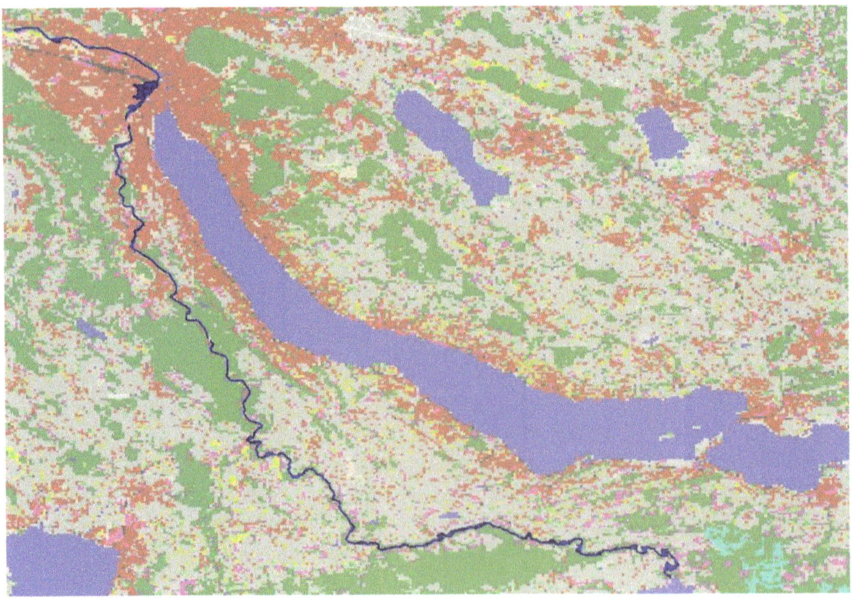

Fig. 14.4 Case for total dam break

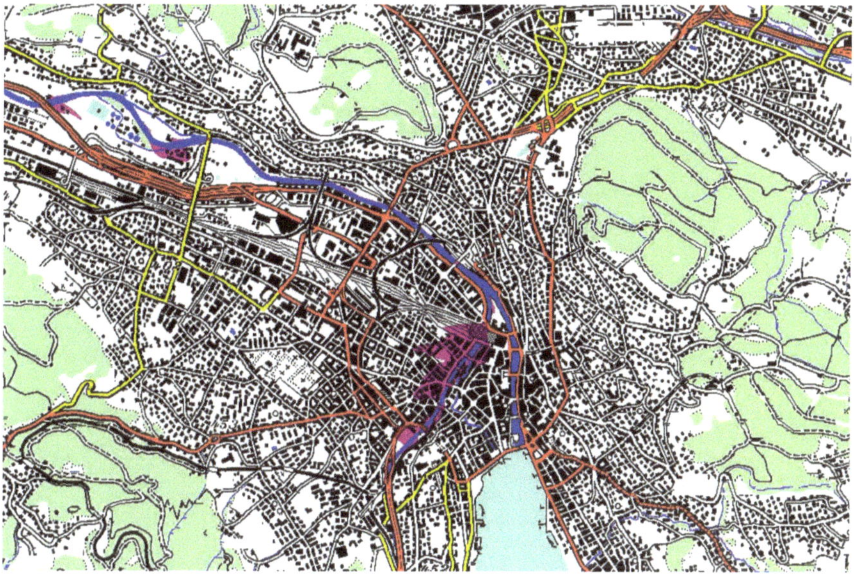

Fig. 14.5 Flooding with Zurich's main downtown

Some initial remarks.

- The *Wasseralarm* document, while sound from the intended standpoint of the Emergency Management Authority, may prove *over-conservative* when seen from the angle of the (Re) Insurer
- While screening models of the kind demonstrated in this memo may provide *useful insights into the relevant phenomenology*, a professional analysis calling the services of hydraulic engineering departments, with a necessary capability of similitude modeling, and downscaling on hydraulic test grounds would be in order, if indeed at stake is to *effectively (re) set insurance premiums.*

Some technical issues associated with present approach.

- Single user application
- Need for availability of GIS data could be easily integrated. Also, maps and satellite pictures can be digitalized and integrated into the tool
- The most time for the consequence analysis is used by (a) appropriate GIS organization and (b) by the digitalizing the channel on a pixel by pixel approach
- The model can be used for dynamic simulations
- It is possible to link GIS data consequence assessment pixel by pixel with another tool for the purposes of, for instance, calculation insured assets at risk
- The model can be extended by the inclusion of loss values per unit.

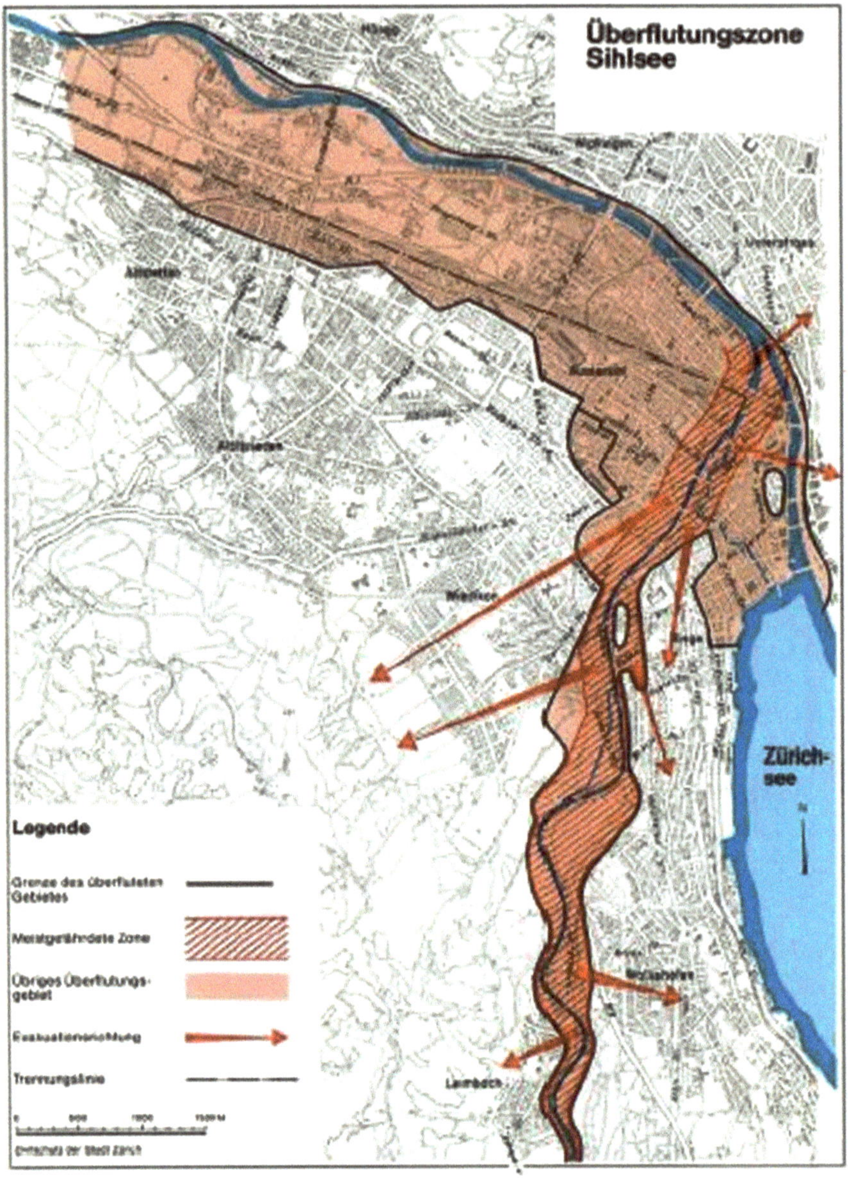

Fig. 14.6 Areas that could be affected

14.3.2 Summary of Discussions

- According to a representative of the *SwissRe* natural catastrophe group, their demand consists in provision of flooding course and extension assessment within 24–72 h. They have their own tools and do not need an additional one. It was confirmed that the shown results exceed the one from the tools applied by the natural catastrophe group.
- According to a representative of product management, there is a need for a portfolio comparison regarding dam break consequences.
- For consequence analysis of tailing dam breaks, there is a demand for a corresponding tool for Risk Engineering Services.
- It was acknowledged that the model result is such as accurate as the GIS resolution. High-resolution GIS data outside from Switzerland are very expensive. KOVERS developed a method to digitize available maps and satellite pictures as well as the interpretation of the color scheme used in maps to indicate for instance, among others, topography or population density.
- Some question mark arose concerning the initial wave front velocity of over 860 m/s within the first seconds after the sudden and complete dam break.
- An ad hoc service as wished by the representative of the natural perils group can only be provided by KOVERS, if and only if GIS data are provided (except for those of Switzerland). Within 3 days, the channel definition, which is very time intense and is done on pixel level and the consequence modeling, can be performed. In all other cases, there must some time provided to gather the corresponding GIS data (approximately 4 weeks).
- According to the statement of the representative of natural perils, they gather in fact worldwide GIS data or those data provided by satellite picture but they do not assess in advance the consequences of dam breaks.
- No time constraints would be present if a portfolio-related consequence analysis indicates a dam break. This suggests, there must be interest in practical tools.

One question raised over the assessment of the 'DAM RISK' tool related to the strikingly high values of what the code has termed 'wave front velocity,' in the initial phase (from a few fractions of a second to perhaps a few seconds) of the flow consecutive to the dam break. A typical data sequence follows (computer printout), to illustrate the matter:

```
Case Sihl
________________________________________________________________

RESERVOIR LENGTH, L (m): 8000
DAM HEIGHT H, (m): 19
CHANNEL SLOPE ANGLE, alpha (deg): -0.83621468e-2
```

FLUID DENSITY, rho (kg/m3): 1406
FLUID DYNAMIC VISCOSITY, miu (Pa.s): 12

TIME SPAN (s): 10200

Physical Data Non-dimensional Data

Time Wavefront Wave Height **Wave Velocity** Time Wavefront
(s) (m) (m) (m/s) tx() xf()

---

1 **867.650** 9.676 **867.650** 0.010255 0.108456 <
2 **1227.042** 9.373 **359.392** 0.020510 0.153380 <
3 1502.813 9.155 275.771 0.030765 0.187852 <
4 1735.299 8.980 232.486 0.041019 0.216912 <
5 1940.123 8.832 204.824 0.051274 0.242515 <
6 2125.299 8.702 185.175 0.061529 0.265662 <
7 2295.585 8.587 170.286 0.071784 0.286948 <
8 2454.083 8.482 158.499 0.082039 0.306760 <
9 2602.949 8.387 148.865 0.092294 0.325369 <
10 2743.749 8.299 140.800 0.102549 0.342969 <
11 2877.668 8.217 133.919 0.112803 0.359708 <
12 3005.626 8.141 127.958 0.123058 0.375703 <
13 3128.355 8.069 122.729 0.133313 0.391044 <
14 3246.447 8.001 118.092 0.143568 0.405806 <
15 3360.392 7.937 113.945 0.153823 0.420049 <
16 3470.598 7.876 110.206 0.164078 0.433825 <
17 3577.411 7.817 106.813 0.174333 0.447176 <
18 3681.125 7.762 103.715 0.184587 0.460141 <
19 3781.996 7.709 100.871 0.194842 0.472750 <
20 3880.247 7.658 98.250 0.205097 0.485031 <
21 3976.070 7.609 95.823 0.215352 0.497009 <
22 4069.637 7.561 93.567 0.225607 0.508705 <
23 4161.101 7.516 91.464 0.235862 0.520138 <
24 4250.597 7.472 89.496 0.246117 0.531325 <
25 4338.248 7.429 87.650 0.256371 0.542281 <
26 4424.162 7.388 85.914 0.266626 0.553020 <
27 4508.439 7.348 84.277 0.276881 0.563555 <
28 4591.170 7.310 82.731 0.287136 0.573896 <
29 4672.436 7.272 81.266 0.297391 0.584054 <
30 4752.312 7.236 79.876 0.307646 0.594039 <
. . . .

```
175 11678.743 5.114 34.928 1.794600 1.459843 <
176 11713.534 5.107 34.791 1.804855 1.464192 <
177 11748.189 5.100 34.655 1.815110 1.468524 <
178 11782.709 5.092 34.520 1.825365 1.472839 <
179 11817.095 5.085 34.386 1.835620 1.477137 <
180 11851.350 5.078 34.254 1.845875 1.481419 <
. . . . . . . . . . . . . . . . . . . . . . . . . . . . . . . . . . . .
```

In understanding the intriguing figures, the following should be considered:

Argument 1. As indicated in the code's online documentation, the model selected for preliminary evaluations of dam break consequences owes to several authors (Nsom et al. 2000; Nsom 2002). An algorithmic transcription of the model, done by the code developers and duly delivered with the code 'Help' sections:

As emphasized in the blued box above, model's constitutive equations, delivering in particular such key-quantities as the *wave front evolving distance from dam*, and the *wave front height*, are not entirely following from evolutionary, partial differential equations of the Fluid Dynamics (Navier–Stokes), but—as explained in the reference paper—also on *correlations derived from scale-model hydrodynamic experiments.* Consequently, *one should not expect a uniform relevance of the figures obtained from correlations over the entire range of physically accessible values of the time.*

To consolidate the algorithm as described (see Sect. 13.2.1 on *Tool description* above), code developers went at some length to reconstruct, using the code, the 4 (four) examples (Figs. 14.7, 14.8, 14.9, and 14.10) reported by Nsom et al. (2000) in their paper. The degree of code-to-reference fitting can be observed on the curves in the right-hand side windows of the reproductions that follow. Observe the input data (i) in the yellow window upper-right, and (ii) in the header of the cyan listing. Note how results are reproduced by the code, in the lower-right window (compare to the upper-right window).

Argument 2. To further clarify the meaning given by the code to the 'wave front velocity,' let us reproduce the first few lines in the listing of *Example #1* above:

```
RESERVOIR LENGTH, L (m): 240
DAM HEIGHT H, (m): 30
SLOPE ANGLE, alpha (deg): 3
Fluid density, rho (kg/m3): 1406
Fluid dynamic viscosity, miu (Pa.s): 12
____________________________________________________________
Non-dimensional Physical
____________________________________________________________
```

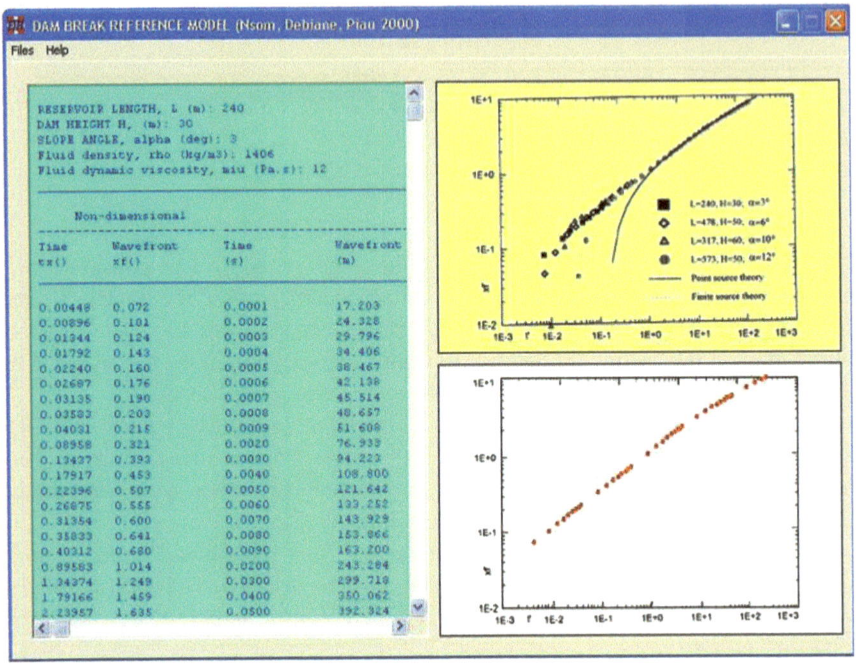

Fig. 14.7 Example # 1, based on Nsom et al. (2000)

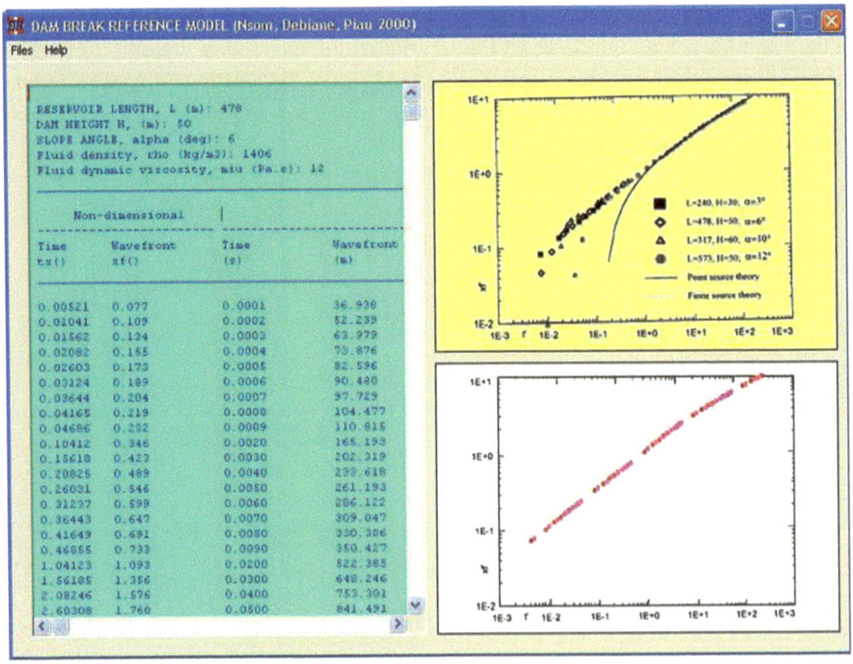

Fig. 14.8 Example # 2, based on Nsom et al. (2000)

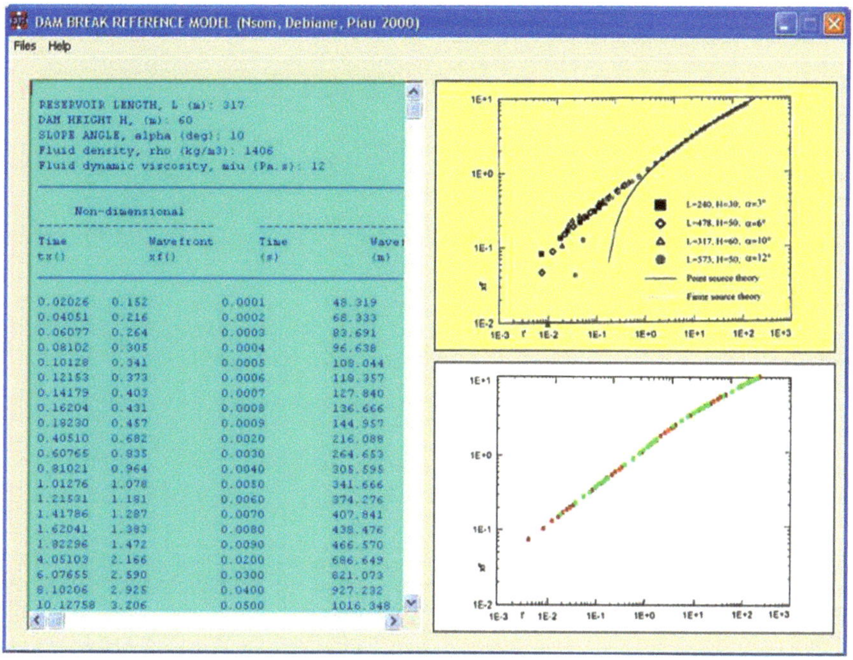

Fig. 14.9 Example # 3, based on Nsom et al. (2000)

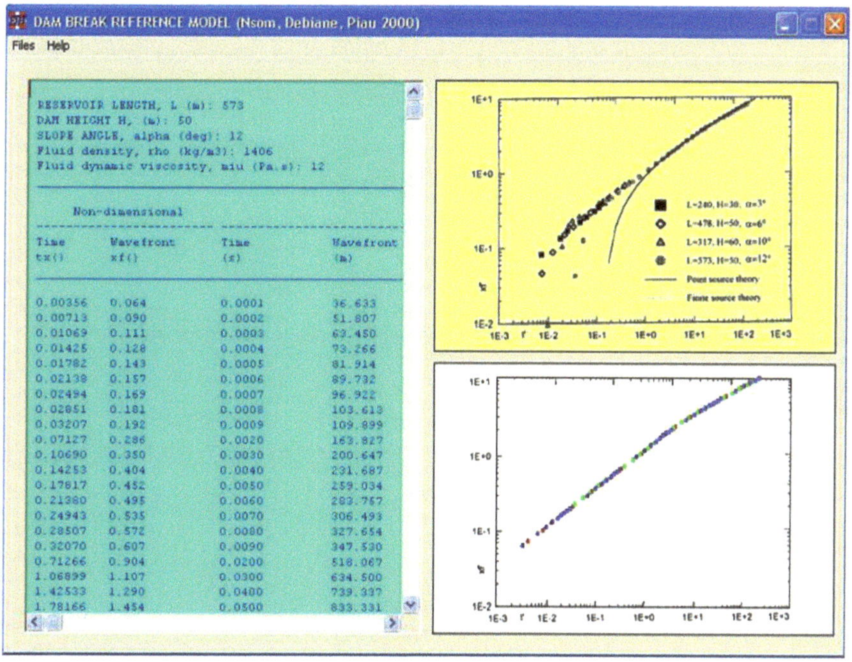

Fig. 14.10 Example # 4, based on Nsom et al. (2000)

```
Time Wavefront Time Wavefront Wave Height Wavefront Velocity
tx() xf() (s) (m) (m) (m/s increment)
________________________________________________________________

1 2 3 4 5 6
________________________________________________________________

0.00448 0.072 0.0001 17.203 15.696 17.203 <
0.00896 0.101 0.0002 24.328 15.357 7.126 <
0.01344 0.124 0.0003 29.796 15.107 5.468 <
0.01792 0.143 0.0004 34.406 14.904 4.609 <
0.02240 0.160 0.0005 38.467 14.730 4.061 <
0.02687 0.176 0.0006 42.138 14.576 3.671 <
0.03135 0.190 0.0007 45.514 14.438 3.376 <
0.03583 0.203 0.0008 48.657 14.312 3.143 <
0.04031 0.215 0.0009 51.608 14.196 2.952 <
. . . . . . . . . . . . . . . . . . . . . . . . . . . . . . . .
```

It is immediately evident that the quantity called, for convenience, 'Wavefront velocity' (column 6) is in effect *a measure of the increment, predicted by the model, of the wave front distance from dam, over the time step employed* (column 4).

In the absence of other indication, the first value of the 'wave front velocity,' (i.e., 17.203) is a mere transcription of the first value issued by the correlational model for the wave front distance. The second value, i.e., 7.126, is the increment from 17.203 to 24.328 m reported in column 4 for the wave front distance, over the time increment given in column 1 (i.e., from 0.00448 to 0.00896 s). In a physical sense, the wave velocity should therefore be:

$$\begin{aligned}\text{Physical wave front velocity} &= (24.328 - 17.203)/(0.00896 - 0.00448)\\ &= 1590.40178\ \text{m/s}\end{aligned}$$

The alternative manner of inferring the wave front velocity—the one reflected in equation:

$$\boldsymbol{Vf} = (\boldsymbol{xf} - \boldsymbol{xf}0) * \boldsymbol{L}\,'\text{Wave front velocity, average over time step, m/s}$$

that can be identified in the algorithm listing above, leads to a close value, namely:

$$\begin{aligned}\text{Physical wave front velocity} &= (0.101 - 0.072) * 240*)/(0.00896 - 0.00448)\\ &= 1553.57142\ m/s\end{aligned}$$

the differences deriving from the different definitions employed (wave front step-wise increment versus true analytical equation). The alternative definitions employed—the definition of convenience by wave front distance increments, and the physical difference obtained from correlations according to the model—are consistent with each other.

Argument 3. While it is important to remember that correlational rules may not cover with uniform accuracy the entire range of values that the physical quantities involved may feature—which is likely to apply to the 'wave front velocity' near their inception (a few seconds), it is also worth noting that high-speed waves are not uncommon in water dynamics. As observed in relation to another form of destructive wave, 'The tidal wave caused by an undersea earthquake in Chile in May 1960, covered the 6000 nm (11,000 km) to New Zealand in about 12 h, travelling at a speed of about 900 km/hr' (250 m/s)! (http://www.seafriends.org.nz/oceano/waves.htm)

The dam break over a wet (river-filled) inclined channel is, in effect, *a wave*. While it may not be supersonic (such high speeds being, most probably, spurious effects of imperfect, correlational, models), *it is sure very fast in its initial stages*. As it subsequently entrains mud, debris etc., it would eventually settle down to a more sedated viscous flow.

Argument 4. Assessors of the 'DAM RISK' experiment should remember that the tool offered under a KT (knowledge transfer) observance *is not about promoting a single dam break model and disfavoring other—possibly superior—models*. Three avenues are already offered online. Beyond these, the project team is definitely open to additional and/or alternative proposals holding promises for a higher performance, from both expert groups and regulatory authorities, and is willing to deploy appropriate efforts to see such alternatives implemented.

14.3.3 Simulation Sequences

The following consequence simulation sequence are in regard to over- and under-conservative model settings (Figs. 14.11 and 14.12). For better comparison with the reference, consequence analysis was integrated in the own consequence representation (Fig. 14.13).

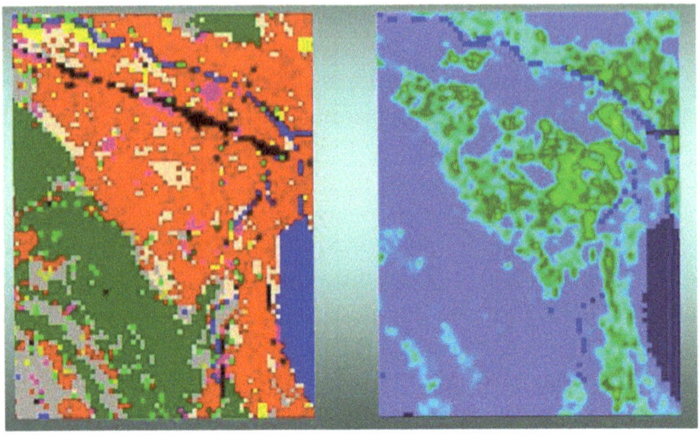

Fig. 14.11 Visual of digital GIS and population type layers

Fig. 14.12 Topographic representations of the digitalized maps

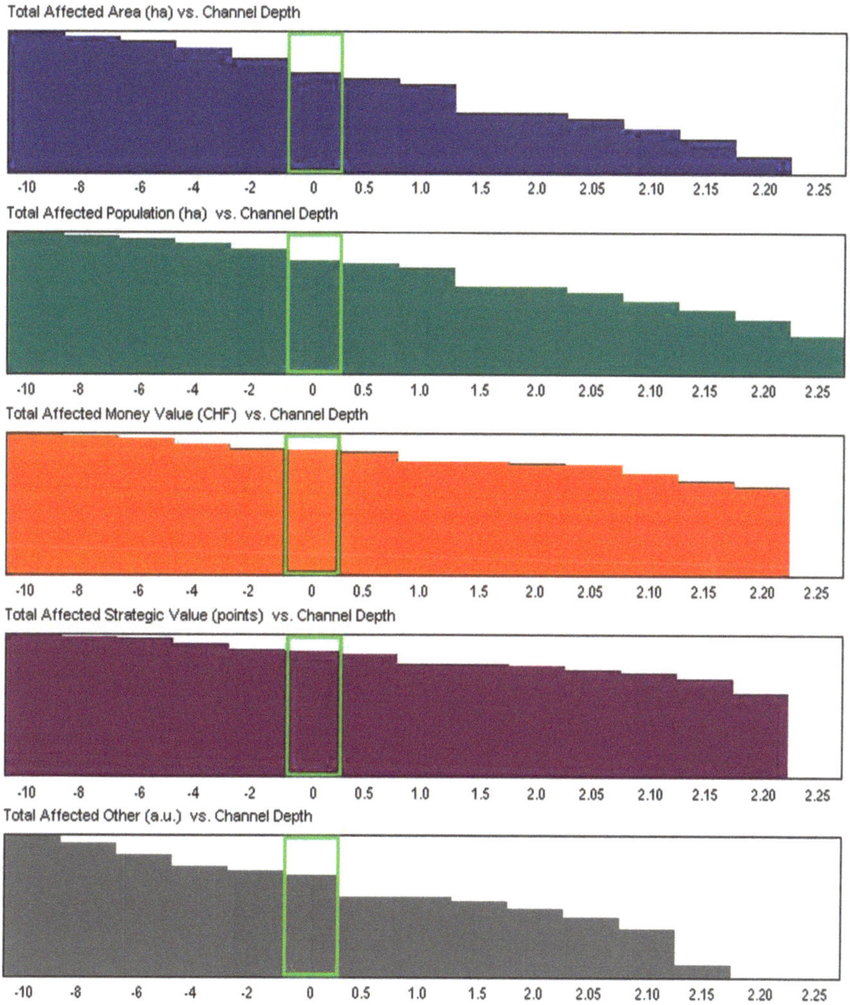

Fig. 14.13 Comparison of different elements of the channel

STATISTICS OF MARKED AREA

Land Type. (ha) (persons) (CHF)		Area Population Money Value Strategic Value Other (points) (a.u.)
Rivers 21.6949		716 2169490 8677960 216.949
Groves 12.2284		293 611420 611420 122.284
HouseD1	77.4992	6742 4.649952e8 46499520 774.992
Arable 37.2347		410 37234700 18617350 372.347

(continued)

(continued)

Land Type. (ha) (persons) (CHF)		Area Population Money Value Strategic Value Other (points) (a.u.)	
Recreatl 28.3835		1079 2.83835e8 24125975 283.835	
ForstO	17.783	36 26674500.0 10669800 177.83	
Resident 32.8422		3350 9.85266e8 32842200 328.422	
Industry 7.639		176 2.2917e9 6875100 76.39	
Garden 5.7767		208 2888350 3466020 57.767	
Orchard 0.3692	12 738400 221520.0 3.692		
Specials 4.4764	224 1.56674e8 3581120 44.764		
FarmPast	0.899	3 674250 629300	8.99
Railways	17.3133	416 1.73133e9 15581970.0	173.133
Lakes 1.9574	4 195740 782960	19.574	
UnPrd 0.8988	4 8988 8988	8.988	

TOTAL AREA accounted (ha): 267
TOTAL POPULATION accounted (persons): 13,671
TOTAL MONEY VALUE accounted (a.u.) 5.98499604e9
TOTAL STRATEGIC VALUE accounted (a.u.) 1.73191203e8
TOTAL OTHER accounted (a.u.) 2669.957
The current map unit values (indices)

Type (CHF/ha)	Money Value (points/ha)	Strategic Value (a.u./ha)	Other
Rivers	1.0e5	0.4e6	10
Groves	5.0e4	5.0e4	10
HouseD1	6.0e6	0.6e6	10
HouseD0	6.0e6	0.6e6	10
Arable	1.0e6	0.5e6	10
Recreatl	1.0e7	0.85e6	10
ForstO	1.5e6	0.6e6	10
Resident	3.0e7	1.0e6	10
Industry	3.0e8	0.9e6	10
Garden	0.5e6	0.6e6	10
Orchard	2.0e6	0.6e6	10
Specials	3.5e7	0.8e6	10
Vineyard	5.0e6	0.6e6	10
FarmPast	0.75e6	0.7e6	10

(continued)

(continued)

Type (CHF/ha)	Money Value (points/ha)	Strategic Value (a.u./ha)	Other
Railways	1.0e8	0.9e6	10
Lakes	1.0e5	0.4e6	10
ForstY	1.25e6	0.6e6	10
UnPrd	1.0e4	1.0e4	10

For the better visualization of both the consequence analysis and the following sequence of simulation results, the KOVERS flood course and extension were digitally overlaid.

CURRENT MAP UNIT VALUES (INDICES)

Land Type (CHF/ha)	Money Value Strategic Value (a.u./ha)	Other (a.u./ha)
Rivers	1.0e5	0.4e6 10
Groves	5.0e4 5.0e4	10
HouseD1	6.0e6	0.6e6 10
HouseD0	6.0e6	0.6e6 10
Arable	1.0e6 0.5e6	10
Recreatl	1.0e7	0.85e6 10
ForstO	1.5e6	0.6e6 10
Resident	3.0e7	1.0e6 10
Industry	3.0e8	0.9e6 10
Garden	0.5e6	0.6e6 10
Orchard	2.0e6	0.6e6 10
Specials	3.5e7	0.8e6 10
Vineyard	5.0e6	0.6e6 10
FarmPast	0.75e6	0.7e6 10
Railways	1.0e8	0.9e6 10
Lakes	1.0e5	0.4e6 10
ForstY	1.25e6	0.6e6 10
UnPrd	1.0e4	1.0e4 10
Misclln	0.0	0.0 0

Abbreviations:

Arable	arable farm land including meadows
HouseD	housing development area
Arable	arable farm land including meadow
Recreatl	recreational areas, e.g. parks, spa
ForstO	Old (Tall) Forest
Resident	dwelling area
FarmPast	farm pastures
Specials	special destination areas

ForstY	young forest
UnPrd	unproductive land
Misclln	other, insignificant

It appears that the most straightforward means to mitigate a flooding crisis are to timely secure a sufficient, positive channeling capability through a decent investment in water management works within the routine urban development planning. As to how to size the channeling—this depends on the *confidence placed on the projections about the maximal height of expected flood wave fronts*. While simple models may give indications in this respect, a full commitment to a capital investment should be preceded by much more profound analytical work and downscaled similitude experiments. To some sizeable extent, the same applies to financial engineering projects in the insurance/reinsurance business.

Given previous information and assuming the following description (Fig. 14.14), sequence of simulations was developed as indicated in Figs. 14.15, 14.16, 14.17, 14.18, 14.19, 14.20, 14.21, and 14.22. Figures 14.15, 14.16, 14.17, and 14.18 take an over-conservative viewpoint, while Figs. 14.19, 14.20, and Fig. 14.21 are under-conservative. Figure 14.22 is optimistic in nature.

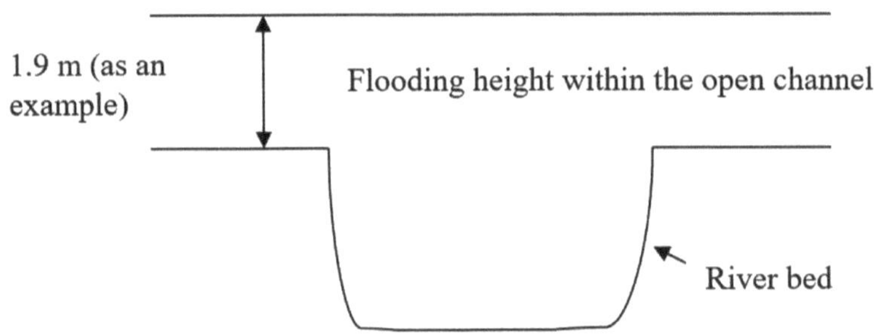

Fig. 14.14 Basic structure of a river and its flooding height

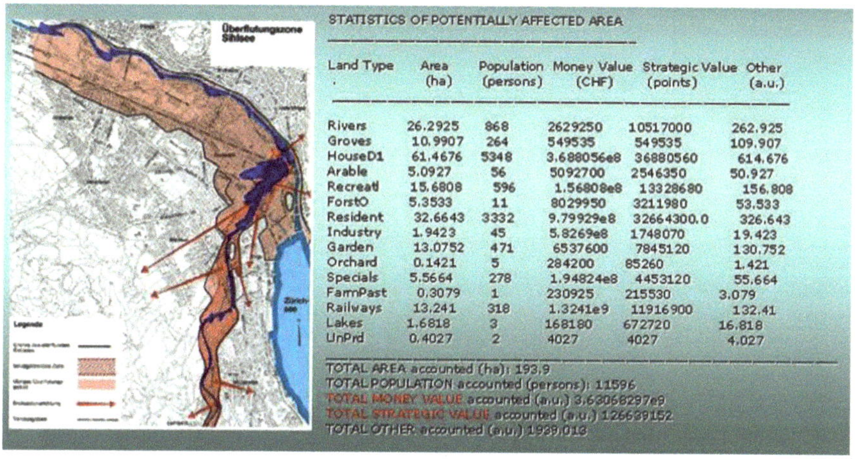

Fig. 14.15 Case #1, Channel depth 0.0 m, GIS effects only

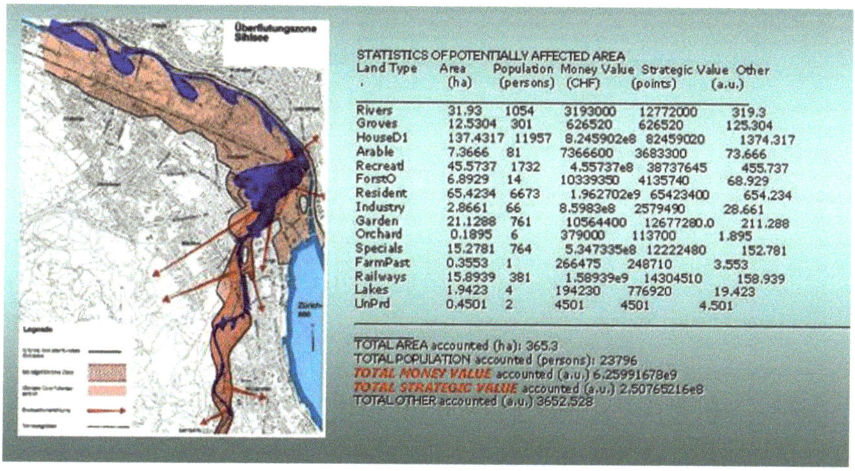

Fig. 14.16 Case #2, Channel depth −2.0 m, Over-conservative

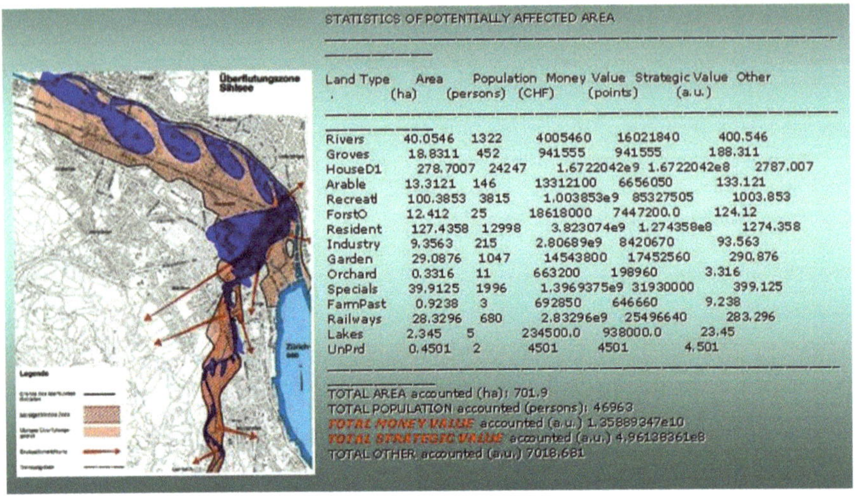

Fig. 14.17 Case #3, Channel depth −6.0 m, Over-conservative

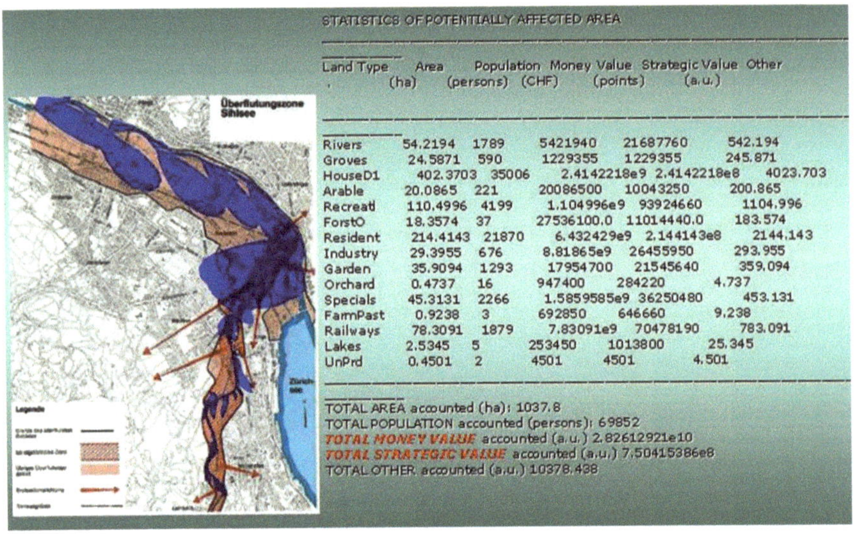

Fig. 14.18 Case #3, Channel depth −10.0 m, Over-conservative

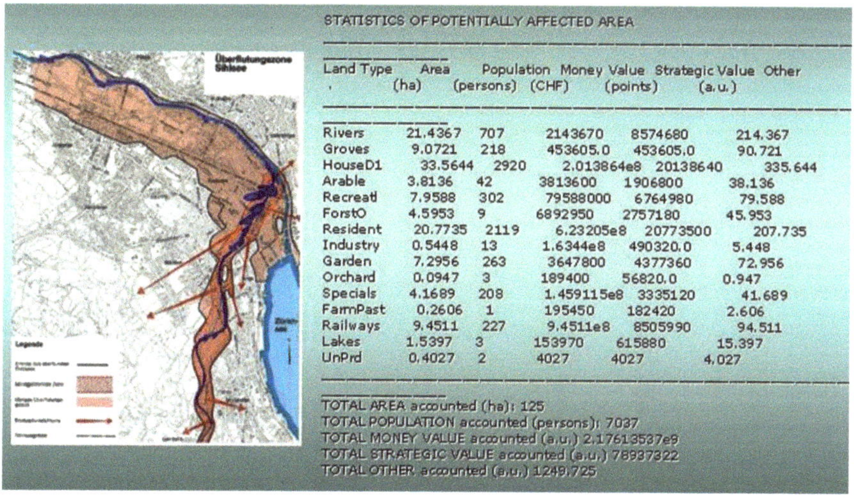

STATISTICS OF POTENTIALLY AFFECTED AREA

Land Type	Area (ha)	Population (persons)	Money Value (CHF)	Strategic Value (points)	Other (a.u.)
Rivers	21.4367	707	2143670	8574680	214.367
Groves	9.0721	218	453605.0	453605.0	90.721
HouseD1	33.5644	2920	2.013864e8	20138640	335.644
Arable	3.8136	42	3813600	1906800	38.136
Recreatl	7.9588	302	79588000	6764980	79.588
ForstO	4.5953	9	6892950	2757180	45.953
Resident	20.7735	2119	6.23205e8	20773500	207.735
Industry	0.5448	13	1.6344e8	490320.0	5.448
Garden	7.2956	263	3647800	4377360	72.956
Orchard	0.0947	3	189400	56820.0	0.947
Specials	4.1689	208	1.459115e8	3335120	41.689
FarmPast	0.2606	1	195450	182420	2.606
Railways	9.4511	227	9.4511e8	8505990	94.511
Lakes	1.5397	3	153970	615880	15.397
UnPrd	0.4027	2	4027	4027	4.027

TOTAL AREA accounted (ha): 125
TOTAL POPULATION accounted (persons): 7037
TOTAL MONEY VALUE accounted (a.u.) 2.17613537e9
TOTAL STRATEGIC VALUE accounted (a.u.) 78937322
TOTAL OTHER accounted (a.u.) 1249.725

Fig. 14.19 Case #4, Channel depth 1.0 m, Under-conservative

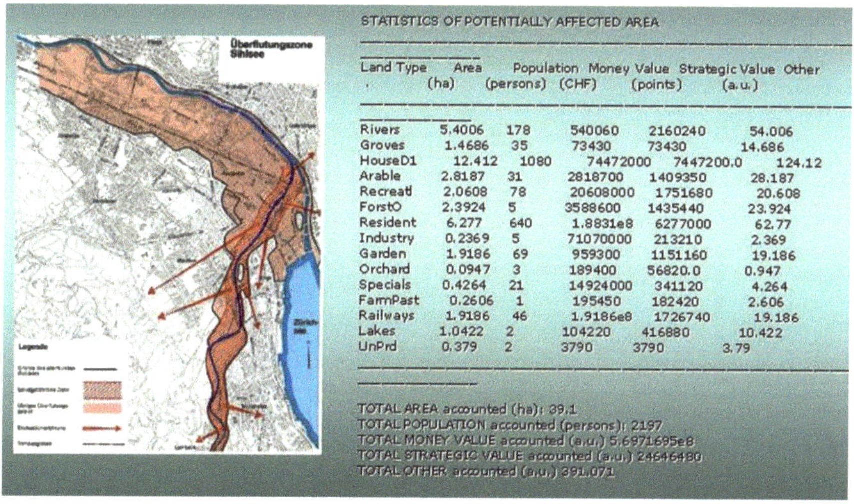

STATISTICS OF POTENTIALLY AFFECTED AREA

Land Type	Area (ha)	Population (persons)	Money Value (CHF)	Strategic Value (points)	Other (a.u.)
Rivers	5.4006	178	540060	2160240	54.006
Groves	1.4686	35	73430	73430	14.686
HouseD1	12.412	1080	74472000	7447200.0	124.12
Arable	2.8187	31	2818700	1409350	28.187
Recreatl	2.0608	78	20608000	1751680	20.608
ForstO	2.3924	5	3588600	1435440	23.924
Resident	6.277	640	1.8831e8	6277000	62.77
Industry	0.2369	5	71070000	213210	2.369
Garden	1.9186	69	959300	1151160	19.186
Orchard	0.0947	3	189400	56820.0	0.947
Specials	0.4264	21	14924000	341120	4.264
FarmPast	0.2606	1	195450	182420	2.606
Railways	1.9186	46	1.9186e8	1726740	19.186
Lakes	1.0422	2	104220	416880	10.422
UnPrd	0.379	2	3790	3790	3.79

TOTAL AREA accounted (ha): 39.1
TOTAL POPULATION accounted (persons): 2197
TOTAL MONEY VALUE accounted (a.u.) 5.6971695e8
TOTAL STRATEGIC VALUE accounted (a.u.) 24646480
TOTAL OTHER accounted (a.u.) 391.071

Fig. 14.20 Case #5, Channel depth 2.0 m, Under-conservative

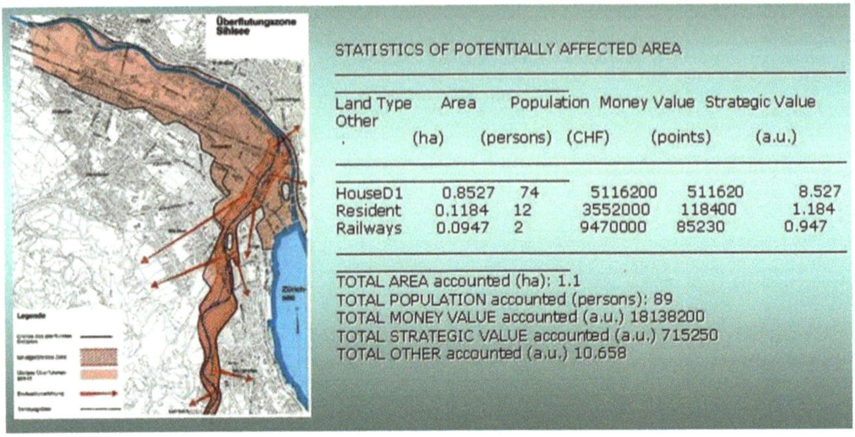

Fig. 14.21 Case #6, Channel depth 2.25 m, Under-conservative

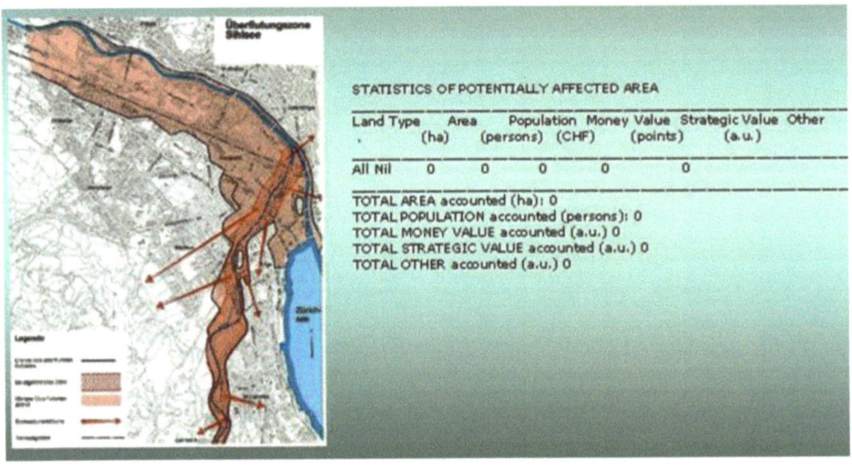

Fig. 14.22 Case #7, Channel depth 2.259 m, Optimistic

References

Nsom, B. (2002). Horizontal viscous dam—break flow: Experiments and theory. *Journal of Hydraulic Engineering*, *128*(5), 543. https://doi.org/10.1061/(ASCE)0733-9429.

Nsom, B., Debiane, K., & Piau, J.-M. (2000). Bed slope effect on the dam break problem. *Journal of Hydraulic Research, 38*(6), 459–464.

Chapter 15
Urban Area Vulnerability Assessment: Cellular Automaton Approach to Airflow Dispersion in Complex Terrains

Abstract This chapter offers insights into the science of the risk and vulnerability with a focus on airflow dispersion approach related to cellular automaton. This discussion, finding itself in the awkward position of having to trade mathematical sophistication and rigor, versus intelligibility and practicality, of the varied stakeholders around, relies heavily on real-life indications associated with challenges of globalization of markets, regionalization of ethnic crucibles, transborder labor, cultural overflows, clashes, and the consecutive 'reactive adaptation' commonly known as terrorism.

15.1 Scientific Precision and Vulnerability

One might argue that scientific anticipations, often relayed in terms of 'projections,' 'prognosis,' 'modeling,' and 'preparedness,' have no place in the landscape of anticipations in which what might happen is a difficult task. Apart from the characteristically large-scale and well-budgeted appearance of the endeavor described the New York Times report (Urbina 2005), worth noting is the attitude that drives the actors. It is aptly reflected in the rhetoric's of one of the project leaders, quoted to have said: 'Our aim is *to begin to understand* how atmospheric dispersion occurs' (author emphasis), which implies a fair recognition of the current limitations in the performance of the atmospheric dispersion models in such complex topographies as the urban areas.

The second line of interest is in the context of a remark, on *how far the expectations about the project output would can go*. Again, in the cited New York Times report, it is noted that '...if a tanker truck carrying toxic gasses crashes downtown or, terrorist releases anthrax in the air, you want to be able to start predicting the places that are downwind.... With computer modeling you can start to figure out whether to tell people to get off the streets immediately or to stay inside. You can also start figuring out where to send the ambulance, police and Fire Department... you can avoid having people running into the plume instead of away from it.'

A.V. Gheorghe et al., *Critical Infrastructures, Key Resources, Key Assets*, Topics in Safety, Risk, Reliability and Quality 34,
https://doi.org/10.1007/978-3-319-69224-1_15

These, we content, are clear-cut and down-to-earth objectives, which of course will not preclude a thorough, heavily scientific, sophisticated investigation behind the scenes. There is no wonder that the stakeholder, seemingly The New York City Hall, has been won over. Of course, the same scenario is applicable to any other location in the world. The technical issue about the afore-stated objectives is—*how easy/difficult is to get the respective answers.* The conventional conduct in the matter is to:

- obtain, via computer modeling, simulations, and field experiments, a prediction on a series of quantities of normative value, recommended by authorities like the American Institute of Chemical Engineers. These might include information on airborne chemical concentration levels known as the *Immediately Dangerous for Life and Health Limit* (IDLH), *Threshold Limit Value* (TLV), and *Short Term Exposure Limit* (STEL)
- have these compared with whatever *Protection Action Guides* (i.e., such as those suggested by U.S. Nuclear Regulatory Commission language) may be in effect, in the targeted legal environment—in this case, the New York State/City health/environmental codes

While there is no doubt that the attempt in this case *will* be made during the exercise envisaged, one may a priori expect a considerable, even disconcerting, divergence in the model predictions as far as the *numbers* (quantity values) are concerned. A far lesser divergence may, on the other hand, be expected in what regards the trends in the air flows' general behavior, these featuring, in particular, (a) the spots of comparatively higher time-integrated concentration of pollutant and (b) the spots of comparatively lower-speed air flow circulation. The obvious expression of these is—output maps laid over City's standard maps or, better, the City's GIS (Geographic Information System layers).

One submits that a fair convergence of the computer models on the items (a) and (b) above is, in effect, (i) reasonably sufficient for the practical purposes emphasized by key-project officials and (ii) comparatively more 'doable,' and less prone to disconcerting divergence and uncertainties, than an approach geared toward hunting for *numbers*, as discussed. Assuming, for the sake of this discussion, that the afore-stated premise is acceptable, the next issue is whether the lower analytical load resulting from it can be paralleled by a comparable reduction in modeling complexity. Indications in the literature and academic practice are that *there is* a positive answer to that, even though some of its aspects are controversial. The response revolves around the notion of 'computational modeling,' having as an outstanding hallmark—the cellular automaton. As evident in its promoters' stands, the said approach involves an entire epistemological motivation and attitude (Gheorghe and Vamanu 2005). Wolfram (1983) submitted that:

> …The ultimate purpose of most scientific investigations is to determine how physical or other systems will behave in particular circumstances. Over the last few years, computer simulation has been emerging as the most effective method in many different cases. The basic approach is to use an algorithm which operates on data in the computer so as to emulate the behavior of the system studied…This algorithm can be considered to provide a

> 'computational model' for the system. The fundamental principle is that the models considered should be as suitable as possible for implementation on digital computers. It is then a matter of scientific analysis to determine whether such models can reproduce the behavior seen in physical and other systems. Such analysis has now been carried out in several cases, and the results are very encouraging.

One may find here, and particularly in the second paragraph quoted, a certain defiance of what a conventionally-educated high-mind means by a 'method.' The accusations of 'empiricism' are met on occasions. Unfortunately, a certain radicalism on behalf of some of the alternative approach promoters who may overstate, in their public stands, both its novelty and efficacy do not help either, in moderating the polemics (e.g., see Glaser and Strauss 1967 on the case for 'grounded theory'). Seeing as a duty of conformity to mention these aspects, the authors here would rather focus, in the sequel, on a limited-liability adherence to systems' computational modeling, in general, and the cellular automaton approach to fluid dynamics, in particular. Our 'attitude' is believed to be honestly reflected in the objective adopted.

> ...to find a PC-based problem solver drawing upon the CA (cellular automaton) methodology, able to deal with the risk and vulnerability-related issues relating to urban areas exposure to emissions of hazardous – chemical and/or radioactive – substances.

15.2 A Computational Model for Air Flows

Undertaken is a general approach in which one starts with a *physical* model to end up with a *computational* model. The following logical descent is in effect:

- The *physical model* thought to be adequate for the purpose described draws upon certain features of the air that warrant some accepted approximations of the Navier-Stokes equation.
- The *computational model*, in turn, draws on a CA-wise interpretation of the chief behavior of the air, indicated by the approximate equations of the physical model.

15.2.1 Accepted Approximations

One seconds the notion that air is, basically, a Navier-Stokes fluid:

$$\frac{\partial u}{\partial t} + (u\nabla)u = f - \frac{\nabla p}{\rho} + \nu\,\Delta\,u \tag{15.1}$$

where

$u(r,t)$ is the space- and time-varying vector field of the fluid velocity
$p(r,t)$ is the local pressure field
$\rho(r,t)$ is fluid's density
f is the vector-density of external forces—such as gravity acting upon the fluid
ν is fluid's kinematic viscosity.

Wolfram (1986) along with other authors admits that air is a negligibly viscous fluid—which dispenses one of the last terms in Eq. (1.1), and only a slightly compressible fluid—which allows taking air density as uniformly equal to 1. Under the standard interpretation of the left-hand side in Eq. (1.1) as the advected fluid's acceleration, a, and also assuming that all external forces are fully compensating each other everywhere and at every time, the physical model equation obtained from Eq. (15.1) reads:

$$a = -\nabla p \tag{15.2}$$

The physical meaning of Eq. (15.2) is, in plain words, what the ancient observers before Evangelista Torricelli have termed as 'Nature's fear of vacuum': *air tends to go where pressure is lower*. This leaves one with a *rule*, instead of an *equation*, thus paving a way to *a rule-based computational model*, the latter being called to just implement the afore-emphasized finding in a manner consistent with computational models' philosophy.

15.3 The Consecutive Rule-Based Model

15.3.1 Terms of Reference

Let us try and see how the implementation of the leading rule above may indeed be made convenient. The first remark is that *any* computer representation of a fluid in motion would ultimately boil down to *pixels-in-motion*. In other words, the governing algorithm would:

(a) take every fluid particle that inherently occupies a given pixel or a 'sprite' memory location region in the trade's slang at moment t_1,
(b) give particle position, speed, and possibly acceleration appropriate increments governed by the same equation/rule (*uniform, or homogeneous updating*) over the time lapse from t_1 to t_2, in consideration of the fluid particles around (*locality of updating*) yet independently, for each and every particle at every time step (*parallel updating*),
(c) relocate the particle at the newly determined pixel/sprite memory region, corresponding to the t_2 time,

(d) cycle the above procedure at user's will.

Thereby and at a closer look, any computer representation of fluid dynamics is (i) relying on a *fixed grid of virtual points*—the only 'space' a computer understands, and (ii) applying a *step-wise, local, homogeneous*, and *parallel* updating of the grid knots by, in fact, virtually 'transporting' the fluid particles across the grid. There is also a matter of the associated terms—parallelism, locality, and homogeneity. In the context of present discussion and in line with previous research, the reader is directed to Rucker and Walker (2017).

Parallelism involves individual cell updates. In a CA grid, updates are performed independently of each other (i.e., we think of all of the updates being done at once). However, and strictly speaking, your computer only updates one cell at a time, but we use a buffer to store the new cell values until a whole screen's worth has been computed to refresh the display. *Locality* has to do with when a cell is updated. The cell's new color value [state] is based solely on the old color values [states] of the cell and of its nearest neighbors. Finally, *homogeneity* deals with the fact that each cell is updated according to the same rules. Typically, the color values [states] of the cell and of its nearest eight neighbors are combined according to some logico-algebraic formula or are used to locate an entry in a preset lookup table.

These terms of reference would, in principle, suffice to support the notion that *any* computer description of a fluid is, in effect, a… *cellular automatonoid—with the 'oid'* suffix left to us by the ancient Greeks to conveniently denominate something that is *alike with* a reference thing, if not necessarily *exactly* the reference thing. To make the leap from an *automatonoid* to a true *automaton*, the final touch is:

(i) make sure one sticks to a (uniformly and parallelly applied) *logical rule*, as opposed to an equation (though frankly, for a computer, between Boolean conditions and the step-wise finite differences, there is only a very thin line); and
(ii) make the rule ASAF (i.e., *As Simple As Feasible*).

There are two additional clauses to the terms of reference above. The *first* follows from the circumstance that true CA implementations may still require too complex rules and geometries, to possibly fit the pattern of a real city—streets, plazas, variously shaped building and all. A consultation of the commonly available literature would indicate that CA wwas successfully demonstrated rather on singular, archetypal objects of the fluid dynamics such as the 'blade,' the 'slab,' and the 'cylinder.' The same stands true for the numerous, and indeed intellectually rewarding implementations of the Chorin's method of 'vortex blobs' (Chorin 1973). This is definitely *not* to say that the respective tools cannot operate at larger scales. The March 2005 New York drill would require modeling on at the size of Manhattan's map — pose quite a challenge for standard CA approach in the sense of Wolfram (1988, 1983).

A second constraint is the common availability of computer power. The quotation below gives an indication of what 'nice-to-haves' may be envisaged, even in a limited-scale CA numerical experiment.

> In our current implementation on a Connection Machine computer with 65,536 processors, lattices of size 4094 × 8192 can be updated at a rate of about 1.0E9 sites per second, allowing the fluid flow patterns around objects to be found interactively up to Reynolds numbers of several hundred (Wolfram 1988).

Therefore, it was found that *an even simpler rule (set of rules) should be employed in order to give* Eq. (15.2) *an implementation counting on the computing power of a standard PC*—expected to ordinarily top-the-desk in a City Hall; a fire department; a police department and the similar.

15.3.2 The Rule

Let us start by reminding the raw form of the rule derived from Eq. (15.2):

'Air tends to go where pressure is lower'

To implement this (trivial) finding that in effect could have equally gone without an equation behind it, a few steps further should be performed:

(a) Based on the well-known pressure equation for ideal gasses, according to which

$$p = n * k_B * T \tag{15.3}$$

where,

p is the gas pressure in P_a
T is the absolute gas temperature in K
k_B is the Boltzmann's constant in P_a*m^3/K, *note that the gas pressure is proportional with the gas particle (molecule) concentration, n* [1/m^3]

(b) Consequently, for the air to 'go where pressure is lower' ***air 'particles' should go where there are few other air particles per unit volume or, more general, in the neighborhood***.
(c) To perform such a feat, ***the particle should, first, become aware of its neighborhood***, be able to somehow discriminate between places where particles are denser and places where particles are scarcer. Once this discrimination is on the record, the particle should, well, go toward the place where particles are scarcer. And, consistent with the CA discipline, *all* particles should perform the said feat parallelly, at every time step.

In view of the above, the practical matters at hand to be solved involve: (i) how can a particle 'become aware,' or otherwise account for its neighbors, (ii) how deep around a particle should look for neighbors, (iii) how far the particle should go, when leaping toward the leaner-in-particles neighborhood, and (iv) how the particle should react to obstacles falling within its investigative radius.

The answers to the above are, in actual fact, the 'rule' constituents. There are virtually many ways to achieve the above targets [i.e., (i) through (iv)]. The reality is that *rule optimization* tends to become an important theme within the focal area of the applied CA research. In the case at hand, the general idea is *to detect the distribution of particles around any given particle.*

A few clarifications are in order: first, the model discussed is limited, in this version, to two dimensions; second, 'particle' should be understood, in the context, in an enlarged sense. For a pure air, 'particle' means an entity that, via its behavior, may show some degree of internal coherence giving it an identifiable, if not always observable, persistence within the flow. Such entities are usually associated with small-scale vertices, a vortex being also termed, by some authors (Chorin 1973; Porthouse and Lewis 1981), as 'air blobs.' In a cloud of pollutant dispersing itself, and advected along with the air flow, one may assimilate particles with pollutant gas 'blobs,' yet also with aerosol particles, and finally, the discrete nature of the fluid's computational model on the one hand, and the fact that PC computer power limitations (memory, speed) preclude too large a number of particles to be accounted for in a parallel (synchronous) manner, both result in an interpretation of the relatively low number of 'particles' employed in the model to cover comparatively extensive spaces, as rather the pollen particles in suspension in water, in a classical Brownian motion experiment, than the water molecules themselves. Model particles would, therefore, present a degree of randomness that reflects/express their being driven by many other, *not* accounted for, fluid particles filling the space in-between the accounted for particles.

Note that none of the conceptual (pre)cautions above would affect the pressure driving of particles, which takes place at a larger scale than the 'blob's scale.' A neighborhood accounting mechanism that would imply the randomness referred to in the preceding paragraphs is described in sequel below (this represents the model's algorithm):

1. Assume n particles thrown in random over a limited 2-D grid consisting of computer screen pixels as knots
2. Assume all particles are advected from left to the right with a uniform velocity, *adv* [pixel/time step].
3. Assume m probing points to investigate the neighborhood of radius r (pixel) of any given particle.
4. Throw the m probing points along m directions diverging from the given particle, spaced at equal angles of 360/m degrees, yet randomly within the 0 to r distance from the particle.

5. Test whether there is/is not a particle, or an obstacle, at probing point positions.
   5.1. If there *is* a particle, ignore case.
   5.2. If there *is not* a particle, then **take case into account by incrementing a counter by 1, and mark case (probing particle) position on record**.
   5.3. If there *is* an obstacle at probing point position, then:
      5.3.1. If the probing point falls on the right-hand side of the given particle, i.e., downwind (see clause 2), then ignore it.
      5.3.2. If the probing point falls on the left-hand side of the given particle, i.e., upwind (the given particle is in the lee of the obstacle), **then take case into account incrementing the counter by 1, and mark case (probing particle) position on record**.
6. Using the valid probing points (number and positions) retained according to clause 5.2 and 5.3.2, **get their weight center, consider it as the new position of the given particle, and move the particle there**, yet in a computer memory area (RAM of HD file as convenient) independent from the area of investigation —call it *store area*, so that the procedure can be performed in parallel (synchronous)manner by toggling between the *work area* and the *store area*.
7. Cycle to step 4, conducting the procedure for all particles in the work area.
8. Refresh now the work area by clearing it and loading in the store area.
9. Repeat the entire process from step 4.

An attempt to graphically illustrate the algorithm at work is given in Fig. 15.1, which is 'computer art,' as opposed to the illustrations in Fig. 15.2, which were obtained through an *ad hoc* code snippet. While more refined or alternative rules may improve the algorithm above, numerical experiments seem to confirm, at this stage, that the current version:

(i) implements with notable efficacy, the key-notion of air going towards the lower pressure areas, both in free-flow and in-flows, constrained by obstacles;
(ii) is flexible as far as tunable parameters (see $\boldsymbol{n}$, $\boldsymbol{m}$, and $\boldsymbol{r}$ in model algorithm above) and, particularly, shape and distribution of obstacles, which makes it a good candidate for an urban area assessment tool;
(iii) is inexpensive as far as required computer power.

15.4 The Computational Results

The following is a selection of sample work cases meant to illustrate the performance of the described rule-based model. Selected for visualization were the *time-integrated concentration* (TIC) of particles; and, occasionally, the *maximal local velocities* (MLV)—both obtained as a color-coded buildup of occupied pixel states. It is common knowledge that TIC and similar quantities (time-integrated powered concentrations) are directly relevant for determining chemical and radiation *doses*,

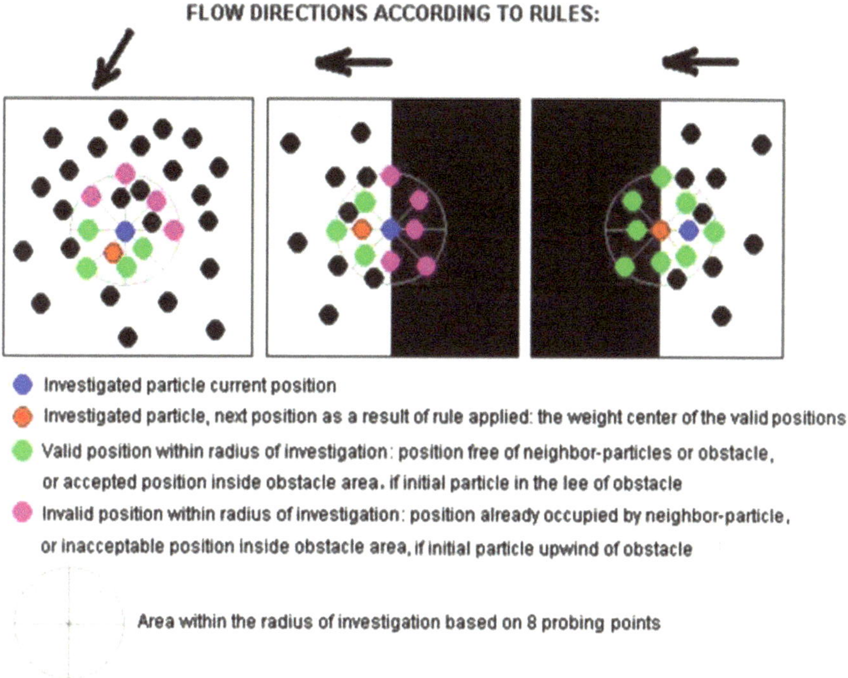

Fig. 15.1 A 'Computer art' representation of model's algorithm

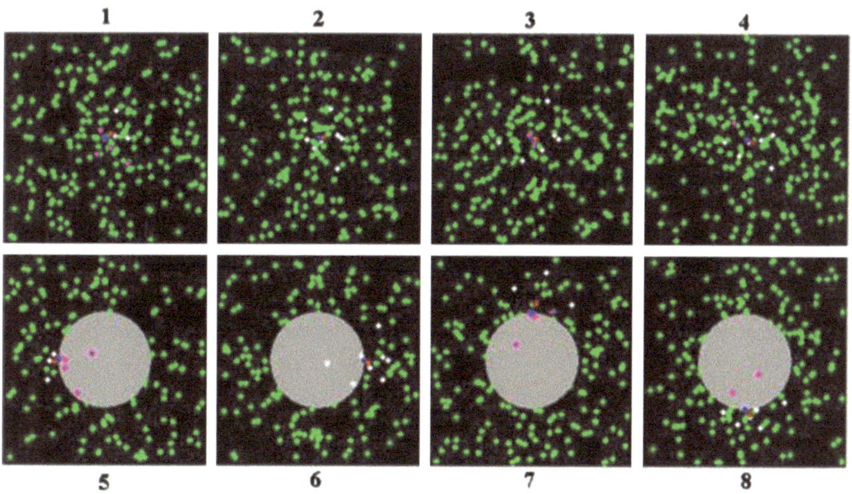

Fig. 15.2 A didactic rendering of the algorithm at work, by a dedicated code snippet

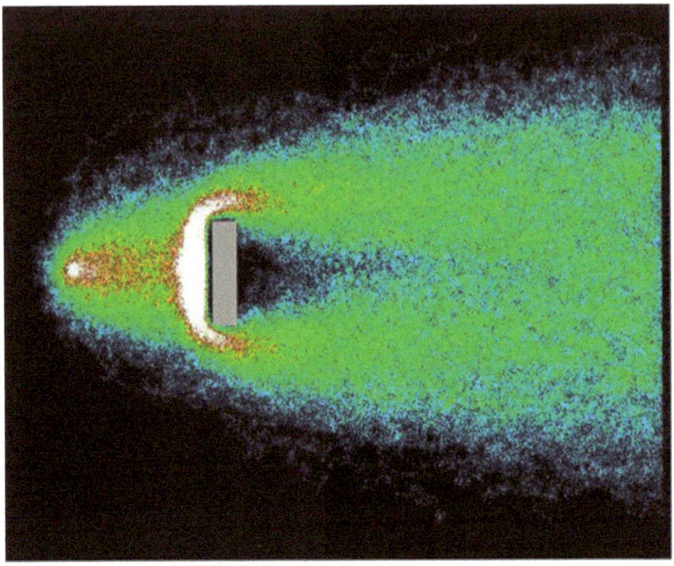

Fig. 15.3 Textbook case of 'the slab.' TIC rendered; 1000 particles, 8 probing points, 20 m-radius probing area; Map width: 632 m

wherefrom health consequences can be derived and quantified via, for instance, the *probit* functional techniques (e.g., see Vamanu et al. 2016 for more information on *probit* functional techniques). In turn, the MLV can be a valuable aid in understanding where aerosol particles tend to pile up on the ground—namely in areas of low-flow velocity (snow-fence effect). Figures 15.3, 15.4, 15.5, 15.6, 15.7, and 15.8 are parallel textbook cases.

One could comment more extensively on various features of the patterns above and their similarity with what wind tunnels and the practice indicate about gas flows. Observe, however, that front wave compressions, cavities form on the lee side of obstacles and even the rudiments of vortex street traces. More important is the utility of the model. For instance, it is possible to develop drills postulating possible terror strikes in large cities involving instantaneous and simultaneous release of toxic gas loads at, for example, different spots with a city. In Figs. 15.9, 15.10, and 15.11, a drill postulating a terror strike consisting of instantaneous and simultaneous release of three toxic gas loads at three different spots downtown the Swiss city of Biel is presented. The results are recorded at 15 min, 30 min, and 1 h into the release. Each release is modeled with 1000 blobs, 8 probing points, and 20-pixel (ca. 100 m) radius probing area. Worth noting is the white 'hot spots' that should primarily draw the attention of the crisis managers, consistently with the response and intervention philosophy of actions associated with critical infrastructure, key resources, and key assets (CIKRKA) and now, Space, Undersea, and Belowground systems.

While the scenario implicit in Figs. 15.9, 15.10, and 15.11 (above) bears relevance to risk analysis, a more vulnerability-oriented case is the one depicted in the

Fig. 15.4 Textbook case of 'the slab.' MLV rendered; 1000 particles, 8 probing points, 20-m-radius probing area; Map width: 632 m, adapted from Vamanu et al. (2010)

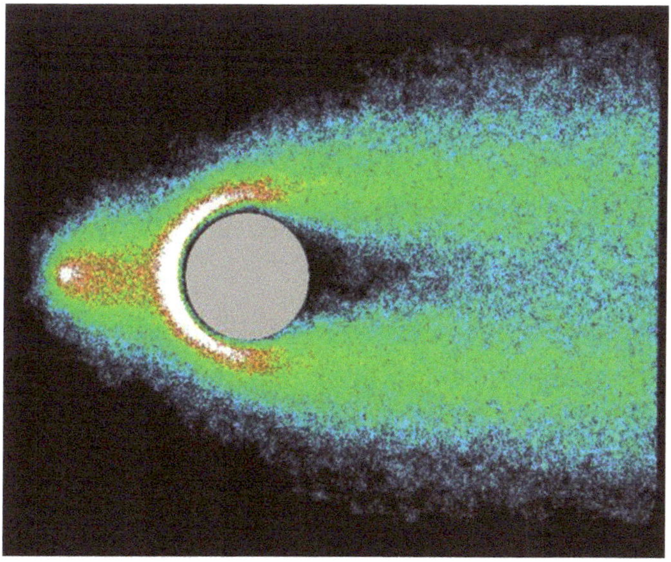

Fig. 15.5 Textbook case of 'the cylinder.' TIC rendered; 1000 particles, 8 probing points, 20-m-radius probing area; Map width: 632 m

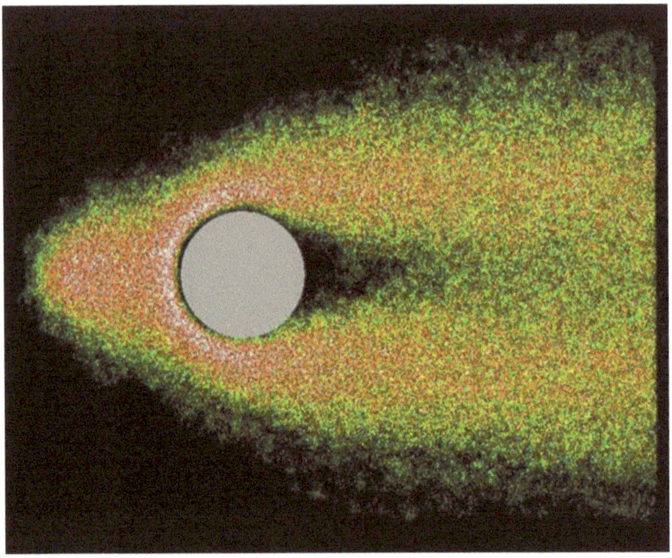

Fig. 15.6 Textbook case of 'the cylinder.' MLV rendered; 1000 particles, 8 probing points, 20-m-radius probing area; Map width: 632 m, adapted from Vamanu et al. (2010)

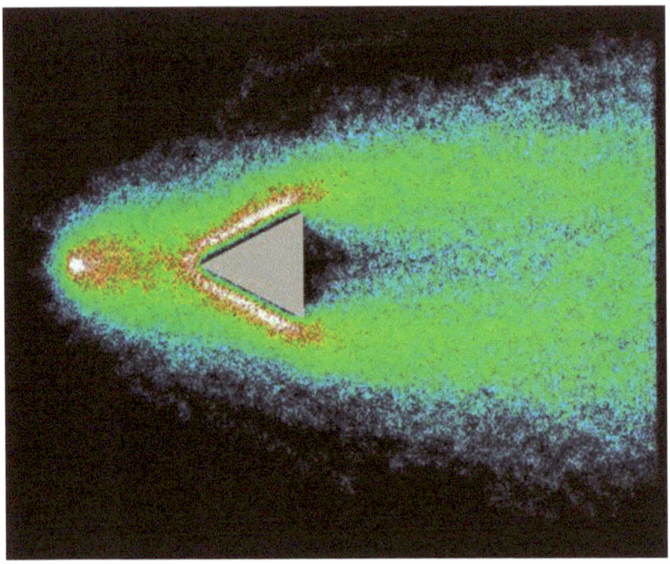

Fig. 15.7 Textbook case of 'the triangular prism.' TIC rendered; 1000 particles, 8 probing points, 20-m-radius probing area; Map width: 632 m, adapted from Vamanu et al. (2010)

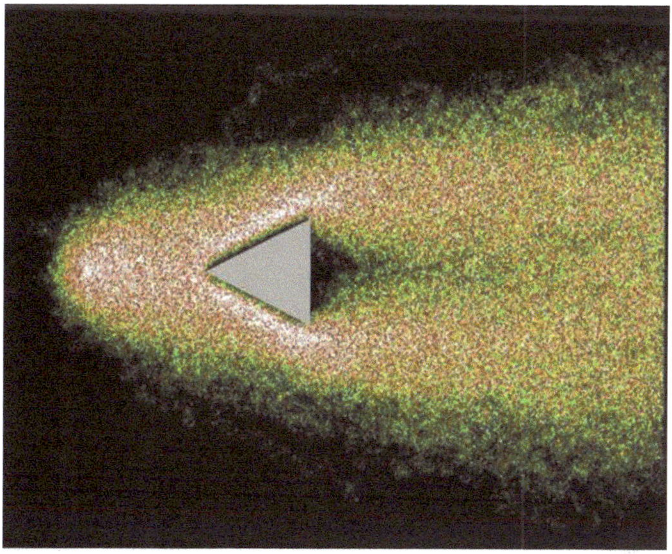

Fig. 15.8 Textbook case of 'the triangular prism.' MLV rendered; 1000 particles, 8 probing points, 20-m-radius probing area; Map width: 632 m, adapted from Vamanu et al. (2010)

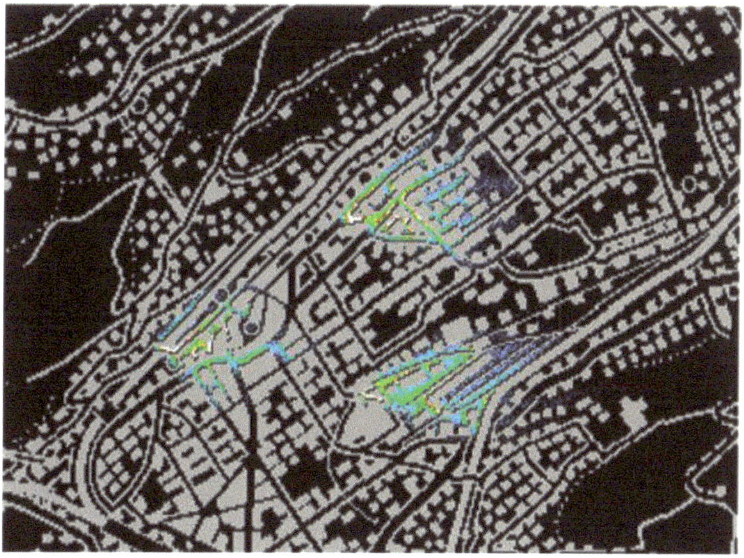

Fig. 15.9 Three instantaneous and simultaneous release in Downtown Biel; Pattern after 15 min, adapted from Vamanu et al. (2010)

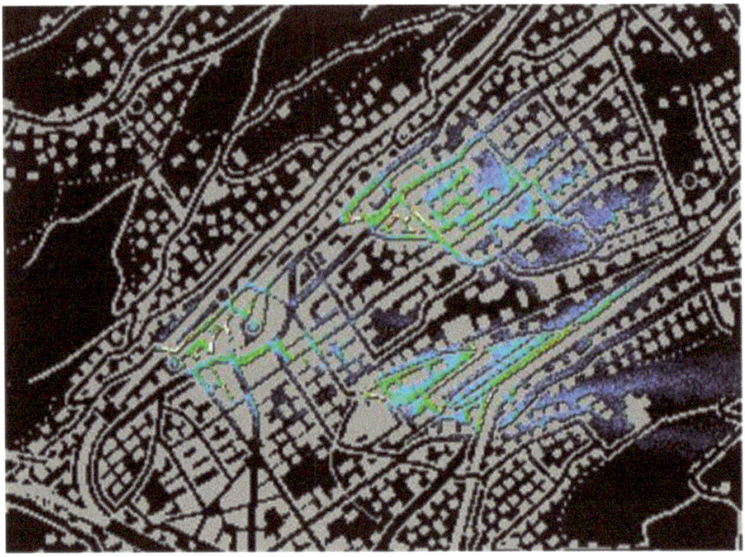

Fig. 15.10 Three instantaneous and simultaneous release in Downtown Biel; pattern after 30 min, adapted from Vamanu et al. (2010)

next following figures (Figs. 15.12 and 15.13). In this case, one searches for 'hot spots' resulting from the exposure of the entire area to an air flow—not necessarily contaminated—that blows uniformly from W to the E. At process outset, particles are randomly distributed over the targeted area. As the flow evolves, areas of higher TIC and lower MLV are forming, which discriminates among the levels of potential exposure.

Assume now that the process is repeated with different orientations of the map with respect to the (fixed) flow direction, left to right and that the wind (advection) is selected according to the multi-annual meteorological statistics of the region, the simplest expression of which is the 'Wind Rose.' Combining the different hot spot patterns obtained, weighted by the respective relative frequency of the meteo data, and perhaps factoring in other elements such as the local population density and age, property value, strategic relevance of the spot, remedial costs etc., one may end up with a meaningful *map of city vulnerability to aggressions for which the air is the vector*.

15.5 Model Calibration

One out of several issues left open at this stage is a consistent calibration of model's control parameters—mainly the appropriate total number of particles, n; and the probing neighborhood radius, r, so that a meaningful correspondence has been established with the conventional atmospheric dispersion theory and practice.

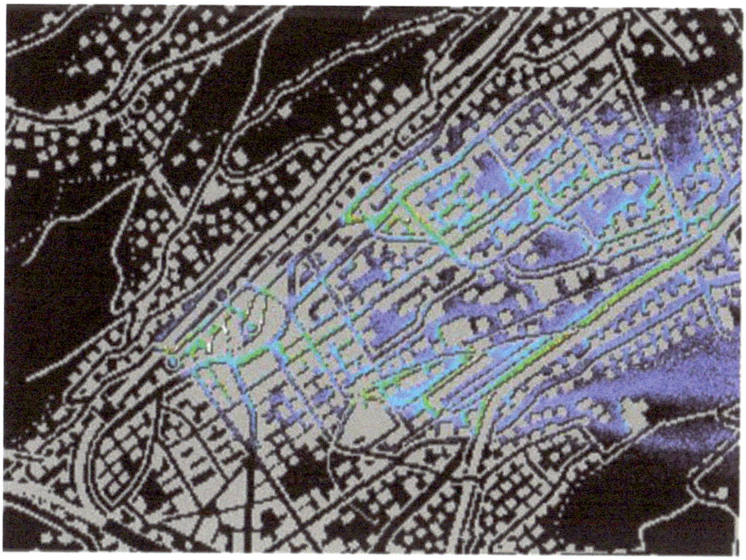

Fig. 15.11 Three instantaneous and simultaneous release in Downtown Biel; pattern after 1 h, adapted from Vamanu et al. (2010)

Fig. 15.12 A possible pattern in a vulnerability-oriented exercise in TIC, adapted from Vamanu et al. (2010)

Preliminary experiments indicate that indeed such a correspondence is feasible. The Karlsruhe-Julich, Klug, Brookhaven, and St. Louis correlations and dispersion coefficients were tested in relation with different n and r, and a systematic is

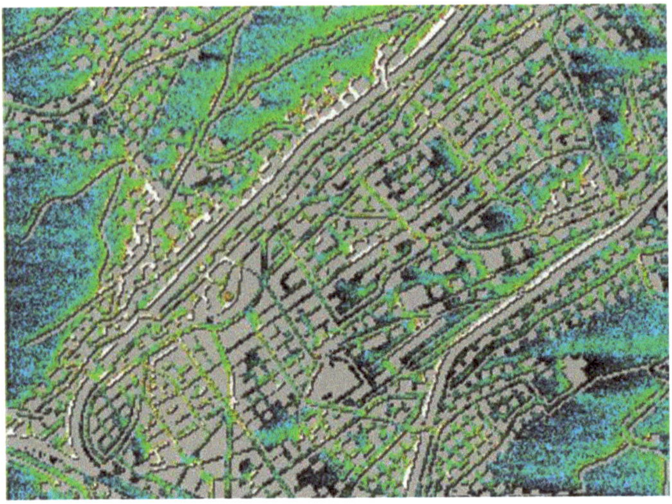

Fig. 15.13 A possible pattern in a vulnerability-oriented exercise in MLV, adapted from Vamanu et al. (2010)

coalescing, to be reported at later time. As an indication of the sensitivity of the model to *r*, the following Figs. 15.14, 15.15, 15.16, and 15.17) render the time-integrated concentration pattern for a release in a flat (no obstacle) terrain, for the same n and r at 20, 10, 5, and 1 m, on a CA grid scaled at 1 m/pixel.

The list of issues pending further investigation also includes, among others, model and implementation-related. Among the model-related is:

- A better accommodation into the model of the wind velocity in relation to urban map 'graphical rugosity'—the distribution of building cross-sections at different scales, so that a correlation to the Reynolds number (Sommerfeld 1908) becomes feasible;
- Exploration of the feasibility of an explicit implementation into the model of the 3rd dimension—the vertical. It is essential to note that the 2-D model described, thus far, offers a 3-rd dimension in an implicit fashion. This is done via a rugosity parameter of a length dimension that would feature the average horizontal cross-section of the building agglomeration. The time step of the advection is set so as to help flow cover, one such characteristic length at each leap forward. In this way, one obtains a kind of 'building tunneling' effect, in the sense given to the word in Quantum Mechanics, which allows some particles to randomly escape being cornered forever in building angles having sharp upwind apertures.
- CA-wise rule optimization, the most urgently needed feature being perhaps a (set of) clause(s) to take into account the actual elevation of the terrain holding the buildings.

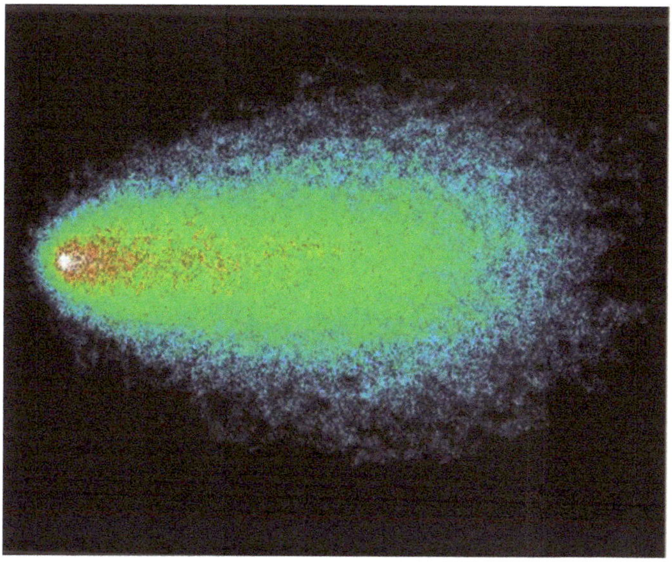

Fig. 15.14 Flat land TIC for r = 20 m, adapted from Vamanu et al. (2010)

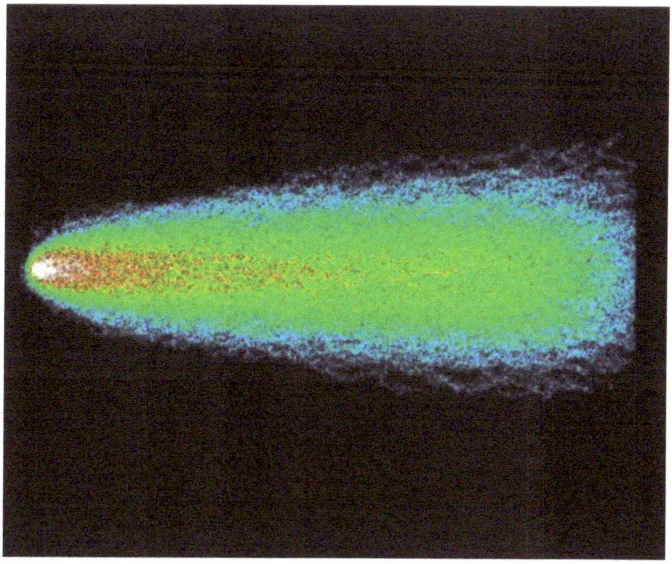

Fig. 15.15 Flat land TIC for r = 10 m, adapted from Vamanu et al. (2010)

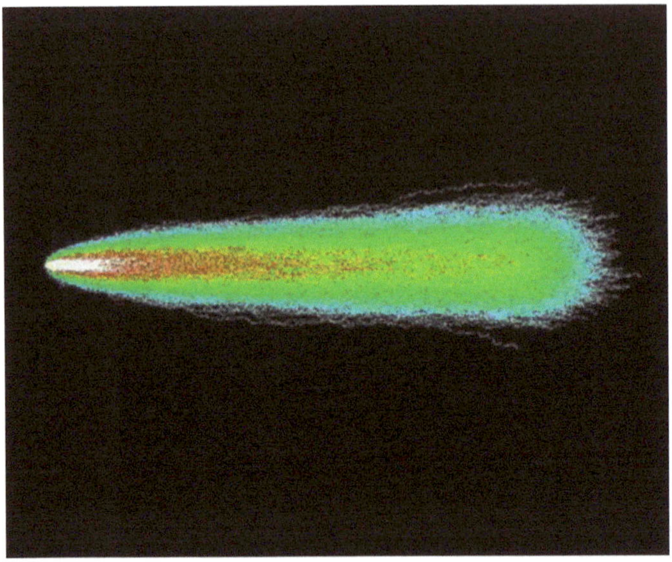

Fig. 15.16 Flat land TIC for r = 5 m, adapted from Vamanu et al. (2010)

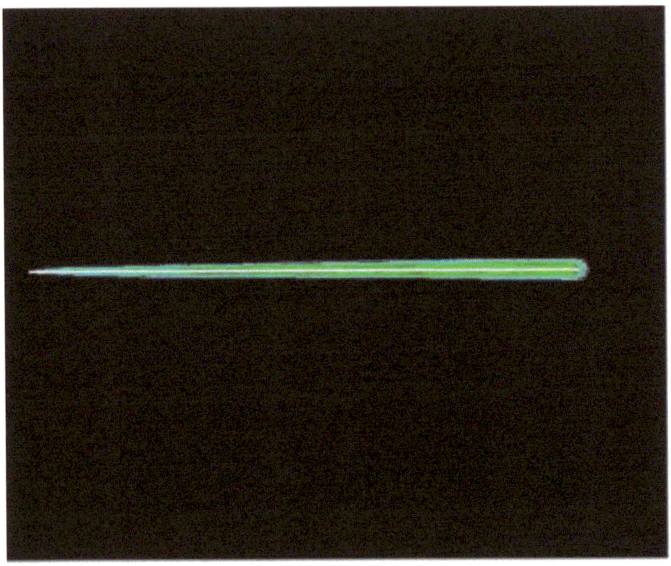

Fig. 15.17 Flat land TIC for r = 1 m, adapted from Vamanu et al. (2010)

- Exploration of the feasibility to apply Chorin's 'vortex blob' approach to 'true' urban agglomerations.

Present authors, for *the purposes of implementation,* suggest improved and more diverse visualization techniques, including color filtering and averaging to reveal hidden regularities in the flows, alpha-blending, and perhaps Direct 3D and other graphical capabilities.

15.6 Remarks

A rule-based computational model of air flows in urban areas is given, based on a physical interpretation of the air as a virtually in viscid, and slightly compressible, fluid. The model draws upon the CA philosophy that assimilates system dynamics to a stepwise, local, homogeneous, and parallel updating of the knot states in a space-grid underlying the allowed freedom of motion of the system. The model construction follows a practical, risk-and-vulnerability-assessment-oriented drive, targeting the simplest rules that may reveal the truths of clear-cut operational value behind an otherwise intrinsically complex physical reality.

Therefore, the chief targets of the investigations supported by the model and the ensuing code are as follows: (i) the 'go/no-go areas' in the event of an accidental or malicious release of hazardous chemicals and radioactive substances such as urban areas—which is a risk-wise assessment approach, and (ii) overall pattern of the most exposed areas in a city, following from the local, multi-annual meteorological statistics—which is a vulnerability-wise assessment approach.

It is important to remember that the associated quantitative information obtained in the process is thought of as indicative in nature, and not essentially affecting the relevance of the qualitative assessment described. Returning to what was presented at the onset of this chapter, it turns out that the work was conveniently motivated by the US Department for Homeland Security project of a comprehensive field experiment with potential pollutants dispersion in Manhattan, New York, conducted in conjunction with varied R&D federal units, academia, and the New York City Hall. In this case and other similar cases, the evidence produced suggests that computational models of the kind, particularly when brought down to PC operability, may prove to be useful in complementing far more elaborated and analytical models and approaches of the traditional Fluid Dynamics. The model confirms in its own right an observation by one experienced author in the field; 'On a small scale, the particle motions appear random. But on a large scale, there is evidence that their average motion corresponds to that expected from a fluid which obeys the usual Navier-Stokes partial differential equations' (Wolfram 1988, p. 90). Moreover, persisting in developing rule-based computational models of fluids may also be warranted by another remark of Wolfram (1988) who suggested that 'some evidence for this comes from the fact that most fluid computations yield results which are accurate to at most the percent level.'

References

Chorin, A. J. (1973). Numerical study of slightly viscous flow. *Journal of Fluid Mechanics, 57*(4), 785–796.

Gheorghe, A. V., & Vamanu, D. V. (2005). A cellular automaton approach to air flow dispersion in urban areas. In G. M. Cojazzi (Ed.), *Proceedings on Systems Analysis for a more Secure World: Application of System Analysis and REMS to Security of Complex Systems* (pp. 369–383). Ispra: European Commission's Joint Research Center.

Glaser, B. G., & Strauss, A. L. (1967). *The discovery of grounded theory: Strategies for qualitative research*. New York, NY: Aldine de Gruyter.

Porthouse, D. T. C., & Lewis, R. I. (1981). Simulation of viscous diffusion for extension of the surface vorticity method to boundary layer and separated flows. *Journal Mechanical Engineering Science, 23*(3), 157–167.

Rucker, R., & Walker, J. (2017). Cellular automata laboratory. Retrieved February 2, 2017, from https://www.fourmilab.ch/cellab/.

Sommerfeld, A. (1908). Ein Beitrag zur hydrodynamischen Erkläerung der turbulenten Flüssigkeitsbewegüngen (A contribution to hydrodynamic explanation of turbulent fluid motions). *International Congress of Mathematicians, 3,* 116–124.

Urbina, I. (2005, February 11). Antiterror test to follow winds and determine airborne paths. *The New York Times*. New York. Retrieved from http://www.nytimes.com/2005/02/11/nyregion/antiterror-test-to-follow-winds-and-determine-airborne-paths.html.

Vamanu, B. I., Gheorghe, A. V., & Katina, P. F. (2016). *Critical infrastructures: Risk and vulnerability assessment in transportation of dangerous goods—Transportation by road and rail* (Vol. 31). Cham, Switzerland: Springer International Publishing.

Vamanu, D. V., Gheorghe, A. V., & Acasandrei, V. T. (2010). Urban vulnerabilities: Think simple. *International Journal of Critical Infrastructures, 6*(3), 304–325.

Wolfram, S. (1983). Statistical mechanics of cellular automata. *Reviews of Modern Physics, 55*(3), 601–644.

Wolfram, S. (1988). *Cellular automaton supercomputing* (No. P/86/9/138) (pp. 87–95). Champaign, IL: University of Illinois at Urbana-Champaign. Retrieved from http://www.stephenwolfram.com/publications/articles/ca/88-cellular/index.html.

Chapter 16
Vulnerability of a Regional Economy in a Global Competition

Abstract In this chapter, a system resilience profile for Germany is introduced. Germany is introduced as a *critical infrastructures ecosystem* of living system of systems landscape using published key information from the German government and the European Union. The produced profile considers Germany in the context of Europe as well as the emerging issues of BREXIT.

16.1 Germany: System Resilience Governance Profile

To introduce the power of a system resilience governance profile, the published key information from the German government and the European Union was partially reused. The focus was on Germany's critical infrastructures ecosystem, because, allegedly, it is the strongest economy in Europe. Figure 16.1 is the profile introducing Germany as *critical infrastructures ecosystem* of a living system of systems landscape (Federal Ministry for Economic Affairs 2010; Federal Ministry for Economic Affairs 2015). The living system of systems Germany is alternatively called—*The Price for Globalization*. If this profile comes under pressure, there will be impacts on the entire system of Europe. 'The Price for Globalization' was chosen as the title of this picture to introduce the systemic dependencies and to show in a simple way who has to pay for all the commitments that politicians make every single day.

If we consider the current vulnerability of Germany, on a high level in a very dynamic and depending economy, we end up with a complex picture. To visualize this kind of complexity, a risk dependency map shows the collaborations, contributions, commitments, and involvement from Germany's point of view on an aggregated level where essential details are still visible. The map is not intended to be comprehensive but covers the currently most important risks and their dependencies. Some of the important figures are product and industry mix, ecosystem dependencies, innovation capabilities (new start-ups, published patents), short- and long-term liabilities, gross domestic product, unemployment rate, and public debt to measure solvability.

A.V. Gheorghe et al., *Critical Infrastructures, Key Resources, Key Assets*, Topics in Safety, Risk, Reliability and Quality 34,
https://doi.org/10.1007/978-3-319-69224-1_16

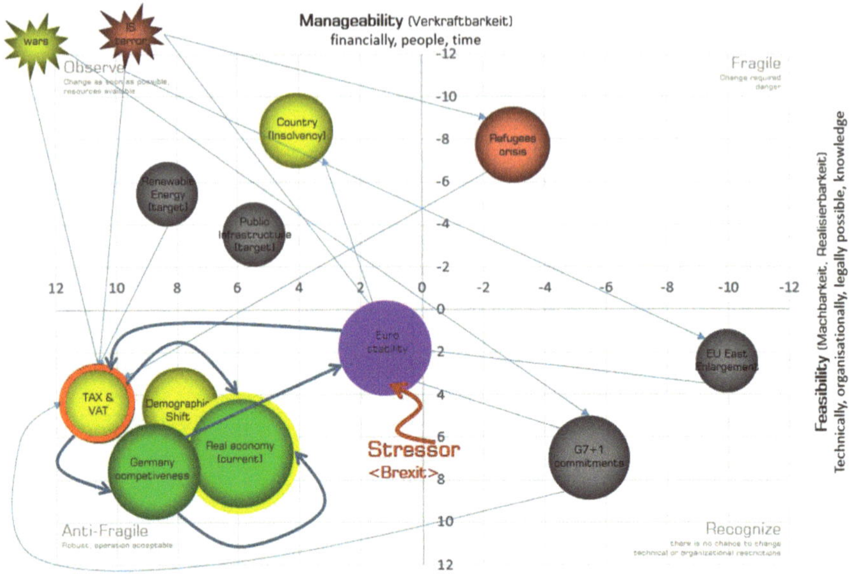

Fig. 16.1 System resilience governance profile Germany

This commitments, collaborations, and contributions can be summarized in groups. Group [A] is introducing some subjects related to the European Union (EU) to highlight EU reform, the Greek and Italy instability, the forthcoming BREXIT and the unknown consequences on both sides, and the emphasis on the solvability of EU member states. [B] summarizing war and terror risks that permanently influence peace, freedom, safety, and notion of democracy, and how the result of human conflicts has created waves of refugees. Regarding risk predictability with this human tragedy, it is simple to see why one must address risk in order to see the difficulty in risk mitigation and why it is important the understand consequences.

One might even speculate that after terror attacks on Berlin, Nice, and Paris, nothing will be the same as before: Risk perception has shifted. Last but not least, unplanned and completely unknown impact of new government administrations. The entire risk landscape, risk assessment, and mitigation strategy will or have to be changed. Therefore, it is important to use an advanced tool like the system resilience governance profile to understand and facing challenges as a whole and not as individual parts. [C] is clustering the unbalanced multi-billion EURO commitments without an action plan. Most of the systems involved are unknown; they take the form of 'system of systems' with conflicting goals; manageability and feasibility are not transparent or balanced; and governance is not defined, and yet, it goes without mentioning, challenges associated with multiple religions, cultures, languages, political systems, time zones, focus, and the values. Without a system resilience governance profile, it can only be said that the problem domain will remain unaddressed, at best we end up with lack of understanding the system and at worse create more risks and vulnerabilities for the common man. [D] is a summary of unsolved local issues, like sustainable social security, health systems, demographic shifts, the shrinking middle class, the constant pressure of and over-regulation in the financial service sector, and inadequate public infrastructure, although megacity trends are generally known. The circle [E] shows who has to pay for all the unmanaged commitments.

16.2 Critical Infrastructures Resilience Governance Profile Germany 2015

Figure 16.1 shows the important and relevancy of visualization regarding system of systems of a critical infrastructure for the purposes of system resilience governance. Terrorism and wars are shown as stressors that could permanently alter system resilience governance profile. For the sake of argument, think of the number of times the responsibility concerning, for example, refugees has shifted and continued to shift. At the same time, a joint assessment does not exist, and, at present, the problem is being 'solved' by Turkey with funds from the EU. By some accounts, the problem is not being resolved. And some might argue, the situation is a ticking time bomb and perhaps more bleak, could turn into a human tragedy. Simultaneously, country-specific persists along with the increasing cost to the EU. But the focus has been shifted to the refugee crisis, and specific country problem is presently out of scope—this is shown in Fig. 16.1 as a shift from the fragile to the observe quadrant.

Permanent change and movement can be recognized between individual quadrants. While the system is clear, feasibility and manageability are assessed and under control; governance is clearly defined; the entire profile with all the systems can be kept stable; and the risks permanently reduced or managed by appropriate measures.

In some systems, the management program is not transparent. It is not clear whether there are enough resources, money, time, and knowledge than available from the required technology (gray bubbles). Also, the governance is not clearly defined. Moreover, in the anti-fragile quadrant, assessment systems are not managed because they are not easy to manage, just like the real economy or the demographic shift. As mentioned, the color of each bubble depends on the system context and the applied, assessed, and monitored risks. The dependency between individual system contexts comes from the aggregated system of systems connectivity or risk dependencies as well as the direction strength and impact. The position of each system context is maintained by the applied, managed, and monitored resources from the roadmap (money, time, people, knowledge, skills, technology, etc.). The size is managed by governance and its susceptibility. In case of an additional stressor like the BREXIT, the fact that many dependencies are systemic, and for some anti-fragile systems, the resources are weaker, the system as a whole is highly unstable and requires high-attention to system resilience governance.

Notice too that the *System Resilience Governance Profile for Germany* shows a profile about the 'ecosystem Germany' in a specific point of time. However, the profile in itself is very dynamic and maintained by the researchers. However, the G7 + 1 commitments made in 2015 are not incorporated and validated because they have not yet been realized. On the one hand, there is a discussion running to enlarge the UE toward the east, while on the other side, countries are planning to leave the union.[1] Both scenarios are missing in the presented profile.

It has been suggested that Germany needs to invest heavily in its public infrastructures including, among others, railroad system, roads, bridges, and public buildings (Federal Ministry for Economic Affairs 2014). This influence also is not processed but will strongly influence the present profile. Furthermore, the energy transition is not fully considered. For this system context, the profile has to be adjusted because the relationships and dependencies are incredibly complex. The system context shown in the anti-fragile quadrant is stable and under control and properly managed but also strongly depends on each other. An essential statement that can be drawn from this profile that the German middle class and the citizen have to pay through TAX and VAT for all investments, commitments, and risks. But already it is clear that the demographic change will cost the taxpayer an incredible amount of money and is already not secured. With the strong tax burden of >40% income tax for individuals and >25% of legal entities, a VAT of 19% is practically no room available.

At this point, it is only fair to ask questions regarding current problems, the related systems, and the approaches being undertaken to understand and govern the different systems. For instance:

- How long before the system (the EU, its member nations) becomes fragile?
- Do we understand system dependencies and interdependencies?
- What are the effects of the ailing € and current zero interest-rate policy which the middle class and the individual taxpayer further unsettled?
- What about the solvability of Germany and the impact on the resilience?

These are, to say the least, tough questions. *However, if one thinks that these are tough questions, then perhaps we ought to wait and first experience the potential consequences of a fragile Europe with crumbling economies.* One need not wait. There are tools that could be used to visualize profiles at the national as well as the EU level. These could include perceptual shift triggered by terrorist, conflict zones, and wars such as the Greece 'insolvency' and the refugee crises. A part of this very dynamic and complex system of systems landscape, critical infrastructure is supporting continuous operation and improvement. Unfortunately, it appears that focus

[1]Think of the BREXIT: Withdrawal of the UK from the European Union. British electorate will address the question again on June 2016 in a referendum on the country's membership, following the passage of the European Union Referendum Act 2015.

is placed on the ‘dominant’ systems such as renewable energy and the rest of systems are left in isolation. A shift is needed toward a ‘system of systems’ approach for governance of critical systems especially since the seemingly isolated events can propagate through the networked interdependencies to cause potential failures (Calida and Katina 2012).

16.3 European Union Resilience Map 2016

The EU, with its multiple cultures, regions, religions, languages, and political systems, is a system with a complex and non-transparent, lethargic, and over-administrated governance. It has insufficient resources and is not managed as a whole, but rather as un-harmonized and manageable parts (individual member states). Not all member states are driven by contribution or are able to contribute because they are not properly managed or solvent.

Figure 16.2 represents a profile for the ecosystem of Germany for a specific point of time. The profile, dynamic in nature, shows the current situation calculated with a certain amount of parameter and visualized with a fuzzy index. The second

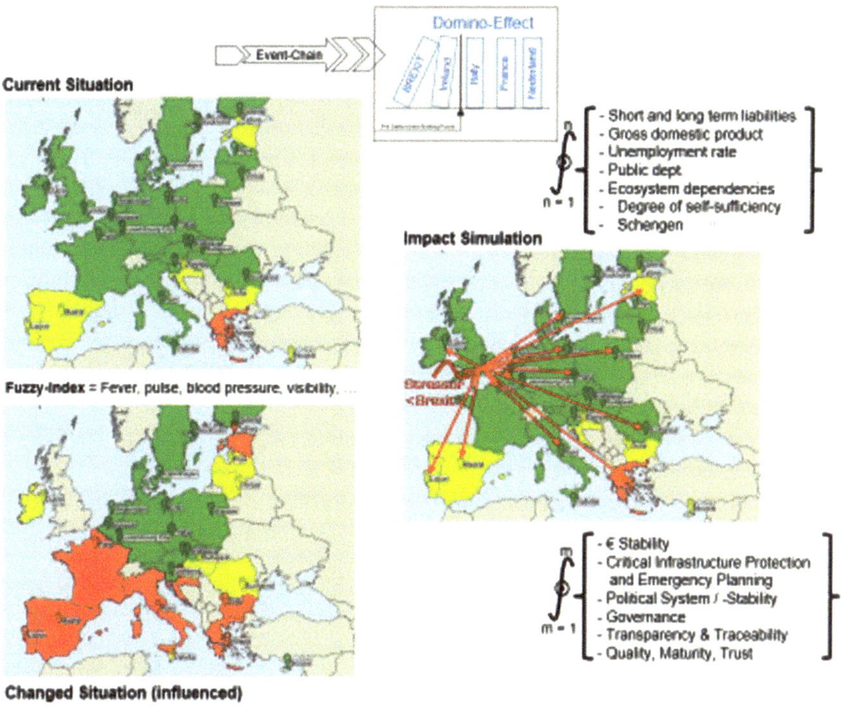

Fig. 16.2 System resilience governance profile visualized as a fuzzy index

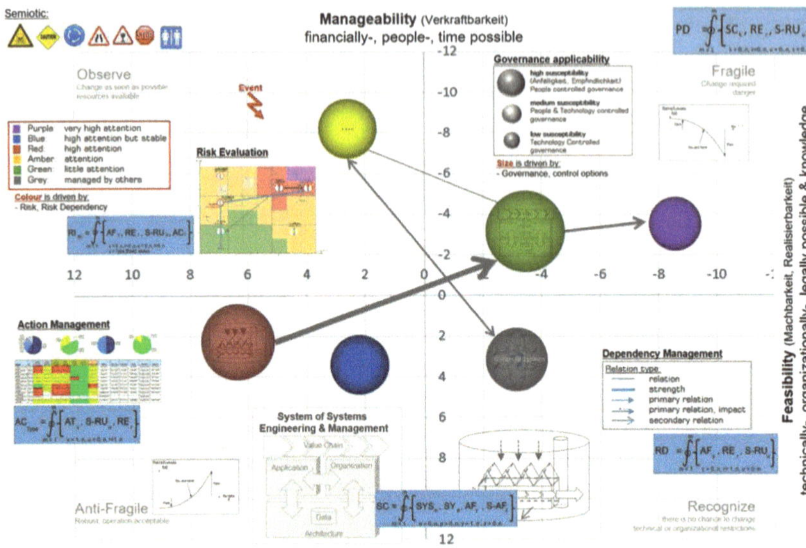

Fig. 16.3 System resilience governance profile

visualization indicates the BREXIT simulation scenario with the calculated impact on each country related based on 'dependency.' The third visualization shows a possible scenario after BREXIT with the visualized consequences to each country based on the calculated new fuzzy index. The fuzzy index is related to the system resilience profile of each country as previously specified in Fig. 16.3.

Again, Fig. 16.2 shows a resilience map for European Union member states in different colors. The color shows the current aggregated system resilience value. Depending on benchmark rules and condition, resilience is evaluated, and the results are visualized in green (fulfilled), amber (incomplete), and red (undefined). In this, the measure of resilience is associated with short- and long-term liabilities, gross domestic product (GDP), unemployment rate, public departments, product-mix, ecosystem dependencies, innovation capabilities (start-up, patent), and degree of self-sufficiency. Further discussions on resilience can be found elsewhere (Gheorghe and Katina 2014).

This rating shows the estimated expected value in case of BREXIT. The map is 100% generated and depends on data quality. This quality involves more than one source, maturity (i.e., repeatable, loaded, and controlled), trust (i.e., substantiated, confirmed, and validated by structures and fact and almost no assumptions). Of course, one must keep in mind that history always the benefit of hindsight, and therefore it is always 20/20. On the other hand, future, because the forecast depends on belief systems, hypotheses, and statistically evaluated numbers, there is always a level of uncertainty. In other words, one can validate historical data, and confirmation is possible, unlike the future. Nonetheless, one can use historical data to a model and to modify the rules and conditions for the purposes of forecasting. With this in mind, historical data associated with official monitoring reports (Dhar 1979; Federal Ministry for Economic Affairs 2013; Federal Ministry for Economic Affairs 2014) were used in the generation of Fig. 16.3 as well as the insights into what could be done to address possible impacts.

In case of additional issues that are not properly managed or stressors, for instance, the BREXIT, the unstable system could collapse driven by domino effects. To the end, present researchers suggest the following:

EU member state should act as individual systems in a system of systems. The system of systems is the EU. This requires understanding basic cybernetic principles of integration of the whole and autonomy of the parts.

Again, the G7 + 1 commitments made in 2015 are not incorporated and validated because the impact has not been realized. This is especially critical some member nations might leave the union, while others are contemplating joining.

16.4 Vulnerability of a Financial System

Undoubtedly, within a system of systems landscape, which forms a country-level economy, there are dependencies within the different critical infrastructures. This is the basis for Fig. 16.4 which represents critical infrastructure for Germany on a certain level of detail and time. Every bubble represents a system (of systems) of critical infrastructure along with its key parties. This figure attempts to illustrate dependencies. Dependencies are relevant in such mappings due to possible effects associated with domino failures. In this case, dependencies (i.e., edges) might be more difficult to handle than system of systems (i.e., nodes) because there are minimum two participants with different observations and perceptions involved or concerned.[2]

The focus on the financial aspects placed in the forefront, for none other than the fact that, in many cases, without money, the world does not seem to operate.[3]

[2]Here is an interesting analogy: If we take it as a given that every second marriage ends in a divorce within 10 years in Europe and considering the 'relation problem,' then we can discern 'systems' and 'relationship.' Consider: if each partner only maintains his/her own body, then who cares for the relation? In terms of systems, this analogy could be used to explain why systems tend to be maintained (better) than relationships. One might even argue that cultivating relationships is getting harder, for instance, when considering the idea 'open-border approach.' This thinking has generated much debate along the lines of culture, religion, and language across European nations.

[3]Also, consider that, if money is missing, no one can pay for shelter or food. Also, credibility and rating of countries are largely based on the financials. Yet research suggests that people and society suggest a need to have the right balance in materialism, technical, social, nutritional, cognitive, spiritual, and environment (Kant 1991; Li 2013).

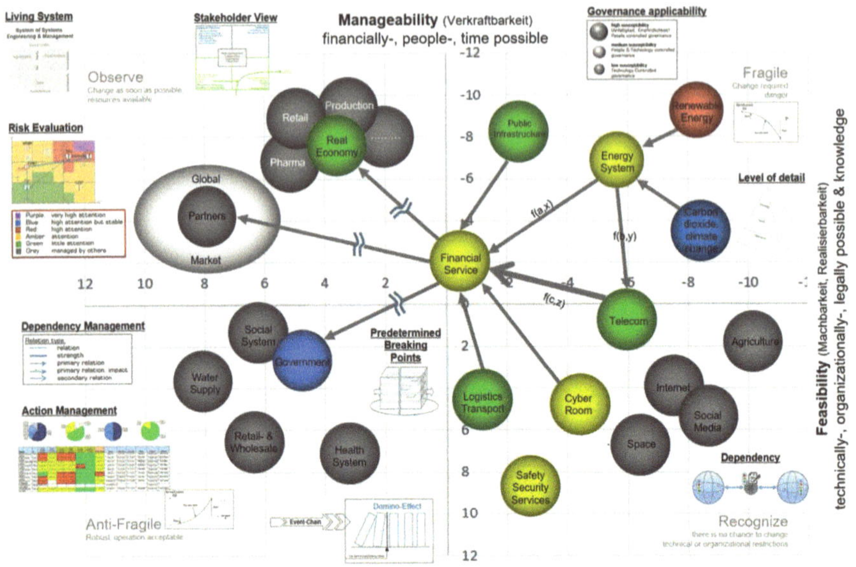

Fig. 16.4 System resilience governance profile—critical infrastructure

The financial service at this system resilience governance profile shown in Fig. 16.4 along with the essential dependencies has a current rating (color, size, position) amber. The color directly depends on details contained with the financials of the selected service industry. Proper modeling of the systems and its relations establishes the details that are manipulated to create aggregations at different levels of the system of interest, be it at the organization, country, or global scale. The expansion of the functions follows the standardized decomposition and aggregation rules for system (of systems) and the underpinning theories in complex systems.

The financial service is the central function of the current running economy and has a present amber risk rating. This risk rating is an aggregation of specific type of risks supported by a risk matrix (i.e., likelihood and impact) and a risk dependency map. The dependencies are based on a specific relation type which indicates strength of the dependency, the direction as well as the impact on the interconnected asset risks. Susceptibility is not worked out and has an acceptable technological and people controlled governance. The position on the map is on the fragile corner. There are many reasons, but consider, missing resources to manage change (manageability) or lack of availability option to change (feasibility).

The fragile corner of the system resilience governance profile indicates an area in which change is required. However, there might not be an option for change because of restrictions associated with technical, organizational, or legal. The maturity is indicated by the x-axis. The y-axis indicates that under the current circumstances, resources (time, money, and people) are not available and therefore, not manageable.

The anti-fragile corner displays system (of systems) where under the current circumstances are in control. This means they are running stable or in an acceptable region, from a feasibility and manageability vantage point, they are in operation. Nevertheless, also in this corner, a system can be under a high risk, a poor quality, and a low susceptibility. The rest of the corners indicate system (of systems) under observation due to lack of resources to fulfill and execute applied actions or recognized system (of systems) due to technical, organizational, or legal aspects of running system of interest. This means challenges cannot be addressed because applied actions cannot be executed at the moment.

The described system resilience governance profile is meant to represent a living system of systems landscape at a specific time and under specific circumstances (i.e., normal, complex situation, any other state). Therefore, how vulnerable a system of system is, is not context-free. It depends on different factors including, among others, attributes, dynamic, rules, and attitude. At the same time, it is possible to create a system of systems visualization for resilience governance profile from available data and then continuously improve the profile with automatically or semi-automatically generated data to improve maturity and quality for the various purposes including management. Authors suggest a period and permanent check to avoid misinterpretation based on wrong or not properly maintained visualizations that could be based on incorrect information.

Figure 16.4 also shows a snapshot of Germany's economy. The profile is developed based on ratings obtained from publicly available information sources published by the government, the European Union, and the World Economic Forum. The financial service system of systems landscape is representing the domestic and foreign banks, institutes, service providers, and payment service organization with an operation in Germany. Additionally, in this landscape are incorporated central bank and regulatory organization. The relation to partners in the cluster Global Market represents the systemic dependencies to other financial service organizations. The real economy represents all type of industry organization established on well-documented aggregation algorithms. Public infrastructure covers all aspect of government-owned elements (e.g., streets, bridges, air-, street-, and ship transports, buildings, and point of interest).

Energy system is the highest aggregation of all energy-relevant infrastructure and sources. The renewable energy system of systems context is summarizing all the infrastructure and sourcing elements to produce and deliver alternative produced energy. The carbon dioxide and climate change system (of systems) are a representation indicating all facts and figures associated with the cluster of interest. In this case, the position undertaken is that global warming and extreme weather events influence energy systems. There is also visible the massive influence of cyber systems to other system of systems. Perhaps this system for many readers is a surprise, but the massive threat of this not precisely possible to define system of systems is incredible.

Additionally, there is a visible triangle among finance, telecommunication, and energy. The three critical infrastructure participants require each other in a highly dependent form. Finally, there is a need to consider relation between government and the financial service and their influence on different layers. The consequence is definitely high for each if some system does not work properly or as expected. Thus, one could argue that the system resilience governance profile, as a leadership instrument, could be used to address many different systems. There are certainly many essential participants as well as their corresponding dependencies. In the present profile, these are many without any rating of impotency or urgency. Moreover, researcher notes that individual system (of systems) vulnerability rating can further be supported by sensitivity analyses or cause-effect simulation to support the message of their current system resilience governance rating.

16.5 Predetermined Breaking Point

Predetermined breaking point is a concept to analyze options to apply rules on explicit points in a model to protect a system for total damage or stop, reduce or avoid domino effects across a network of dependencies in a system of systems landscape. Authors submit that it is possible to arrive at a comprehensive description of a complex situation, if a complex situation is properly documented along with the creation of a risk dependency map and qualified risk scenarios. Such a documentation can be used to validate system behavior and to outline system capabilities. If loops are visible in the risk dependency maps, the system description can be used to identify or calculate predetermined breaking points to reduce system damage or to reduce fragility.

In addition, such a documentation could serve as a basis for protecting a system from damage or collapse through examination of predetermined breaking points in the form of stopping the cycle, loop, or spiral. As in the real world, definitions and the change of predetermined breaking points can also be applied to the system of systems engineering. This concept supports preventive install of algorithms for protection of system damage. The predetermined breaking points also support loops interruption so that no spirals effect violated the system as well as interruption of relations that are excessively large at predefined locations. The assumption is that such measures aid in system protection and thus aid in reduction of possible impacts.

16.6 Remarks

It is certainly possible to create a resilience profile for systems. Using basic publicly available data, a system can be modeled as a living system of systems along with a consideration of issues that could affect the system. A predetermined breaking point analyses could then be applied to every layer of the system and its dependencies to identify options. Predetermined breaking points depend on attribute, behavior, and capabilities of every single artifact. Since predetermined breaking points, as used in the model, can be used to create relations among different systems (e.g., financial service, real economy, partners, and government), it can be used as a basis to create protection measures to prevent systemic driven domino effects.

References

Calida, B. Y., & Katina, P. F. (2012). Regional industries as critical infrastructures: A tale of two modern cities. *International Journal of Critical Infrastructures, 8*(1), 74–90.

Dhar, S.B. (1979). Power system long-range decision analysis under fuzzy environment. *IEEE Transactions on Power Apparatus and Systems, PAS-98*(2), 585–596.

Federal Ministry for Economic Affairs (2010). *Energy efficiency—Made in Germany: Energy efficiency in industry, building service technology and transport.* München, Germany: Federal Ministry for Economic Affairs and Energy. Retrieved from http://www.efficiency-from-germany.info/ENEFF/Redaktion/EN/Downloads/Publikationen/energy_efficiency_made_in_germany.pdf?__blob=publicationFile&v=4.

Federal Ministry for Economic Affairs (2013). *Monitoring—Report: Digital economy 2013, Digitalization and the new working world.* München, Germany: Federal Ministry for Economic Affairs and Energy. Retrieved from http://ftp.zew.de/pub/zew-docs/gutachten/Monitoring_Report_2013_EN_Shortversion.pdf.

Federal Ministry for Economic Affairs. (2014). *Monitoring—Report: Digital economy 2014, ICT as innovation driver.* München, Germany: Federal Ministry for Economic Affairs and Energy. Retrieved from http://www.zew.de/en/publikationen/monitoring-report-digital-economy-2014-ict-as-innovation-driver/?cHash=361546c1ac3e8c9b39c2e5529047eba9.

Federal Ministry for Economic Affairs. (2015). *An electricity market for Germany's energy transition: White Paper by the Federal Ministry for Economic Affairs and Energy.* München, Germany: Federal Ministry for Economic Affairs and Energy. Retrieved from http://www.bmwi.de/Redaktion/EN/Publikationen/whitepaper-electricity-market.pdf?__blob=publicationFile&v=6.

Gheorghe, A. V., & Katina, P. F. (2014). Editorial: Resiliency and engineering systems—Research trends and challenges. *International Journal of Critical Infrastructures, 10*(3/4), 193–199.

Kant, I. (1991). *The metaphysics of morals.* (trans: Gregor, M.J.). Cambridge, UK: Cambridge University Press.

Li, J. (2013). The visible hand: From struggling survival to viable sustainability. In *Proceedings of the 56th Annual Meeting of the ISSS.* (pp. 1–19). San Jose, CA: International Society for the Systems Sciences. Retrieved from http://journals.isss.org/index.php/proceedings56th/article/view/1959.

Chapter 17
The Postface—Toward Space, Undersea, and Belowground Governance

Abstract The purpose of this chapter is to provide a forward-looking summary to critical space, undersea, and underground systems. Proposed areas of research at the methodology, epistemology, ontology, and nature of man are then presented.

17.1 A Summation

The present authors, thus far, have managed to illustrate a need for new and innovative approaches to immerging risks and opportunities in space, undersea, and underground by extending the concept of critical infrastructures. Emphasis has been placed on governance of critical and complex system (of systems) using several models grounded in physics and other fields involving concepts of risk, vulnerability, resilience, fragility, and perception to address risk, be it malicious, technical, or natural (Fig. 17.1).

However, at the most fundamental level, any rigorous research needs to establish a paradigm for which knowledge claims can be contrasted (Churchman 1968; Warfield 1976). Despite this claim, the literature suggests that there is not one widely accepted approach to knowledge claim (Burrell and Morgan 1979; Flood and Carson 1993). As it turns out, this is a discussion related to philosophy and certainly worth exploring given the current topic of CIKRKA and certainly not exclude is open bazaar of space, undersea, and belowground systems.

If one takes the view of Burrell and Morgan (1979) and extensions of Flood and Carson (1993), then the key issues are ontology, epistemology, methodology, and nature of human beings as they relate to knowledge. Ontology deals with how an observer views reality.

Epistemology deals with how one obtains and communicates knowledge. Nature of man deals with how man is described in relation to environment/systems. Methodology deals attempts to investigate and obtain knowledge in the world we find ourselves.

A.V. Gheorghe et al., *Critical Infrastructures, Key Resources, Key Assets*, Topics in Safety, Risk, Reliability and Quality 34,
https://doi.org/10.1007/978-3-319-69224-1_17

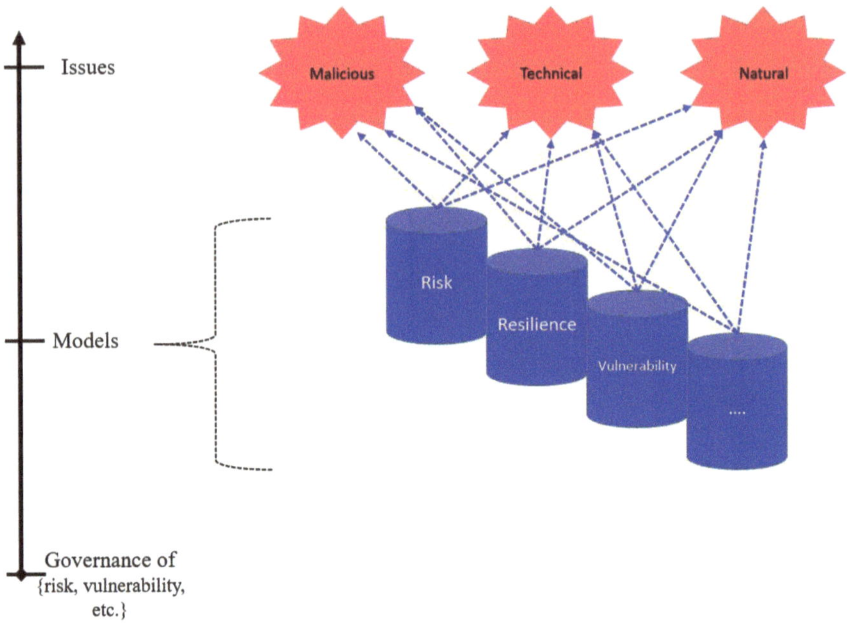

Fig. 17.1 Relating issues, models, and governance

17.1.1 Methodology

Jackson (1991) suggests that a methodology involves 'procedures for gaining knowledge about systems and structured processes involved in intervening in and changing systems' (p. 134). Following Burrell and Morgan (1979), methodological approaches can be categorized into two opposing extremes of *idiographic* and *nomothetic*. An idiographic view of a methodology supports subjectivity in research of complex systems as suggested by Flood and Carson (1993, p. 248):

> …the principal concern is to understand the way an individual creates, modifies, and interprets the world. The experiences are seen as unique and particular to the individual rather than general and universal. An external reality is questioned. An emphasis is placed on the relativistic nature of the world to such an extent that it may be perceived as not amenable to study using the ground rules of the natural sciences. Understanding can be obtained only by acquiring firsthand knowledge of the subject under investigation.

The opposing view of methodology—nomothetic—supports the traditional scientific method and its reductionist approach to address problematic issues

(Churchman 1968, 1971) and is described as (Flood and Carson 1993, pp. 247–248):

> …analyze relationships and regularities between the elements of which the world is com posed…identification of the elements and the way relationships can be expressed. The methodological issues are concepts themselves, their measurement, and identification of underlying themes. In essence, there is search for universal laws that govern the reality that is being observed. Methodologies are based on systematic process and technique.

There is no shortage of methodologies that can be used in intervening and changing systems. These include as Systems Analysis, Systems Engineering, Operations Research, Complex System Governance, Critical Systems Heuristics, Interactive Planning, Organizational Cybernetics, Organizational Learning, Sociotechnical Systems, Soft Systems Methodology, Strategic Assumption Surfacing and Testing, Systems Dynamics, Systems of Systems Engineering Methodology, and Total Systems Intervention.[1] Of interests are two important issues: Each methodological approach is developed and grounded in certain core conceptual foundations, and the 'selection of a method is based on the context of problematic situation and purpose of analysis' (Katina 2015). This begs the question, *what is the basis for methodologies for CIKRKA and the open bazaar of space, undersea, and belowground systems?* The response is rather simple:

> *A need to address pressing issues in critical infrastructures, space, undersea, and belowground, including and certainly not limited to risk, fragility, vulnerability, resilience, and perception. Current methodologies, as suggested in present research, are not sufficient in addressing present, emerging, and future issues.*

The suggested methodologies (e.g., QVA) are grounded in physics and other fields with proved utility, and thus, present research tends to lean toward nomothetic approach to methodology. However, the research also includes aspects of idiographic view of a methodology, in as much as the presented methodologies embrace subjectivity in research of critical and complex system (of systems) in space, undersea, and belowground. This is an issue well-addressed by the inclusion of 'risk perception' and 'safety culture' in different systems.

---

[1]The reader is directed elsewhere for proponents of these methodologies as well as notes on classifications, descriptions, advantages, and disadvantages (Jackson 2003; Katina 2015; Katina and Calida 2017).

17.1.2 Epistemology

An epistemological aspect of research deals with how a researcher (i.e., a system observer) begins to understand problematic situations and communicate knowledge to fellow researchers or observers. This dimension provides the form of knowledge, how knowledge is acquired, and what is considered to be 'true' or 'false' (Burrell and Morgan 1979). There are two opposite extremes of epistemology: positivism and anti-positivism. A ***positivistic*** approach to research indicates that 'knowledge is hard, real, and capable of being transmitted in a tangible form' (Flood and Carson 1993, p. 247). This stance of epistemology supports the idea that it is possible to 'explain and predict what happens in the social world by searching for regularities and causal relationships between its constituent elements…[and] that the growth of knowledge is essentially a cumulative process in which new insights are added to existing stock of knowledge and false hypotheses eliminated' (Burrell and Morgan 1979, p. 5).

In contrast, ***anti-positivism*** approach to research opposes positivism's view of knowledge as a hard, concrete, and tangible. This approach does not search for 'laws or underlying regularities in the social affairs…[but supports] that one can only 'understand' by occupying the frame of reference of the participant in action' (Burrell and Morgan 1979, p. 5). In anti-positivism, 'knowledge is soft, more subjective, spiritual, or even transcendental—based on experience, insight, and essentially of a personal nature' (Flood and Carson 1993, p. 247).

The scarcity of literature on the topic of space, undersea, and belowground in the context of critical infrastructures and the related concepts, for instance, fragility, render present research anti-positivistic in nature. Certainly, this is the case when '…people hold different views on (a) whether there is a problem [with space, undersea, and belowground systems], and if they agree there is, (b) what the problem [with such systems] is' (Vennix 1996, p. 13). That being said, the present textbook, itself, along with the research it contains is positivistic in nature since it contains 'knowledge that is hard, real, and capable of being transmitted in a tangible form.' (Flood and Carson 1993).

17.1.3 Ontology

An ontological aspect of research deals with the existence of entities and how such entities can be grouped based on similarities and differences. Moreover, ontology can also describe how 'an observer views the nature of reality or how concretely the external world might be understood' (Katina et al. 2014, p. 49). Two opposite extremes of ontology are realism and nominalism. Based on Burrell and Morgan (1979) and extrapolations from Flood and Carson (1993), *realism* is captured as 'external to the individual imposing itself on individual consciousness; it is a given "out there"' (p. 247). Realism suggests that reality is objective in nature.

On the other hand, *nominalism* describes reality as a product of individual consciousness. More significantly, nominalism ascribes to the assumption of individual cognition. Under nominalism, Burrell and Morgan (1979) note that 'the social world external to individual cognition is made up of nothing more than names, concepts and labels which are used to structure reality' (p. 4). The utility of 'concepts,' 'labels,' and 'names' is based on the convenience they offer as tools that can be used to make sense and describe reality (Flood and Carson 1993).

One might argue that present research fits more toward nominalistic view of the nature of reality in which present researchers have conceived the nature, development, and interpretation of issues in CIKRKA, space, undersea, and belowground as well as potential approach to address the related issues. And present researchers may tend to agree since the presented ideas are emerging. However, in as much as these ideas are partially dependent on cognition of observers, one must not be mistaken to assume that threats and risks in space, undersea, and belowground a mire fantasy of present researchers; recall *black swans* (see **Appendix F**)?

17.1.4 Nature of Human Beings

The final dimension of research consideration is the nature of human beings. This aspect is essential since it provides a stance on man and his activities in society. It has been suggested that two opposite extremes of *determinism* and *voluntarism* can describe the nature of human beings (Burrell and Morgan 1979; Flood and Carson 1993). A *deterministic* view of human beings suggests that a researcher views human beings as 'mechanistic, determined by situations in the external world; human beings and their experiences are products of their environment; they are conditioned by external circumstances' (Flood and Carson 1993, p. 247).

On the other hand, *voluntarism* one could take the view that human beings are 'completely autonomous and free-willed' (Burrell and Morgan 1979, p. 6) and that therefore they have a 'creative role [in their environment] and [can] create their environment' (Flood and Carson 1993, p. 247). Burrell and Morgan's (1979) research also indicates that to the extent that social theories are concerned with human activities, a theory must be disposed to either implicitly or explicitly to one of these viewpoints or an intermediate that can be used to address human activities.

At this point in this research, it should be evident that present authors took humans as being voluntaristic. They are endowed with the ability to do something regarding issues in critical infrastructures, key resources, and key assets as well as space, undersea, and belowground. In effect, they are responsible for the development of methodologies, methods, framework, models, and techniques that shape research and intervene the present and future humanity landscape. A summary of philosophical issues related to present research is provided in Fig. 17.2.

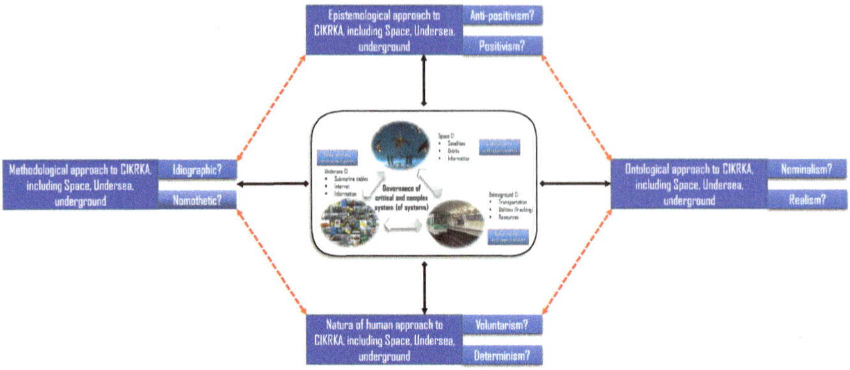

Fig. 17.2 Philosophical issues informing research into open bazaar of space, undersea, and belowground systems

17.2 Research Agenda

In the preface of the present text, we audacity to offer a thought commonly attributed to Albert Einstein: *we cannot solve our problems with the same thinking we used when we created them.* In the present text, we have attempted to offer a viewpoint, perhaps a misguided one, that could be used to address emerging issues in critical infrastructures, key resources, and key assets. More importantly, however, is the need to apply the presented ideas and models to advance the stated fields and push the boundaries of critical infrastructures into the open fields of space, undersea, and belowground.

Under these considerations, it is only at this point that one might realize that there are many more questions than answers. These questions ought to become (Table 17.1) the center of debates, discussions, and further research. This is essential if our society is to avoid the continued path and direction outlined in the outset of present research: *continuing moral decay, vulnerable infrastructure systems, susceptible and fragile space systems, and crumbling economies.*

Arguably, these research questions might be addressed at multiple levels including specific system of interests such as satellites, organizations, regions, state, and on a global scale. Additionally, these questions are meant to establish the value propositions for entities and organizations that might be involved and/or interested in space, undersea, and belowground critical systems, including addressing risk and the related topics.

There might be research interest in sustainability of space, undersea, and belowground systems. This might involve a need for implementation of policies, processes, and practices that meet the needs of the present without compromising the ability of future generations to meet their own needs. Certainly, attainment of sustainability involves understanding of systems in question, human and environment, and the complex interactions among such systems. And as suggested by John

Table 17.1 Suggested research areas for critical space, undersea, and belowground systems

Area of research	Suggested research issues
Methodology	• What methodological approaches could be undertaken to address emerging issues in space, undersea, and belowground as critical systems? • How can a methodology be classified with respect to the conceptual, philosophical, and theoretical grounding upon which it is based? • What guidance can be developed to establish the compatibility of a methodology for the context within which it will be deployed? • In addition to concepts of risk, vulnerability, resilience, fragility, and perception, what are other concepts that could be added to governance frameworks? • What measures might be developed for selection, design, execution, and evaluation of methodologies? • How can one quantify and visualize the suggested concepts of interest? • What is the form of 'governance' for space, undersea, and belowground critical systems?
Epistemology	• What are the necessary and sufficient conditions of knowledge in space, undersea, and belowground as critical systems? • What are the sources of such knowledge? What is the structure, and what limits of such knowledge? • What is the degree to which knowledge is temporally, contextually, or worldview dependent? • How can knowledge be classified such that enabling and constraining implications for thinking–decision–action manifestations can be examined? • What is the criteria for justified belief and what makes justified beliefs, justified? • Is such justification internal or external to one's own mind?
Ontology	• What is the ontology for open bazaar of critical space, undersea, and belowground systems? • What ontology can be developed for the engineering of critical space, undersea, and belowground systems? • How can ontological perspective be made explicit, represented, and factored into system development? • What approaches can be established to ensure the ontology is current with domain knowledge and term use? • To what degree does the ontological perspective influence system design, execution, development, and interpretation? • How can one provide sufficient specificity and concept coverage for the domain of interest, thus minimizing the content completeness problem? • What are the implications for observer independence versus observer dependence in attributions concerning of the nature of systems? • What can be done to ensure the ontology can support its use cases? • What is the role of emerging science-based engineering domains, for instance, cyber-physical systems (CPS) and Internet-of-Things (IoT) in addressing issues in the emerging domain of space–undersea–belowground critical systems?

(continued)

Table 17.1 (continued)

Area of research	Suggested research issues
Human nature	• What is the role of humans in engineering of space, undersea, and belowground systems? • How can the ideological diversity within groups be determined and impacts on productivity and creativity be assessed? • What are the implications of 'voluntarism' and 'determinism' on managing emerging issues including risks, threats, and opportunities in critical space, undersea, and belowground systems? • How can individual ideological preferences be identified and assessed with respect to their influence on system design/execution? • What approaches could be undertaken to address propensity and human inclinations for destructive behaviors and their perceptions? '*persons of interests*' anyone?

Casti, one of the viable approaches is 'loosening up the tightly bound interconnections…[and]…sustainability is a delicate balancing act calling upon us to remain on the narrow path between organizational and chaos, simplicity and complexity' (Casti 2012, p. 46).

Finally, there might be a need for consideration of the role of emerging science-based engineering developments, for instance, cyber-physical systems (CPS), Internet-of-Things (IoT), and modeling, simulation, and analysis (MS&A), in addressing issues in the emerging domain of space–undersea–belowground critical systems. For instance, serious gaming approach from MS&A could be used 'enables experts to obtain a thorough understanding of the complexity and interdependency of the system while offering a platform to experiment with various strategies and scenarios' (Ancel 2011). This would suggest a serious gaming approach for risk assessment, as a methodological approach to assist in space, undersea, and belowground infrastructure data elicitation and planning. Such considerations could be the basis for addressing potential cybersecurity issues, interdependencies, and monitoring and diagnostics of systems in question. There is certainly a variety of research questions associated with present research. These questions and emerging risks—man-made, technical, or natural—suggest a being on the verge of a need for 'sustainable system models' for the open bazaar of space, undersea, and belowground critical systems.

17.3 Final Remarks

Three related concepts are known: *critical infrastructures*—so vital and ubiquitous that their incapacity or destruction would not only affect the security and social welfare of any nation, but also cascade [send disruptive waves] across borders; *key resources*—publicly or privately controlled resources essential to the minimal operations of the economy and government; and *key assets*—'alone may not be vital to the continuity of critical services on a national scale, but an attack on any one of them could produce, in the worst case, significant loss of life and/or public health and safety consequences.' To this, we can add, more less, start considering a discussion onto risks, threats, and opportunities in *space, undersea*, and *below-ground*. This chapter suggests a need for development of conceptual foundations as well as accompanying methodologies, epistemologies, ontologies as well as the consideration of man. In this initial phase, researchers focused on the extreme ends of knowledge claims, and this should not be taken to suggest that 'there is no need' for the middle or hybrid approaches. That could not be further from the intended purpose.

References

Ancel, E. (2011). *A systemic approach to next generation infrastructure data elicitation and planning using serious gaming methods* (Ph.D.). Old Dominion University, United States—Virginia.

Burrell, G., & Morgan, G. (1979). *Sociological paradigms and organisational analysis*. Burlington, VT: Ashgate Publishing.

Casti, J. (2012). *X-Events: Complexity overload and the collapse of everything*. New York, NY: William Morrow.

Churchman, C. W. (1968). *Challenge to reason*. New York, NY: McGraw-Hill.

Churchman, C. W. (1971). *The design of inquiring systems*. New York, NY: Basic Books.

Flood, R. L., & Carson, E. R. (1993). *Dealing with complexity: An introduction to the theory and application of systems science*. New York: Plenum Press.

Jackson, M. C. (1991). *Systems methodology for the management sciences*. New York, NY: Plenum Press.

Jackson, M. C. (2003). *Systems thinking: Creative holism for managers*. Chichester, UK: Wiley.

Katina, P. F. (2015). *Systems theory-based construct for identifying metasystem pathologies for complex system governance* (Ph.D.). Old Dominion University, United States—Virginia.

Katina, P. F., & Calida, B. Y. (2017). Complex system analysis for engineering of systemic failures. In M. Hopkins (Ed.), *Systems engineering: Concepts, tools and applications* (pp. 105–132). New York: NY: Nova Science Publishers.

Katina, P. F., Keating, C. B., & Jaradat, R. M. (2014). System requirements engineering in complex situations. *Requirements engineering, 19*(1), 45–62.

Vennix, J. (1996). *Group model building: Facilitating team learning using system dynamics* (1st ed.). Chichester, UK: Wiley.

Warfield, J. N. (1976). *Societal systems: Planning, policy and complexity*. New York, NY: Wiley-Interscience.

Appendices

This section was developed for inclusion of selective technical aspects and detailed aspects of several works presented in present book. The reader is invited to consult this section for description of terms and concepts used in the present book. The reader can also use this appendix as a reference for research efforts related to the presented terms and concepts.

Appendix A: Hierarchical Holographic Vulnerability Assessment

The complexity of critical infrastructures and its related security issues calls for a holistic approach. Various modeling techniques of complex systems are in rapid development and facilitate vulnerability assessment and management with good theoretical and practical foundation. The present appendix places emphasis on Hierarchical Holographic Modeling (HHM), which is used in stepwise approach within the framework of parsing the vulnerability concept, hazards and accident scenarios identification, and vulnerability management. The proposed framework can serve as generic vulnerability assessment platform and leaves the potential to be further developed with application cases.

Methodology and Scientific Contribution

Song (2005) provides a clear-cut differentiation of risk and vulnerability as suggested explicitly in Fig. A.1. The following summary is provided for amplification:

- Risk and vulnerability are all hazard-oriented concepts.
- Hazards are multiform.
- Due to the multifaceted and hierarchical characteristics of systems (especially large-scaled complex systems), the associated vulnerabilities are diversified.
- Based on two contributors-hazard and vulnerability, risks are also multifarious.

A.V. Gheorghe et al., *Critical Infrastructures, Key Resources, Key Assets*, Topics in Safety, Risk, Reliability and Quality 34,
https://doi.org/10.1007/978-3-319-69224-1

- Without vulnerability study, risk assessment and management is incomplete and sometimes inaccurate (misleading).
- To assess and manage the security and survivability of systems, risk and vulnerability are two very close-linked study entities, but play the different roles, respectively.

Based on these noticeable points, one can conclude: (i) vulnerability assessment and management are necessary and important for system protection, (ii) risk and vulnerability study all need systematic and holistic approach, and (iii) risk analysis methodologies may be adjusted or modified to apply in vulnerability analysis.

Compared with vulnerability studies, risk studies are relatively mature. A number of methodologies and approaches have been developed and successfully used in this field. Particularly, in terms of systematic and holistic philosophy, a methodological framework RFRM (Risk Filtering, Ranking, and Management) for risk assessment and management has been developed (Haimes et al. 2002).

RFRM captures six risk assessment and management questions in a step-by-step procedure to identify, prioritize, and manage risks. It builds on HHM to identify the possible sources of risks, then filters and ranks these sources, and in risk management, HHM can be used to identify all possible risk management options (Haimes et al. 2004). Obviously, the core of RFRM framework is HHM, which reveals the multifarious nature of systems (especially large-scaled systems), provides a holographic view of a modeled system, so it is capable of identifying most, if not all, major source of risk and renders possibility of holistic, systematic system risk assessment and management.

The HHM is a holistic methodological approach aimed at capturing and representing the essence of inherent diverse characteristics and attributes of a system—its multiple aspects, perspectives, facets, views, dimensions, and hierarchies. Since vulnerabilities share the same nature with systems assessed, it is reasonable to use HHM in vulnerability study. And by analogy, the vulnerability assessment and management methodological framework (i.e., Hierarchical Holographic Vulnerability Assessment—HHVA) can be constructed as a scientific contribution comparing with risk assessment and management methodological framework. This comparison is illustrated in Fig. A.2.

Goal and the Overview of HHVA

The goals of the methodology framework are as follows:

1. better understand the system, its elements, and their interdependencies,
2. holistically identify hazards (threats) the system could expose to,
3. systematically point out and assess vulnerabilities,
4. develop policy options against these vulnerabilities,
5. and filter, ranking and recommend policy options

The overview of proposed new methodological framework for 'Hierarchical Holographic Vulnerability Assessment' is illustrated in Fig. A.3 and explained in the sequel.

Assessment Procedure

Step I: Parsing the vulnerability concept with HHM

Based on the definition of vulnerability concept and the method of HHM, it is possible to decompose vulnerability concept into two head topics and the related hierarchic subtopics Fig. A.4), namely analyzing and assessing vulnerability from two aspects: susceptibility and resilience, which comprise of several hierarchical components, respectively. This HHM outlines the consideration scope of vulnerability assessment.

Understandably, a system can always encounter various hazards (threats), and if a system is unable to 'handle' such hazards, unwanted harms or situations can arise in terms of losses, and we say then the system is vulnerable to the hazards. In another word, developing a clear understanding of hazards (threats) is a fundamental element of vulnerability assessment and management. It makes little sense to talk about a system's vulnerability without specifying the hazards to which it is vulnerable, since facing particular hazard various vulnerabilities manifest. Thus, it is important, at first step of HHVA, to identify all possible hazards (threats) that the system can expose to or can access to the system. To characterize these hazards (threats), their developing trends and the ways in which vulnerabilities are exploited should be conducted in this task. For simplification, in following context, we use only term hazard.

Step II: Identification of hazards and scenarios through hierarchical holographic modeling

Most, if not all, source of hazards can be identified through HHM methodology. Hazards can be formed from internal factors that are inside system boundary such as component failures and operator errors and external factors that are outside the system boundaries, such as terrorism and nature-triggering events. The overview here does not aim at giving a complete description of all possible hazards factors, but rather highlight some important types, see Fig. A.5.

It is constructive to identify the two basic structural components of HHM: head topic and subtopic. Head topic constitutes the major visions, views, concepts, perspectives, and decomposition, such as those depicted in Fig. A.5. These are the eight major perspectives related to hazards. They are structure, technical, interdependencies, geography, organization, management, temporal, and societal.

Second are the subtopics which provide a more detailed classification. Each subtopic class corresponds to a class of successive hazard, namely if the subtopic goes wrong, it will induce hazards. Central to HHM framework is the ability to branch out from each of the decompositions or considerations and explore the connectedness and ramifications within all other perspectives, and it also assumes an iterative approach to provide structure for identifying hazards. Thus, by its

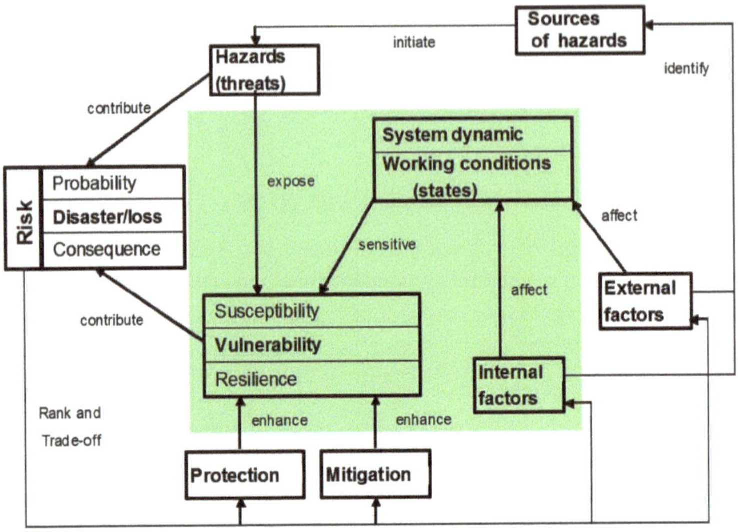

Fig. A.1 A model of hazards to a vulnerable system results in risks, adapted from Song (2005)

nature and construction of 'iterated cross-reference' among all decompositions, the HHM methodology identifies a comprehensive set of hazards and also provides a same thinking path to generate all possible scenarios, which is defined as a sequence of potential events.

An example related to the hazards induced by vision dependencies referencing to structure vision is depicted in Fig. A.6. For instance, through examination of the system interconnections and the consideration of their configurations simultaneously, the corresponding hazards can be recognized and identified in systems components, and the environment. A reverse example is shown in Fig. A.7. Variations in the hierarchical representation indicate some scenarios that hazards of failures or malfunctions of any infrastructures, systems, components, and configurations result in some accidental events which may through multifold interconnections (for further discussion on interdependencies, see Rinaldi et al. 2001; Katina et al. 2014), interconnection give rise to a chain of consequential events and simultaneously develop new hazards. The interfaces between the structure components are critical factors for the evolution of hazards.

Usually, the temporal vision will not be used as primary decomposition, but the reference of other vision (decomposition). It is important to involve temporal vision in the vulnerability assessment, because temporal vision articulates the change and evolution of system vulnerabilities. For example, in Fig. A.8, possible hazards to the system derived from organization factors, that is, *decision-making, policy, communication, finance*. Emergency response and recover, particularly, are related to temporal factors that need to be identified. In Fig. A.9, three hierarchical references are represented, which indicate various societal factors that can influence

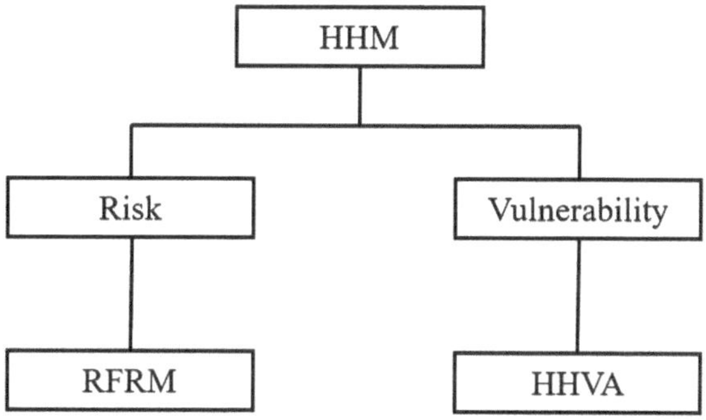

Fig. A.2 Based on HHM, comparing with risk, a systematic and holistic methodological framework can be developed for vulnerability, modified from Song (2005)

system vulnerability and imply multiple hazards related to these societal factors in technical vision from different structure components.

For instance, ‘human’ in ‘societal’ vision can be related to various economic, political, market, and individual reasons that could be used to suggest willful attacks to any components of the system structure. These hazards can affect physical, cyber, process, or functional multiple technical aspects of these components. Associated with the identification of hazards, the correspondent scenarios (i.e., sequence of potential events) can be generated. Additionally, for generating attack scenarios in detail, another HHM of modes of attack can be constructed.

Step III: Scenario filtering based on scope, consequence, and level of decision-making

Following Step II, the identified hazards can be overwhelming! Clearly, not all these hazards and corresponding scenarios can be of immediate and simultaneous concern to all levels of decision making and at all times.

At this step of scenario, filtering is needed. This filtering looks at hazard factors and filters according to the interests and responsibilities of specific decision-maker(s). The filtering criteria at this step include the decision-making level, the scope [i.e., what hazards are of prime important to the decision-maker(s)], and the consequences of scenarios [i.e., what situations are most unwanted to the decision-maker(s)]. The filtering in Step III is achieved on the bases of expert experience and knowledge of the nature, function, and operation of the system being studied and the role and responsibility of the decision-maker(s). For example, the American Petroleum Institute suggests that hazards could involve willful attacks. The correspondent factor of this hazard is human, including terrorists, activist, disgruntled employees, and criminals (API and NPRA 2003). In SINTEF Energy Research’s report (Doorman et al. 2004), three types of consequences (i.e., high price, curtailment, and blackout) are described as the unwanted situations and are the only ones related to hazards under consideration.

Step IV: Identification of vulnerable points through tracking scenarios

In this step, the remaining subset of scenarios are examined again. We know various vulnerabilities in the system facing multiform hazards generate diversified scenarios. Tracking the potential realization of each scenario various, vulnerable points can be identified. For this simulation is useful technology.

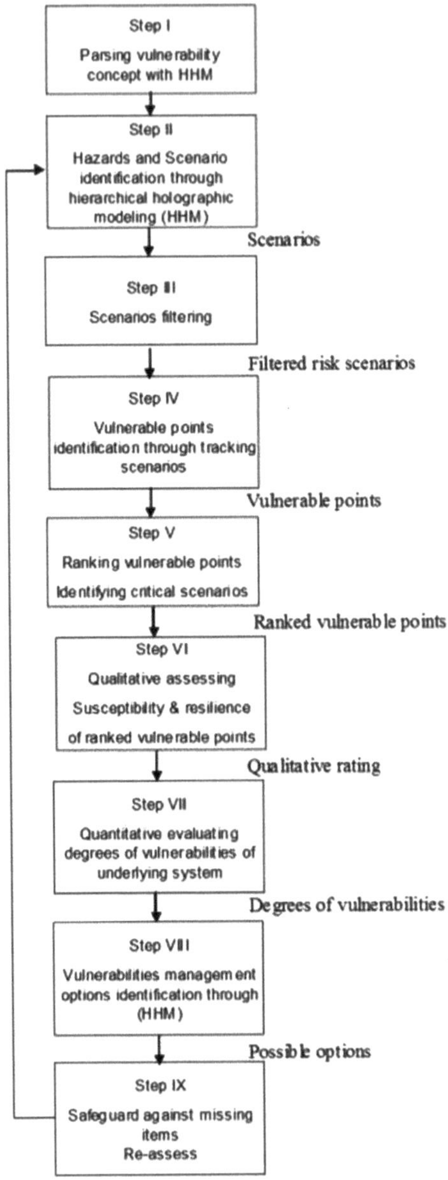

Fig. A.3 HHVA methodology, adapted from Song (2005)

For simplicity, a simplified tracking process is suggested as depicted in Fig. A.10. Layer 1 represents the 1-order vulnerable points which cause the initial accident events and forming the new hazards; layer 2 denotes the 2-order vulnerable points which facing new hazards induce the further accident events and hazards; other layers may be deduced by analogy. In this simplified case, the scenarios can

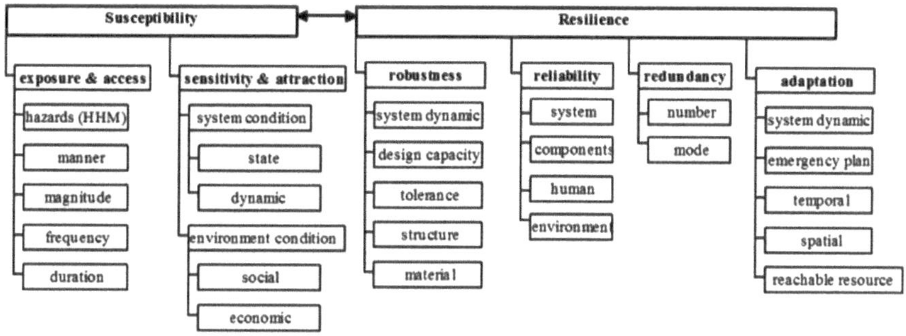

Fig. A.4 Parsing vulnerability of a system by HHM, adapted from Song (2005)

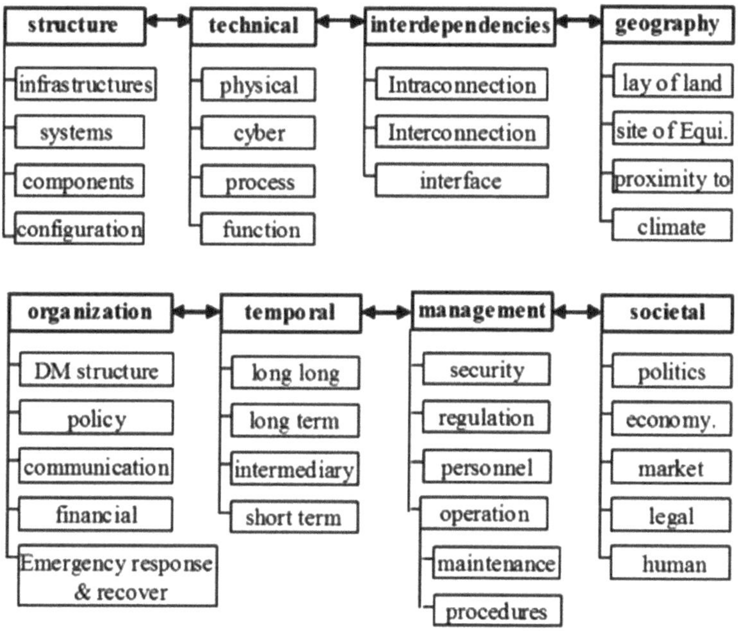

Fig. A.5 A high-level classification of hazards and scenarios for HHM, adapted from Song (2005)

through 2- or 3-order vulnerable points lead to three unwanted situations X, Y, and Z, e.g., ①-②-X, and ④-⑤-⑥-X-Y-Z.

Step V: Ranking the vulnerable points

Ranking the vulnerable points is mainly based on two criteria: consequence and likelihood, and other subsidiary criteria can be applied depending on the concrete applications. In this case, one can define consequence as a number of accident events (include unwanted situations) of a vulnerable point can lead to, and determine it by the number of output arrows from the circle of the vulnerable point (see Fig. A.10); likelihood as the chance of a vulnerable point induced accident event or frequency of a vulnerable point involving in the risk-scenarios, and is determined by the number of input arrows to the circle of the vulnerable point. The adopted methodology is based on the back-tracking philosophy of dynamic programming. Taking Fig. A.10 as an example, the ranking procedure is described as:

- Rank scenarios groups are *firstly* done in order to identify the vulnerable points which can directly lead to unwanted situations. These are depicted by circles labeled with number 2, 6, and 8 in Fig. A.10. *Secondly*, we compare the consequences that these vulnerable points can induced to rank these vulnerable points. It is obvious that the vulnerable point labeled with number 6 can cause three unwanted situations, which denotes with three output arrows to X, Y, Z, respectively, and the consequence is (X + Y + Z). Vulnerable point 6 is the most critical point. Due to the supposition, vulnerable points 2 and 8 bring the same consequences (X + Z), but by the likelihood criteria, it is easy to see that the probability of accident events happening through the vulnerable point 2 (three input arrows) is more than the vulnerable points 8 (one input arrows). Thus, the vulnerable point 2 is more critical than the vulnerable point 8.

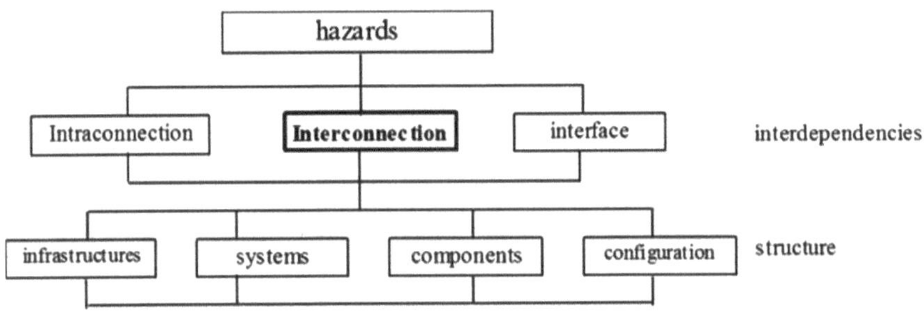

Fig. A.6 Interdependencies vision reference structure vision for identification of hazards, adapted from Song (2005)

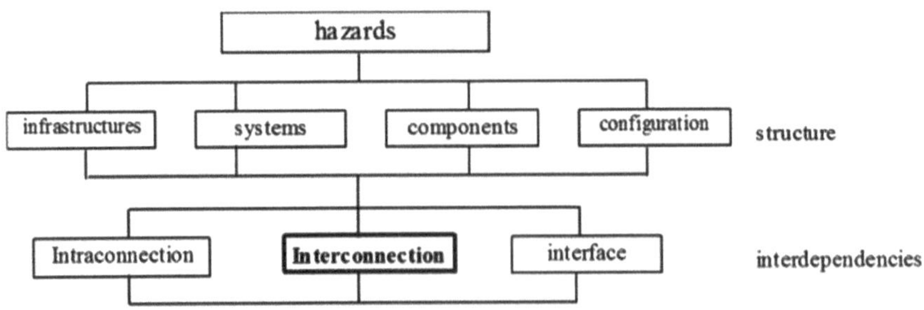

Fig. A.7 Structure vision reference interdependencies vision for identification of hazards, adapted from Song (2005)

Considering another case, if the vulnerable point 8 can cause unwanted situation not only Z but also Y (dot and dash line), based on consequence criteria, we can get a reversed ranking for 2 and 8. *Thirdly*, identify scenarios involving these direct vulnerable points, respectively, and rank the scenarios groups based on the sequence of the criticality of the concerned vulnerability points. For example, in Fig. A.10, scenarios group ①-⑥, ④-⑤-⑥, ⑦-⑤-⑥, are ranked at first position as most critical scenarios.

- Rank the vulnerable points to realize the critical scenarios based on likelihood criterion. For example, as suggested in Fig. A.10, vulnerable points ①, ⑤, ⑦ have the same likelihood, thus have the same ranking. To further rank these vulnerable points, one needs to compare the consequences they can induce as well as related factors of expensiveness of the equipment, temporal factor, or order of impact. For example, ①-⑥ only through two order impacts leads to unwanted situation, but ⑦-⑤-⑥ through three order impacts, thus, ① should be ranked higher than ⑦.

This ranking process offers a number of advantages, including:

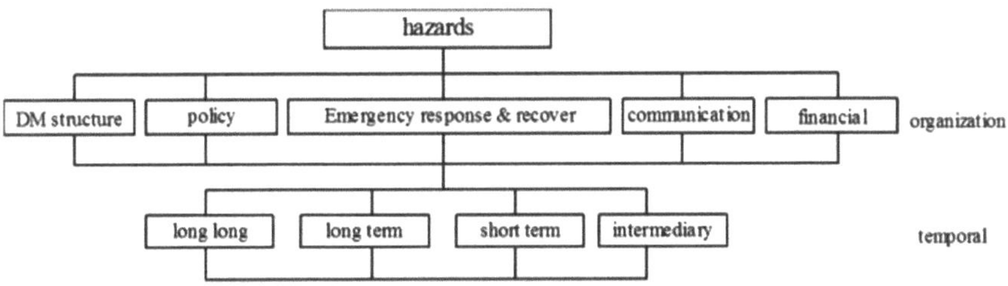

Fig. A.8 Organization vision reference temporal vision for identification of hazards, adapted from Song (2005)

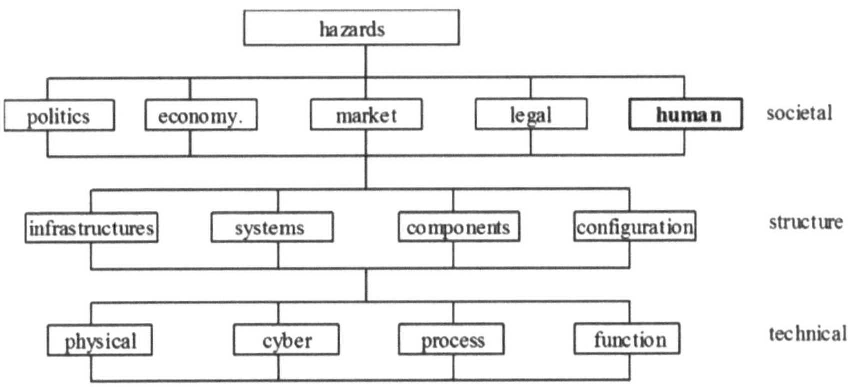

Fig. A.9 Societal vision reference structure and technical vision for identification of hazards, adapted from Song (2005)

1. Identifying most critical scenarios related to the concerned unwanted situations
2. Ranking the vulnerable points to provide a foundation for allocation of limited resource and establishment of mitigation plan
3. Determine the dominating vulnerabilities of the system for the assessment steps that follow

Step VI: Analyzing and assessing the susceptibility and resilience of the ranked vulnerable points

Sensitivity and attractiveness are a surrogate measure for likelihood of the happening of accident events. For sensitivity, one needs to examine if the vulnerable points are on critical stress under various internal or external unfavorable conditions. For attractiveness, one supposes that the vulnerable points are targets for the adversaries. Similarly, various system and environment conditions are examined to determine the targets' value from the adversary's perspective. These conditions include the operation state, dynamic of the underlying system, social, and the economic situations system of interest. To reflect the defensive properties of the underlying system, the resilience (i.e., as described in Fig. A.4), namely the subtopics robustness, reliability, redundancy, and adaptation, will be explored.

As an aid to this reflection, a set of criteria relating to the resilience can be generated from branches of the four subtopics in Fig. A.4. All existing vulnerability management capacities should be identified in this step. It may be helpful to rate the vulnerable points being examined as either '*high*,' '*medium,*' or '*low*' against each criterion and then to use this combination of rating to judge the ability of the vulnerable points to the hazards. These criteria are intended to be generally applicable but the user may of course modify them to suit the specific system under study.

The qualitative assessment of susceptibility and resilience of the vulnerable points involving in the filtered critical scenarios can be applied in the next step for quantitative assessment of the degree of vulnerabilities and the acceptability of the vulnerabilities, through different quantified criteria or mechanism in terms of the different requirement and the application in different models or approaches.

Up to this point, Step I through Step V, attempts have been made to respond to issues of 'what could be vulnerable? why are they vulnerable?' and identified, ranked, and qualitatively assessed the existed vulnerable points in the system. All these steps are guided by HHM of Fig. A.3. In next step, there becomes a need for information, knowledge, and results gained from previous steps to assess the degree of various system vulnerabilities quantitatively.

Step VII: Evaluating system vulnerabilities and the degree of vulnerabilities with multi-approaches

During the ranking process in Step V, we have ranked scenario groups and the related vulnerable points. By investigating the realization of high-ranked scenarios and corresponding vulnerable points, it is easy to understand at which aspects a

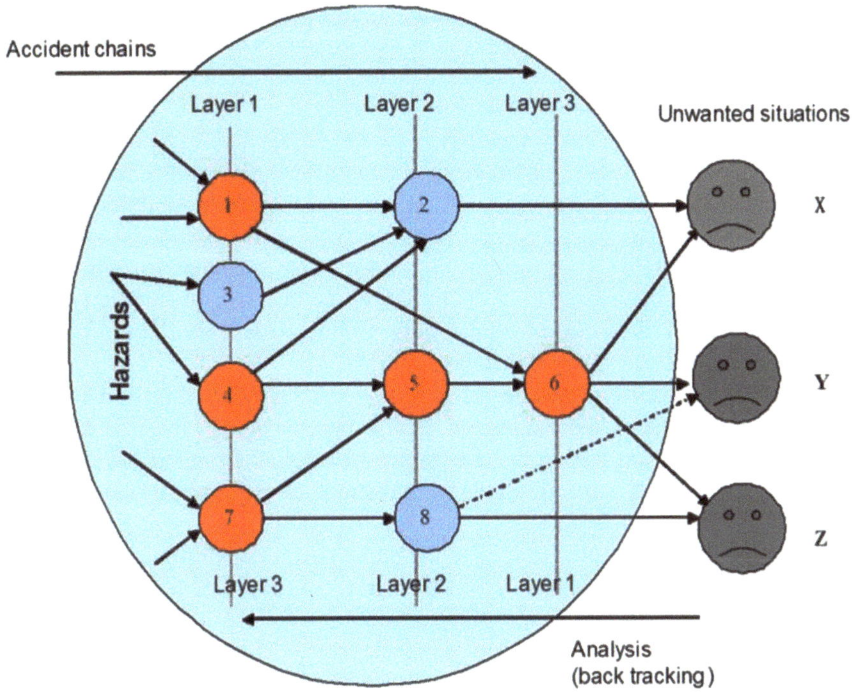

Fig. A.10 Vulnerable points identification through tracking scenarios and rank through back tracking, adapted from Song (2005)

system is most vulnerable to the related hazards (i.e., to determine what dominated vulnerabilities, which can lead to unwanted situations, the system has).

Because subsystems, components, and people along with the levels of organization and management in a large-scale system cannot share the same kind vulnerability, it is impractical to assess the vulnerability of a system only at certain facet and with same method or model. In author's context, we suppose a large-scale complex man-made system generally may have mainly 6 categories vulnerabilities (Table A.1):

To evaluate these six categories of vulnerability items, different model or method should be used specifically and complementarily. There is no shortage of vulnerability assessment methods (e.g., see Vamanu et al. 2016; Song 2005). However, none of these give a complete picture of large-scale system vulnerabilities profile or level. Thus, it is reasonable and valuable to integrate a number of these methods into the holistic vulnerability evaluation of a system, for solving the problem from multi-perspective.

Haimes (2008) suggests that HHM can have holistic dual representation. First is a holistic investigative paradigm. Second is a holistically, hierarchically, multiple objectively, mathematical methodology. Exploiting the inherent synergy of HHM's duality provides the necessary theoretical, methodological, and practical foundation for any large-scale systems. HHM provides multiple perspectives, or review, of a given problem, referred to as hierarchical holographic sub-models (HHS). Each perspective has its own unique qualities, issues, limitations, and factors that may require a particular approach to modeling and analysis.

Being consistent with this philosophy, a system vulnerabilities evaluation HHM is constructed in Fig. A.11. The corresponding ‘model and method list’ box for each category of vulnerability can be filled out continuously by multi-disciplinary researchers with time.

An example HHM is shown in Fig. A.12. Four quantitative vulnerability assessment model/methods (i.e., BOX1 through BOX4) are illustrations for the evaluation of the 6 categories of vulnerability. Correspondent brief introductions of these four methods are given as below:

BOX1: A method is proposed by to assess system vulnerability in terms of probabilities that the system undergoes significant changes, or even disintegrates under progressive stress (Gheorghe and Vamanu 2005). This method can be used in analysis and assessing ‘operation-induced vulnerability’ in aspect of the nonlinear dynamic cooperation processes.

In this method, the system is seen as the phase portrait of a critical process, and the process is modeled via a set of ordinary differential (or difference) equation. The dynamic variables y_i are nonlinear coupled, and the coupling intensity is expressed by coefficients, represented as control parameter or exchange constant.

For a designed (or expected) operational pattern, i.e., with a set of well-defined coefficients, implies certain phase portrait. Facing the hazards that the fluctuations of control parameters (subject to disturbance or engineered variations), the phase portrait may change its structure (fixed point, limit circle, strange attractor, unbounded state or chaos), which defines the vulnerability that the system has inadequate resilience to adapt to changes in hazards.

This method is based on nonlinear dynamics and chaos theory, and applies the simulation method to accomplish the system probability resilience analysis and assessment. Probability of keep same phase portrait indicates system topologic identity resilience; probability of keep bound state denotes system structure resilience (at the edge of system disintegration). Additionally, it is pointed out that system sensitivity to the changes of phase portrait depends on the position of original design in basin of state solution space. A generic software named BIZ has been developed to support the realization of this method.

BOX2: A theory-based indicator method to quantify the vulnerability of a complex multicomponent system, as well as to dynamically monitor the time evolvement of vulnerability as the indicators change (Gheorghe and Vamanu 2004b). The associated generic software is developed and named AIDRAM. Under this method, vulnerability is defined as system’s virtual openness to lose its design functions, and/or structural integrity, and/or identity under the combined interplay of two sets of factors: *U* and *V*. U represents risk-featuring factors/indicators which feature the risks that internal to the system. V represents management response-featuring factors which feature the capability of the system’s management to react/respond to internal developments within the system. Such factors feature the ambient in which the system evolves; they are mainly external.

System's measurable/monitored indicators (parameters) are aggregated basing on the fuzzy theory to form the two control variables U and V are, respectively:

$$U(X_1, X_2 \ldots X_n) = \min\left\{1, \left(X_1^P + X_2^P + \cdots + X_n^P\right)^{\frac{1}{P}}\right\}$$

where,

X_i are the normalized indicators obtained from the physical indicators Y_i as:

$X_i = A \log_{10}(Y_i) + B$, $i = 1, 2, \ldots, n$. A similar set of equations would give $V(X_1, X_2, \ldots, X_n)$.

To model system state is inspired by reference frameworks in classical Statistical Physics such as the Bragg–Williams approximation of the Ising model, feed from the alternative interpretations by Thom and Zeeman, of the Stability problem in Systems Theory (see Vamanu et al. 2016). The obtained state equation is:

$$th\left(\frac{(U * \zeta + V)}{\theta}\right) = 2\zeta$$

where U and V are two state variables, and ζ is membership fraction which denotes the ratio of state of system members. θ is a temperature parameter which represent the proper degree of choice of system indicators from 'expert judgment.'

The characteristic's topological foil of the system state equation reflects catastrophe theory, and its planar projection defines the U-V operational region (U >=0, V >=0). In ***BOX2***, the lowest picture shows this U-V plane; red triangle indicates the system unstable, thereby featuring a high vulnerability; the yellow region indicates the critical unstable vulnerability; the white region indicates the system stable, thereby featuring low vulnerability; the boundary line between red and yellow region is named cusp line; the system vulnerability index or degree is defined as;

$$V_{\text{scale}} = 100\left(\frac{1 - D}{15}\right)$$

where D is the distance of the system state (U, V) in the operation region to the cusp line, the range is from 0 to 100.

This method/model can be used in HHVA to analyze and assess the vulnerabilities induced by structure stability, management, organization, respectively, or their compound. The vulnerable points involving in the scenarios of corresponding vulnerability categories can be chosen as indictors, and their quantization is based on the result of qualitative assessment in Step VI.

BOX3: A model based on graphs concepts is to attach a metrics to complexity—internal connectivity, of multi-component systems (Gheorghe and Vamanu 2004a). The capacity of a system made of people, hardware, software, and organizational

Table A.1 Vulnerability categories

Vulnerability category	Description
Structure stability-induced vulnerability	Structure indicates various components that constitute the system, including subsystems, components, and their configurations. Structure stability-induced vulnerability means diversified physical, cyber, organizational failures, or unfavorable changes in structure components induce structure unstable, increasing the chances that a will not keep the assigned function and operation pattern and thus possible leading to unwanted situations
Complexity-induced vulnerability	Taking complexity as multi-components within or between the large-scaled systems with intricate interdependencies, complexity-induced vulnerability is taken to describe a situation in which high interdependency and interconnection could create right conditions for a small defect or accident initiated at one point to high chances to propagate throughout the system and escalate into unwanted situations
Operation-induced vulnerability	Operations of a system include: (1) the cooperation (i.e., resource and function dynamic assignment and nonlinear interactions), with other interdependent systems or within interconnected multi-components in the system and (2) various maintenance, procedure(process), and emergence action factors within the system. Miss-cooperation can affect the resilience of the system, and through nonlinear interactions, a determined small change can lead to unexpected emergence, making system at critical situation. And many major accidents occur either during maintenance and procedure operation or because of inadequate or faulty-executed maintenance and procedure control
Geography-induced vulnerability	Geography is a determinant of climate and primary disadvantage environment controls on a system; it is a determinant as to which natural factors pose hazards to a system. Geography-induced vulnerability is also an important path leading to unwanted situations
Organization induced vulnerability	Organization factors comprise decision-making structure, policy and regulation establishment, emergency communication and response, etc. Fallible decision or outdated policy and regulation can threat the survivability of a system, and inadequate organization can cause system breakdown
Management-induced vulnerability	Management mainly implies the security, personnel, operational, and financial management. Absence of detection and control for the security issues increase the system susceptibility. Inadequate personnel education and appointment can lead 'sharp end' of the system functions. Unreasonable resource allocation reduces the system resilience to the related hazards

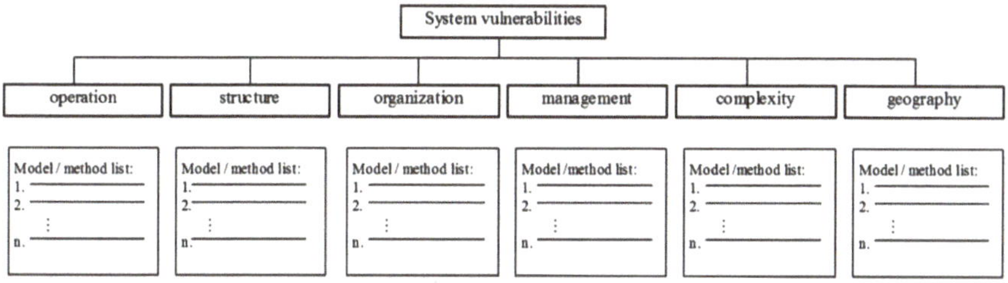

Fig. A.11 HHM of system vulnerability and their corresponding assessing models/methods

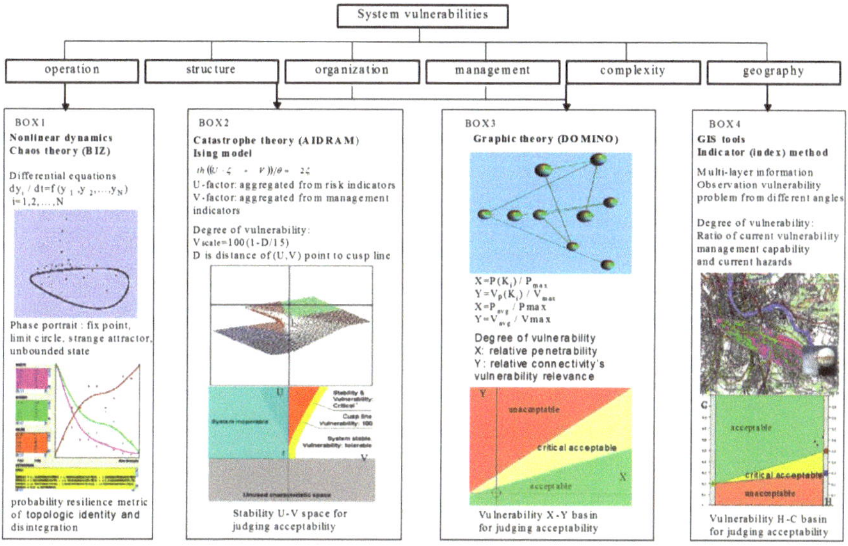

Fig. A.12 An example HHM of system vulnerabilities and their corresponding assessing models/methods, adapted from Song (2005)

and management procedures to be penetrated is defined as the vulnerability of such a system. The associated software for this method is DOMINO.

Three fundamental elements are used to represent system: *Knots*, *links,* and *features*. Knots as components of a system and the subjects of the analysis. Links that connect the knots in the sense that exchange/trade information, energy, and/or substance. Features are meant to characterize the knots, with quantitative vulnerability assessment through values and weights. It is nature to use this model in HHVA just by choosing the related vulnerable points as knots, and quantifying the result of qualitative assessment of vulnerable points in Step VI as features of the knots. With this model, to characterize a system's vulnerability in terms of its complexity, consider two parameters:

1. *System's penetrability*: a quality that may have as its metrics the (average) number of knots that can be accessed staring from a (any) given knot in the system
2. *The connectivity's vulnerability relevance*: depending on the penetrability defined in 1, yet also on the vulnerability relevance grades assigned to knots features.

The related mathematic models are for signaling extremes or vulnerability spikes:

$$X = P(K_i)/P_{\max}$$
$$Y = V_p(K_i)/V_{\max\, relevance}$$

Where,

$P(K_i)$	is number of distinct knots that can be accessed from K_i
$P_{\max}$	is the number of knots
$V_p\ (K_i)$	is the search-path vulnerability relevance of knots K_i
$V_{\max}$	is maximum possible vulnerability

- qualifying a system's connectivity (complexity) overall vulnerability relevance:

$$X = P_{avg}/P_{\max}$$
$$Y = V_{avg}/V_{\max}$$

Where,

P_{avg}	is average system penetrability, per knot
V_{avg}	is average vulnerability relevance or knot of system

In the X–Y plane, the appraisal of 'vulnerability tolerance' can be conducted. The X–Y space (shown in lowest picture in BOX3) is divided into 3 basins: basin of acceptable vulnerability, basin of critical vulnerability, and basin of unacceptable vulnerability.

BOX4: GIS combines with indicator (index) method to analyze and assess vulnerability induced by geography. In the GIS multilayer, information can be achieved. This makes it possible to look at vulnerability problem from a different angle.

Working in HHVA, one can select vulnerable points involving the related scenarios as indicators. Based on information acquired from GIS, one can then analyze and evaluate the hazards and management capacity to corresponding vulnerable points, and, through specific aggregating method (criteria), obtain the index of hazard and index of capability. The ratio of index of management capacity to index of hazard may be used to determine degree of vulnerability. This can be done in H (hazard)-C(capacity) plane in which three basins (shown in lowest picture in BOX4) are divided. With the index pair (H,C), one could then evaluate the acceptability of the vulnerability induced by geography.

Step VIII: Vulnerability Management

In this step, one turns attention to vulnerability management and asks: *what can be done to reduce these vulnerabilities?* Clearly, this question should put one in a creative mode. Having evaluated the acceptance of system vulnerabilities, and quantified the degree of system vulnerabilities, one needs to go back to Step V to review the selected critical scenarios as well as the ranking of related vulnerable points and do some relevant adjustment or re-ranking.

Since one knows the system and the major vulnerable points, at this point, there is need to create options for actions, by asking: *What design modifications, operational changes or internal and external factors (see Fig.* A.3*) adjustment and control could we make that would reduce the vulnerabilities of this vulnerable points?* Particularly, in this step, a new HHM resembling Fig. A.3 can be built from multi-respective to identify all possible protection and mitigation measures. At this point, authors will not discuss a new HHM since it is understood that each system is different.

Having set forth these options, it is not possible to shift back to analytics through a number of questions: First, *how much would we reduce the vulnerabilities, and are they acceptable?* And, *how much would it costs to implement (one or more of) these options?*

Re-doing Step VI and Step VIII, one is able to move back and forth and arrive at a set of cost-effective options. At this point, one needs to recall that options have been evaluated against the filtered set of scenarios remaining at the end of Step II; thus, in the next step, one needs to take another look at the effect these options might have on the scenarios that were previously filtered out.

Step IX: Safeguard against missing critical items

Reducing the initial large number of scenarios to a reasonably smaller number of scenarios at the completion of Step III may inadvertently filter out scenarios that could become important if the proposed options were actually implemented. Also, in a dynamic world, early indicator of newly emerging critical threats or hazards should not be overlooked (Andriani and McKelvey 2011; Calida and Katina 2012; Haimes et al. 2002). Following the completion of Step VIII, which generates and aids in selection of vulnerabilities management policy options and their associated trade-offs, we ask the question: *How robust has the policy selection and scenario filtering process been?* Step IX is then aimed at providing added assurance that the proposed HHVA methodology creates flexible reaction plans if indictors signal the emergence of new or heretofore undetected critical items. In particular, in Step IX of the analysis, we:

1. Ascertain the extent to which the vulnerabilities management options developed in Step VIII affect or are being affected by any of the scenarios discarded in Step III to VI. That is, in light of the interdependencies within the scenarios, one must evaluate the proposed management policy options against the scenarios previously filtered out
2. Revise as appropriate the vulnerabilities management options developed in Step VIII in light of what is learned in step 1 (above). This serves to enable further refinement of vulnerabilities management options
3. Detailed deployment of Step IX is mostly driven by the specific characteristics of the system. The main guiding principle in this step is the cascading effects due to the intra- and interdependencies that may have been overlooked during the filtering and ranking process in Step III to VI.

A Summary

The proposed HHVA approach is a holistic and dynamic methodology for dealing emerging hazards and threats. In HHM, the analysis is never considered finished since new items could be added or revisited. Modification and adaptation of HHVA to the changing environment and the increasing knowledge that stakeholders acquired with time and to the changing of stakeholders' own choices and requirements will all contribute to the driving forces of HHVA development.

In simple words, HHVA undergoes 'explore—feedback—adaptation—explore' the open chain of processes for continuous improvement on, in this case, *critical infrastructures*. Key value of this approach is to help the stakeholders to be cognizant of the vulnerability status and its dynamics, and to facilitate communication and cooperation on safety policy making and implementation.

The ideas of vulnerability assessment and management can be taken as an extension and complementary of risk assessment and management. And while it matters that one differentiates risk and vulnerability, there can even be a greater value on the consideration of degrees of vulnerability, which will inherently affect vulnerability communication as well as management approaches. Moreover, researchers suggest there can be utility in explicitly linking present research to topics of resilience, fragility, and even governance. The proposed framework: *Hierarchical Holographic Vulnerability Assessment* (Song 2005) captures a set of questions including:

- What can be vulnerable?
- Why are they vulnerable?
- How vulnerable are they?
- What are the vulnerabilities of the system? Are they acceptable?
- What can be done to reduce vulnerabilities?
- What are the trade-offs of the options in terms of costs, benefits?

And is a step-by-step procedure to identify, prioritize, and manage vulnerabilities. A concerned researcher will take topic of risk, resilience, fragility, governance, etc., and see the influence that have on the framework, management, and the system of in the face of complex and dynamic operational environment.

References

Andriani, P., & McKelvey, B. (2011). Using scale-free processes to explain punctuated-change in management-relevant phenomena. *International Journal of Complexity in Leadership and Management, 1*(3), 211–251.

API, & NPRA. (2003). *Security vulnerability assessment methodology for the petroleum and petrochemical industries*. Washington: DC: American Petroluem Institute.

Calida, B. Y., & Katina, P. F. (2012). Regional industries as critical infrastructures: A tale of two modern cities. *International Journal of Critical Infrastructures, 8*(1), 74–90.

Doorman, G., Kjølle, G., Uhlen, K., Huse, E. S., & Flatabø, N. (2004). *Vulnerability of the Nordic Power System* (No. TR F5962). Sem Sælandsvei 11: SINTEF Energy Research.

Gheorghe, A. V., & Vamanu, D. V. (2004a). Complexity induced vulnerability. *International Journal of Critical Infrastructures, 1*(1), 76–84.

Gheorghe, A. V., & Vamanu, D. V. (2004b). Towards QVA—Quantitative Vulnerability Assessment: A generic practical model. *Journal of Risk Research, 7*(6), 613–628.

Gheorghe, A. V., & Vamanu, D. V. (2005). Reading vulnerability in phase portraits: An exercise in probabilistic resilience assessment. *International Journal of Critical Infrastructures, 1*(4), 312–329.

Haimes, Y. Y. (2008). *Risk modeling, assessment, and management*. Hoboken: John Wiley & Sons, Inc.

Haimes, Y. Y., Kaplan, S., & Lambert, J. H. (2002). Risk filtering, ranking, and management framework using hierarchical holographic modeling. *Risk Analysis: An Official Publication of the Society for Risk Analysis, 22*(2), 383–397.

Katina, P. F., Pinto, C. A., Bradley, J. M., & Hester, P. T. (2014). Interdependency-induced risk with applications to healthcare. *International Journal of Critical Infrastructure Protection, 7* (1), 12–26.

Rinaldi, S. M., Peerenboom, J., & Kelly, T. K. (2001). Identifying, understanding, and analyzing critical infrastructure interdependencies. *IEEE Control Systems, 21*(6), 11–25.

Song, C. (2005). *A methodological framework for vulnerability assessment for critical infrastructure systems, hierarchical holographic vulnerability assessment (HHVA)* (Thesis). ETH Zürich, Zürich.

Vamanu, B. I., Gheorghe, A. V., & Katina, P. F. (2016). *Critical infrastructures: Risk and vulnerability assessment in transportation of dangerous goods—Transportation by road and rail* (Vol. 31). Cham, Switzerland: Springer International Publishing.

Appendix B: Complex System Governance

The purpose of this appendix is to provide an overview of *Complex System Governance*,[1] as an emerging field grounded in management cybernetics, systems thinking, and governance to help practitioners more effectively deal with increasing complex systems.

Background

Thought this book, especially Chaps. 2 and 3, references to 'governance' is made. However, up to this point, we had not delved into the meaning of the term: *Governance*. Governance, as defined in a recent report of the World Bank, is "the process through which state and nonstate actors interact to design and implement policies within a given set of formal and informal rules that shape and are shaped by power" (The World Bank 2017, p. 3). This definition suggests that the notion of governance involves elements of power, policy, people, and outcomes. Moreover, governance can take different forms. This can be the case when one refers to international bodies, national state institutions, local government agencies, community, and business associations (The World Bank 2017). However, one cannot be naïve to assume that there is a clear distinction among the listed forms of governance, especially given the elements of ambiguity, complexity, emergence, interdependence, and uncertainty that characterize the current operational environment. Hence, there are always overlaps among different systems, and in present research, different critical systems along with their actors.

There is a variety of reasoning for involving on 'governance' research, and that is an issue for specific systems. In the present case, the focus is: (i) establishing a 'systems' view of the current complex problem domain facing practitioners, (ii) articulating the nature, role, and functions for complex system governance as a compelling response to avoid negative implications of *system drift*, and (iii) identifying the high-level approach and value for engaging in complex system governance development.

In the hopes of facilitating a conversation and potential further interest in complex system governance (CSG), this appendix discuses CSG problem space, primer for CSG including detailed concept definition, paradigm, and functions. It also includes an overview of 3 critical stages for CSG development, values

[1]This appendix is made available by kind contributions of Dr. Charles B. Keating and his colleagues in the 'CSG Learning Community' at Old Dominion University.

associated with CSG engagement, and, finally, offers essential perspectives as well as cautions for CSG development.

A fundamental assumption for CSG development is acceptance that the system of interest (e.g., organization, entity) is in fact a 'complex' system. This assumption is critical for purposes of design, analysis, operation, maintenance, and evolution. This assumption is not innocuous. It invokes a level of thinking, language, and worldview that transcends approaches that are not truly systems-based in their orientation. This in no way demeans non-systems-based approaches or their capacity to make improvements in a complex system (Keating 2015). In contrast, CSG offers a purposeful, 'holistic,' and comprehensive approach to complex system development.

This development is tempered, based on the degree to which a system is capable of engaging feasible activities. As we unfold CSG, we cannot lose sight of the inherent difficulty in improving individual capacity, organizational competence, and systems to more effectively deal with the increasing complexity they face. If there were already universal approaches available to accomplish this, there would be no need to study and improve CSG. Unfortunately, such an approach does not currently exist. Thus, CSG is emerging as a response to systems in which governance is not functioning well on its own.

CSG has been previously identified as a framework to guide design, assessment, and evolution of *nine essential functions* that are required to sustain and evolve system performance (Keating et al. 2014; Keating and Katina 2015). A detailed discussion of these functions is provided later in this appendix. However, important to the present discussion is that the CSG functions enable systems and their practitioners (owners, operators, designers, performers) to excel in the midst of constant flux, disorder, and environmental turbulence. So, our question becomes, *what is it about our current systems that suggest CSG development should be considered?*

Arguably, and in many cases, many of our 'systems' are developed over time through processes of 'accretion' or 'self-organization.' Accretion is a process whereby elements are added in a piecemeal fashion until the whole system appears fragmented and no longer makes sense. Self-organization involves letting system structure and resulting behavior develop with minimal design oversight. This can produce results that may or may not be consistent with expectations or desirable performance. The result of either of these system development processes, accretion or self-organization, can and often do result in systems that fail to meet performance expectations. In effect, development is not *purposeful*, resulting in a condition we refer to as *'system drift.'* Just as a powerless ship drifts along its intended course subject to uncontrollable currents, so too can our systems experience drift resulting from development by accretion or self-organization. *System drift* symbolizes a system that is subject to the unintended consequences that accrue in the absence of a purposefully executed design. In the end, *system drift* describes a condition all too familiar to practitioners who must navigate systems through the increasingly complex environment, while confronting seemingly intractable issues on a daily basis. CSG is a coherent response to *system drift.*

CSG is focused on providing practitioners with perspective, methods, and tools to better understand and deal with complexities they must routinely confront. In essence, CSG helps avoid *system drift* through purposeful design, similar to a ship changing heading or speed to compensate for the effects of wind or current. Figure B.1 depicts five critical realities that practitioners responsible for modern complex systems must face. The ability to effectively respond to these realities will separate the high-performance systems from the 'also ran' systems in the future. We might hope that this situation would only be a temporary aberration from normal. Unfortunately, these conditions are not likely to subside in the near or distant future. Instead, they are more likely to intensify. Practitioners responsible for systems must adjust to thrive in this 'new normal' reality. Those who do not shift the level of decision, action, and understanding in response, in the best-case scenario, will likely experience *system drift* firsthand. In the worst-case scenario, they are likely to experience outright failure.

CSG is built on the foundations of two primary fields, *Systems Theory* and *Management Cybernetics*. Systems Theory provides a set of axioms and propositions that define structure, behavior, development, and performance of complex systems (Adams et al. 2014, Hester and Adams 2014). Just as the laws of physics (e.g., gravity) are immutable, so too are the propositions of Systems Theory. *All complex systems are subject to the axioms and propositions of systems.* Systems propositions create the absolute conditions that define system performance. Violation of these propositions comes with real consequences (e.g., diminished performance). Not acknowledging systems propositions does not preclude their impact on systems. Systems Theory supports CSG development by providing a focus on integration and coordination of complex systems.

Management Cybernetics is described as 'the science of effective organization' (Beer 1979). This field, and its corresponding model, supports understanding of communication and control in complex systems. These two historically proven and insightful fields have not been in the mainstream, or made easily accessible, for practitioners who must deal with increasingly complex systems and their associated problems. CSG brings the power of these two fields together, in a novel way, for the first time.

CSG is one of many systems-based approaches (e.g., see Katina and Calida 2017; Keating et al. 2016) designed to better deal with complexity and what we referred to earlier as *'system drift.' System drift* denotes systems that, irrespective of the noblest intentions, have either never been properly designed or whose execution continually fails to meet desired performance expectations. In short, these *'drifting'* systems fall short of delivering minimal value expected, much less producing high performance. We do not need to look far to see examples of drifting systems. In fact, it would be a rarity for one not to be impacted by systems in drift in any given day. Consider the following examples: (1) launching of a new Enterprise Resource Planning initiative that collapses due to emergent incompatibilities with existing systems, (2) a costly crisis from discovery of noncompliance to a regulatory requirement that has been in existence for several years but never identified, or (3) introduction of a new purchasing policy that achieves intended reductions in supplier costs but increases overall costs due to resulting schedule delays. Unfortunately, the impacts of *system drift* are not limited to increased costs. These drifting systems have considerable associated human cost. These human costs are borne by those that must suffer through these drifting systems by compensating for their ineffectiveness. CSG supports thinking, decision, and action to proactively and purposefully address *system drift.* Ultimately, CSG is intended to reduce the high human costs characteristic of these systems in drift.

Systems-based approaches, such as CSG, and the systems thinking upon which they are founded, are certainly not 'new' in trying to address what we described as *system drift.* In fact, the foundations of systems' thinking have been traced as far back as the ancient Chinese work *The I Ching*—translated as 'Book of Change' dated as far as 400 B.C. (Wilhelm 1967) that noted the dynamic nature of changing relationships among elements. Additionally, the central philosophical tenet of systems' thinking, *holism*, can be traced back to the writings of Aristotle, who suggested that 'the whole is more than the sum of its parts' (Aristotle 2002). Thus, approaches based in systems thinking and *'holism'* are not new and have historically represented a significant step toward dealing with system drift. However, what is new in bringing CSG applied research to the problem domain is the fusion of *Systems Theory* and *Management Cybernetics* to provide practitioners with perspective, supporting methods, and tools to confront drifting systems. This practitioner-focused CSG research seeks to increase capabilities for better understanding, decision, and action in dealing with complex systems and their associated problems. CSG seeks to increase effectiveness in dealing with *system drift.*

Problem Domain for Complex System Governance

The problems facing practitioners in modern systems appear to be intractable given the apparent ineffectiveness of the responses provided to address them. These problems continue to proliferate into all aspects of human endeavor and the systems designed to orchestrate those endeavors. They are not the privilege, or curse, of any particular field or sector (energy, utilities, health care, transportation, commerce, defense, security, services), as none are immune to the effects of this problem domain. Problems stemming from this domain do not have a precise cause–effect relationship that would make understanding and resolution easy. In fact, they are more likely products of a 'circular causality,' where the precise singular determination of cause is doubtful (Goodkind et al. 2011; Komljenovic et al. 2017; Korzybski 1994; von Foerster et al. 1953). Instead, these problems are consistent with the notion of Ackoff's (1981) 'messes' (interrelated sets of problems that are not well formulated, understood, or easily resolved) and Rittel and Webber's (1973) 'wicked problems' (problems that are intractable with current levels of thinking, decision, action, and interpretation). This problem domain is likely to continue, and perhaps accelerate, as we continue to grapple with twenty-first century complex systems and their problems.

Arguably, complex systems and their associated problems have been in existence long before the twenty-first century. However, the landscape for modern systems has changed appreciably into a much more ‘complex problem space.’ This problem space (Fig. B.1) is marked by difficulties encountered across the holistic range of technical, organizational, managerial, human, social, information, political, and policy issues. The different aspects of this ‘new normal’ complex problem space have been previously established (Jaradat and Keating 2014; Keating and Katina 2011; Keating 2014) as being characterized by conditions identified in Fig. B.2.

While this listing is not presented as exhaustive, it illustrates two important points. First, the issues emanating from this domain continue without consistent resolution methods. Thus, there is certainly room for new thinking and derivative approaches to address this domain. Second, the conditions identified are not likely to recede in the future. In essence, this domain represents the ‘new normal’ for the practitioners dealing with complex systems.

As mentioned earlier, these conditions are not the privilege of any particular system or sector. As an illustrative example, we have selected the water utilities sector to demonstrate the pervasive nature of the complex system problem domain. For instance, Fig. B.3 is a compilation of challenges facing the water utilities sector compiled from several sources (Baranowski et al. 2012; EPA 2015; Naphade et al. 2011). As evident from the circumstances marking the water utilities sector, we can certainly extrapolate those to the challenges in complex system problem domain. In addition, we can also project the majority to a wider array of enterprises, sectors, and systems facing similar circumstances.

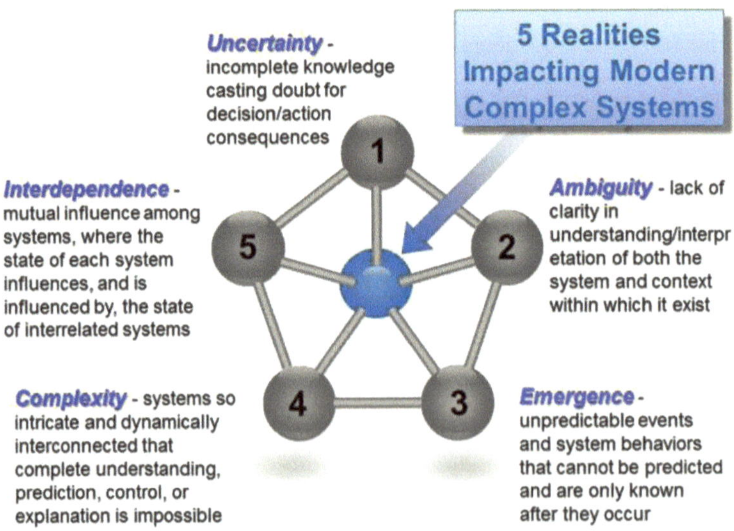

Fig. B.1 Realities for practitioners in modern complex systems

Fig. B.2 Challenges for practitioners in the complex system problem domain

Fig. B.3 Challenges facing the water utilities sector

Effectiveness in dealing with these problem domains beckons for individuals and organizations capable of engaging in a different level of thinking, decision, and action to produce alternative paths forward. As one response, CSG is proposed as an emerging field to enable practitioners to build capabilities to better diagnose and effectively respond to deeper-level systemic issues that impede system performance. Thus, CSG seeks to identify and 'design through' fundamental system issues such as those identified earlier (Fig. B.2). Unfortunately, these issues exist at deep tacit levels and appear only as symptomatic at the surface. Thus, efforts to address the problems at the surface level, although providing temporary 'fixes,' continually fail to resolve the deeper fundamental system issues. This deeper fundamental system-level resolution is necessary to preclude recurrence of the symptomatic issue in another superficial form.

Continual treatment of symptomatic conditions contributes to *'system drift'* by focusing on temporary correction of deficiencies at a superficial level. Unfortunately, this correction behavior is endemic to modern systems, fostering *'system superheros'*. These *'system superheros'* are recognizable as individuals who resolve surface symptoms (crises) through brute force and knowing how to navigate problematic systems. However, this behavior for error correction fails to address underlying systemic inadequacies or pathologies (Davidz 2017; Katina 2015; Troncale 2013), instead opting to reactively focus on apparent resolution that only serves to mask deeper systemic inadequacies. This is not to disparage the hard work and noble efforts of practitioners who become skilled at compensating for poorly designed and executed systems (system superheroes). On the contrary, we seek to draw attention to the liabilities of dependence of *'system superheroes'* to resolve 'crises' invoked by faulty systems. One ought to ask three important questions of systems that operate in the *'system superhero'* reactive problem resolution mode:

1. Is the existence of *'system superhero'* behavior masking more fundamental deficiencies in the underlying system?
2. Is reliance on *'system superheroes'* unsustainable, creating conditions for an eventual system collapse?
3. What happens when the *'system superheroes'* get overwhelmed, tired, retire, or just leave?

While CSG cannot claim to eliminate the existence of *system superheroes,* it does provide an opportunity to address underlying systemic deficiencies that this behavior masks. And perhaps, if not making them obsolete, at least reducing reliance on them for system performance.

CSG is certainly not portrayed as a 'panacea' to singularly guarantee success with the present and future twenty-first century problems facing organizations and their systems. However, it does offer a compelling argument as an approach to generate alternative thinking, decision, and action to address system problems. In addition, CSG can foster enhanced collaboration and partnerships across a system. This includes supporting:

- A 'total systems view' based in a holistic perspective
- Effective communication with multiple stakeholders through more explicit system understanding and representations
- Development of systems-based leadership skills that enhance capabilities for dealing with increasingly complex systems
- Increasing the likelihood of achieving expected performance

Again, while CSG is not a singular remedy to produce better performing systems, it does provide a solid complementary set of methods, tools, and thinking to enhance practice. Having established the basis for the problem domain of concern for CSG, we shift our focus to elaborate details of the emerging field. While a comprehensive treatment of the emerging CSG field (Keating et al. 2014) is beyond this appendix, the present focus is on providing the fundamental basis for the field. The intent here is offering a concise overview of the field, including the perspective of CSG, a high-level development approach, and the expected value from engaging in CSG development.

Complex System Governance: A Concise Overview

Succinctly stated, CSG paradigm is :

> *From a systems theoretic conceptual foundation, a set of nine interrelated functions is enacted through mechanisms. These mechanisms invoke metasystem governance to produce the communication, control, coordination, and integration essential to ensure continued system viability*

At first glance, the situation for dealing with complex systems and their constituent problems appears bleak. However, CSG is offered as an approach that can provide insights and a fruitful path forward. CSG has been presented as *"Design, execution, and evolution of the metasystem functions necessary to provide control, communication, coordination, and integration of a complex system"* (Keating et al. 2014, p. 274). At a high level, the following elements of the definition are elaborated as an essential foundation:

- ***Design***—purposeful and deliberate arrangement of the governance system to achieve desirable system performance and behavior.
- ***Execution***—performance of the system design within the unique system context, subject to emergent conditions stemming from interactions within the system and between the system and its external environment.
- ***Evolution***—the change of the governance system in response to internal and external shifts as well as revised trajectory.
- ***Metasystem***—the set of nine interrelated higher-level functions that provide for governance of a complex system.
- ***Control***—invoking the minimal constraints necessary to ensure desirable levels of performance and maintenance of system trajectory, in the midst of internally or externally generated perturbations of the system.
- ***Communication***—the flow, transduction, and processing of information within and external to the system, that provides for consistency in decisions, actions, interpretations, and knowledge creation made with respect to the system.

- ***Coordination***—providing for interactions (relationships) between constituent entities within the system, and between the system and external entities, such that unnecessary instabilities are avoided.
- ***Integration***—continuous maintenance of system integrity. This requires a dynamic balance between autonomy of constituent entities and the interdependence of those entities to form a coherent whole. This interdependence produces the system identity (uniqueness) that exists beyond the identities of the individual constituents.
- ***Complex system***—a set of bounded interdependent entities forming a whole in pursuit of a common purpose to produce value beyond that which individual entities are capable.

Instrumental to the formulation of CSG is the unique role of the 'metasystem.' The metasystem construct brings several important considerations for the CSG paradigm development, including:

1. The metasystem operates at a logical level beyond the elements that it must integrate
2. The metasystem has been conceptually grounded in the foundations of systems theory and management cybernetics
3. A metasystem is a set of interrelated functions—which only specify 'what' must be achieved for continuing system viability (existence), not 'how' those functions are to be achieved
4. The metasystem functions must be performed if a system is to remain viable—this does not preclude the possibility that a system may be poorly performing, yet still continue to be viable (exist)
5. A metasystem can be purposefully designed, executed, and maintained, or left to its own (self-organizing) development.

There is no one right answer with respect to metasystem design and development, just the level of system performance that either meets desired expectations or falls short. In addition, the metasystem functions are enacted through mechanisms, or devices that permit performance of the particular functions (e.g., a leadership council). These mechanisms must also be compatible with the context and supporting infrastructure within which the metasystem is embedded. Fig. B.4 identifies the primary organization of the CSG paradigm, including the central role of the metasystem.

Critical to understanding the metasystem is the particular positioning of the metasystem in relationship to the environment, context, and system of interest (Fig. B.5). The following descriptions are provided to focus our discussion:

- ***Environment***: The aggregate of all surroundings and conditions within which a system operates. It influences and is influenced by a system.
- ***Context***: The circumstances, factors, patterns, conditions, or trends within which a system is embedded. It acts to constrain or enable the system.

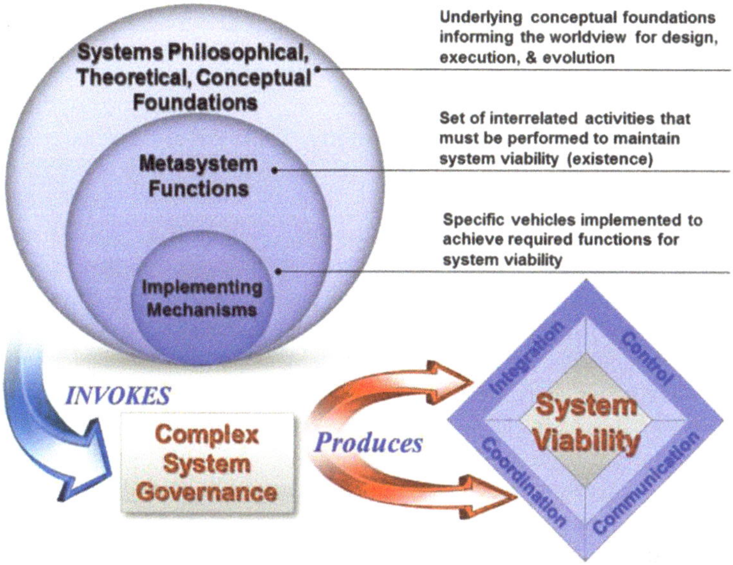

Fig. B.4 CSG paradigm

- ***System(s)***: The set of interrelated elements that are subject to immutable system axioms and propositions and are governed to produce that which is of value and consumed external to the system.
- ***Metasystem***: The set of functions that are invoked through mechanisms to govern a system such that viability (existence) is maintained

There are four important points concerning the relationship of these four elements. First, the metasystem, system, and context are embedded in the larger environment. This implies that separation can only occur through a process of 'abstraction.' The process of abstraction not only is essential for analysis, but also carries with it inevitable abstraction errors. There is no perfect abstraction, meaning all abstractions have some level of error. Care must be taken to account for the choices (assumptions, judgments) made for abstraction. Second, the metasystem exists as meta (beyond/above) to the system(s) that it seeks to govern. While it serves the objective of purposeful analysis and development, the metasystem simply imposes a viewpoint to examine the interconnected mechanisms that perform the functions necessary for integration, coordination, communication, and control for the system. Therefore, the metasystem is a construct that allows organization of mechanisms by essential functions to support analysis of a complex system. Third, the system is separated from the environment by the system boundary. The system boundary is established by the criteria that define inclusion and exclusion with respect to what constitutes the system. Although the boundary is imposed for purposes of analysis, care must be taken to be conscious of both the initial establishment as well as shifts in the boundary conditions over time. Fourth, the system and metasystem are embedded within the context. In essence, the context acknowledges conditions that are more closely coupled to the system/metasystem than those in the environment (e.g., system leadership style).

The separation of the environment, context, system, and metasystem is for convenience and permits analysis. In reality, these four elements exist as an inseparable whole. The separation of these elements always requires judgments. Judgments of boundaries, relevant aspects of the environment, contextual definition, and articulation of the metasystem are always subject to 'abstraction error.' Therefore, CSG requires purposeful decisions with respect to abstraction of the context, system(s), and metasystem from the environment (Fig. B.5).

The fundamental foundation for CSG is found in systems, including the philosophical, theoretical, and conceptual underpinnings that serve as a grounding for the field. The metasystem is a construct that defines the set of 9 interrelated functions that act to provide governance for a complex system (Fig. B.6). Also, central to the metasystem are the communication channels that act, through their own mechanisms, to provide for the interface within the metasystem, and between the metasystem and governed systems. These *communication channels* help enact metasystem functions and are captured in Table B.1 and also denoted in Fig. B.6.

The nine metasystem functions include:

- ***Policy and Identity—Metasystem Five (M5)***—focused on overall steering and trajectory for the system. Maintains identity and balance between current and future focus.
- ***System context—Metasystem Five Star (M5*)***—focused on the specific context within which the metasystem is embedded. Context is the set of circumstances, factors, conditions, or patterns that enable or constrain execution of the system.

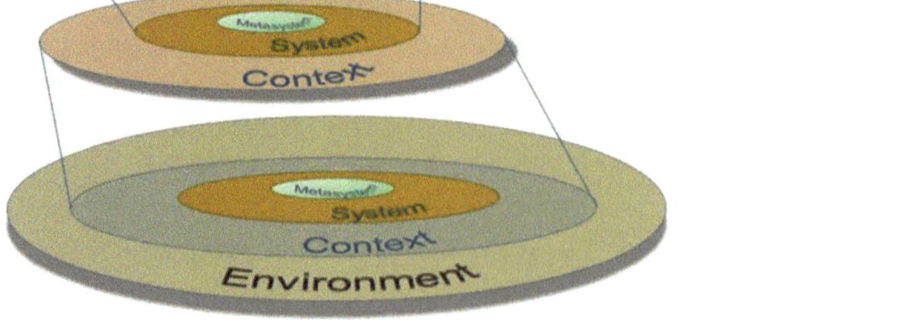

Fig. B.5 Relationship among elements of environment, context, system, and metasystem

Table B.1 Communication Channels for CSG

Communications channel and responsibility	CSG Metasystem role
Command (Metasystem 5)	• Provides non-negotiable direction to the metasystem and governed systems • Primarily from the Metasystem 5 and disseminated throughout the system
Resource bargain/ Accountability (Metasystem 3)	• Determines and allocates the resources (manpower, material, money, information, support) to governed systems • Defines performance levels, responsibilities, and accountability for governed systems • Primarily an interface between Metasystem 3 to the governed systems
Operations (Metasystem 3)	• Provides for the routine interface focused on near term operational focus • Concentrated on direction for system production (products, services, processes, information) consumed external to the system • Primarily an interface between Metasystem 3 and governed systems
Coordination (Metasystem 2)	• Provides for metasystem and governed systems balance and stability • Ensures that information concerning decisions and actions necessary to prevent disturbances are shared within the Metasystem and governed systems • Primarily a channel designed and executed by Metasystem 2
Audit (Metasystem 3)*	• Provides routine and sporadic feedback concerning operational performance • Investigation and reporting on problematic performance issues within the system • Primarily a Metasystem 3* channel for communicating between Metasystem 3 and governed systems concerning performance issues
Algedonic (Metasystem 5)	• Provides a 'bypass' of all channels when the integrity of the system is threatened • Compels instant alert to crisis or potentially catastrophic situations• for the system • Directed to Metasystem 5 from anywhere in the metasystem or governed systems
Environmental Scanning (Metasystem 4')	• Provides design for sensing of the external environment • Identifies environmental patterns, • Activities, or events with system implications • Provided for access throughout the metasystem as well as governed systems
Dialog (Metasystem 5')	• Provides for examination of system decisions, actions, and interpretations for consistency with system purpose and identity • Directed to Metasystem 5 from anywhere in the metasystem or governed systems

(continued)

Table B.1 (continued)

Communications channel and responsibility	CSG Metasystem role
Learning (Metasystem 4)*	• Provides detection and correction of error within the metasystem as well as governed systems, focused on system design issues as opposed to execution • Directed to Metasystem 4* from anywhere in the metasystem or governed systems
Informing (Metasystem 2)	• Provides for flow and access to routine information in the metasystem or between the metasystem and governed systems • Access provided to entire metasystem and governed systems

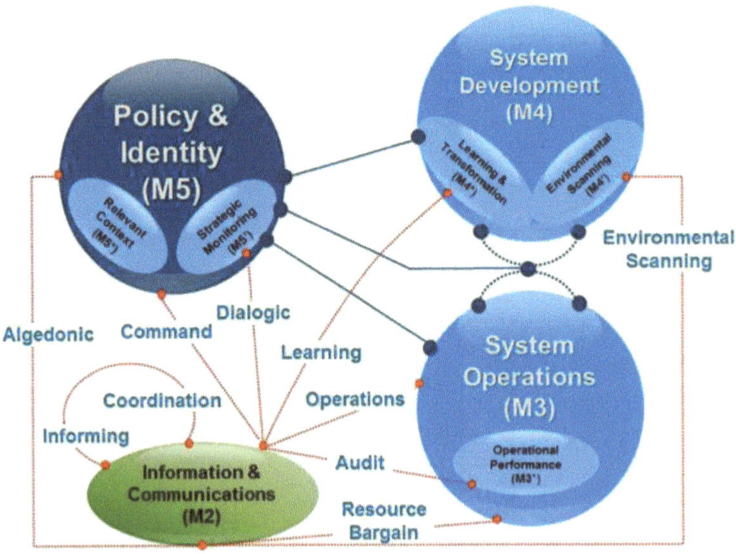

Fig. B.6 Nine interrelated functions that form the metasystem

- ***Strategic System Monitoring—Metasystem Five Prime (M5')***—focused on oversight of the system performance indicators at a strategic level, identifying performance that exceeds or fails to meet established expectations.
- ***System Development—Metasystem Four (M4)***—maintains the models of the current and future system, concentrating on the long-range development of the system to ensure future viability.
- ***Learning and Transformation—Metasystem Four Star (M4*)***—focused on facilitation of learning based on correction of design errors in the metasystem functions and planning for transformation of the metasystem.
- ***Environmental Scanning—Metasystem Four Prime (M4')***—designs, deploys, monitors, and communicates sensing of the environment for trends, patterns, or events with implications for both present and future system viability

- ***System Operations—Metasystem Three (M3)***—focused on the day-to-day execution of the metasystem to ensure that the overall system maintains established performance levels.
- ***Operational Performance—Metasystem Three Star (M3*)***—monitors system performance to identify and assess aberrant conditions, exceeded thresholds, or anomalies.
- ***Information and Communications—Metasystem Two (M2)***—designs, establishes, and maintains the flow of information and consistent interpretation of exchanges (communication channels) necessary to execute metasystem functions.

The ten communication channels are adapted from the work of Beer (1979, 1981, 1985) and extensions of Keating and Morin (2001):

Implementing mechanisms is the final element that forms a CSG triad (Fig. B.7), complementing *Conceptual Foundations* and *Metasystem Functions (and their corresponding communication channels)*. *Conceptual Foundations* help to explain and understand *'why'* systems behave and perform as they do, based on the axioms and propositions of Systems Theory and Management Cybernetics. These axioms and propositions are immutable and cannot be negotiated away. The consequences for violation of the propositions are real and will impact system viability. The *Metasystem Functions and their communication channels* identify *'what'* must be achieved to ensure continued system viability.

ALL systems must perform the metasystem functions at a minimal level to maintain viability. However, viability is not a 'guarantee' of performance excellence. On the contrary, viability simply assures is that the system continues to exist. There are degrees of viability, the minimal of which is existence. *Implementing Mechanisms* are the specific vehicles (e.g., processes, procedures, activities, practices, plans, artifacts, values/beliefs, customs, mores) that implement metasystem governance functions and their communication channels for a specific system of interest. These mechanisms may be explicit/tacit, formal/informal, routine/

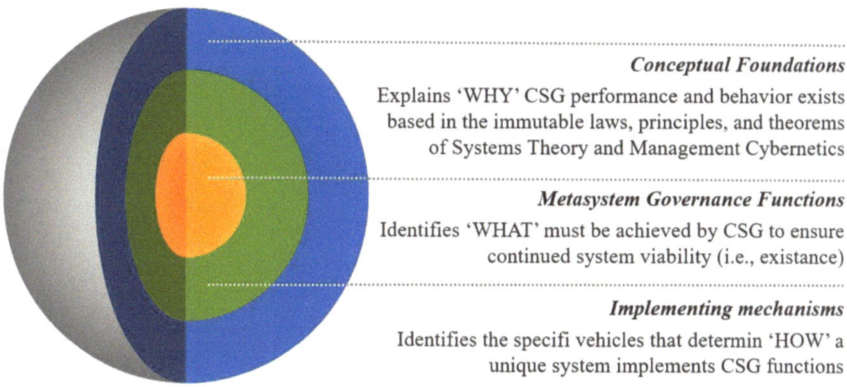

Fig. B.7 Triad of the CSG paradigm

non-routine, effective/ineffective, or rational/irrational. However, all mechanisms can be articulated in relationship to the metasystem governance functions and corresponding communication channels they support.

Approach for Complex System Governance Development

Detailed account of CSG methodology for development is beyond the scope of present effort and is articulated elsewhere (Keating and Katina 2016). Thus, a high-level approach to the three primary stages of CSG development is provided. It is important to note that the CSG development approach is applicable to 'any' system chosen as the *system of interest.* However, just as each system is unique and exists in a unique context, the specific approach to CSG development must be tailored to appreciate the uniqueness of the particular system of interest. For succinctness, the three primary stages of CSG development, as depicted in Fig. B.8, are *Initialization, Readiness Level Assessment,* and *Governance Development.*

Stage 1: Initialization

This stage is the most critical in CSG development. It establishes a reference point for the current state of CSG for a system. There are two primary aspects of initialization. First, the metasystem is *'framed'*, which involves definition of: (a) the metasystem functions, (b) the mechanisms that execute those functions, (c) the interrelationships [governance architecture] among those mechanisms, and (d) the current state of

Fig. B.8 Three stages of CSG development

performance of metasystem functions and corresponding mechanisms [including pathologies limiting performance]. Second, the '*context*' (circumstances, factors, conditions, patterns, or trends that enable or constrain the metasystem) within which the system in focus operates is articulated. This includes establishing the clear lines of demarcation between the system, its context, and the environment. In essence, articulation of system context involves identification pathological conditions enabling or constraining the system of interest (Katina 2016). The product from this stage is a CSG profile. This profile represents the current state of CSG and the context for the system of interest.

Stage 2: Readiness Level Assessment

This stage establishes a classification for the existing CSG which is represented by a profile. The profile establishes the current effectiveness level for governance and identifies the current state (readiness level) for engaging different levels of development difficulty for the CSG. Thus, the readiness level informs what can be reasonably undertaken as feasible activities to improve the CSG. In effect, the *Governance Readiness Level* (GRL) helps to set reasonable expectation thresholds for what can be successfully accomplished to improve the state of governance from an established baseline (current effectiveness level). The GRL is a function of both the current state of the metasystem and the context within which the metasystem operates.

Stage 3: Governance Development

This stage is directed to identification, prioritization, implementation, and monitoring of activities to enhance CSG. It is important to note that the CSG functions are already being performed if a system exists. However, the mere existence of a system does not guarantee high performance. Existence may range from 'barely alive' to 'thriving.' The selection of activities to improve governance is conducted against the backdrop of the initialization and readiness-level assessment. The initialization provides the particular 'gaps,' deficiencies, or opportunities that exist for CSG development within the specific context. In contrast, the Governance Readiness Level (GRL) establishes the maturity of the evolving CSG. Therefore, the GRL identifies the difficulty level of CSG development activities which can be undertaken with a reasonable chance for successful completion. This is followed by Governance Development which is focused on the selection, planning, execution, and evaluation of activities to improve CSG, and ultimately enhance the GRL.

While these stages of CSG development are presented with a clear degree of separation, in the reality of development, their separation is not clear-cut. On the contrary, their execution is more likely to be emergent and overlapping, particularly as CSG follows a cyclic development path.

Implications and Value for CSG Development

Ultimately, CSG offers significant contributions to help practitioners address some of the most vexing current, as well as future, system problems they must confront. CSG is not suggested as a panacea for all problems facing societal enterprises and systems. Instead, CSG is advocated as an emerging field with significant opportunity to provide value in the following areas:

- ***Enhance capacity*** of individual practitioners to engage in the level of systems thinking necessary to more effectively deal with the entire range of complex system problems. These problems are a byproduct of modern enterprises and their systems. Effectiveness is achieved through development and propagation of CSG language, methods, and tools to assist practitioners in their efforts to design, analyze, execute, and evolve complex systems and their associated problems.
- ***Develop competencies*** at the organizational level for dealing with complex systems and their derivative problems. This involves generation of knowledge, development of skills, and fostering abilities beyond the individual level to embrace problems holistically. For CSG, holism suggests competency development that expands beyond narrow technology-centric infusions. Instead, enhanced competencies that span the entire range of sociotechnical considerations endemic to complex systems are an outcome from CSG engagement.
- ***Understand infrastructure compatibility*** necessary to support systems-based endeavors. This compatibility is necessary to formulate contextually consistent approaches to problems, create conditions necessary for governance system stability, and produce coherent decisions, actions, and interpretations at the individual and organizational levels. The most exceptional system solutions, absent compatible supporting infrastructure, are destined to outright fail in the worst-case scenario and underachieve in the best-case scenario.
- ***Governance Effectiveness Level Identification*** is a direct byproduct from the initial assessment of the existing state and performance of the CSG. In addition, the set of 'unique' indicators developed for a specific system of interest can provide a baseline that can be used to longitudinally establish the continuous progression of governance improvement. In effect, the degree of improvement stemming from initiatives undertaken to improve CSG can be established. Therefore, the state and shifts in governance can be monitored.
- ***Governance Readiness Level Identification*** can help establish a feasible set of initiatives that can be undertaken with a higher probability of successful achievement. This does not minimize the degree of CSG discovered inadequacies that might exist in a system. However, it does take into account the current sophistication in system governance, as well as the limiting/enabling context, that will influence what can be reasonably taken on with confidence of success. Minimally, exploration of the CSG readiness level can provide new insights into past successes/failures as well as cautions for impending future endeavors.

- ***Explicit Models for Understanding*** generated through CSG efforts can provide insights into the structural relationships, context, and systemic deficiencies that exist for a system of interest. These insights can accrue regardless of whether or not specific actions to address issues are initiated. The models can be constructed without system modification. Therefore, alternative decisions, actions, and interpretations can be selectively engaged based on consideration of insights and understanding generated through modeling efforts.
- ***Purposeful Governance Development*** through focused design, analysis, and evolution of the CSG functions necessary to maintain system viability. While all viable (existing) systems perform the CSG functions, it is rare that they are purposefully articulated, examined, or developed in a comprehensive fashion. Purposeful CSG development can produce a 'blueprint' against which development can be achieved by design, rather than serendipity. This includes establishment of the set of 'dashboard indicators' for CSG performance. These performance indicators exist beyond more 'traditional' measures of system/organizational performance.
- ***Coherent Decision Support*** can be achieved by the 'big picture' view of the governance landscape. This includes identification of highest leverage strategic impact areas and their interrelationship to the larger CSG performance gaps. Thus, decisions for resource allocation can be better targeted. This allows steering away from activities that are simply 'intriguing' without demonstrating the highest substantial benefit to the larger 'systemic' governance concerns. Considering CSG development priorities, low contribution efforts can be eliminated, or resources shifted appropriately.
- ***Rigorous Guided 'Self-Study'*** into CSG can provide significant insights into how the system actually functions. Although enterprises and their systems function routinely and successfully on a daily basis, as a matter of course practitioners are not particularly skilled nor do they engage in deep reflection as to why, how, and what they do from a systems point of view. The gains to be made by reflective self-examination, from a systemic point of view, can reveal insights far beyond traditional methods of examination (e.g., Strategic Planning, SWOT analysis). Thus, practitioners can examine a different level of analysis through 'self-study' and experience insights in a 'safe-to-fail' setting. Additionally, self-study might suggest the level of education/training that might be necessary for individuals and the organization to increase individual capacity and organizational competence for systems thinking.

Ultimately, CSG seeks to increase the probability of achieving desirable system performance (viability, growth, etc.) in the flux of a turbulent environment. There are multiple opportunities to accrue value from CSG efforts. CSG development value can span practitioner, enterprise, support infrastructure, context, and system levels. The specific achievement of value is certainly dependent on the degree to which an individual, enterprise, or system is willing to engage in the development effort. However, value is not limited to an 'all or nothing' application of CSG. There is much to be gained in even small endeavors within the larger framework of

CSG development. For instance, training in systems thinking can provide insights and improve the capacity of system practitioners to 'think systemically.' Thus, they may be better prepared to understand the sources of systemic issues and enhance the potential for alternative responses to more effectively deal with increasing complexity.

The Promise and Cautions of CSG Development

The CSG perspective developed in this appendix offers significant contributions to help address some of the most vexing problems faced by practitioners (owners, operators, designers, performers) responsible for governance of modern systems. While the specific details of CSG development are beyond the scope of this document, we have tried to 'make the case' for the promise of the emerging CSG field.

The development of the metasystem governance functions and context is central to CSG development. CSG development is an evolution, not an onetime effort. It requires simultaneous development of: (1) *individual capacity* to engage in holistic systemic thinking and action necessary to implement CSG, (2) *organizational competency* for governance that focuses on generation of knowledge, skills, abilities at a level beyond individuals to collectively engage deeper analysis, design, and evolution of the governance system, and (3) *compatibility of support infrastructures* that are capable of supporting and reinforcing governance development. As such, CSG development is a *continuous process* that persistently improves the level of governance effectiveness and the Governance Readiness Level through purposeful action.

In essence, at a most fundamental level, governance development can occur through three distinct processes, *accretion*—where new elements, activities, or modifications occur in a piecemeal fashion, *self-organization*—where the relationships and activities are allowed to 'take their own unfettered' course of development, or *purposeful design*—where the design and execution of CSG is pursued with deliberate intention. CSG is targeted to purposeful design.

However, the pursuit of CSG Development as purposeful design should not be entered into lightly. Unfortunately, CSG development has limitations, as does any systems-based approach, in dealing with complex systems and their associated problems. In realistic caution for CSG development pursuit, consider 7 important points:

1. ***CSG development must involve the individuals who own the system (i.e., accountable for system performance) and responsible to ensure that the system continues to develop such that viability is maintained***. CSG development pursuit without engagement of these individuals is unlikely to achieve anticipated results. There is no shortcut for system practitioners—CSG responsibility cannot be relegated
2. ***Individual capacity, organizational competence, and infrastructure compatibility to engage in systemic thinking/action will determine the degree to which system governance can enhance system performance***. Without a commensurate effort to understand the impacts, and necessity to include their development, these three areas can severely limit CSG developmental achievement.

3. ***Feasible actions to improve the governance system are a function of the degree of engagement, resources, will, and the existing state of 'governance readiness' for the enterprise***. Realization of 'full potential' for CSG development requires alignment of all of these elements. Outcome-expectation desires that are incongruent with investments of time, energy, and resources are likely to produce disappointing results.
4. ***The design for comprehensive governance development is fallible and must be continually adjusted***. It is naïve to engage in CSG development assuming that action outcomes can be known in advance. Instead, care must be taken to understand that the design for CSG development cannot be static. CSG development must adjust in response to changes in the system itself, the external environment, and the context within which CSG is embedded. The *rate of change for CSG development* design must minimally keep pace with the rate of change in the system, external environment, and context.
5. ***The nature of CSG development is evolutionary rather than revolutionary***. Therefore, the implementation of CSG development requires 'the long view' and patience. Expectations for CSG development must be appreciative of the current state of governance effectiveness and Governance Readiness Level. These will dictate what level of system improvement might be feasibly engaged over the near and long term.
6. ***In essence, CSG development is a protracted 'self-study' of the system of interest, enacted through a new set of lenses, corresponding language, methods, and tools***. New thinking requires new language, which can produce alternative decision, action, and interpretation in route to pursuit of different outcomes (system performance levels). The willingness to engage in protracted self-study is essential for realization of the benefits of CSG development. There is no shortcut to the reflective self-study required.
7. ***Engaging governance development is not a trivial endeavor***. It is hard work, requiring significant investment of resources, patience to take the 'long view,' and sacrifice of instant gratification for sustainable longer term performance improvement. Superficial CSG efforts are not likely to produce desirable or sustainable results, and in fact may make matters worse.

These cautions are provided to ensure that practitioners considering CSG development are aware of what CSG development entails. This does not suggest that elements of CSG development (e.g., improvement in individual systems thinking capacity) will not be beneficial. However, what can be achieved by CSG development must be consistent with the commitment invested in development efforts. This is the work of the 'owners' of enterprises and their systems.

References

Ackoff, R. L. (1981). The art and science of mess management. *Interfaces*, *11*(1), 20–26.

Adams, K. M., Hester, P. T., Bradley, J. M., Meyers, T. J., & Keating, C. B. (2014). Systems theory as the foundation for understanding systems. *Systems Engineering*, *17*(1), 112–123.

Aristotle. (2002). *Metaphysics: Book H—Form and being at work*. (J. Sachs, Trans.) (2nd ed.). Santa Fe, CA: Green Lion Press.

Baranowski, C., Ampleman, M., Whitler, J., & Posner, A. (2012). EPA's Climate Ready Water Utilities (CRWU) Initiative. *Proceedings of the Water Environment Federation*, *2012*(1), 600–604.

Beer, S. (1979). *The heart of the enterprise*. New York, NY: John Wiley & Sons.

Beer, S. (1981). *The brain of the firm: The managerial cybernetics of organization*. Chichester, UK: Wiley.

Beer, S. (1985). *Diagnosing the system for organizations*. Oxford, UK: Oxford University Press.

Davidz, H. L. (2017). Systems engineering pathology: Leveraging science to characterize dysfunction. In *Annual INOSE International Workshop*. Los Angeles: INCOSE. Retrieved from www.incose.org/IW2017

EPA. (2015). *Adaptation strategies guide for water utilities* (No. EPA 817-K-15–001). Washington: DC: United States Environmental Protection Agency.

Goodkind, J. R., Githinji, A., & Isakson, B. (2011). Reducing health disparities experienced by refugees resettled in urban areas: A community-based transdisciplinary intervention model. In M. Kirst, N. Schaefer-McDaniel, S. Hwang, & P. O'Campo (Eds.), *Converging Disciplines* (pp. 41–55). New York, NY: Springer New York.

Hester, P. T., & Adams, K. M. (2014). *Systemic thinking: Fundamentals for understanding problems and messes*. New York, NY: Springer Berlin Heidelberg.

Jaradat, R. M., & Keating, C. B. (2014). Fragility of oil as a critical infrastructure problem. *International Journal of Critical Infrastructure Protection*, *7*(2), 86–99.

Katina, P. F. (2015). *Systems theory-based construct for identifying metasystem pathologies for complex system governance* (Ph.D.). Old Dominion University, United States—Virginia.

Katina, P. F. (2016). Metasystem pathologies (M-Path) method: Phases and procedures. *Journal of Management Development*, *35*(10), 1287–1301.

Katina, P. F., & Calida, B. Y. (2017). Complex system analysis for engineering of systemic failures. In M. Hopkins (Ed.), *Systems Engineering: Concepts, Tools and Applications* (pp. 105–132). New York: NY: Nova Science Publishers.

Keating, C. B. (2014). Governance implications for meeting challenges in the system of systems engineering field. In *2014 9th International Conference on System of Systems Engineering (SOSE)* (pp. 154–159). Adelaide, Australia: IEEE. https://doi.org/10.1109/SYSOSE

Keating, C. B. (2015). *White Paper: Complex System Governance: Confronting System Drift—Unpublished manuscript*. Norfolk: VA: National Centers for System of Systems Engineering.

Keating, C. B., Calida, B. Y., Jaradat, R. M., & Katina, P. F. (2016). Systems thinking. In J. V. Farr, S. J. Gandhi, & D. N. Merino (Eds.), *The Engineering Management Handbook* (2nd ed., pp. 281–316). Rolla: MO: The American Society of Engineering Management.

Keating, C. B., & Katina, P. F. (2011). Systems of systems engineering: Prospects and challenges for the emerging field. *International Journal of System of Systems Engineering*, 2(2/3), 234–256.

Keating, C. B., & Katina, P. F. (2015). Editorial: Foundational perspectives for the emerging complex system governance field. *International Journal of System of Systems Engineering*, *6*(1/2), 1–14.

Keating, C. B., & Katina, P. F. (2016). Complex system governance development: A first generation methodology. *International Journal of System of Systems Engineering*, *7*(1/2/3), 43–74.

Keating, C. B., Katina, P. F., & Bradley, J. M. (2014). Complex system governance: concept, challenges, and emerging research. *International Journal of System of Systems Engineering*, *5*(3), 263–288.
Keating, C. B., & Morin, M. (2001). An approach for systems analysis of patient care operations. *The Journal of Nursing Administration*, *31*(7–8), 355–363.
Komljenovic, D., Loiselle, G., & Kumral, M. (2017). Organization: A new focus on mine safety improvement in a complex operational and business environment. *International Journal of Mining Science and Technology*, *27*(4), 617–625. https://doi.org/10.1016/j.ijmst.2017.05.006
Korzybski, A. (1994). *Science and sanity: An introduction to non-Aristotelian systems and general semantics*. New York, NY: Wiley.
Naphade, M., Banavar, G., Harrison, C., Paraszczak, J., & Morris, R. (2011). Smarter Cities and Their Innovation Challenges. *Computer*, *44*(6), 32–39. https://doi.org/10.1109/MC.2011.187
Rittel, H. W. J., & Webber, M. M. (1973). Dilemmas in a general theory of planning. *Policy Sciences*, *4*(2), 155–169.
The World Bank. (2017). *World development report 2017: Governance and the law* (No. 112303). Washington: DC: The World Bank. Retrieved from http://elibrary.worldbank.org/doi/book/10.1596/978-1-4648-0950-7
Troncale, L. (2013). Systems processes and pathologies: Creating an integrated framework for systems science. *INCOSE International Symposium*, *23*(1), 1330–1353.
von Foerster, H., Mead, M., & Teuber, H. L. (1953). *Cybernetics: Circular causal and feedback mechanisms in biological and social systems*. New York, NY: Josiah Macy, Jr Foundation.
Wilhelm, H. (Ed.). (1967). *The I Ching, or, Book of Changes*. (C. F. Baynes, Trans.) (3rd ed.). Lincolnwood: Princeton University Press.

Appendix C: An Expert-Oriented Tutorial for an State of a System with Many Bi-Stable Entities

This appendix contains an expert-oriented tutorial for arriving at the equation for a system with many bi-state systems/entities. This appendix is necessary as, on occasions, students, practitioners, as well as researchers, have expressed discomfort with the deductive algebraic/calculus flow that takes one equation and manipulates into until arriving at the right and compact analytic solution for the equation of state of systems with multi-component systems with bi-stable entities (Vamanu et al. 2016).

The Quantitative Vulnerability Assessment (QVA) model was first introduced in Gheorghe and Vamanu (2004) and offers an equation of state of systems with multi-component systems with bi-stable entities; namely

$$\tanh\left(\frac{u\zeta + v}{\Theta}\right) = 2\zeta \tag{C.1}$$

The tutorial that follows, although presumptuous about readers' level of mathematical proficiency and inelegant by all academic standards, is the authors' honest attempt to meet such concerns.

1. Consider a dynamic system made of a large number, M, of bi-stable entities. Assume that M_1 entities are, at a given moment in time, in 'State 1' (e.g., the normal state or functional state; and that M_2 entities are in 'State 2' (e.g., the abnormal state or dysfunctional state). This forms the basis for Eq. (C.2):

$$M = M_1 + M_2 \tag{C.2}$$

It comes natural to contend that the dynamics of the system consists, at the elemental (or 'atomic') level, of an entity collapsing from State 1 into a State 2—which takes the number M_1 down to $M_1 - 1$ and M_2 to $M_2 + 1$; or conversely—ascending from State 2 up into State 1—which takes M_2 to $M_2 - 1$ and M_1 to $M_1 + 1$. The fundamental lack of knowledge on when such an act takes place induces the common recourse to probabilities: assume, therefore, that $w_{21}(M_1, M_2)$ is the probability of a State 1-to-State 2 transition, whereas $w_{12}(M_1, M_2)$ is the probability of a State 2 to State 1 transition, where both probabilities are some functions of M_1 and M_2 (see Eq. C.3).

$$(M_1 - 1, M_2 + 1) \underset{w_{12}}{\overset{\rightarrow w_{21}}{\longleftarrow}} (M_1, M_2) \underset{w_{21}}{\overset{w_{12}}{\longrightarrow}} (M_1 + 1, M_2 - 1) \tag{C.3}$$

What the D.3 scheme graphically depicts, a customary routine in Physics translates into a 'master equation' for the distribution function, $f(M_1, M_2; t)$, of the probabilistic process described in Eq. C.4—a quantity that depends, as intuitively expected, on the current populations M_1 and M_2, while also varying with the time, t.

$$\begin{aligned} \frac{\partial f(M_1, M_2; t)}{\partial t} &= w_{21} f(M_1 - 1, M_2 + 1) + w_{12} f(M_1 + 1, M_2 - 1) \\ &\quad - (w_{21} + w_{12}) f(M_1, M_2) \end{aligned} \tag{C.4}$$

In plain words, Eq. C.4 tells us that the variation in time of the distribution function—the left-hand side of the master equation—covers, in the integrative manner provided by the concept,

(i) the acts of transitions from State 1 to State 2—the 1st term on right-hand side of Eq. C.4;
(ii) the acts of transitions from State 2 to State 1—the 2nd term on right-hand side of Eq. C.4; and
(iii) would naturally depend also on the current state of the system, appropriately characterized by the (M_1, M_2) pair of numbers.

2. To make the master equation useful, one has to take it to a form amenable to an algebraic handling. First, one operates a change of variables (Eq. C.5):

$$\zeta = \frac{1}{2}\frac{M_1 - M_2}{M_1 + M_2} = \frac{M_1 - M_2}{2M} \tag{C.5}$$

To see the meaning in it, let us take the system to its limits: indeed, if one assumes that *all* entities have somehow got into State 1 ('functional'), therefore making $M_1 = M$ and, by way of consequence, making $M_2 = 0$ in Eq. (C.5), the variable ζ takes the value 1/2. Conversely, if one assumes that all entities get in State 2 ('dysfunctional'), then, by making $M_2 = M$ and, of course, $M_1 = 0$, variable ζ becomes −1/2 (scheme 6). In-between the extremes, variable ζ would indeed work as a telling *measure of system functionality*, opposing the functional population of entities, M_1 to the dysfunctional population, M_2.

$$M_1 = M \rightarrow \zeta = \frac{1}{2} \tag{C.6}$$

$$M_2 = M \rightarrow \zeta = -\frac{1}{2}$$

To take full advantage of the change of variable (C.5), the immediate obvious consequences are worth noting:

$$\begin{cases} M_1 + M_2 = M \\ M_1 - M_2 = 2M\zeta \end{cases} \tag{C.7}$$

$$\begin{aligned} M_1 &= \frac{M}{2} + M\zeta = M\left(\frac{1}{2} + \zeta\right) \\ M_2 &= \frac{M}{2} - M\zeta = M\left(\frac{1}{2} - \zeta\right) \end{aligned} \tag{C.8}$$

$$\begin{aligned} M_1 \mp 1 &= M\left(\frac{1}{2} + \zeta\right) \mp 1 = M\left(\frac{1}{2} + \zeta \mp \frac{1}{M}\right) \\ M_2 \mp 1 &= M\left(\frac{1}{2} - \zeta\right) \mp 1 = M\left(\frac{1}{2} - \zeta \mp \frac{1}{M}\right) \end{aligned} \tag{C.9}$$

Upon these, one reaches the level of convenience that indeed justifies the change of variable: all states of the system—the *current*, i.e., (M_1, M_2); the *functionally-depleted*, i.e., $(M_1 - 1, M_2 + 1)$; and *functionally-enriched*, i.e., $(M_1 + 1, M_2 - 1)$ can be algebraically represented by a single (as opposed to two) variable, ζ, along with the constant M—the total population of entities in the system (Eq. C.10):

$$\begin{array}{ll} (M_1, M_2) & \zeta \\ (M_1-1, M_2+1) & \zeta^- = \frac{(M_1-1)-(M_2+1)}{2M} = \frac{M_1-M_2}{2M} - \frac{1}{M} = \zeta - \frac{1}{M} \\ (M_1+1, M_2-1) & \zeta^+ = \frac{(M_1+1)-(M_2-1)}{2M} = \frac{M_1-M_2}{2M} + \frac{1}{M} = \zeta + \frac{1}{M} \end{array} \tag{C.10}$$

On using the notation ζ, ζ^-, and ζ^+, the master equation can be rewritten as:

$$\frac{\partial f(\zeta)}{\partial t} = w_{21}(\zeta^-)f(\zeta^-) + w_{12}(\zeta^+)f(\zeta^+) - (w_{21}+w_{12})f(\zeta) \tag{C.11}$$

which, explicitly, reads:

$$\begin{aligned} \frac{\partial f(\zeta)}{\partial t} = {} & w_{21}\left(\zeta - \frac{1}{M}\right) f\left(\zeta - \frac{1}{M}\right) + w_{12}\left(\zeta + \frac{1}{M}\right) f\left(\zeta + \frac{1}{M}\right) \\ & - (w_{21}+w_{12})f(\zeta) \end{aligned} \tag{C.12}$$

Featuring a single variable, ζ, Eq. (C.12) is now ready for more comprehensive interpretations.

3. Of a first evidence is the fact that, according to the original assumption that the number M of system constituents is large, all functions involving its inverse, $1/M$—a small quantity—in the right-hand side of Eq. (C.12) may be expressed by standard, convergent series expansions than could safely be cut off at their terms in the 2nd order of $1/M$:

$$\begin{aligned} \frac{\partial f(\zeta)}{\partial t} = {} & \left(w_{21} - \frac{1}{M}\frac{\partial w_{21}}{\partial \zeta} + \frac{1}{2M^2}\frac{\partial^2 w_{21}}{\partial \zeta^2}\right)\left(f - \frac{1}{M}\frac{\partial f}{\partial \zeta} + \frac{1}{2M^2}\frac{\partial^2 f}{\partial \zeta^2}\right) \\ & + \left(w_{12} + \frac{1}{M}\frac{\partial w_{12}}{\partial \zeta} + \frac{1}{2M^2}\frac{\partial^2 w_{12}}{\partial \zeta^2}\right)\left(f + \frac{1}{M}\frac{\partial f}{\partial \zeta} + \frac{1}{2M^2}\frac{\partial^2 f}{\partial \zeta^2}\right) \\ & - (w_{21}+w_{12})f \end{aligned} \tag{C.13}$$

Straightforward multiplications in the right-hand side of Eq. (C.13), followed by ignoring all resulting terms of an order greater than 2 in $1/M$ (i.e., $1/M^3$ or $1/M^4$) and a re-grouping of the resulting terms, will now make the contributions to the time-partial-derivative of the distribution function f in the left-hand side be arranged by their order in $1/M$, as indicated in the chain of equalities (Eq. C.14), next:

$$
\begin{aligned}
\frac{\partial f(\zeta)}{\partial t} &= -(w_{21}-w_{12})\frac{1}{M}\frac{\partial f}{\partial \zeta} + (w_{21}+w_{12})\frac{1}{2M^2}\frac{\partial^2 f}{\partial \zeta^2} \\
&\quad - f\frac{1}{M}\left(\frac{\partial w_{21}}{\partial \zeta} - \frac{\partial w_{12}}{\partial \zeta}\right) + \frac{1}{M^2}\frac{\partial f}{\partial \zeta}\left(\frac{\partial w_{21}}{\partial \zeta} + \frac{\partial w_{12}}{\partial \zeta}\right) + f\frac{1}{2M^2}\left(\frac{\partial^2 w_{21}}{\partial \zeta^2} + \frac{\partial^2 w_{12}}{\partial \zeta^2}\right) \\
&= -\frac{1}{M}\frac{\partial}{\partial \zeta}[(w_{21}-w_{12})f] + \frac{1}{M}f\frac{\partial}{\partial \zeta}(w_{21}-w_{12}) - \frac{1}{M}f\frac{\partial}{\partial \zeta}(w_{21}-w_{12}) \\
&\quad + \frac{1}{2M^2}\left[(w_{21}+w_{12})\frac{\partial^2 f}{\partial \zeta^2} + 2\frac{\partial f}{\partial \zeta}\frac{\partial}{\partial \zeta}(w_{21}+w_{12}) + f\frac{\partial^2}{\partial \zeta^2}(w_{21}+w_{12})\right] \\
&= -\frac{1}{M}\frac{\partial}{\partial \zeta}[(w_{21}-w_{12})f] + \frac{1}{2M^2}\frac{\partial^2}{\partial \zeta^2}[(w_{21}+w_{12})f] \\
&= -\frac{\partial}{\partial \zeta}\left[\frac{1}{M}(w_{21}-w_{12})f - \frac{1}{2M^2}\frac{\partial}{\partial \zeta}\left((w_{21}+w_{12})\frac{\partial f}{\partial \zeta}\right)\right]
\end{aligned}
\tag{C.14}
$$

To make the long story short, Eq. (C.14) may now be written as:

$$
\frac{\partial f}{\partial t} + \frac{\partial J}{\partial \zeta} = 0 \tag{C.15}
$$

which is a manner of evidencing a quantity, J, known in Physics as a 'current':

$$
J = \frac{1}{M}(w_{21}-w_{12})f - \frac{1}{2M^2}\frac{\partial}{\partial \zeta}[(w_{21}+w_{12})f] \tag{C.16}
$$

In keeping with the same tongue/thinking, Eq. (C.15) indicates that the probability distribution function f is subject, in the inner dynamics of the system, to a 'law of conservation.'

4. It is time now for assumptions that transcends the mere Algebra: the system is assumed to find itself in a *stationary state*, (i.e., a state in which the probability distribution function, f, does not vary in time) which reads:

$$
\frac{\partial f}{\partial t} = 0 \tag{C.17}
$$

By virtue of Eq. (C.15), the condition in Eq. (C.17) automatically entails that the 'current' J does not vary with the 'system operability fraction'—as we have termed z, which in turn reads:

$$
\frac{\partial J}{\partial \zeta} = 0 \tag{C.18}
$$

Further on, the 1st derivative of the current J being nil entails that the 'current' J itself assumes a constant value, in the stationary state of the system:

$$J = \text{constant} \tag{C.19}$$

and, moreover, nothing would prevent us to take this arbitrary constant as being zero.
So, let us have a recap of the last reasoning:

$$\frac{\partial f}{\partial t} = 0 \rightarrow \frac{\partial J}{\partial \zeta} = 0, \; J = \text{constant}, \quad \text{constant} = 0 \tag{C.20}$$

5. We are now back to some algebra: given the expression in Eq. (C.16) of the 'current' J, the condition $J = 0$ (see Eq. C.20) reads in fact:

$$(w_{21} - w_{12})f = \frac{1}{2M}\frac{\mathbf{d}}{\mathbf{d}\zeta}[(w_{21} + w_{12})f] \tag{C.21}$$

Note that the system's stationary condition assumed allows us to replace partial derivatives by straight derivatives.
Observe now that we can equally write Eq. (C.21) as:

$$\frac{w_{21} - w_{12}}{w_{21} + w_{12}}(w_{21} + w_{12})f = \frac{1}{2M}\frac{\mathbf{d}}{\mathbf{d}\zeta}[(w_{21} + w_{12})f] \tag{C.22}$$

The petty trick of multiplying the left-hand side of Eq. (C.22) by $(w_{21} + w_{12})/(w_{21} + w_{12})$—which actually means by 1—will prove more useful than one may first realize: indeed, if we introduce now a new function, g, relating to f as:

$$g = (w_{21} + w_{12})f \tag{C.23}$$

then we can easily rewrite Eq. (C.22) as:

$$2M\frac{w_{21} - w_{12}}{w_{21} + w_{12}}g = \frac{\mathbf{d}g}{\mathbf{d}\zeta} \tag{C.24}$$

or, which is the same thing, as:

$$\frac{\mathbf{d}g}{\mathbf{d}\zeta} = 2M\frac{w_{21} - w_{12}}{w_{21} + w_{12}}g \tag{C.25}$$

And thus, we have ourselves an ordinary differential equation describing the stationary state of the system.
We are now three-steps-away from integral solution of this equation. It goes like this:
Step 1—change places of g and $\mathbf{d}\zeta$ in Eq. (C.25):

$$\frac{\mathbf{d}g}{g} = 2M\frac{w_{21} - w_{12}}{w_{21} + w_{12}}\mathbf{d}\zeta$$

Step 2—integrate both members in the Step 1 results in Eq. (C.26); recall your math:

- the primitive of $1/g$ is $\ln g$;
- $\ln A - \ln B = \ln (A/B)$;
- if $\ln g = \boldsymbol{C}$
  then $g = e^{\boldsymbol{C}}$

Note also that $\boldsymbol{C}$ is a constant that will remain inconsequential in the further reasoning.

$$\int_{-\frac{1}{2}}^{\zeta} \frac{dg}{g} = 2M \int_{-\frac{1}{2}}^{\zeta} \frac{w_{21}(k) - w_{12}(k)}{w_{21}(k) + w_{12}(k)}\mathbf{d}k$$

$$\ln\, g(k)\Big|_{-\frac{1}{2}}^{\zeta} = 2M \int_{-\frac{1}{2}}^{\zeta} \frac{w_{21}(k) - w_{12}(k)}{w_{21}(k) + w_{12}(k)}\mathbf{d}k$$

$$\ln\, g(\zeta) - \ln\, g\left(-\frac{1}{2}\right) = 2M \int_{-\frac{1}{2}}^{\zeta} \frac{w_{21}(k) - w_{12}(k)}{w_{21}(k) + w_{12}(k)}\mathbf{d}k$$

$$\ln\, g\left(-\frac{1}{2}\right) = \ln \boldsymbol{C}$$

$$\ln g\,(\zeta) - \ln \boldsymbol{C} = 2M \int_{-\frac{1}{2}}^{\zeta} \frac{w_{21}(k) - w_{12}(k)}{w_{21}(k) + w_{12}(k)}\mathbf{d}k$$

$$ln\frac{g(\zeta)}{\boldsymbol{C}} = 2M \int_{-\frac{1}{2}}^{\zeta} \frac{w_{21}(k) - w_{12}(k)}{w_{21}(k) + w_{12}(k)}\mathbf{d}k$$

Step 3

$$g = \boldsymbol{C}e^{2M\int_{-\frac{1}{2}}^{\zeta}\frac{w_{21}(k)-w_{12}(k)}{w_{21}(k)+w_{12}(k)}\mathbf{d}k}$$

At last, given the definition in Eq. (C.23) of function g, one obtains the target-function f as:

$$f(\zeta) = \boldsymbol{C}\cdot\frac{e^{2M\int_{-\frac{1}{2}}^{\zeta}\frac{w_{21}(k)-w_{12}(k)}{w_{21}(k)+w_{12}(k)}\mathbf{d}k}}{(w_{21}(\zeta)+w_{12}(\zeta))} \tag{C.26}$$

6. To get further on, one has to employ some more Physics (alas!..). An additional, yet intuitively natural assumption is that one has to look for *the extremal surface of the probability density of states f*, which would allow one *to detect the areas of maximal probability of system's real behavior.* In math language, looking for extremes of a function is to force its first derivative with respect to the relevant variable—in our case ζ—to zero:

$$\frac{\mathbf{d}f(\zeta)}{\mathbf{d}\zeta} = 0 \tag{C.27}$$

with f given by Eq. (C.26).
Performing the derivative of $f(\zeta)$ given by Eq. (C.26) is textbook material. A full account of the operation may look like this:

$$\frac{\mathbf{d}f(\zeta)}{\mathbf{d}\zeta} = 0 \rightarrow \frac{d}{d\zeta}\left[\frac{e^{2M\int_{-\frac{1}{2}}^{\zeta}\frac{w_{21}(k)-w_{12}(k)}{w_{21}(k)+w_{12}(k)}\mathbf{d}k}}{w_{21}(\zeta)+w_{12}(\zeta)}\right] = 0 \tag{C.28}$$

$$\begin{aligned}&\frac{\mathbf{d}}{\mathbf{d}\zeta}\left[e^{2M\int_{-\frac{1}{2}}^{\zeta}\frac{w_{21}(k)-w_{12}(k)}{w_{21}(k)+w_{12}(k)}\mathbf{d}k}\right]\cdot\frac{1}{w_{21}(\zeta)+w_{12}(\zeta)}\\&\quad+e^{2M\int_{-\frac{1}{2}}^{\zeta}\frac{w_{21}(k)-w_{12}(k)}{w_{21}(k)+w_{12}(k)}\mathbf{d}k}\cdot\frac{\mathbf{d}}{\mathbf{d}\zeta}\left[\frac{1}{w_{21}(\zeta)+w_{12}(\zeta)}\right] = 0\end{aligned} \tag{C.29}$$

$$\frac{\mathbf{d}}{\mathbf{d}\zeta}\left[e^{2M\int_{-\frac{1}{2}}^{\zeta}\frac{w_{21}(k)-w_{12}(k)}{w_{21}(k)+w_{12}(k)}\mathbf{d}k}\right]=\frac{\mathbf{d}}{\mathbf{d}\zeta}\left[2M\int_{-\frac{1}{2}}^{\zeta}\frac{w_{21}(k)-w_{12}(k)}{w_{21}(k)+w_{12}(k)}\mathbf{d}k\right]\cdot e^{2M\int_{-\frac{1}{2}}^{\zeta}\frac{w_{21}(k)-w_{12}(k)}{w_{21}(k)+w_{12}(k)}\mathbf{d}k}$$
$$=\frac{\mathbf{d}}{\mathbf{d}\zeta}\left[2M\int_{-\frac{1}{2}}^{\zeta}\frac{w_{21}(k)-w_{12}(k)}{w_{21}(k)+w_{12}(k)}\mathbf{d}k\right]$$
$$=2M\frac{w_{21}(\zeta)-w_{12}(\zeta)}{w_{21}(\zeta)+w_{12}(\zeta)} \tag{C.30}$$

$$\frac{\mathbf{d}}{\mathbf{d}\zeta}\left[\frac{1}{w_{21}(\zeta)+w_{12}(\zeta)}\right]=-\frac{\frac{\mathbf{d}}{\mathbf{d}\zeta}\left[w_{21(\zeta)}+w_{12(\zeta)}\right]}{\left[w_{21}(\zeta)+w_{12}(\zeta)\right]^2}=-\frac{w'_{21}(\zeta)+w'_{12}(\zeta)}{\left[w_{21}(\zeta)+w_{12}(\zeta)\right]^2} \tag{C.31}$$

$$\left[2M\frac{w_{21}(\zeta)-w_{12}(\zeta)}{w_{21}(\zeta)+w_{12}(\zeta)}\cdot\frac{1}{w_{21}(\zeta)+w_{12}(\zeta)}-\frac{w'_{21}(\zeta)+w'_{12}(\zeta)}{\left[w_{21}(\zeta)+w_{12}(\zeta)\right]^2}\right]e^{2M\int_{-\frac{1}{2}}^{\zeta}\frac{w_{21}(k)-w_{12}(k)}{w_{21}(k)+w_{12}(k)}\mathbf{d}k}=0 \tag{C.32}$$

$$2M\frac{w_{21}(\zeta)-w_{12}(\zeta)}{\left[w_{21}(\zeta)+w_{12}(\zeta)\right]^2}-\frac{w'_{21}(\zeta)+w'_{12}(\zeta)}{\left[w_{21}(\zeta)+w_{12}(\zeta)\right]^2}=0 \tag{C.33}$$

$$\frac{w'_{21}(\zeta)+w'_{12}(\zeta)}{w_{21}(\zeta)-w_{12}(\zeta)}=2M \tag{C.34}$$

In the equations above, apostrophes (') indicate first derivatives of the functions $w_{12}(\zeta)$ and $w_{21}(\zeta)$ with respect to ζ.

7. At this stage, an analytic look into how the transition probability functions $w_{12}(\zeta)$ and $w_{21}(\zeta)$ may look like can no longer be avoided. The solution is again inspired by standard Statistical Physics: in physical systems of binary-state entities such as Ising, or Heisenberg magnets, where magnetic moments are carried by ½ 'spins,' holding either ½ or—½ values, the transition probabilities are assumed to depend on system's state variable ζ as follows:

$$\begin{cases} w_{12}=wM_1e^{-\frac{u\zeta+v}{\theta}}=wM\left(\frac{1}{2}+\zeta\right)e^{-\frac{u\zeta+v}{\theta}} \\ w_{21}=wM_2e^{\frac{u\zeta+v}{\theta}}=wM\left(\frac{1}{2}-\zeta\right)e^{\frac{u\zeta+v}{\theta}} \end{cases} \tag{C.35}$$

with already known notations, there are notable exceptions involving parameters u, v and θ in the exponentials.

To avoid becoming completely parochial, let us confine ourselves to loosely saying that u, that multiplies the 'system functionality variable' ζ, is *a measure of the intensity of interaction between any two, closeset-neighbouring, entities in the system*; whereas v is *a measure of the interaction of entities with influences outside the system*, known as 'fields.' For more on these, we redirect the reader to the main in Sect. 2.2 (Chap. 2) and the accompanying references. On the other hand, θ is *some measure of system's 'temperature'*—that again should be understood in the context as suggested in Sect. 2.2 in Chap. 2.
And now, back to elementary calculus: taking the first derivative of $w_{12}(\zeta)$ and $w_{21}(\zeta)$ yields:

$$\begin{cases} w'_{12} = wM\left[1 - \frac{u}{\theta}\left(\frac{1}{2} + \zeta\right)\right]e^{-\frac{u\zeta + v}{\theta}} \\ w'_{21} = wM\left[-1 + \frac{u}{\theta}\left(\frac{1}{2} - \zeta\right)\right]e^{\frac{u\zeta + v}{\theta}} \end{cases} \tag{C.36}$$

Taking the expressions of w'_{12}, w'_{21} from Eq. (C.36) and the expressions of w_{21} and w_{12} from Eq. (C.35) into the Eq. (C.34) of system's surface of extremal probability one has:

$$\frac{wM\left[-1 + \frac{u}{\theta}\left(\frac{1}{2} - \zeta\right)\right]e^{\frac{u\zeta + v}{\theta}} + wM\left[1 - \frac{u}{\theta}\left(\frac{1}{2} + \zeta\right)\right]e^{-\frac{u\zeta + v}{\theta}}}{wM\left(\frac{1}{2} - \zeta\right)e^{\frac{u\zeta + v}{\theta}} - wM\left(\frac{1}{2} + \zeta\right)e^{-\frac{u\zeta + v}{\theta}}} = 2M \tag{C.37}$$

Processing the fraction in (C.37) by simplifications and terms re-grouping is, again, textbook stuff. Here it is:

$$\frac{\left[-1 + \frac{u}{\theta}\left(\frac{1}{2} - \zeta\right)\right]e^{\frac{u\zeta + v}{\theta}} + \left[1 - \frac{u}{\theta}\left(\frac{1}{2} + \zeta\right)\right]e^{-\frac{u\zeta + v}{\theta}}}{\left(\frac{1}{2} - \zeta\right)e^{\frac{u\zeta + v}{\theta}} - \left(\frac{1}{2} + \zeta\right)e^{-\frac{u\zeta + v}{\theta}}} = 2M \tag{C.38}$$

$$\frac{\frac{u}{\theta}\left[\left(\frac{1}{2} - \zeta\right)e^{\frac{u\zeta + v}{\theta}} - \left(\frac{1}{2} + \zeta\right)e^{-\frac{u\zeta + v}{\theta}}\right] - \left[e^{\frac{u\zeta + v}{\theta}} - e^{-\frac{u\zeta + v}{\theta}}\right]}{\left(\frac{1}{2} - \zeta\right)e^{\frac{u\zeta + v}{\theta}} - \left(\frac{1}{2} + \zeta\right)e^{-\frac{u\zeta + v}{\theta}}} = 2M \tag{C.39}$$

$$\frac{u}{\theta} - \frac{\left[e^{\frac{u\zeta + v}{\theta}} - e^{-\frac{u\zeta + v}{\theta}}\right]}{\left(\frac{1}{2} - \zeta\right)e^{\frac{u\zeta + v}{\theta}} - \left(\frac{1}{2} + \zeta\right)e^{-\frac{u\zeta + v}{\theta}}} = 2M \tag{C.40}$$

$$\frac{\left[\left(\frac{1}{2} - \zeta\right)e^{\frac{u\zeta + v}{\theta}} - \left(\frac{1}{2} + \zeta\right)e^{-\frac{u\zeta + v}{\theta}}\right]}{\left[e^{\frac{u\zeta + v}{\theta}} - e^{-\frac{u\zeta + v}{\theta}}\right]} = \frac{1}{\frac{u}{\theta} - 2M} \tag{C.41}$$

$$\frac{\frac{1}{2}\left[e^{\frac{u\zeta+v}{\theta}}-e^{-\frac{u\zeta+v}{\theta}}\right]-\zeta\left[e^{\frac{u\zeta+v}{\theta}}+e^{-\frac{u\zeta+v}{\theta}}\right]}{\left[e^{\frac{u\zeta+v}{\theta}}-e^{-\frac{u\zeta+v}{\theta}}\right]}=\frac{1}{\frac{u}{\theta}-2M} \tag{C.42}$$

$$\frac{1}{2}-\zeta\frac{e^{\frac{u\zeta+v}{\theta}}+e^{-\frac{u\zeta+v}{\theta}}}{e^{\frac{u\zeta+v}{\theta}}-e^{-\frac{u\zeta+v}{\theta}}}=\frac{1}{\frac{u}{\theta}-2M}$$

$$\frac{1}{2}-\zeta\coth\left(\frac{u\zeta+v}{\theta}\right)=\frac{1}{\frac{u}{\theta}-2M} \tag{C.43}$$

$$\coth\left(\frac{u\zeta+v}{\theta}\right)=\left(\frac{1}{2}-\frac{1}{\frac{u}{\theta}-2M}\right)\frac{1}{\zeta}$$

8. The last step into the chores requires substantive observation that, in the right-hand side of Eq. (C.43), the 2nd fraction in parenthesis has the large number M as its denominator, overwhelming—at nonzero 'temperatures' θ, a zero-temperature being inconceivable anyway, the u/θ term, which makes the respective fraction negligible when compared with ½—the 1st in the same parenthesis. If we ignore the said small fraction, this leaves us with:

$$\coth\left(\frac{u\zeta+v}{\theta}\right)\sim\frac{1}{2\zeta} \tag{C.44}$$

A simple inversion of the quantities in Eq. (C.44) now gives:

$$\boxed{\tanh\left(\tfrac{u\zeta+v}{\theta}\right)=2\zeta} \tag{C.45}$$

where we take the liberty of making the equality categorical, which, in the light of the arguments displayed in this appendix, is believed to be defendable.

The readers may wish to recognize the result, Eq. (C.45), as the equation of state of the system of binary-state entities discussed in Sect. 2.2 of this book.

A final remark: neither the book authors, nor their associates quoted in relation with the subject of this appendix, are claiming or have ever claimed to originate the way of thinking and methodological clues regarding how one arrives at the equation for system state with many bi-stable entities. Those assets belong to such highly noted (and duly quoted) predecessors like Haken and Weidlich in Synergetics, Thom in the Theory of Catastrophes; Bragg, Williams, Ising, and Heisenberg in Physics; and many others as noted in the References. Our only feat was to make the solutions work for our purposes.

References

Gheorghe, A. V., & Vamanu, D. V. (2004). Towards QVA—Quantitative Vulnerability Assessment: A generic practical model. *Journal of Risk Research*, 7(6), 613–628.
Vamanu, B. I., Gheorghe, A. V., & Katina, P. F. (2016). *Critical infrastructures: Risk and vulnerability assessment in transportation of dangerous goods—Transportation by road and rail* (Vol. 31). Cham, Switzerland: Springer International Publishing.

Appendix D: The Mix Game

In Sect. 2.2.1, the basic assumptions of the QVA model were stated. In an attempt to make this 'fun,' researchers developed a 'mix game' approach that elaborates on the outlined concepts. This ad-hoc, and simplified model conceived for the objective-oriented design of optimal primary energy mixes, has set to deliberately avoid the arcane of both the textbook linear, and nonlinear, programming technology, while however retaining its basic philosophy that, as these authors believe, can be summarized as:

Pursue your target while observing the given constraints

Instead of betting on consecrated mathematical techniques, the gaming has chosen to draw on the implicit ability of fast, mechanical computation to sort out many thousands of random throws (i.e., values assigned to hosts of variables), by their degree of compliance with numerical and logical (Boolean) constraints.

The master plan of the approach goes as follows:

Organize the Variables

Let

$$\mathrm{mix}(i,j); i = 1, 2, \ldots, n\mathrm{Mix}, \tag{D.1}$$

be an array of features of nMix energy sources. Note that: (i) '*energy source*' means, in the context, a pair consisting of a primary energy species (e.g., natural gas, oil, coal) and a conversion technology to an end-use form—electric power, or heat-and-work, and (ii) '*features*' designate, again in the context, physical quantities attached to 'energy sources' on both the input side—as variables feeding a multi-attribute source assessment, and the output side—as containers for assessment results the aggregation of which would result in a *choice*—of 'the best' energy mix, analyst's (gamer's) criteria/preferences considered.

For the case in point, the energy sources and features have emerged as a compilation of data adopted in the EU reference document (European Commission 2007). The sources read as follows (Table D.1):

where $i = 1, 2, \ldots, 19$ is the first array index in Eq. (D.1).

The features are grouped in the following manner (Table D.2), the indices in the first column being the j-index in the mix(i, j) array, of Eq. (D.1).

The mix assessment also requires additional data. This data is provided in Table D.3:

In fact, the composition of the data palette above is reflective of an assessment philosophy that looks for a primary energy mix: (i) starting from the end-use demand, (ii) observing a discipline of not exceeding a certain annual rate of depletion, of proven reserves, and (ii) compliant with national commitments, at political level, concerning GHG emissions while also giving away an allowance to other, non-energy-related activities prone to increment the GHG environmental burden.

Table D.1 Energy sources

1	Natural gas, open cycle gas turbine (OCGT)
2	Natural gas, combined cycle gas turbine (CCGT)
3	Natural gas, heat-and-work (HW)
4	Oil, diesel engine
5	Oil, heat-and-work (HW)
6	Coal, pulverized fuel with flue gas desulphurisation (PF)
7	Coal, circulating fluidized bed combustion (CFBC)
8	Coal, integrated gasification combined cycle (IGCC)
9	Coal, heat-and-work (HW)
10	Nuclear, light water reactor (LWR)
11	Biomass, generation plant
12	Biomass, heat-and-work (H/W)
13	Wind, offshore
14	Wind, onshore
15	Hydro, large
16	Hydro, small (< 10 MW)
17	Solar, photovoltaic (PHV)
18	Solar, heat-and-work (HW)
19	Geothermal, heat (H)

Table D.2 Features of mix

Index		Features
On the input side:	0	Source is (value = 1), or is not (value = 0), selected in the mix;
	1	Source addresses electric power as an end use (value = 0), or heat-and-work (value = 1)
	2	Market price of energy delivered by source, in euro/toe (ton oil equivalent);
	3	Cost of energy delivered by source, in euro/toe
	4	The efficiency of energy conversion from primary to the end use
	5	The specific Greenhouse Gas (GHG) Emissions, in tCO_2 eq/toe (ton of carbon dioxide equivalent emission, per ton of oil equivalent energy generated)
On the output side	6	Electric power supply secured by the 'source,' in toe/year
	7	Heat-and-work supply secured by the 'source,' in toe/year
	8	Energy sales secured by the source, in euro/year
	9	Energy costs incurred by the source, in euro/year
	10	Profit (raw) obtained (sales—costs) from the source, in euro/year
	11	Pollution entailed by operating the source, intCO_2 eq/year

Table D.3 Additional data for the mix

Quantity		Variable name
Energy demand data	Population (persons)	Population[a]
	Total demand (toe/year), of which:	Demand
	Electric power (% of total)	Demandpower
	Heat-and-work (% of total)	Demandheat
	Nonenergy uses (% of total)	Demandother
	Demand satisfaction target (% covered by supply)	DemandTarget
Reserve-driven consumption ceilings (independent of conversion technology)	Natural gas	Stress$(1)
	Oil	Stress$(2)
	Coal	Stress$(3)
	Nuclear	Stress$(4)
	Biomass	Stress$(5)
	Wind	Stress$(6)
	Hydro	Stress$(7)
	Solar	Stress$(8)
	Geothermal	Stress$(9)
Other constraints	GHG national target ceiling (tCO2 eq/yr)	GHGtarget
	Allowance for non-energy polluters (% of target)	Allowance

[a]This variable is idle in the current model version

Design an Algorithm Implementing the Adopted Mindset

The main steps are:

Step (1) Initialize data

```
maxProfit = 0 'maximum profit, of a game session
maxSales = 0 'maximum sales, of a game session
minCosts = val("1.0e30") 'minimum costs, of a game session
minPollution = val("1.0e30") 'minimum pollution, of a game session
minImports = val("1.0e30") 'minimum imports, of a game session
```

Step (2) Set a number of iterations, niter, for a game session; default is 10.

Step (3) Loop throughout iterations:

```
for jiter = 1 to niter 'start with iteration #1 and end with iteration #niter
gosub [plan] 'the through-and-optimize routine
```

Step (4) As the `[plan]` routine has delivered the current iteration's 'best' findings—maximal or minimal, as appropriate, namely

```
'profit—maximum,
'sales—maximum,
'costs—minimum,
'totImportCost—minimum of the total costs of imports required to complement the
'supply from domestic resources, and
'pollution—minimum
```

proceed to retaining the extremes that would enable a summary of options, by the end of the iteration:

```
if profit > maxProfit then
maxProfit = profit
profitIter = jiter
if sales > maxSales then
maxSales = sales
salesIter = jiter
if costs < minCosts then
minCosts = costs
costsIter = jiter
if totImportCost < minImports then
minImports = totImportCost
importIter = jiter
if pollution < minPollution then
```

minPollution = pollution
pollutionIter = jiter

Step (5) Display the results of the current iteration.
Step (6) Go for the next iteration:

next jiter

Step (7) Summarize.

All iterations are so designed as to *maximize profit* while observing the constraints set by the admitted limited capability to satisfy the demand; the physical/economical limitations in drawing upon the domestic reserves; the committed targets on energy import ceiling; and the politically committed environmental (GHG) targets—not necessarily in this order. However, given the stochastic nature of the game, of any number of iterations several may catch the attention by the end of a game session, apart from 'the highest-profit mix choice.' These include: (i) the high-profit-*highest-sales* mix choice, (ii) the high-profit-*lowest-costs* mix choice, (iii) the high-profit-*lowest pollution* mix choice, and (iv) the high-profit-*lowest-imports* mix choice, all deserving, in game authors' opinion, an interest in their own right. In consideration of this, a game session obliges by listing all choices as mentioned.

The Game Key

Key to the entire process is the `[plan]` throw-and-optimize routine. It works on a number of iterations (i.e., throws) of its own. The respective variable is `n`, and it is hard-coded, in this version to `n = 25000`. One has:

```
[plan]
n=25000 'Set the number of throws to 250000
Emax=0 'Initialize a variable holding the highest profit, to 0

for jn=1 to n 'Start throwing...

for k=1 to 9 'Initialize the reserves depletion container
stress(k,1)=0 (see the stress$(k), k = 1, 2, ..., 9 vector in Table 3
next

pSupply=0 'Initialize cumulative power supply, of mix
hSupply=0 'Initialize cumulative heat-and-work supply, of mix
sales=0 'Initialize cumulative sales, of mix
costs=0 'Initialize cumulative costs, of mix
profit=0 'Initialize cumulative profit, of mix
pollution=0 'Initialize cumulative pollution, of mix

E=0 'Initialize trial profit variable to 0

'The natural steering factor in code's 'throws' (random, trial allocations per mix)
'is the demand. Using the inputs one then computes, first:
pDemand=demand*demandpower*0.01
hDemand=demand*demandheat*0.01
oDemand=demand*demandother*0.01

for i=1 to nmix 'Loop throughout the mix
if mix(i,0) then 'Assess only components that were user-selected in the mix
if mix(i,1)=0 then 'In the sequel, all quantities are per year:
mix(i,6)= rnd(1)*2.0*pDemand/nPower 'trial power supply, as a fraction of demand
pSupply=pSupply+mix(i,6) 'cumulate...
else
mix(i,7)= rnd(1)*2.0*hDemand/nHeat 'trial heat supply, as a fraction of demand
hSupply=hSupply+mix(i,7) 'cumulate...
end if
mix(i,8)=(mix(i,6)+mix(i,7))*mix(i,2) 'resulting trial sales
sales=sales+mix(i,8) 'cumulate...
mix(i,9)=(mix(i,6)+mix(i,7))*mix(i,3) 'resulting trial costs
costs=costs+mix(i,9) 'cumulate
mix(i,10)=mix(i,8)-mix(i,9) 'resulting profit
profit=profit+mix(i,10) 'cumulate
mix(i,11)=(mix(i,6)+mix(i,7))*mix(i,5) 'resulting pollution
pollution=pollution+mix(i,11) 'cumulate...

for k=1 to 9 'compute cumulated stress from
if instr(mix$(i),stress$(k))>0 then 'trial supply, on domestic reserves
if mix(i,1)=0 then stress(k,1)=stress(k,1)+mix(i,6)
if mix(i,1)=1 then stress(k,1)=stress(k,1)+mix(i,7)
end if
next k
end if

next i 'close the mix loop per throw

E=profit 'set optimization variable to the profit/year

okstress=1 'determine whether reserves
for k=1 to 9 'constraints are observed
if stress(k,1)>stress(k,0) then okstress=0
next
```

```
okpollution=1 'determine whether GHG
if pollution>GHGtarget*(1-allowance/100) then okpollution=0 'targets are observed

pAvgPrice=0 'compute an average market price,
mm=0 'over mix components ('sources')
for m=1 to 19
if mix(m,0) then
mm=mm+1
pAvgPrice=pAvgPrice+mix(m,2)

end if
next
pAvgPrice=pAvgPrice/mm 'the mix-averaged price

pImport=(pDemand-pSupply) 'power import needed, physical
hImport=(hDemand-hSupply) 'heat-and-work-oriented import needed, physical
totImport=pImport+hImport 'total imports needed, physical
pImportCost=pImport*pAvgPrice 'power imports cost
hImportCost=sImport*pAvgPrice 'heat-and-work-oriented imports cost
totImportCost=pImportCost+hImportCost 'total cost of annual imports required to
demandCost=demand*pAvgPrice 'fully meet the demand
okdemand=1

'now test whether the costs of imports exceed the Demand Satisfaction Target (v. Table 3)
if totImportCost/demandCost>demandTarget/100 then okdemand=0

'...And the moment of truth:
' - if the error in meeting the current Maximum Profit Target, Emax, is larger than
' 1/100 of the current profit, E, then reset the Maximum Profit 'Target to the
' current profit value, and continue to throw, within the preset limit of '25000
' throws;
' - if, otherwise, the error falls below 0.01 of the current profit, then exit the
' 250000-throws loop, and exit the routine, leaving to the main code to print the
' current iteration results and go for the next iteration. This reads:

if okstress=1 and okpollution=1 and okdemand=1 and E>=Emax then
if abs(E-Emax)>=0.01*E then
Emax=E
else
exit for
end if
end if

next jn
return
```

The inequality from the computer output above was proved, in numerical experiments, to ensure an acceptable convergence of the process, so that conclusive mix options could be obtained without open-loop incidents, despite the random nature of the initializing (i.e., the throwing) drill.

Limitations, Caveats

To be sure, the model behind the game is no more and probably far less conducive to *absolute* maxima/minima than what is described in Chap. 2, Sect. 2.1.1 or advanced linear/nonlinear-programming models. However, it does produce extremal values which, even if relative, are convincing enough about the notion that designing proper energy mixes as a part of energy strategic planning. While hard to control in every detail of the multi-attribute process, and sometimes rippled by random perturbations, given the physical or human in origin nature, and perhaps also requiring a twist of *good instincts* for gambling, the model is, nevertheless, an attainable goal.

An underlying message to all those who have found some pleasure in playing the Mix Game is that the science and art of *Modeling, Simulation and Visualization* (MS&V) is increasingly becoming a must in the risky business of designing sustainable futures—energy and all.

Reference

European Commission. (2007). *An energy policy for Europe. Communication from the Commission to the European Council and the European Parliament* (No. {SEC(2007) 12}). Brussels, Belgium: Commission of the European Communities.

Appendix E: Systems Pathologies

Webster's New Explorer Encyclopedic Dictionary suggests that the term *pathology* is derived from two ancient Greek terms: *pathos* (i.e., suffering, experiencing and emotions) and *logia* (i.e., the study of) (Merriam-Webster 2006). Etymologically, this terms "has historically been related to attempts to understand observed symptoms and determining the cause of disease and death through dissection" (Katina 2015a, p. 248). However, there is an increasing trend to apply the underlying concepts to other fields. See, for example, Barnard (1946) in management theory and organizational studies, Dery (1984), in policy analysis, Beer (1984) and Ríos (2012) in management cybernetics, Bobba et al. (2007) in computer systems, Sheptycki (2004) intelligent systems, Keating and Katina (2012) in system of systems and Davidz (2017) and Troncale (2011a, 2011b, 2013) in systems engineering. In this appendix, a listing of pathologies developed as part of doctoral research (Katina 2015b, 2016a, 2016b) addressing utility of systems theory in problem formulation is provided. This list of pathologies could be incorporated into different models that inform (including of pathologies) governance of critical and complex system (of systems) in space, undersea, and belowground. Certainly, this research would include aspects of identifying and diagnosis (including pathogenesis) and development of treatment against pathologies.

Systems Theory-Based Pathologies

Pathology of complementarity. A situation in which an organization ignores other perspectives/models that are not entirely compatible with the established-predominate perspectives including missions, goals, and objectives. An organization in this case mistakenly assumes that there is only one 'right' perspective.

Pathology of diminishing returns. A condition in which management mistakenly assumes that increasing number of workforce increases the productivity of the organization as a whole without expanding the landscape of operations.

Pathology of requisite hierarchy. A situation in which the regulatory body of an organization is not well-designed to match the variety of the organization. This pathology is evident in situations where variety of the system is higher than what the regulatory body can handle.

Pathology of requisite knowledge. A situation in which an organization simply has a bad regulator. A bad regular for an organization is simply a regulator that is not well-informed of the relevant facts that enable organizational viability.

Pathology of requisite parsimony. A condition in which a system fails because the human element of the organization has assumed more activities than what can reasonably be handled. The number is limited to seven plus or minus two.

Pathology of requisite saliency. A condition in which organization productivity is reduced due to having undifferentiated importance of organizational missions and objectives. This pathology is related to having *spurious saliency* (i.e., the organization is emphasizing the wrong elements, out of proportion to what they deserve), *unproductive emulation* (i.e., members of the organization might be behaving as those who help create rather than resolve problems), and *having a cultural lag* (i.e., not operating using a common established knowledge base).

Pathology of requisite variety. Specifically addressing channel linking the regulator and system, this is a situation in which the regulatory entity of an organization has insufficient capacity to address the variety of the system.

Pathology of adaptation. A situation in which neither the internal structures of a system are able to change in response to external disturbances, nor system being able to lessen environmental changes affecting it.

Pathology of autonomy. A situation in which a subsystem does not have the ability to act as an independent agent without the constraints of a higher system. Autonomy in this case might include being able to make decisions and taking actions.

Pathology of balance of tensions. A situation in which the system lacks a governing structure that can relive tension among different subsystems/elements. The governing structure (i.e., the metasystem structure) can be used to balance tensions along the dimensions of (1) independence of subsystems and missions of the whole,

(2) structured design and self-organization, and (3) maintaining stability and allowing for change commensurate with unpredictability in the system/environment.

Pathology of basins of stability. A condition in which system's stability is reduced because of the inability to recognize different system configurations and their periods of transitions. It has been suggested that complex systems have three configurations: *order*, *chaos*, and *transition phase*. Each configuration requires different resources and produces different consequences.

Pathology of buffering. A condition in which a system lacks surplus resources. In essence, the system is being operated without slack. In this case, slack is reserve and might be defined as 'capacity in excess of immediate needs.'

Pathology of circular causality. A situation in which a traditional (linear) causality model of thinking is applied without recognizing the intricate interactions in subsystems of a complex system. Under the traditional model of thinking, event A is directly related to B (i.e., causes) and in turn B causes C. Emphasis is placed on finding single causes while ignoring a multitude of other factors.

Pathology of consequent production. A condition in which there is failure to focus on the underlying structure of the system causing the outcomes/outputs, desired or otherwise. The focus should be on attempting to (re)calibrate the structures of the system in order to produce an improved product or service.

Pathology of cybernetic stability. A condition in which a system lacks a sufficient number of external connections. This is a like a freestanding structure. It has been suggested that an increased the number of connections makes a system more stable and easily adaptive.

Pathology of darkness. A situation in which a system is operated upon under the assumption that all its relevant aspects including behaviors are known.

Pathology of dialecticism. A condition in which a system lacks the ability to detect errors and learn. More specifically, this condition involves the lack of means to correct errors through single loop where reflection is made on what is good/bad about operations.

Pathology of emergence. A condition in which management assumes behaviors of the system whole can be directly inferred from properties of subsystems, independent of subsystem interaction. In this case, management fails to recognize that complex systems exhibit behaviors beyond those of the individual subsystems.

Pathology of environmental-modification. A condition in which a system fails to negotiate its environment. As indicated by the pathology of adaptation, systems can either change themselves or change the environment. The pathology of environmental-modification places more emphasis on the efforts undertaken to influence the environment of the system.

Pathology of equifinality. A situation in which a system is operated with a belief that there exists only one approach/method to achieve a final desired state —

including goals and missions. There might indeed be one approach/solution, however, the issue at hand is whether other alternative approaches can be examined and taken into consideration.

Pathology of equivocation. A situation in which communication channels of a system are inefficient in delivering intended signal (i.e., messages) from one point to the next. In delivering messages (i.e., information), the sender may wish to conceal the meaning so that only the intended receiver can decipher and understand its meaning. In a secret system, the receiver is able to understand the meaning.

Pathology of eudemony. A situation in which precedence is placed on the financial profitability of a system above any other measures. This situation involves ignoring import measures that are desirable in describing overall well-being since they not easily quantifiable. Specifically, the literature suggests that the overall well-being of a system, including people and the society at large, is related to having the right balance in material, technical, physical, social, nutritional, cognitive, spiritual, and environment.

Pathology of events of low probability. A situation in which a complex system is expected to accommodate all scenarios including those of low probability. More specifically this pathological condition indicates that it's an error to attempt to be all things to all people at all times.

Pathology of feedback. A situation in which a system lacks the means to improve its behaviors because of insufficient scanning processes. Scanning processes provide the basis for bringing the system close to a desired state.

Pathology of flatness. A situation in which the structure of governance is an inverted pyramid. This is a situation in which there is a 'larger the number of administrators relative to that of producers.'

Pathology of frame of reference. A situation in which a system lacks standards by which it can be judged. In this case, a standard is not a sufficient measure for the truth of the judgment but it is a reliable indication of how the system and its elements are.

Pathology of hierarchy. A situation in which a system lacks a basic structure of a hierarchy. A hierarchy provides a regulatory structure that enables 'organization' of the system to generate desired system performance/behavior.

Pathology of high-flux. A situation in which the rate of arrival of resources to systems is less than failures. Related to recovery time, the pathology of high-flux suggests the need to have resources arrive as soon as failure occurs. The lag in arrival of resource has implications on system stability.

Pathology of holism. A situation in which the management assumes a mode of operation suggesting that behaviors of an integrated system are possessed in its subsystem parts. This pathology is different from the pathology of emergence in that it suggests that understanding of a system cannot be maintained past a particular point of reduction. Under the pathological condition of holism, there are system properties (i.e., behaviors) that cannot be deduced from parts; likewise, there are subsystem behaviors that cannot be deduced from the system.

Pathology of homeorhesis. A situation in which a system lacks mechanisms to guide and enable it to return it to a pre-set path or trajectory following an environmental disturbance.

Pathology of homeostasis. A situation in which a system lacks monitoring mechanisms that can be used to alert of any external changes affecting system's essential internal variables. Systems can use negative feedback to reduce fluctuations in the output caused by the environment.

Pathology of internal elaboration. A condition in which the management style creates silos due to overemphasis on development of policies and procedures of subsystems and people management.

Pathology of iteration. A situation in which a system lacks means to account for continuous comparison of first iteration to the norm to discover errors. Similar to a continuous process that keeps comparing actual state and the desired state of the system, the iteration process provides the means to measure errors in a timely manner.

Pathology of least effort. A situation in which a system attempts to move forward by selection of a path of high resistance. Started differently, this is a situation in which a system pursues its goals using methods and tools that are deemed inefficient.

Pathology of maximum power. A situation in which a system lacks ability to maximize its production through increased capacity for intake and transformation rate. Failure to be able to keep up with demand.

Pathology of minimal critical specification. A situation in which a system is managed by prescribing detailed account of what must be done and how it must be done. In managing complex systems, it is recommended to minimal specifications.

Pathology of multifinality. Involves the notion of experience and the fact that humans have a tendency to draw premature conclusions regarding complex situations that they have previously experienced. Consequently, it is an error for one to anticipate the same results using the same approach; outcomes might vary widely based on subtle situational differences.

Pathology of omnivory. A situation in which system's internal structures (i.e., pathways) cannot be modified to increase their ability to intake a diverse number of resources. Systems that can take in a diverse number of resources are more stable since a decline on one of the resources will not affect the system.

Pathology of organizational closure. A situation in which a system lacks a critical part in the structure that provides closure.

Pathology of over-specialization. A situation in which a system is so specialized that it cannot afford to change.

Pathology of Pareto. A condition in which significant efforts are undertaken to alter the '80/20 production curve.' This pathology steams from assuming the existence of a 'causal-interrelationships' evident in simple systems.

Pathology of patchiness. A situation in which a system lacks ability to increase diversity in terms of consumption of resources from the environment. This pathology does not apply situations where the environment has only one resource. Counter to the *pathology of omnivory* which primarily addresses diversification of internal structures, patchiness pathology addresses complex system failure to 'acquire' a taste for different resources such that 'if one set of resources declines, there will not be any other to take their place.'

Pathology of polystability. A circumstance in which a system is managed as if system-level equilibrium is similar to its subsystems. Subsystems have their own equilibriums which are different from that of the system.

Pathology of redundancy of potential command. A condition where subsystems entities lack the 'freedom' to decide and act on behalf on the system.

Pathology of redundancy of resources. A condition in which a system is designed and operated under the assumption of optimum efficiency. Under this condition, the allocated resources, for example, might be exactly what is needed—no more no less. In other words, critical redundant resources are not provided.

Pathology of relaxation time. A situation in which a system experiences too many changes at the same time. When a system is continuously bombarded with many changes, it becomes incapable of processing or assimilating any of the changes and becomes chaotic.

Pathology of resilience. A situation in which a system, when it experiences a disturbance, has no ability to quickly return to its previous configuration.

Pathology of robustness. A situation in which a system lacks the ability to use simple and/or complex mechanisms to withstand environmental changes without modifying the system.

Pathology of safe environment. A situation in which a system fails to create a permanently stable environment.

Pathology of satisficing. A condition in which the management team of a system searches for the best possible solution (i.e., optimization) instead of searching for good-enough solution (i.e., satisficing).

Pathology of self-organization. A condition in which management fails to work with the self-organizing tendencies of complex systems. This condition happens when an organizing structure limits autonomy of its subsystems by using global patterns to influence local interactions.

Pathology of separatibility. A situation in which subsystems are tightly coupled together such that a small disturbance is reflected throughout the entire system. In other words, the tight coupling in a large number of subsystems along with positive feedback creates the right conditions for a single breakdown in one of the subsystems to have a major effect on other subsystems and the system as a whole.

Pathology of steady state. A condition in which one focuses on the steady state of a system whole while ignoring steady states of subsystems. This is an error since a system cannot be in a steady state if any of its subsystems are not in steady states.

Pathology of sub-optimization. This condition elaborates on several other pathologies including emergence, holism, and satisficing. It suggests that independent improvement of subsystems does not always improve the performance of the integrated system whole.

Pathology of subsidiarity. A situation in which local issues need to always be solved by a higher authority. A local issue is a subsystem issue and a local authority must solve it.

Pathology of system context. An attempt to address systemic issues (or systems) independent of the context in which they are embedded. It is impossible to understand and draw the meaning of system independent of its context.

Pathology of the first cybernetic control. A situation in which system lacks ability to compare system behavior against a set standard. When the comparison is done, the system might lack mechanism to enable corrective measures and actions to be undertaken.

Pathology of the Red Queen. A condition in which a system fails to survive because of its inability to compete with other systems in the same environment. This goes beyond the idea of adapting, evolving and proliferation inasmuch as they relate to gaining a competitive advantage. It relates to the idea of simply surviving inasmuch as surviving means that an organization takes all the running it can do, just to stay in the same place.

Pathology of the second cybernetic control. Similar to the *first cybernetic control pathology* and addressing control in terms of communication, this pathology suggests that a system might go out of control if its communication elements are incapable of providing sufficient regulations to address variety. In this case, communication provides regulations that enable the system to address any disturbances that might impede the system.

Pathology of the third cybernetic control. This is a grave warning against tinkering with an unbroken system. It states that a system can only be brought into control (i.e., a more preferred state), if and only if it has gone out of control.

Pathology of transcendence. The assumption that stability and viability in complex systems can only be achieved within the confines of reality as defined and understood within the objective realm of 'scientific' approach.

Pathology of ultra-stability. A condition in which a system can fend off knows and anticipated disturbances but it is not sufficiently designed to fend off unknown disturbances without changing its internal structures; stability at a logically higher level.

Pathology of undifferentiated coding. This pathology deals with the issue of 'objectivity' and 'subjectivity' in understanding issues affecting systems. More specifically, this pathology is a situation in which reality and knowledge are directly attributed to observable results such that anything that does not involve human sensors such as eyes, ears, and touch is not valued.

Pathology of unity. A situation in which a system lacks an integrated system purpose or having an identity that is not easily distinguishable from other systems.

Pathology of viability. Concerned with failure to balance two related elements: subsystem autonomy and integration of the whole and system stability and system adaptation.

Shannon-Hartley's channel capacity pathology. This pathology has to do with the ability of a communication channel to transmit different messages without channel modification. A well-designed communication channel accounts for noise (i.e., any factor in the process that works against the predictability of the outcome of the communication process) in transmission.

Gödel's incompleteness pathology. Operating a system upon the assumption that the traditional terms of discourse/frame of reference is both consistent and complete. Any given frame of reference/framework is always incomplete.

Pathology of information redundancy. A situation in which little and perhaps insufficient effort is dedicated to reducing error in information transmission. More specifically, it suggests that transmission of information (i.e., communication) can be enhanced through making redundancy of transmitted messages.

Pathology of morphogenesis. A situation in which a system fails to remain stable after creating a new and radically different structure (system) elaborating on the existing structures as conditioned by morphocatalyst influencing the system.

Pathology of morphostasis. A condition in which stability of an organization is reduced by resisting change; preferring the *status quo*.

Pathology of Pareto optimality. A situation in which a measure, for instance, allocation of resources, is undertaken to improve one part of a system and is believed to have no adverse effects on other parts of systems. In welfare economics, it has been shown that it is not possible to make one part of the system better without making another part worse-off.

Pathology of purposive behaviorism. A situation in which the purpose of the system is unguided and primarily based on intended results as opposed to what the system produces.

Pathology of recursiveness. A violation of the *theorem of system recursion* defined as a condition in which system in question is incapable of defining itself as a viable system containing viable systems and being contained in a viable system.

Pathology of reification. A situation in which reality is distorted because of confusing abstract ideas to concrete entities. Young's (1964) words make it more apparent: this pathology occurs when "an analytic or abstract relationship [is treated] as though it were a concrete entity" (p. 109).

Pathology of genesis of structure. Addresses the need to initiate and maintain communications among forming structures in a system.

Pathology of synchronicity. A situation in which phenomena about a system appears to be meaningfully related but is ignored since it is impossible to be explained in terms of causality-language.

Pathology of communication. The receiver of information is unable to receive information as intended by the sender. Communication is broadly defined as 'all of the procedures by which one mind may affect another.'

Pathology of control. A condition that emerges out of having ineffective control mechanisms. It has been suggested that control is what 'permits the system to adapt and remain viable.'

Pathology of dynamic equilibrium. A situation in which system expected performance is reduced due to imbalance in interactions with external systems.

Pathology of punctuated equilibrium. A situation in which the long periods of stasis (i.e., relative calmness) become the basis for a potentially catastrophic event.

Pathology of sociotechnicality. A condition in which organization has a preference for either the social (i.e., soft/human) aspects or the technical (i.e., technology in the workplace) aspect of an organization but not both.

Pathology of system boundary. A situation in which a boundary (i.e., line of demarcation) of a system is fuzzily defined. A line of demarcation provides minimum description distinguishing a system from its environment.

Pathology of system environment. Concerned with understanding the relationship between system and its environment. A complement to *pathology of boundary*, this involves a failure to understand a line of demarcation distinguishing environment from system.

References

Barnard, C. I. (1946). Functions and pathology of status systems in formal organizations. In W. F. Whyte (Ed.), *Industry and Society* (pp. 46–83). New York, NY: McGraw-Hill.

Beer, S. (1984). The viable system model: Its provenance, development, methodology and pathology. *The Journal of the Operational Research Society*, *35*(1), 7–25.

Bobba, J., Moore, K. E., Volos, H., Yen, L., Hill, M. D., Swift, M. M., & Wood, D. A. (2007). Performance pathologies in hardware transactional memory. In C. Scheideler & P. Gibbons, (Eds.), *Proceedings of the 34th annual international symposium on Computer architecture* (pp. 81–91). San Diego: CA: ACM.

Davidz, H. L. (2017). Systems engineering pathology: Leveraging science to characterize dysfunction. In *Annual INOSE International Workshop*. Los Angeles: INCOSE. Retrieved from www.incose.org/IW2017

Dery, D. (1984). *Problem definition in policy analysis*. Lawrence, KS: University Press of Kansas.

Katina, P. F. (2015a). Emerging systems theory–based pathologies for governance of complex systems. *International Journal of System of Systems Engineering*, *6*(1/2), 144–159.

Katina, P. F. (2015b). *Systems theory-based construct for identifying metasystem pathologies for complex system governance* (Ph.D.). Old Dominion University, United States—Virginia.

Katina, P. F. (2016a). Metasystem pathologies (M-Path) method: Phases and procedures. *Journal of Management Development*, *35*(10), 1287–1301.

Katina, P. F. (2016b). Systems theory as a foundation for discovery of pathologies for complex system problem formulation. In A. J. Masys (Ed.), *Applications of Systems Thinking and Soft Operations Research in Managing Complexity* (pp. 227–267). Geneva, Switzerland: Springer International Publishing.

Keating, C. B., & Katina, P. F. (2012). Prevalence of pathologies in systems of systems. *International Journal of System of Systems Engineering*, *3*(3/4), 243–267.

Merriam-Webster. (2006). *Webster's new explorer encyclopedic dictionary*. Springfield, MA: Federal Street Press.

Ríos, J. P. (2012). *Design and diagnosis for sustainable organizations: The viable system method*. New York, NY: Springer Berlin Heidelberg.

Sheptycki, J. (2004). Organizational pathologies in police intelligence systems: Some contributions to the lexicon of intelligence-led policing. *European Journal of Criminology*, *1*(3), 307–332.

Troncale, L. (2011a). Can a Theory that Integrates the Natural Systems Sciences Help Systems Engineering of Defense against Security Threats? In *2011 Eighth International Conference on Information Technology: New Generations (ITNG)* (pp. 947–952). Las Vegas, NV. https://doi.org/10.1109/ITNG.2011.213

Troncale, L. (2011b). Would a rigorous knowledge base in systems pathology add significantly to the systems engineering portfolio. In *CSER'11 Proceedings, Conference on Systems Engineering Research, April 14–16* (pp. 1–17). Redondo Beach, CA. Retrieved from http://lentroncale.com/wp-content/uploads/2011/12/SysPath-CSER11-Troncale-article.pdf

Troncale, L. (2013). Systems processes and pathologies: Creating an integrated framework for systems science. *INCOSE International Symposium*, *23*(1), 1330–1353.

Young, O. R. (1964). A survey of general systems theory. *General Systems*, *9*, 61–80.

Appendix F: Lexicographical Threat Index

In a typical approach to threat, the norm is to consider ABC: *Atomic*, *Biological*, and *Chemical*. However, this is a narrow scope of threats and can leave entities exposed and therefore vulnerable to a slew of threats.

Threat Categories

Three broad categories are suggested: *ABC threat spectrum*, *infrared-ultraviolet threat spectrum*, and *black swan threat spectrum*. A summary of these threats is presented in Fig. F.1. ABC threats are conventional threats. These threats include atomic, biological, chemical, and the rest of the lexicon of the alphabet: drugs, epidemics, finance, global warming, h/entropy (CI), information security, job loss, NEO, Piracy, etc. These might require traditional risk management approaches.

The infrared-ultraviolet threat spectrum addressing threats that might emerge, for example, as a result of implementing a new technology (e.g., Autonomous Aircraft Systems), methodology, and the likes.

The Black Swans threat spectrum are threats that only become visible once they are revealed. These events are often labeled as *low probability* but *high consequences*, or as John Casti calls them, 'x-events,' events so rare that they are unthinkable and often dismissed and labeled as fictitious and improbable (Casti 2012). These can be classified into:

- Ambiguous Threats: These are threats/or benefit depending on the perspective, more or less, the interpretation of the meaning. Think, for instance, the benefit and threat of Nanotechnology
- Completely Unexpected Threats: this spectrum of threats includes events that one thinks cannot happen, until they happen. Think of, for instance, the 9/11 attacks.

Potential Research Questions

For those must deal with risk, well, perhaps everyone, and especially those that must deal with governance of critical and complex system (of systems) in space, undersea, and belowground, it becomes obvious that the traditional 'definitions' of phenomenon and noumenon are changing. For example, one cannot think of risk in terms

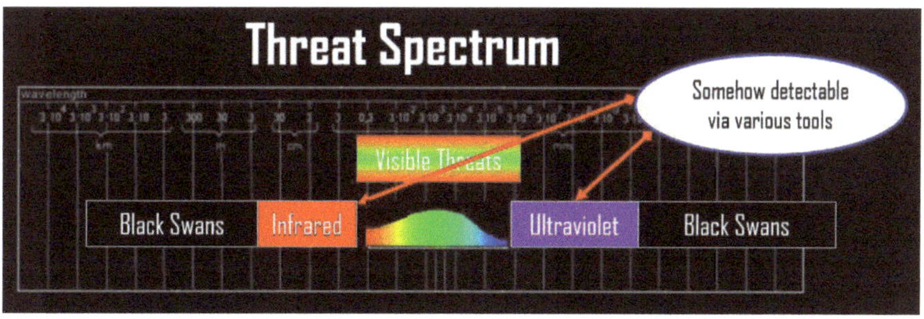

Fig. F.1 Three broad categories of threats

of only probability and consequence without a consideration, for example, vulnerability, perception, and other-related concepts. Inevitably, this changes how might approach, for example, 'risk formulation' (Katina et al. 2014). Perhaps it's time to think in terms of the changing society trends, threat in terms of threat 2.0, and the likes (Fig. F.2). Regarding emerging areas space, undersea, and belowground, present authors suggest that much of the threat are invisible to human kind (Fig. F.3).

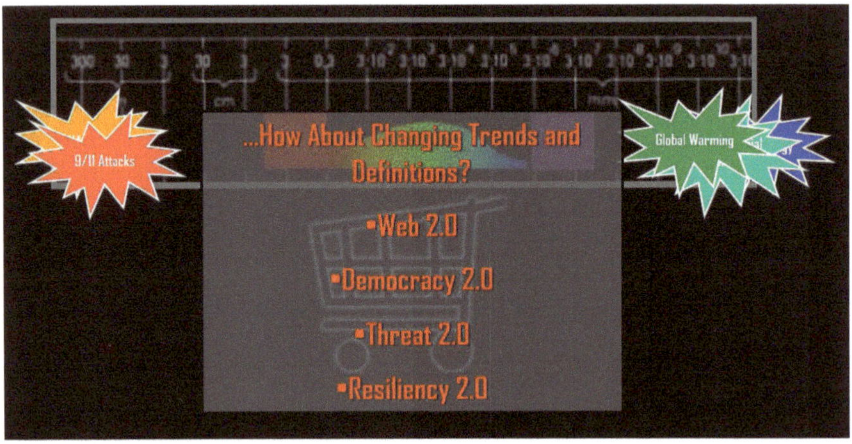

Fig. F.2 Revealing threats over time

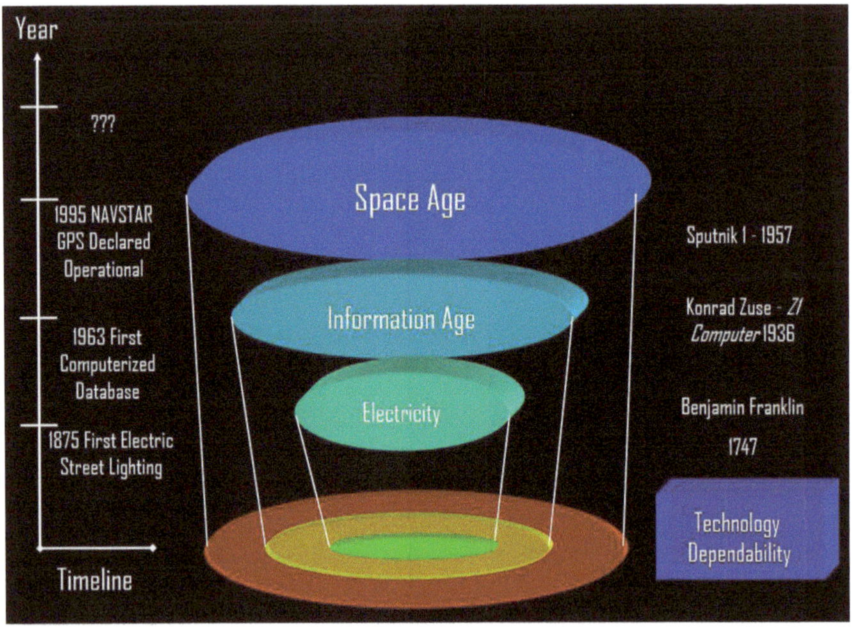

Fig. F.3 A fertile new domain of threat

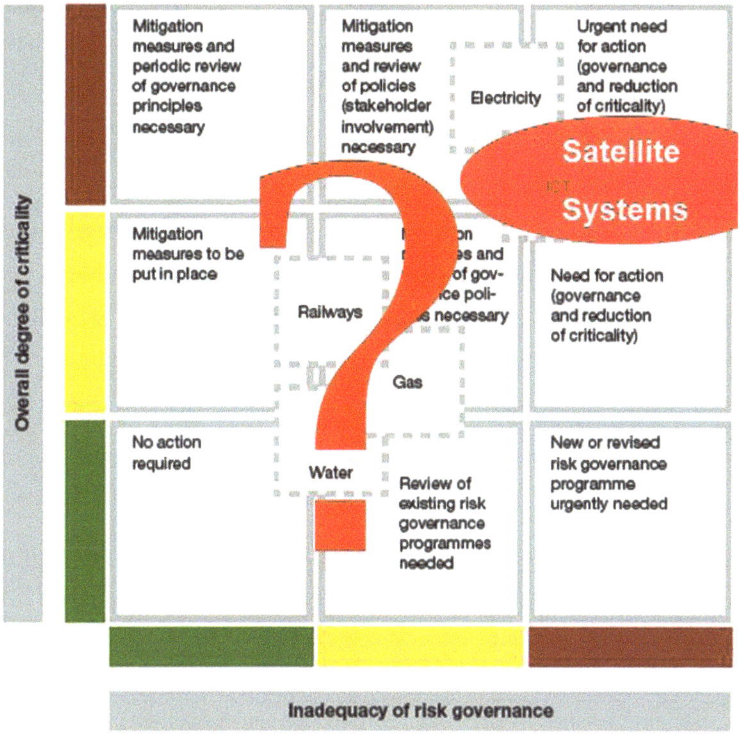

Fig. F.4 Examining phenomena using degree of criticality and adequacy of risk governance, modified from Gheorghe (2009)

A dashboard approach to governance was earlier suggested in Chap. 2, Sect. 2.2.2. A complementary approach might involve identification of 'degree of criticality' and contrasted with 'adequacy of risk governance' as suggested in Fig. F.4. In related efforts, especially in regard to measuring cyberthreats, one might look at issues of trusted display, authenticity, freshness, communication channels, as well as the trifecta of integrity, privacy, and secrecy. Certainly, such discussions are part of the system of systems involving *socionomics, infranomics*, *economics*.

A Cautionary Tale

Here is a rather bleak picture of our current approach as suggested by Gheorghe (2009) to a group of senior executives:

> *'We are a crude form of life right now, in the evolutionary shape our civilization...*
>
> *Really, we're not even civilized yet...*

So, as the world joins together, and we are through with military regimes, prisons, torture, hunger, poverty, deprivation...

When that is gone, that will be the beginning of the civilized world...

... We are not there yet.'

References

Casti, J. (2012). X-Events: Complexity overload and the collapse of everything. New York, NY: William Morrow.

Gheorghe, A. V. (2009, March). *Privacy enhanced technology: A landscape of threats*. Presented at the Private client: Threat Lexicographical, Norfolk, VA—United States.

Katina, P. F., Pinto, C. A., Bradley, J. M., & Hester, P. T. (2014). Interdependency-induced risk with applications to healthcare. *International Journal of Critical Infrastructure Protection, 7* (1), 12–26. https://doi.org/10.1016/j.ijcip.2014.01.005

Appendix G: Introductory Notes on VULPET

The purpose of this appendix is to introduce the reader to a software platform developed to consolidate and enhance the ability of contractor to understand, diagnose, and predict abnormal occurrences of risks in petrochemical enterprises. The software platform considers a range of issues including technical, business, and managerial.

The What, Who, How of VULPET

VULPET is a software platform use in assessing vulnerabilities in the petrochemical industry. It was developed by the appointment of *Swiss Re* by an alliance of independent contractors. The product is meant *to consolidate and enhance Contractor's ability to understand, diagnose, and predict abnormal occurrences in the field of the risks and safety of petrochemical enterprise—technical, business, and managerial aspects considered on a comparable footing, and in an integrative manner*.

The VULPET project was originated and managed by *BC2 Basel*, and was technically designed and executed by *KOVERS-KT*, a dedicated research and knowledge transfer unit within the *Laboratory of Safety Analysis* (*LSA*) of *the Swiss Federal Institute of Technology, ETH-Zurich* (Zürich, Switzerland).

VULPET is designed as an analyst's toolkit assembling traditional and less conventional methods to approach a quantitative risk and vulnerability assessment. On the traditional line, *the Risk Assessment Matrix* (ASMAT) method—by now a well-established analytical tool within *SwissRe*'s corps of engineers—is given a convenient and expeditious software implementation. Less conventional is, on the other hand, an approach originating in ETH/LSA academic publications to the quantitative vulnerability analysis, centered on the notion of system stability and drawing upon a physical analogy provided by the theory of cooperative phenomena in Statistical Physics. What both approaches have in common is *the shared motivation to provide a framework for a comparative categorizing of reinsurance targets—in this case petrochemical plants—using an as limited as feasible set of numbers, and synoptic charts.* Placing the two methods on the same platform was believed to having the advantage of offering the analyst *a fair choice*, between a tool he/she had already accepted and is accustomed to, and a complementary approach that, if perhaps thought-provoking, may yet need time to be digested and appraised.

Caveats

VULPET is, essentially, a research product the notion of which was adopted as a result of brainstorming sessions conducted within the *Swiss Re Engineering Services* community, on an intellectual drive to keep an open mind on various developments and trends in the field of Systems Engineering and other disciplines that have the potential of being supportive of an ever more adequate, accurate, and performant analysis of the complex phenomenology underlying the threats and challenges to which the reinsurance targets, as well the reinsurers themselves, are exposed. As such, the product claims no access to the absolute truth of the matter, nor does it endeavor to cover all possible angles in the attempt to provide quantitative methods to express vulnerability in a single, or a few, numbers and charts. In a practical sense, VULPET is just another tool on the SwissRe's analyst's desktop. In an epistemological sense, on the other hand, it is a live evidence on the feasibility to enhance the scope, ways, and means of the reinsurance strategy and overall business—which is believed to be a line of action to be further pursued.

Upon these caveats*, the product is provided as is, with no liabilities either assumed or accepted by the proponents and developers.* The first formal version of VULPET was open to comments and feedback and continues to be improved. This is the case for ergonomic shortcomings and bugs are *a priori* to be expected, as with all software products of a certain degree of complexity. Error reports as well as 'nice to have' issues are, therefore, welcome.

Overview

VULPET is designed on the following working philosophy:

(a) Describe the condition and performance of a petrochemical plant seen as a technological and business entity by one, or several, alternative, *sets of indicators* that may imply *potential system deficiencies*, and *management deficiencies*.

(b) Based on methodologically grounded criteria, aggregate the indicators such as to eventually obtain

   (i) a single *risk index* (the traditional ASMAT method), or a single *vulnerability index* (the quantitative vulnerability assessment method: VASMAT, proposed as a complement to ASMAT); and

   (ii) appropriate additional *synoptics* (ASMAT), or chart-embedded *matrices* (VASMAT), supportive for a comparative assessment of plant risk/vulnerability performance.

The software is consequently designed *as a turntable-wise user interface* (Fig. G.1) dispatching analyst's interests among a variety of natural functions to be expected. These include:

1. *The Risk Assessment Engine*, operating a short-hand implementation of SwissRe's traditional ASMAT method.
2. *The Vulnerability Assessment Engine*, featuring:

   2.1. The interactive definition of vulnerability-relevant sets of indicators.
   2.2. The interactive setting of interdependence relationships between indicators.
   2.3. A data library on the reinsurance targets—corporations, divisions, and plants dwelling in the petrochemical business—and the respective data acquisition and management unit.
   2.4. The *Vulnerability Engine* itself, producing, out of sets of indicators, *Vulnerability Indexes*, and *Vulnerability Matrices* spanned by a *System Deficiency Index,* and a *Management Deficiency Index.*

A *flowchart* and a *folder structure* reflective of the above terms of reference are attached to this overview (Fig. G.2). The software companion manual provides substantive information on ASMAT method in 'Section 3,' VULPET method *per se* in 'Section 4,' a 'Getting Started' tour over the VULPET modules in 'Section 5,' and an appendix holding a structured collection of the source code listings, making VULPET a potential 'open source project' for SwissRe's IT Division or an authorized third party.

Fig. G.1 VulPet interface

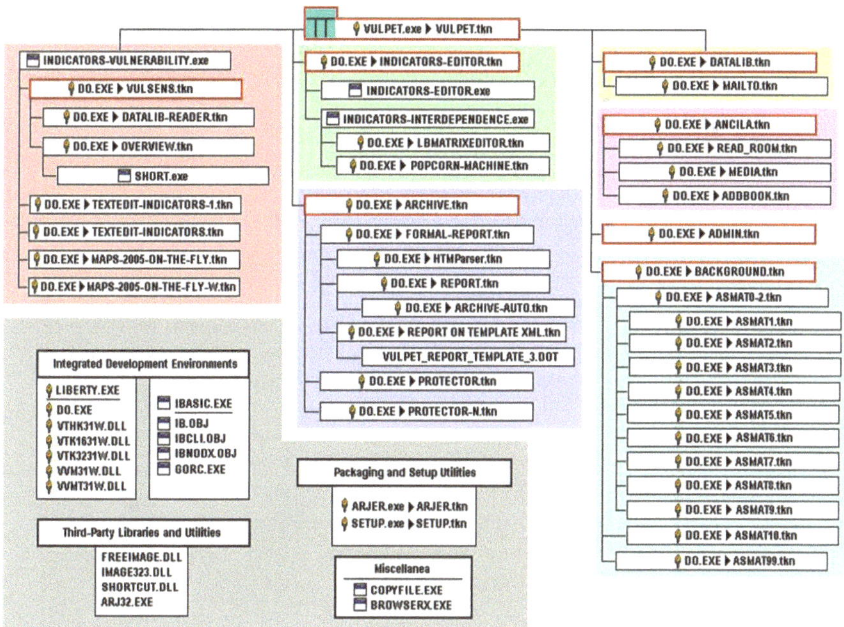

Fig. G.2 A basic structure of the VulPet software, adapted from Gheorghe et al. (2006)

The VULPET Folders

The software in organized into various folders as described below:

ARCHIVE	Stores sets of files making up the assessment results, generated by the vulnerability assessment engine, VULSENS.TKN. Any single assessment file set revolves around the Interim (Intermediate) Assessment Report, including files:	
	yyyy-mm-dd_hh-mm-ss.htm	the *Interim Assessment Report* serving as raw source for the VULPET Formal Report; readable from inside VULPET (module ARCHIVE.tkn)
	yyyy-mm-dd_hh-mm-ss-copy.htm	copy of the Interim Report file, adjusted for independent reading from outside VULPET by virtually any browser
	yyyy-mm-dd_hh-mm-ss.idt	case identifier
	yyyy-mm-dd_hh-mm-ss.cas	alternate case identifier
	yyyy-mm-dd_hh-mm-ss%overview-1.jpg	scoring overview table #1
	yyyy-mm-dd_hh-mm-ss%overview-2.jpg	scoring overview table #2
	yyyy-mm-dd_hh-mm-ss%frame1.jpg	'dashboard' frame #1
	yyyy-mm-dd_hh-mm-ss%frame2.jpg	'dashboard' frame #2 (if any)
	yyyy-mm-dd_hh-mm-ss%frame3.jpg	'dashboard' frame #3 (if any)
	yyyy-mm-dd_hh-mm-ss.jpg	case's *Vulnerability Assessment Matrix* (the S-M Matrix)
	with *yyyy-mm-dd_hh-mm-ss* the computer date and time of assessment, in the format: *year-month-day_hour-minute-second*	
ASMAT	Holds text and graphics files (input, output, support) pertaining to the VULPET implementation of the standard, accepted *Risk Assessment Matrix Method*	
BIN	Holds fixed graphics and text requisites employed in running VULPET; includes bitmaps, cursors, icons, .htm prototypes, miscellanea. May still include files that have lost relevance over the code's development history and/or are still believed of possible use in presentations, etc. Outstanding is file *VULPET_Report_Template_3.dot*, the template for the runtime generation of VULPET Formal Reports (VULPET_Report_Template_3.doc)	
FINALS	Stores *VULPET Formal Reports*. Each report holds an independent subfolder in its own right. Subfolders revolve around	
	(i) file **VULPET.doc**—the standalone Formal Report, exportable outside	
	VULPET *as is* (not needing the other files assistance);	
	(ii) file **VULPET.htm**—the HTML version of the Final Report, needing the	
	assistance of several other files in the subfolder	
	A number of files serve in the generation of the reports, and may find good use as independent texts and/or graphics, in presentations	
Foundations		

(continued)

(continued)

	Holds intermediate and final graphics that has served in the generation of the VULPET Manual. Also holds file *VULPET_Indicator_Matrix_V1.XLS*, that can be summoned at runtime to assist risk engineers in the way of a notepad, from within the Vulnerability Assessment engine
HTM	Serves as workspace for the code to generate Formal Reports. Cleared out at the inception of every report-generation process, it is however left untouched till the next round, primarily for checkout purposes
Indicators	The indicator data repository. Holds indicator lists (*.IND); indicator interdependence matrices (*.MTX); indicator scoring directions and explanations (*.RTF); and a few other incidental runtime requisites
Manual	The VULPET Manual and requisites for its runtime reading
Media	Documentary files, going with VULPET's *SUPPORT—MEDIA* section
Read room	Documentary files, going with VULPET's SUPPORT—*PAPERS* section
Scoring	A repository for saving, and retrieving indicator scores (*.SRE)
Systems	A repository for VULPET 'systems,' i.e., cuspidal topological folds in the (S, M, z) space (*System Deficiency, Management Deficiency, Operability Fraction*). A 'system' is essentially defined by its *temperature* (see VULPET model) and attaches a set of files: .BMP, .CSP, .DAT, .SYT
VULPET_DATA_LIBRARY	The repository of files holding information on companies, divisions, and plants. For details, see file *The VULPET Data Library—A Briefing.doc*, in this folder
VULPET_DATA_LIBRARY.BIN	The repository of *prototype* files serving in the interactive enhancement of VULPET's library on companies, divisions, and plants (see above). While such 'template' files are plain texts, the files generated with these are encrypted
Workspace	Work space for the code, handling indicator scoring directions and explanations called at runtime (.RTF files). *Do not tamper with*
Workspace-1	Work space for the code, handling user recommendations provided at runtime (.RTF files), to be included in the Interim and Formal Reports
	Do not tamper with

Reference

Gheorghe, A. V., Kröger, W., & Capaul, B. (2006). *VulPet: Vulnerability assessment of petrochemical plants [Computer software]*. ETH—Zurich: Laboratorium fur Siecherheist Analytik (LSA) KOVERS-KT.

Glossary

This section contains a glossary of terms for the present knowledge domain as articulated in the *National Infrastructure Protection Plan (NIPP) 2013: Partnering for Critical Infrastructure Security and Resilience* (USDHS 2013). The listed terms have been used in the present text or closed related to the domain knowledge. In a general sense, explanations of concepts relevant to present topic are provided. A reader might also use this section as a reference to material found elsewhere

All Hazards The term 'all hazards' means a threat or an incident, natural, or man-made that warrants action to protect life, property, the environment, and public health or safety, and to minimize disruptions of government, social, or economic activities. It includes natural disasters, cyber incidents, industrial accidents, pandemics, acts of terrorism, sabotage, and destructive criminal activity targeting critical infrastructure

Asset Person, structure, facility, information, material, or process that has value

Business Continuity Activities performed by an organization to ensure that during and after a disaster the organization's essential functions are maintained uninterrupted, or are resumed with minimal disruption

Consequence The effect of an event, incident, or occurrence, including the number of deaths, injuries, and other human health impacts along with economic impacts both direct and indirect and other negative outcomes to society

Control Systems Computer-based systems used within many infrastructure and industries to monitor and control sensitive processes and physical functions. These systems typically collect measurement and operational data from the field, process and display the information, and relay control commands to local or remote equipment or human–machine interfaces (operators). Examples of types of control systems include SCADA systems, Process Control Systems, and Distributed Control Systems

A.V. Gheorghe et al., *Critical Infrastructures, Key Resources, Key Assets*, Topics in Safety, Risk, Reliability and Quality 34,
https://doi.org/10.1007/978-3-319-69224-1

Critical Infrastructure Systems and assets, whether physical or virtual, so vital to the USA that the incapacity or destruction of such systems and assets would have a debilitating impact on security, national economic security, national public health or safety, or any combination of those matters

Critical Infrastructure Community Critical infrastructure owners and operators, both public and private; Federal departments and agencies; regional entities; SLTT governments; and other organizations from the private and nonprofit sectors with a role in securing and strengthening the resilience of the Nation's critical infrastructure and/or promoting practices and ideas for doing so.

Critical Infrastructure Cross-Sector Council Private sector council that comprises the chairs and vice chairs of the SCCs. This council coordinates cross-sector issues, initiatives, and interdependencies to support critical infrastructure security and resilience.

Critical Infrastructure Information ***(CII)*** Information that is not customarily in the public domain and is related to the security of critical infrastructure or protected systems. CII consists of records and information concerning any of the following:

- Actual, potential, or threatened interference with, attack on, compromise of, or incapacitation of critical infrastructure or protected systems by either physical or computer-based attack or other similar conduct (including the misuse of or unauthorized access to all types of communications and data transmission systems) that violates Federal, State, or local law; harms the interstate commerce of the United States; or threatens public health or safety.
- The ability of any critical infrastructure or protected system to resist such interference, compromise, or incapacitation, including any planned or past assessment, projection, or estimate of the vulnerability of critical infrastructure or a protected system, including security testing, risk evaluation, risk management planning, or risk audit.
- Any planned or past operational problem or solution regarding critical infrastructure or protected systems, including repair, recovery, insurance, or continuity, to the extent that it is related to such interference, compromise, or incapacitation.

Critical Infrastructure Owners and Operators Those entities responsible for day-to-day operation and investment of a particular critical infrastructure entity

Critical Infrastructure Partner Those Federal and SLTT governmental entities, public and private sector owners and operators and representative organizations, regional organizations and coalitions, academic and professional entities, and certain not-for-profit and private volunteer organizations that share responsibility for securing and strengthening the resilience of the Nation's critical infrastructure

Critical Infrastructure Partnership Advisory Council (CIPAC) Council established by DHS under 6 U.S.C. §451 to facilitate effective interaction and coordination of critical infrastructure activities among the Federal Government; the private sector; and SLTT governments.

Critical Infrastructure Risk Management Framework A planning and decision-making framework that outlines the process for setting goals and objectives, identifying infrastructure, assessing risks, implementing risk management activities, and measuring effectiveness to inform continuous improvement in critical infrastructure security and resilience

Cybersecurity The prevention of damage to, unauthorized use of, or exploitation of, and, if needed, the restoration of electronic information and communications systems and the information contained therein to ensure confidentiality, integrity, and availability; includes protection and restoration, when needed, of information networks and wireline, wireless, satellite, public safety answering points, and 911 communications systems and control systems.

Cyber System Any combination of facilities, equipment, personnel, procedures, and communications integrated to provide cyber services; examples include business systems, control systems, and access control systems.

Dependency The one-directional reliance of an asset, system, network, or collection thereof—within or across sectors—on an input, interaction, or other requirement from other sources in order to function properly

Executive Order 13636 Executive Order that calls for the Federal Government to closely coordinate with critical infrastructure owners and operators to improve cybersecurity information sharing; develop a technology-neutral cybersecurity framework; and promote and incentivize the adoption of strong cybersecurity practices.

Emergency Support Functions (ESF) The primary, but not exclusive, Federal coordinating structures for building, sustaining, and delivering the response core capabilities. ESFs are vital for responding to Stafford Act incidents but also may be used for other incidents

Function Service, process, capability, or operation performed by an asset, system, network, or organization

Fusion Center A State and major urban area focal point for the receipt, analysis, gathering, and sharing of threat-related information between the Federal Government, SLTT, and private sector partners

Government Coordinating Council (GCC) The government counterpart to the Sector Coordinating Council for each sector, established to enable interagency and intergovernmental coordination; comprises representatives across various levels of government (Federal and SLTT) as appropriate to the risk and operational landscape of each sector.

Hazard Natural or man-made source or cause of harm or difficulty

Incident An occurrence, caused by either human action or natural phenomenon, that may cause harm and require action, which can include major disasters, emergencies, terrorist attacks, terrorist threats, wild and urban fires, floods, hazardous materials spills, nuclear accidents, aircraft accidents, earthquakes, hurricanes, tornadoes, tropical storms, war-related disasters, public health and medical emergencies, cyber-attacks, cyber failure/accident, and other occurrences requiring an emergency response

Information Sharing and Analysis Centers (ISACs) Operational entities formed by critical infrastructure owners and operators to gather, analyze, appropriately sanitize, and disseminate intelligence and information related to critical infrastructure. ISACs provide 24/7 threat warning and incident reporting capabilities and have the ability to reach and share information within their sectors, between sectors, and among government and private sector stakeholders

Information Sharing and Analysis Organization Any formal or informal entity or collaboration created or employed by public or private sector organizations, for purposes of:

(a) Gathering and analyzing critical infrastructure information in order to better understand security problems and interdependencies related to critical infrastructure and protected systems, so as to ensure the availability, integrity, and reliability thereof;
(b) Communicating or disclosing critical infrastructure information to help prevent, detect, mitigate, or recover from the effects of an interference, compromise, or an incapacitation problem related to critical infrastructure or protected systems; and
(c) Voluntarily disseminating critical infrastructure information to its members, State, local, and Federal Governments, or any other entities that may be of assistance in carrying out the purposes specified in subparagraphs (a) and (b).

Infrastructure The framework of interdependent networks and systems comprising identifiable industries, institutions (including people and procedures), and distribution capabilities that provide a reliable flow of products and services essential to the defense and economic security of the United States, the smooth functioning of government at all levels, and society as a whole; consistent with the definition in the Homeland Security Act, infrastructure includes physical, cyber, and/or human elements.

Interdependency Mutually reliant relationship between entities (objects, individuals, or groups); the degree of interdependency does not need to be equal in both directions.

Joint Terrorism Task Forces (JTTFs) FBI-led local task forces of highly trained Federal, State, and local law enforcement and intelligence agencies established to collect terrorism-related intelligence and conduct investigations. The local

FBI JTTFs receive and resolve reports of possible terrorism activity submitted by private industry partners and the public

Mitigation Capabilities necessary to reduce loss of life and property by lessening the impact of disasters

National Cyber Investigative Joint Task Force The multi-agency national focal point for coordinating, integrating, and sharing pertinent information related to cyber threat investigations, with representation from Federal agencies, including DHS, and from State, local, and international law enforcement partners

National Cybersecurity and Communications Integration Center The national cyber critical infrastructure center, as designated by the Secretary of Homeland Security, which secures Federal civilian agencies in cyberspace; provides support and expertise to private sector partners and SLTT entities; coordinates with international partners; and coordinates the Federal Government mitigation and recovery efforts for significant cyber and communications incidents.

National Infrastructure Coordinating Center The national physical critical infrastructure center, as designated by the Secretary of Homeland Security, which coordinates a national network dedicated to the security and resilience of critical infrastructure of the United States by providing 24/7 situational awareness through information sharing, and fostering a unity of effort

National Operations Center A DHS 24/7 operations center responsible for providing real-time situational awareness and monitoring of the homeland, coordinating incident response activities, and, in conjunction with the Office of Intelligence and Analysis, issuing advisories and bulletins concerning threats to homeland security, as well as specific protective measures

National Preparedness The actions taken to plan, organize, equip, train, and exercise to build and sustain the capabilities necessary to prevent, protect against, mitigate the effects of, respond to, and recover from those threats that pose the greatest risk to the security of the Nation

Network A group of components that share information or interact with each other to perform a function

Partnership Close cooperation between parties having common interests in achieving a shared vision

Presidential Policy Directive 8 (PPD-8) Facilitates an integrated, all-of-Nation approach to national preparedness for the threats that pose the greatest risk to the security of the Nation, including acts of terrorism, cyber-attacks, pandemics, and catastrophic natural disasters; directs the Federal Government to develop a national preparedness system to build and improve the capabilities necessary to maintain national preparedness across the five mission areas covered in the PPD: prevention, protection, mitigation, response, and recovery.

Presidential Policy Directive 21 (PPD-21) Aims to clarify roles and responsibilities across the Federal Government and establish a more effective partnership with owners and operators and SLTT entities to enhance the security and resilience of critical infrastructure

Prevention Those capabilities necessary to avoid, prevent, or stop a threatened or actual act of terrorism

Protected Critical Infrastructure Information (PCII) All critical infrastructure information that has been properly submitted and validated pursuant to the Critical Infrastructure Information Act and implementing directive; all information submitted to the PCII Program Office or designee with an express statement is presumed to be PCII until the PCII Program Office determines otherwise.

Protection Those capabilities necessary to secure the homeland against acts of terrorism and man-made or natural disasters

Recovery Those capabilities necessary to assist communities affected by an incident to recover effectively, including, but not limited to, rebuilding infrastructure systems; providing adequate interim and long-term housing for survivors; restoring health, social, and community services; promoting economic development; and restoring natural and cultural resources.

Recovery Support Functions (RSF) Coordinating structures for key functional areas of assistance during recovery operations; RSFs support local governments by facilitating problem solving, improving access to resources, and fostering coordination among State and Federal agencies, nongovernmental partners, and stakeholders.

Regional Entities and interests spanning geographic areas ranging from large multi-State areas to metropolitan areas and varying by organizational structure and key initiatives, yet fostering engagement and collaboration between critical infrastructure owners and operators, government, and other key stakeholders within the given location

Regional Consortium Coordinating Council Comprises regional groups and coalitions around the country engaged in various initiatives to advance critical infrastructure security and resilience in the public and private sectors

Resilience The ability to prepare for and adapt to changing conditions and withstand and recover rapidly from disruptions; includes the ability to withstand and recover from deliberate attacks, accidents, or naturally occurring threats or incidents.

Response Capabilities necessary to save lives, protect property and the environment, and meet basic human needs after an incident has occurred

Risk The potential for an unwanted outcome resulting from an incident, event, or occurrence, as determined by its likelihood and the associated consequences

Risk-Informed Decision Making The determination of a course of action predicated on the assessment of risk, the expected impact of that course of action on that risk, and other relevant factors

Sector A logical collection of assets, systems, or networks that provide a common function to the economy, government, or society; the *National Plan* addresses 16 critical infrastructure sectors, as identified in PPD-21.

Sector Coordinating Council (SCC) The private sector counterpart to the GCC, these councils are self-organized, self-run, and self-governed organizations that are representative of a spectrum of key stakeholders within a sector; serve as principal entry points for the government to collaborate with each sector for developing and coordinating a wide range of critical infrastructure security and resilience activities and issues.

Sector-Specific Agency (SSA) A Federal department or agency designated by PPD-21 with responsibility for providing institutional knowledge and specialized expertise as well as leading, facilitating, or supporting the security and resilience programs and associated activities of its designated critical infrastructure sector in the all-hazards environment

Sector-Specific Plans (SSP) Planning documents that complement and tailor application of the *National Plan* to the specific characteristics and risk landscape of each critical infrastructure sector; developed by the SSAs in close collaboration with the SCCs and other sector partners.

Secure/Security Reducing the risk to critical infrastructure by physical means or defens[ive] cyber measures to intrusions, attacks, or the effects of natural or man-made disasters

Steady State The posture for routine, normal, day-to-day operations as contrasted with temporary periods of heightened alert or real-time response to threats or incidents

System Any combination of facilities, equipment, personnel, procedures, and communications integrated for a specific purpose

Terrorism Premeditated threat or act of violence against noncombatant persons, property, and environmental or economic targets to induce fear, intimidate, coerce, or affect a government, the civilian population, or any segment thereof, in furtherance of political, social, ideological, or religious objectives

Threat A natural or man-made occurrence, individual, entity, or action that has or indicates the potential to harm life, information, operations, the environment, and/or property

Threat and Hazard Identification and Risk Assessment (THIRA) A tool that allows a regional, State, or urban area jurisdiction to understand its threats and hazards and how the impacts may vary according to time of occurrence, season, location, and other community factors. This knowledge helps a jurisdiction establish informed and defensible capability targets for preparedness

Value Proposition A statement that outlines the business and national interest in critical infrastructure security and resilience actions and articulates the benefits gained by partners through collaborating in the mechanisms described in the *National Plan*

Vulnerability A physical feature or operational attribute that renders an entity open to exploitation or susceptible to a given hazard

Reference USDHS. (2013). *NIPP 2013: Partnering for critical infrastructure security and resilience*. Washington, D.C.: U.S. Dept. of Homeland Security. Retrieved from www.dhs.gov/xlibrary/assets/nipp-ssp-national-monuments-icons.pdf

Index

A.V. Gheorghe et al., *Critical Infrastructures, Key Resources, Key Assets*, Topics in Safety, Risk, Reliability and Quality 34,
https://doi.org/10.1007/978-3-319-69224-1

FSC
www.fsc.org
MIX
Papier | Fördert gute Waldnutzung
FSC® C083411

Kolibri 360 GmbH
Ferdinand-Jühlke-Straße 7
99095 Erfurt, Deutschland
produktsicherheit@kolibri360.de